EXTENDED
Mathematics
for Cambridge IGCSE

Audrey Simpson

CAMBRIDGE UNIVERSITY PRESS

CAMBRIDGE UNIVERSITY PRESS

Cambridge, New York, Melbourne, Madrid, Cape Town, Singapore,
São Paulo, Delhi, Dubai, Tokyo, Mexico City

Cambridge University Press
4381/4 Ansari Road, Daryaganj, Delhi 110002, India

www.cambridge.org
Information on this title: www.cambridge.org/9780521186032

© Cambridge University Press 2011

This publication is in copyright. Subject to statutory exception
and to the provisions of relevant collective licensing agreements,
no reproduction of any part may take place without the written
permission of Cambridge University Press.

First Published 2011

Printed in India at Replika Press Pvt. Ltd., Kundli, Haryana

A catalogue for this publication is available from the British Library

ISBN 978-0-521-18603-2 Paperback

Cambridge University Press has no responsibility for the persistence or
accuracy of URLs for external or third-party internet websites referred to in
this publication, and does not guarantee that any content on such websites is,
or will remain, accurate or appropriate. Information regarding prices, travel
timetables and other factual information given in this work are correct at
the time of first printing but Cambridge University Press does not guarantee
the accuracy of such information thereafter.

Contents

Introduction	vii
Acknowledgements	ix
Note	x

1 Real Numbers — 1
- Core Skills — 1
- More about Sets — 2
- Defining a Set — 3
- Notation and Symbols used with Sets — 4
- More Notation and Symbols — 5
- Venn Diagrams — 7
- More about Square Roots and Irrational Numbers — 12
- Fractions, Decimals and Percentages — 13
- Mixed Exercise — 14
- Examination Questions — 14

2 Algebra I — 20
- Core Skills — 20
- Expanding Products of Algebraic Factors — 21
- Factorising Quadratic Expressions — 23
- Factorising a Difference of Squares — 27
- Factorising by Pairing — 27
- Factorising Systematically — 28
- More about Indices — 29
- Algebraic Fractions — 32
- Mixed Exercise — 39
- Examination Questions — 40

3 Working with Numbers — 44
- Core Skills — 44
- Upper and Lower Bounds in Calculations — 46
- Increase and Decrease by a Given Ratio — 47
- Reverse Percentages — 48
- Mixed Exercise — 50
- Examination Questions — 51

4 Algebra II — 56
- Core Skills — 56
- More Equations — 57
- Quadratic Equations — 58
- More Sequences — 68
- More Simultaneous Equations — 70
- Inequalities — 73
- Variation — 76
- Rearranging Formulae — 78
- Mixed Exercise — 81
- Examination Questions — 82

5 Geometry and Shape — 88
- Core Skills — 88
- Symmetry in Three-Dimensional Shapes — 90
- Further Circle Facts — 93
- Investigations — 96
- Finding the Centre of a Circle — 98
- Irregular Polygons — 100
- Mixed Exercise — 101
- Examination Questions — 103

6 Algebra and Graphs — 109
- Core Skills — 109
- Distance/Time Graphs — 112
- Speed/Time Graphs — 114
- More Graphs of Curves — 120
- Straight Line Segments — 129
- Mixed Exercise — 133
- Examination Questions — 135

7 Length, Area and Volume — 146
- Core Skills — 146
- Arc Lengths in Circles — 147
- Sector Areas in Circles — 148
- More Volumes and Surface Areas of Solids — 148
- Curved Surface Areas — 149
- Volumes — 150
- Similar Shapes — 152
- Similar Triangles — 155
- Mixed Exercise — 160
- Examination Questions — 161

8 Further Algebra — 168

- Beginning Matrices — 168
- Operations on Matrices — 169
- Multiplication of Matrices — 171
- The Zero Matrix — 176
- The Identity Matrix and Inverse Matrices — 177
- Functions — 183
- Inverse Functions — 188
- Composite Functions — 190
- Graphs of Inequalities — 193
- Linear Programming — 200
- Mixed Exercise — 204
- Examination Questions — 205

9 Trigonometry — 214

- Core Skills — 214
- Angles of Elevation and Depression — 215
- Trigonometry with Angles between 0° and 180° — 217
- The Sine Rule — 222
- The Cosine Rule — 228
- Area of a Triangle — 233
- 3-D Trigonometry — 235
- Mixed Exercise — 239
- Examination Questions — 241

10 Transformations, Vectors and Matrices — 253

- Core Skills — 253
- More about Vectors — 256
- More about Transformations — 268
- Matrices and Transformations — 279
- Mixed Exercise — 291
- Examination Questions — 293

11 Statistics — 308

- Core Skills — 308
- Histograms — 309
- The Mean from a Grouped Frequency Table — 318
- Cumulative Frequency — 320
- Quartiles, Inter-Quartile Range and Percentiles — 322
- Mixed Exercise — 330
- Examination Questions — 332

12 Probability — 342
- Core Skills — 342
- Combined Events — 342
- Tree Diagrams — 347
- Venn Diagrams — 351
- Using Set Notation — 352
- Mixed Exercise — 356
- Examination Questions — 358

Revision and Examination Hints — 365

Answer key — 368

Index — 439

Introduction

Extended Mathematics for Cambridge IGCSE continues the work started in *Core Mathematics for Cambridge IGCSE,* preparing students for the Cambridge IGCSE Mathematics Extended Level examination from University of Cambridge International Examinations. It can be used in one of two ways.

- This book can form the second year of a two-year course, following the study of the Core Book in the first year.
- It can also be studied in conjunction with the Core Book, by following the order in which the studies should be studied as outlined on the next page. Each topic is started in the Core Book and continues in more depth in the Extended Book.

The structure of both books enables students to work through at their own pace, checking their progress as they go. Each topic is carefully introduced and followed by worked examples and exercises. The books are designed to be worked through sequentially and the entire Extended syllabus is covered by both books taken together.

A final section alerts the reader to common mistakes made by students in the Examinations, and lists particular items that need to be committed to memory.

If you choose to study the Core Book and the Extended Book together, we recommend that you read the chapters in the order shown below.

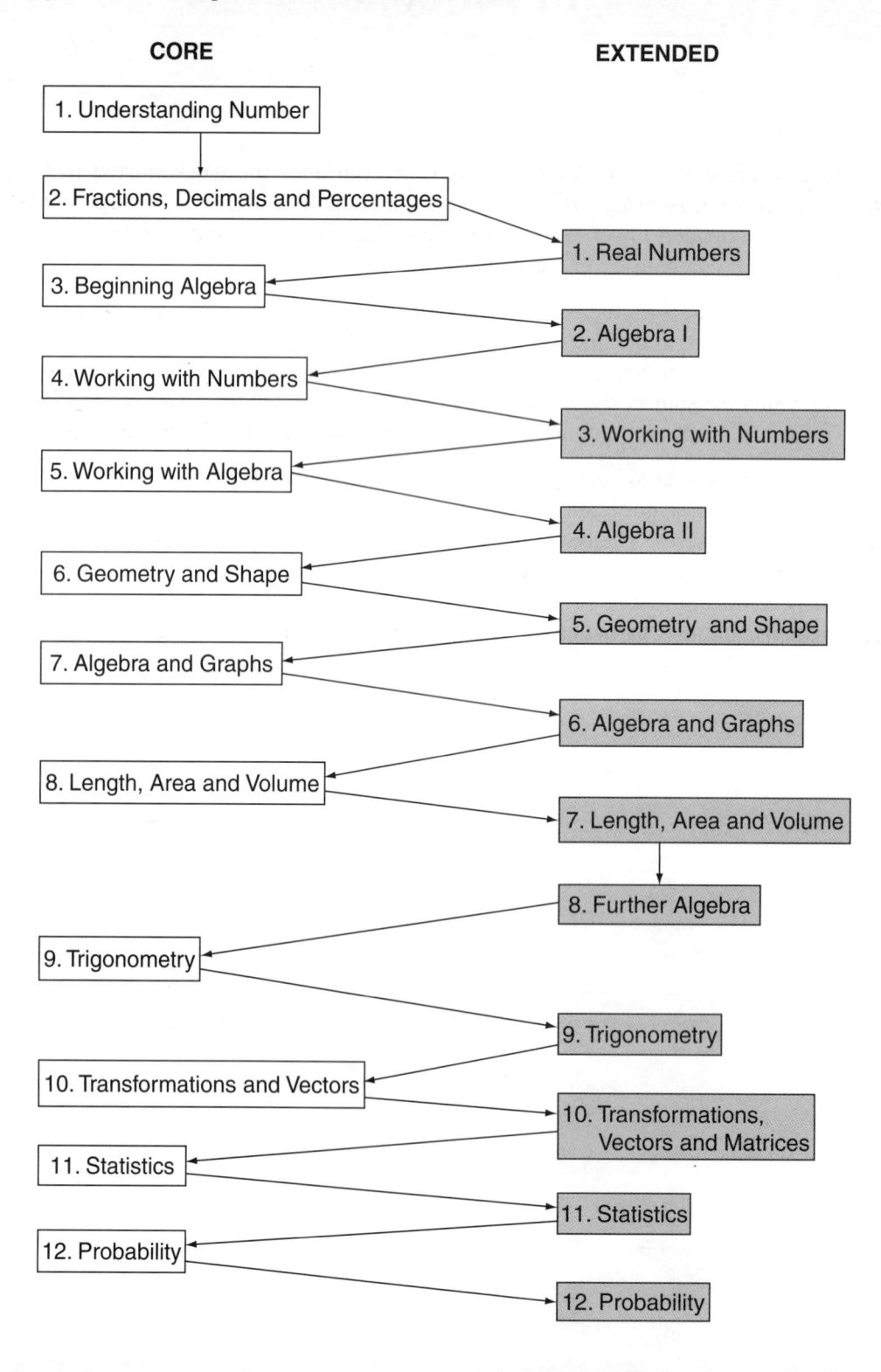

CORE

1. Understanding Number
2. Fractions, Decimals and Percentages
3. Beginning Algebra
4. Working with Numbers
5. Working with Algebra
6. Geometry and Shape
7. Algebra and Graphs
8. Length, Area and Volume
9. Trigonometry
10. Transformations and Vectors
11. Statistics
12. Probability

EXTENDED

1. Real Numbers
2. Algebra I
3. Working with Numbers
4. Algebra II
5. Geometry and Shape
6. Algebra and Graphs
7. Length, Area and Volume
8. Further Algebra
9. Trigonometry
10. Transformations, Vectors and Matrices
11. Statistics
12. Probability

Acknowledgements

I would like to thank Professor Gordon Kirby for his invaluable advice and encouragement. I am also grateful for his efforts to check my work patiently for errors, both mathematical and stylistic.

I am also indebted to my sister, Pat Victor, for the times she sorted out frustrating problems with both my computer and the software needed for the production of the manuscript.

Audrey Simpson

Note

The currencies used in this book are:
- 1 dollar ($), worth 100 cents
- 1 rupee [Re (singular) or Rs (plural)], worth 100 paise (singular paisa)
- 1 pound (£), worth 100 pence
- 1 euro (€), worth 100 cents.

Unless greater accuracy is required, answers to money questions which are not whole numbers, should be given to 2 decimal places.

For example, $12.7 should be given as $12.70, which is read 'twelve dollars and seventy cents'.

Chapter 1

Real Numbers

The set of Real Numbers includes all the numbers we need in everyday life. Your course for *Extended Mathematics for Cambridge IGCSE* is based entirely on the set of real numbers. You may wonder what other numbers there could possibly be. If you study Mathematics further you will eventually meet the set of Imaginary Numbers, which are the square roots of negative numbers. As you know, it is not possible to multiply a number by itself to make a negative number, so the square root of a negative number does not exist. However, Mathematicians need to work with the square roots of negative numbers, so they are called imaginary numbers. Imaginary numbers form the basis of a very interesting branch of Mathematics and you might like to find out more about them.

This Chapter follows on from Chapters 1 and 2 in *Core Mathematics for Cambridge IGCSE*. Before starting the new work you should work through the Core Skills exercise and check your answers. It is essential that you can complete the exercise with confidence. If you are unsure of anything you will need to go back to Chapters 1 and 2 in the Core Book.

Core Skills

1. 100 π 22.5   1   9   18   5   24   6   0 $\sqrt{2}$ $\sqrt{25}$ 2 $\frac{2}{7}$ 49

   From the list above select
   (a) the natural numbers
   (b) the integers
   (c) the rational numbers
   (d) the irrational numbers
   (e) the prime numbers
   (f) the factors of 45
   (g) the multiples of 6
   (h) the cube root of 8
   (i) the square of 7
   (j) the square root of 81

2. (a) List the set of factors of 144.
   (b) List the set of prime factors of 144.
   (c) Write 144 as a product of its prime factors.

3. Find the Highest Common Factor of 198 and 110.

4. Find the Lowest Common Multiple of 15 and 18.

5. Arrange the following in order of size, starting with the smallest.
   0.513   0.5006   0.52   0.5111

6. Insert a correct symbol between each of the following pairs of numbers.

   $\sqrt{64}, 2^3$ $\sqrt{64}, 3^2$ 19, 18

7. Write the following in standard form
   (a) 60137.1   (b) 0.005401
8. Use the correct order of working to calculate the following
   (a) $20 - 4 \times 5$ (b) $15 \div (5 - 2)$ (c) $(6 + 2) \div 2 - 3$
9. List the set of prime numbers between 20 and 40.
10. Write $\dfrac{37}{3}$ as a mixed number.
11. Write $5\dfrac{3}{5}$ as an improper fraction.
12. Fill in the spaces in this set of equivalent fractions
    $$\dfrac{}{75} = \dfrac{1}{3} = \dfrac{4}{}$$
13. Without using a calculator, calculate the following
    (a) $2\dfrac{2}{7} - \dfrac{4}{7}$ (b) $2\dfrac{2}{7} \times \dfrac{4}{7}$ (c) $2\dfrac{2}{7} \div \dfrac{4}{7}$
14. (a) Change 0.71 to a percentage.
    (b) Change 15% to a fraction.
    (c) Change $\dfrac{7}{50}$ to a decimal.
15. Find 16% of 54.
16. Find 28 as a percentage of 63.
17. Find a fraction that lies between $\dfrac{4}{7}$ and $\dfrac{5}{7}$.

More about Sets

We have already studied some sets of numbers. These were the sets of natural numbers, integers, rational numbers, real numbers and prime numbers.

A set is a collection of objects, ideas or numbers that can be clearly defined.

Clearly defined means that we can tell with certainty whether or not something belongs to the set.

For example,
   (a) The set of colours of the rainbow is {red, orange, yellow, green, blue, indigo and violet}. Brown is a colour, but it does not belong to this set.
   (b) The set of domestic pets may be quite large, but it would not include a brontosaurus! However a brontosaurus would be included in the set of dinosaurs.
   (c) The set of factors of 10 would include 1, 2, 5 and 10, but not 4.

In each of these cases, the sets are well defined because it is possible to tell whether or not something belongs to the set.

Defining a Set

There are various ways of defining a set.

- We can **list** the members of the set in curly brackets.
  The set of factors of 10 = {1, 2, 5, 10}
- If there is no end to the set, we use dots to represent 'and so on'.
  The set of natural numbers = {1, 2, 3, 4, 5, 6, 7, ...}
- If there is an end to the set, but it is too long to list all the numbers, we can still use dots, but include one or more of the last numbers in the set.
  The set of even numbers between 1 and 99 = {2, 4, 6, 8, 10, ..., 96, 98}
- Sets can also be defined by a description.
  For example, {planets in our solar system} or {capital cities of the world}.
- Sets are often labelled by using capital letters. The letters N, Z, Q and R are reserved for the sets we have already met, namely natural numbers, integers, rational numbers and real numbers.
- If we use a capital letter to represent another set we must define it clearly.
  For example,
  $$P = \{\text{the prime numbers between 1 and 20}\}$$
  or
  $$B = \{\text{boys in your class}\}$$

Example 1

(a) Write down a description of each of the following sets (there could be more than one suitable description).
  (i) {1, 3, 5, 7}   (ii) {5, 10, 15, 20, 25, ...}   (iii) {1, 2, 3, 4, 6, 12}

(b) List the following sets
  (i) The set of square numbers less than 30.
  (ii) The set of vowels in the English alphabet.
  (iii) The set of natural numbers less than 80.
  (iv) The set of cube numbers.

Answer 1

(a) (i) The set of odd numbers between 0 and 8, or simply {odd numbers between 0 and 8}
  (ii) The set of multiples of 5
  (iii) The set of factors of 12

(b) (i) {1, 4, 9, 16, 25}
  (ii) {a, e, i, o, u}
  (iii) {1, 2, 3, 4, 5, 6, 7, ..., 76, 77, 78, 79}
  (iv) {1, 8, 27, 64, 125, ...}

Notation and Symbols used with Sets

The members of the set are also called **elements** of the set.

If we define a set C = {primary colours}, then the elements of the set C are red, blue and green. The number of elements or members in the set is referred to as **n(C)**.

In this case, n(C) = 3. Of course n(Z) (the number of elements in the set of integers) is undefined, or infinitely large, because the integers go on forever.

The Greek letter $\in$ is used to mean 'is a member of' or 'is an element of'.

So Monday $\in$ {days of the week}, should be read as 'Monday is an element of the set of days of the week'. If this letter has a cross through it, i.e. $\notin$, it reads 'is not an element of'.

Hence, April $\notin$ {days of the week}.

We also need an 'empty' set. An **empty set** has no members.

For example, {even prime numbers *greater* than 2} is an empty set because 2 is the *only* even prime number. The symbol used for the empty set is $\emptyset$.

In some textbooks, you may see the empty set written as { }, which is an alternative for $\emptyset$.

Notice that we cannot use {0} as the empty set, because it has a member, which is zero.

So far the symbols we have met are:

Number of elements in set A	n(A)
'is an element of'	$\in$
'is not an element of'	$\notin$
The empty set	$\emptyset$

Exercise 1.1

1. List the elements of these sets
   (a) {square numbers between 10 and 40}
   (b) {months of the year beginning with M}
   (c) {natural numbers $\leqslant$ 10}

2. Describe these sets
   (a) {1, 3, 5, 7, 9}
   (b) {Monday, Tuesday, Wednesday, Thursday, Friday, Saturday, Sunday}
   (c) {a, b, c, d, e, ..., x, y, z}

3. Which of the following are empty sets?
   (a) {people over 5 metres tall}   (b) {birds that swim}
   (c) {0}   (d) {integers between 1.1 and 1.9}

4. Insert the correct symbol in the following statements.
   (a) 1 ... {prime numbers}   (b) 1000 ... {even numbers}

5. A = {letters used in the word 'mathematics'}.
   (a) List set A (b) Find n(A)

More Notation and Symbols

You will find that when we talk about sets we need to limit the possibilities in some way. For example, if you were to make up a school team to play football, it would be no good including students from another school. The team would be chosen from the set of all the students in *your* school. This is called the **universal set**. The symbol used to denote the universal set is $\mathscr{E}$.

So we could have: $\mathscr{E}$ = {students in your school}

F = {students in your school's football team}

M = {students in your maths class}

Both F and M belong to the universal set $\mathscr{E}$.

Another situation we need to be able to describe is that some students may be in your maths class *and* in the school football team. This is called an **intersection of sets**, and has the symbol $\cap$.

NOTE: This symbol, $\cap$, could be likened to a bridge between the two sets.

We write {students *both* in the school football team *and* in your maths class} = $F \cap M$.
If there is no one in your maths class who is also in the football team, then $F \cap M = \emptyset$. In this case, the intersection of F and M has no members, so it is the empty set.

It is perhaps easier to see this using sets with numbers.
We will use the following sets:

$$A = \{1, 2, 3, 4, 5, 6, 7, 8, 9\}$$
$$B = \{10, 11, 12, 13, 14, 15\}$$
$$C = \{2, 4, 6, 8, 10, 12\}$$
$$D = \{1, 3, 5, 7\}$$

From these sets it will be seen that:

$A \cap C = \{2, 4, 6, 8\}$ because 2, 4, 6 and 8 are in both A and C

and $A \cap B = \emptyset$

What about $A \cap D$?

$$A \cap D = \{1, 3, 5, 7\} = D$$

The whole of set D is in set A.

We say that D is a proper subset of A, and we use the symbol $\subset$, so $D \subset A$. You will usually hear a proper subset being referred to as just a subset. Strictly speaking a subset is different from a proper subset, which can be shown using the following example.

V = {vowels in the English language)

E = {letters of the English language}

L = {a, e, i, o, u}

F = {a, b, c}

V is a proper subset of E because there are elements in E that are *not* in V. However, V could be thought of *either* as equal to L, or as a subset of L, so V is not a *proper* subset of L.

Therefore, V is a proper subset of E. $\qquad V \subset E$

V is a subset of L (and V is equal to L). $\qquad V \subseteq L$

NOTE: Think of $\subseteq$ as 'is a subset or equal to' in the same way as $\leqslant$ means 'less than or equal to'. Also think of '$\subset$ is a proper subset', and '$<$ is strictly less than'.

The same symbol with a line through it $\not\subset$ means 'is not a proper subset of'.

V is not a proper subset of L. $\qquad V \not\subset L$

We can also say: $\qquad F \subset E$

$\qquad F \not\subset L$

and $\qquad F \not\subseteq L$

Suppose we wanted to use the whole of A and B. This would be called the union of A and B, and we would use the symbol $\cup$.

NOTE: You could think of this as a U, standing for 'Union'.

$$A \cup B = \{1, 2, 3, 4, 5, 6, 7, 8, 9, 10, 11, 12, 13, 14, 15\}$$
$$A \cup C = \{1, 2, 3, 4, 5, 6, 7, 8, 9, 10, 12\}$$

Notice that the elements that appear in both sets, are not listed twice.

We can now add to our list of symbols.

Universal set	$\mathscr{E}$
Intersection	$\cap$
Proper subset	$\subset$
Subset	$\subseteq$
Not a subset	$\not\subset$
Union	$\cup$

Example 2

$F = \{10, 20, 30, 40\} \qquad G = \{11, 13, 17, 19\}$

$H = \{10, 11, 12, 13\} \qquad J = \{11, 13\}$

Use these sets to answer the questions.

(a) List
  (i) $G \cap H$ (ii) $F \cup H$

(b) J is a subset of two of the sets. Which are those two sets?

(c) Suggest a suitable universal set for F, G, H and J.

(d) What can you say about $F \cap J$?

Answer 2

(a) (i) $G \cap H = \{11, 13\}$ (notice that this is also equal to J)

    (ii) $F \cup H = \{10, 11, 12, 13, 20, 30, 40\}$

(b) $J \subset G$ and $J \subset H$

(c) A universal set could be N (the set of natural numbers), or we could restrict it more, say
$\mathscr{E} = \{10, 11, 12, 13, 14, 15, ..., 38, 39, 40\}$
or $\mathscr{E}$ = {natural numbers between 10 and 40 inclusive}
(d) $F \cap J = \emptyset$

There is one more symbol left to add to the list.
Look at these sets:
$$\mathscr{E} = \{1, 2, 3, 4, 5\}$$
$$A = \{1, 4\}$$
$$B = \{2, 4, 5\}$$
The new symbol is the letter for the set followed by a dash, for example, A'.
This means everything in the universal set but not in set A. It is called the **complement** of A.
So $A' = \{2, 3, 5\}$ and $B' = \{1, 3\}$.
Since $A \cap B = \{4\}$
then $(A \cap B)' = \{1, 2, 3, 5\}$
In the same way, can you show that $(A \cup B)' = \{3\}$?

Exercise 1.2

$\mathscr{E} = \{3, 4, 5, 6, 7, 8, 9\}$
$A = \{4, 5, 6\}$
B = {odd numbers between 2 and 8}
C = {square numbers $\leqslant 10$}

Using the above sets,
1. List
   (a) B
   (b) $A \cap B$
   (c) $A \cup B$
   (d) A'
   (e) the complement of B
   (f) $A' \cap B$
   (g) $(A \cup B)'$
2. List the intersection of B and C.
3. List the union of A and C.
4. Write down n(C).
5. Find n($A \cup C$).
6. Find n(B').

Venn Diagrams

A Venn diagram is a very useful method of visualising the relationship between sets. A rectangular box is drawn to show the universal set, and within this, other shapes, usually ovals or circles, represent sets within the universal set. If there is an intersection between sets, it will be shown by an overlap of the ovals or circles.
The following example should help you to understand the concept of a Venn diagram.

8 Extended Mathematics for Cambridge IGCSE

Example 3

$\mathscr{E} = \{1, 2, 3, 4, 5, 6, 7, 8, 9, 10\}$

$A = \{\text{even numbers}\}$

$B = \{1, 2, 3, 4, 5\}$

$C = \{6, 7, 8\}$

(a) Draw a Venn diagram to represent the above sets. Show the elements in each set.

(b) List the following sets.

    (i) $A \cap B$ (ii) $A \cup B$ (iii) $A \cup C$ (iv) $(A \cup C)'$

    (v) $A \cup B \cup C$ (vi) $(A \cup B \cup C)'$ (vii) $B \cap C$

Answer 3

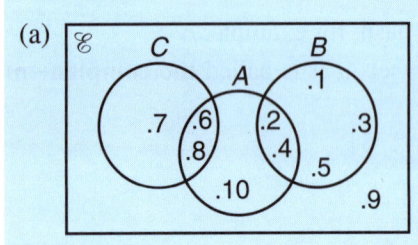

(a) [Venn diagram showing sets C, A, B within universal set ℰ, with elements placed appropriately]

NOTE: The symbol ($\mathscr{E}$) for the Universal Set may be placed either inside or outside the rectangle. Either is correct. You may find different symbols used for the Universal Set in other books, but it should always be clear what is intended.

(b)   (i) $A \cap B = \{2, 4\}$ (ii) $A \cup B = \{1, 2, 3, 4, 5, 6, 8, 10\}$

    (iii) $A \cup C = \{2, 4, 6, 7, 8, 10\}$ (iv) $(A \cup C)' = \{1, 3, 5, 9\}$

    (v) $A \cup B \cup C = \{1, 2, 3, 4, 5, 6, 7, 8, 10\}$ (vi) $(A \cup B \cup C)' = \{9\}$

    (vii) $B \cap C = \emptyset$

NOTE: You may find that shading or hatching can help to make the distinction between union and intersection between sets in the Venn diagram more clear.

If you look at the three Venn diagrams given in the diagrams below, you will see that in the first, one set has been hatched in a different direction from the other.

In the second Venn diagram, the two sets show an intersection, where the hatching appears in both directions (cross hatching).

In the third Venn diagram, the union is shown by all the shaded areas, in either direction, taken together.

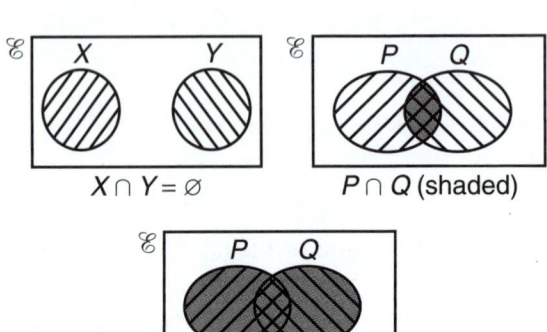

$X \cap Y = \emptyset$ $P \cap Q$ (shaded)

$P \cup Q$ (shaded)

Example 4

(a) Draw three identical Venn diagrams to illustrate the following sets.
   $\mathscr{E}$ = {students in your school}
   C = {students in your class}
   B = {students who come to school by bus}
(b) (i) In the first diagram shade and clearly label $B \cap C$.
   (ii) In the next diagram shade and clearly label $B \cup C$.
   (iii) In the last diagram shade and clearly label $C \cap B'$.
(c) How would you describe $B \cap C$ and $C \cap B$ in words?

Answer 4

(a) and (b)

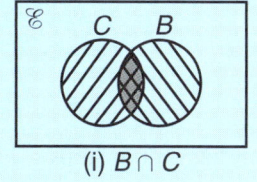

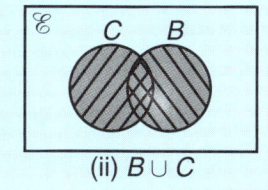

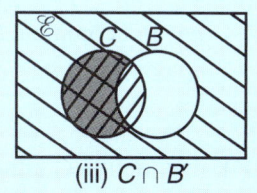

(i) $B \cap C$ (ii) $B \cup C$ (iii) $C \cap B'$

(c) $B \cap C$ = {students in the class who come to school by bus}
   $C \cap B'$ = {students in the class who do not come to school by bus}

You should be able to see what a powerful tool Venn diagrams can be in the study of sets. We will use them again later in the course.

Study the next four diagrams carefully to see how hatching and shading can clarify the required areas of Venn diagrams.

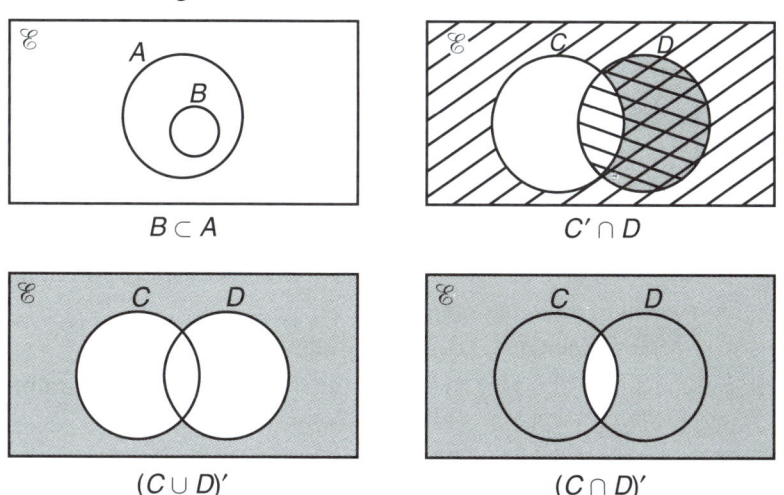

$B \subset A$ $C' \cap D$

$(C \cup D)'$ $(C \cap D)'$

NOTE: The hatching used above is to help you understand which area of the Venn diagram is required. In your examination you should only shade the area specified in the question.

Venn diagrams can also show the number of elements in each set as the next example shows.

Example 5

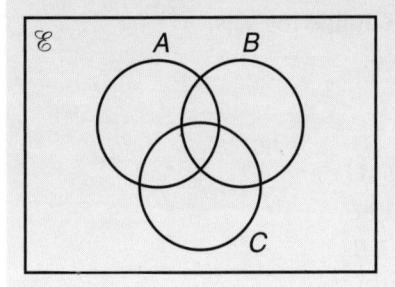

You are given the following information.

n($\mathscr{E}$) = 30, n(A) = 11, n(B) = 12, n(C) = 10,

n(A ∩ B ∩ C) = 3, n(A ∩ B) = 5, n(A ∩ C) = 4, n(B ∩ C) = 6

(a) Copy and complete the Venn diagram.
(b) Find (i) n(A ∪ B ∪ C)′ (ii) n(A ∪ B)

Answer 5

(a) Using the information given, it is possible to fill in the Venn diagram in steps.

STEP 1: n(A ∩ B ∩ C) = 3, so 3 goes right in the centre of the diagram.

STEP 2: n(A ∩ B) = 5, so take away the 3 that is already there, which leaves 2 to go in the rest of the intersection.

n(A ∩ C) = 4 and n(B ∩ C) = 6 means we can fill in the other two intersections in the same way. (See the diagram below for steps 1 and 2)

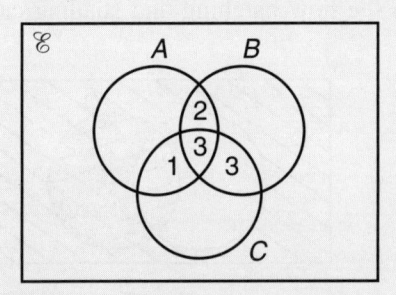

STEP 3: n(A) = 11 so take away the 1, 2 and 3 which are already in A, leaving 5 to be written in the remaining part of A.
Use n(B) = 12 and n(C) = 10 in the same way.

STEP 4: Find n(A ∪ B ∪ C) by adding all the numbers in your diagram.

n(A ∪ B ∪ C) = 5 + 1 + 3 + 2 + 4 + 3 + 3 = 21

It was given that n($\mathscr{E}$) = 30, so n(A ∪ B ∪ C)′ = 30 − 21 = 9

Write in the 9 to complete the Venn diagram. (See the diagram below for step 4.)

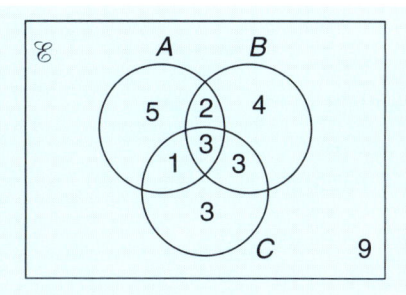

(b) (i) $n(A \cup B \cup C)' = 9$ (ii) $n(A \cup B) = 5 + 1 + 2 + 3 + 4 + 3 = 18$
The answers to these and many other similar questions can now be easily obtained from the diagram.

Exercise 1.3

1. Draw Venn diagrams to illustrate the following.
   (a) $\mathscr{E} = N$, $P =$ {prime numbers less than 10}, $E = \{2, 4, 6\}$
   (Mark and label each element in its correct place.)
   (b) $\mathscr{E} = Z$, $S =$ {square numbers}, $E =$ {even numbers}
   (The individual elements cannot be labelled because there are an infinite number of elements.)

2. Copy these Venn diagrams and in each case, shade the required areas.
   (a) B'
   (b) $A \cup B$
   (c) $A \cap B'$
   (d) $(A \cap B)'$

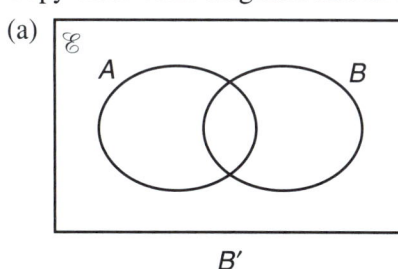

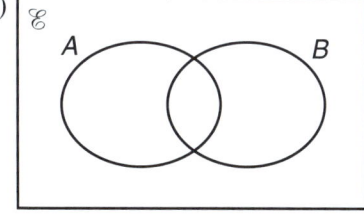

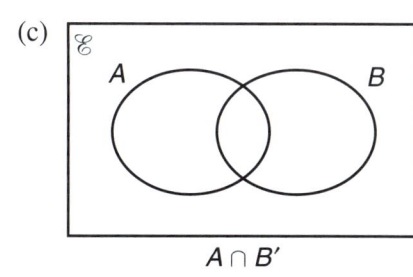

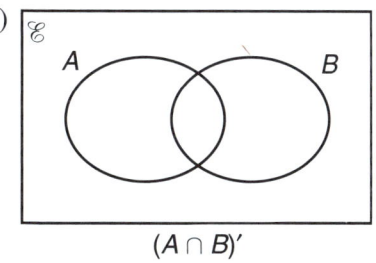

3. Draw Venn diagrams to illustrate each of the following.
   (a) $A \cap B = \emptyset$ (b) $A \subset B$

4. Draw a Venn diagram to illustrate the following.
   $\mathscr{E} =$ {natural numbers from 1 to 10}
   $P = \{1, 2, 3, 4, 5, 6, 7, 8\}$

$R = \{2, 4, 6, 8\}$
$S = \{1, 2, 3, 4\}$
Mark and label the correct position for each element in the diagram.

5. (a) Draw a Venn diagram to show two sets, A and B, where $B \subset A$.
   (b) Given that $n(\mathscr{E}) = 48$, $n(A) = 25$ and $n(B) = 10$, find
       (i) $n(A \cap B')$ (ii) $n(A')$
   (c) Complete the Venn diagram showing the numbers in each area. Check that all your numbers add up to 48.

6. (a) Draw a Venn diagram showing two sets, P and S, with an intersection.
   (b) Given that $n(\mathscr{E}) = 20$, $n(P) = 7$, $n(S) = 16$ and $n(P \cup S)' = 0$, find $n(P \cap S)$.
   (c) Complete the diagram.

NOTE: Since $n(P \cup S)' = 0$, all 20 of the elements must lie in P and S, but adding 7 and 16 makes too many, so how many must be in the intersection?

More about Square Roots and Irrational Numbers

In this section of this chapter we will see how we can find the square root of any perfect square, or express irrational square roots in their simplest form, without using a calculator. There are two possible methods.

(a) In the first, we express the number, as a product of its prime factors (by using a factor tree if necessary).

For example, $\quad 900 = 2 \times 2 \times 3 \times 3 \times 5 \times 5$

so $\quad \sqrt{900} = \sqrt{2 \times 2 \times 3 \times 3 \times 5 \times 5}$

This is the same as $\quad \sqrt{2} \times \sqrt{2} \times \sqrt{3} \times \sqrt{3} \times \sqrt{5} \times \sqrt{5}$

We know $\quad \sqrt{2} \times \sqrt{2} = \sqrt{4} = 2$, $\sqrt{3} \times \sqrt{3} = 3$ and $\sqrt{5} \times \sqrt{5} = 5$

This means $\quad \sqrt{900} = \sqrt{2 \times 2 \times 3 \times 3 \times 5 \times 5} = 2 \times 3 \times 5 = 30$

so $\quad \sqrt{900} = 30$.

This is also the best method for finding cube roots.

$\sqrt[3]{216} = \sqrt[3]{2 \times 2 \times 2 \times 3 \times 3 \times 3} = \sqrt[3]{2 \times 2 \times 2} \times \sqrt[3]{3 \times 3 \times 3} = 2 \times 3 = 6$

(b) In the second method, we look for factors of the number that are perfect squares themselves, such as 4, 9, 16, 25 and so on.

So $\quad \sqrt{900} = \sqrt{4 \times 9 \times 25} = 2 \times 3 \times 5 = 30$

Either method will also work for numbers whose square roots are irrational.
For example, by the first method,

$\sqrt{450} = \sqrt{2 \times 3 \times 3 \times 5 \times 5} = \sqrt{2} \times \sqrt{3 \times 3} \times \sqrt{5 \times 5} = \sqrt{2} \times 3 \times 5 = 15 \times \sqrt{2}$

and $15 \times \sqrt{2}$ is usually written $15\sqrt{2}$.

By the second method,
$$\sqrt{450} = \sqrt{9 \times 25 \times 2} = 3 \times 5 \times \sqrt{2} = 15\sqrt{2}.$$

This is as far as you can go easily without using a calculator.
Square roots that are irrational are also called **surds**. The number $15\sqrt{2}$ is a surd because $\sqrt{2}$ is irrational.

Example 6
(a) Find $\sqrt{225}$ without using a calculator.
(b) Simplify $\sqrt{432}$ by writing it as a surd, in terms of $\sqrt{2}$ or $\sqrt{3}$ or $\sqrt{5}$.

Answer 6
(a) $\sqrt{225} = \sqrt{3 \times 3 \times 5 \times 5} = \sqrt{9 \times 25} = 3 \times 5 = 15$
(b) $\sqrt{432} = \sqrt{2 \times 2 \times 2 \times 2 \times 3 \times 3 \times 3} = \sqrt{16 \times 9 \times 3}$
    $4 \times 3 \times \sqrt{3} = 12\sqrt{3}$

Exercise 1.4
1. Without using a calculator (and showing all your working) find the following.
   (a) $\sqrt{784}$ (b) $\sqrt{1600}$ (c) $\sqrt{625}$
2. Express the following square roots as surds in terms of $\sqrt{2}$, $\sqrt{3}$ or $\sqrt{5}$.
   (a) $\sqrt{180}$ (b) $\sqrt{98}$ (c) $\sqrt{192}$
3. In each case, show that the following square roots are either rational or irrational.
   (a) $\sqrt{50}$ (b) $\sqrt{144}$ (c) $\sqrt{45}$

Fractions, Decimals and Percentages

The basic work on fractions, decimals and percentages was covered in *Core Mathematics for Cambridge IGCSE*. The questions you will be asked at Extended Level may be just a little more difficult. Work through the next exercise and check your answers before attempting the final exercise.

Exercise 1.5
1. Write 2.23 as
   (a) a mixed number,   (b) an improper fraction.
2. Arrange in order of size, starting with the smallest
   0.21% $\dfrac{21}{500}$ 2×10^{-3} $\dfrac{1}{250}$
3. (a) Calculate 65 as a percentage of 50.
   (b) Calculate 38.2 as a percentage of 40.
   (c) Calculate 6 as a percentage of 1200.

4. Arrange the following in order of size, starting with the smallest.

$\dfrac{22}{41}, \dfrac{47}{99}$ and $\dfrac{53}{79}$

NOTE: Use your calculator to change each fraction to a decimal for comparison.

5. Find a fraction halfway between
   (a) $\dfrac{17}{20}$ and $\dfrac{9}{10}$ (b) $\dfrac{4}{5}$ and $\dfrac{5}{6}$

Exercise 1.6

Mixed Exercise

1. $\mathscr{E}$ = {a, b, c, d, e, f, g, h} L = {a, c, e} M = {b, c, d, e, f}
   (a) Draw a Venn Diagram showing these sets. Shade L in one direction and M in another.
   (b) List
      (i) $L \cap M$ (ii) $L \cup M$ (iii) L' (iv) M' (v) $(L \cap M)'$
   (c) Draw another Venn diagram, Shade L in one direction and M' in another.
   (d) List
      (i) $L \cap M'$ (ii) $L \cup M'$

2. List the integers from −2 to 2 inclusive. (Inclusive means including −2 and 2.)

3. Describe the following
   (a) {1, 8, 27, 64}   (b) {1, 4, 9, 16, 25}

4. Simplify as far as possible
   (a) $\sqrt{4096}$ (b) $\sqrt{2450}$

5. Which of the following are irrational numbers?
   (a) $\sqrt{500}$ (b) $\sqrt{121}$

6. You are given the following information.
   $\mathscr{E}$ = {integers from −10 and 10 inclusive}
   F = {2, 4, 6}
   G = {−2, −1, 0, 1, 2}
   H = {1, 2, 3, 4, 5, 6, 7, 8}
   By first drawing a Venn diagram, find
   (a) n($F \cap G$)   (b) n($F \cup G$)   (c) n($F \cap G \cap H$)
   (d) n($F \cup G \cup H$)   (e) n($F \cup G \cup H$)'

Examination Questions

7. From the list of numbers $\dfrac{22}{7}$, π, $\sqrt{14}$, $\sqrt{16}$, 27.4, $\dfrac{65}{13}$ write down
   (a) one integer,
   (b) one irrational number.

(0580/02 Oct/Nov 2004 q 3)

8. $\mathscr{E} = \{40, 41, 42, 43, 44, 45, 46, 47, 48, 49\}$
   $A = \{\text{prime numbers}\}$ $\qquad$ $B = \{\text{odd numbers}\}$
   (a) Place the ten numbers in the correct places on the Venn diagram.

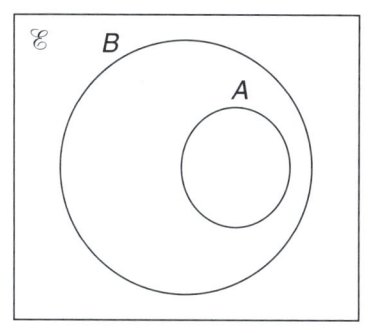

   (b) State the value of n($B \cap A'$)

   (0580/02 Oct/Nov 2004 q 11)

9. $\mathscr{E} = \left\{-2\frac{1}{2}, -1, \sqrt{2}, 3.5, \sqrt{30}, \sqrt{36}\right\}$

   $X = \{\text{integers}\}$
   $Y = \{\text{irrational numbers}\}$
   List the members of $\qquad$ (a) X, $\qquad$ (b) Y.

   (0580/02 Oct/Nov 2003 q 5)

10. (a) Express in set notation, as simply as possible, the subset shaded in the Venn diagram.

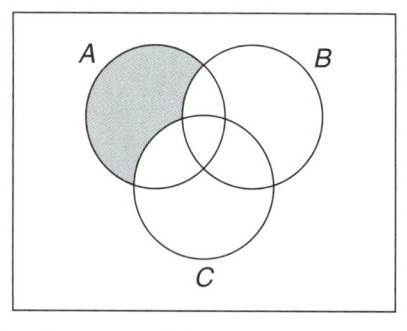

    (b) It is given that n($\mathscr{E}$) = 20 and P and S are sets such that n(P) = 7 and n(S) = 16. Find the smallest possible value of n($P \cap S$).

    (4024/01 Oct/Nov 2004 q 18 a and d)

11. In a group of language students, 24 studied Spanish, 23 studied French and 15 studied German. 12 studied Spanish and French, 10 studied German and French, 6 studied Spanish and German and 4 studied all three languages. By drawing a Venn diagram or otherwise, calculate the number of students who studied
    (a) both Spanish and French, but not German,
    (b) only one language.

    (4024/01 May/June 2003 q 18 a)

12. Rearrange the quantities in order with the smallest first.

    $\frac{1}{8}\%$, $\frac{3}{2500}$, 0.00126

    (0580/02 Oct/Nov 2003 q 4)

13. Lin scored 18 marks in a test and Jon scored 12 marks. Calculate Lin's mark as a percentage of Jon's mark.

    (0580/02 May/June 2008 q 3)

14. 0.0008 8×10^{-5} 0.8% $\frac{1}{125000}$

    Write the above numbers in order, smallest first.

    (0580/02 Oct/Nov 2006 q 6)

15. (a) Shade the region $A \cap B$.

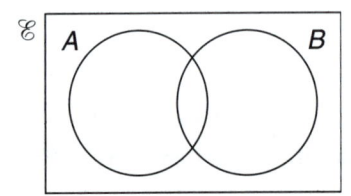

    (b) Shade the region $(A \cup B)'$.

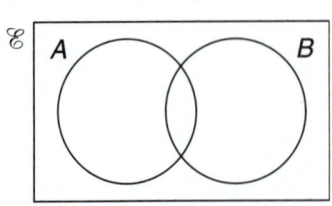

    (c) Shade the complement of set B.

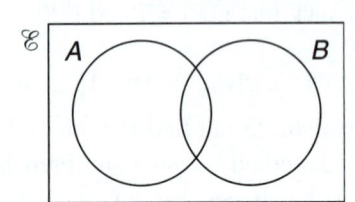

    (0580/92 Oct/Nov 2006 q 11)

16. Write down the next prime number after 89.

    (0580/02 May/June 2006 q 2)

17. n(A) = 18, n(B) = 11 and n(A∪ B)' = 0.
   (a) Label the Venn diagram to show the set A and B where n(A ∪ B) = 18.
       Write down the number of elements in each region.

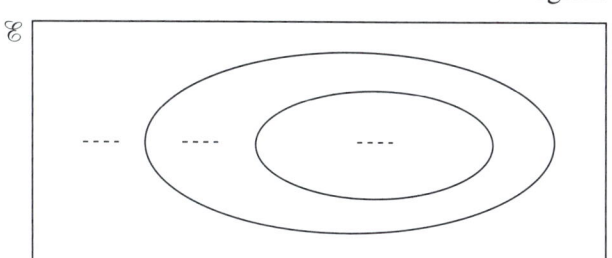

   (b) Draw another Venn diagram to show the sets A and B where n(A ∪ B) = 29.
       Write down the number of elements in each region.

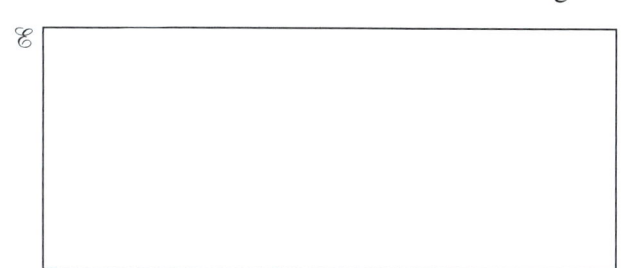

(0580/02 May/June 2006 q 17)

18. On the Venn diagrams shade the regions
    (a) $A' \cap C'$,

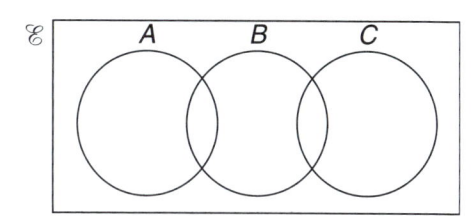

    (b) $(A \cup C) \cap B$.

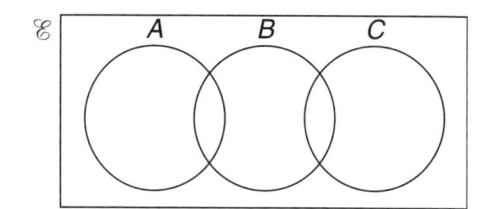

(0580/02 May/June 2007 q 8)

19. n($\mathscr{E}$) = 21, n($A \cup B$) = 19, n($A \cap B'$) = 8 and n(A) = 12.
Complete the Venn diagram to show this information.

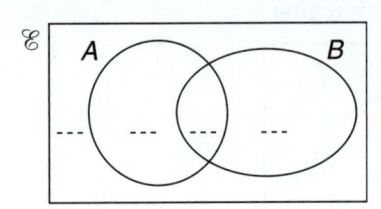

(0580/02 May/June 2005 q 11)

20. In a survey, 100 students were asked if they liked basketball (B), football (F) and swimming (S).
The Venn diagram shows the results.

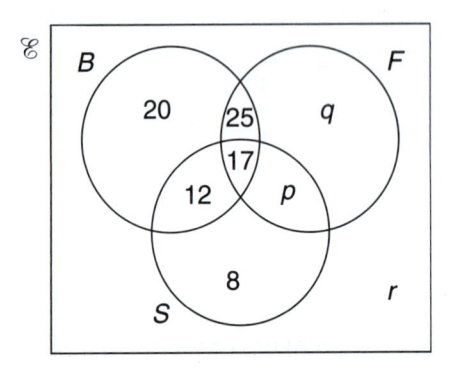

42 students liked swimming.
40 students liked exactly one sport.
(a) Find the value of p, q and r.
(b) How many students like
    (i) all three sports,
    (ii) basketball and swimming but not football?
(c) Find
    (i) n(B)   (ii) n($(B \cup F) \cap S'$)

(0580/04 Oct/Nov 2008 q 9 a, b and c)

21. Write the following in order of size, smallest first.

$\dfrac{399}{401}$ $\dfrac{689}{701}$ $\dfrac{598}{601}$

(0580/02 May/June 2008 q 6)

22. A and B are sets.
    Write the following sets in their simplest form.
    (a) $A \cap A'$.
    (b) $A \cup A'$.
    (c) $(A \cap B) \cup (A \cap B')$.

    (0580/02 Oct/Nov 2007 q 12)

23. Work out the value of $1 + \dfrac{2}{3 + \dfrac{4}{5+6}}$

    (0580/21 Oct/Nov 2008 q 3)

Chapter 2

Algebra I

Algebra is a vital tool in Mathematics. It is essential that you understand the language of algebra, both the significant words such as 'equation' and 'evaluate', and the shorthand notation that mathematicians use to communicate their ideas.
Make sure that you can complete the Core Skills exercise before you attempt to continue with the rest of the chapter. If necessary refer back to Chapter 3 in *Core Mathematics for Cambridge IGCSE* to refresh your memory.

Core Skills

1. $y = 2x + 3$ $2x + 5y - x - 1$ $a^2 = b^2 + c^2$
   From the selection above, choose
   (a) a formula,   (b) an equation,   (c) an expression.
2. From the expression, $2x - x^2 + 3xy - 6 - 4xy$ choose
   (a) a term in x,   (b) a constant term,
   (c) the coefficient of the term in x^2,   (d) two like terms.
3. Calculate
   (a) $2 - 1 + 5 - 6 - 3$ (b) $-7 + 3 - 4 + {}^-1$
   (c) ${}^-2 + {}^-2 - {}^-2$ (d) $3 - {}^-4$
4. Calculate
   (a) $2 \times {}^-4$ (b) ${}^-5 \times {}^-6$ (c) $2 \div {}^-4$
   (d) ${}^-5 \div {}^-6$ (e) $(-1)^3$ (f) $(-2)^4$
5. Simplify
   (a) $a^2 \times a^3$ (b) $x^3 \times x^5 \times x$ (c) $b^5 \div b^2$
   (d) $(c^2)^5$ (e) $(ab^2)^3$
6. Simplify where possible
   (a) $(-x)^2 + (-x)^3$ (b) $-6x \div -2y$
   (c) $xy + x^2y^2$ (d) ${}^-2a \times {}^-3b \div 6a$
7. Simplify
   (a) $x^4 \times y^5 \times y^2 \times x^3$ (b) $2x^3 \div 3x^2 \times 9x^5$
   (c) $x^{11} \div x^7 \times x^8$ (d) $(x^2y^5)^6$

8. Simplify the following expressions
   (a) $2x - 3y - 5x + y$
   (b) $a + b + c - {}^-a + {}^-b - {}^-c$
   (c) $pqr - qpr$
   (d) $ab + 3a - 4b - 2ab + 4b + a$
9. Evaluate the following expressions by substituting $x = 2$, $y = {}^-3$ and $z = {}^-4$
   (a) $xyz + 1$
   (b) $x^2 + y^2 + z^2$
   (c) $xy - yz + zx$
   (d) $x^2 + y^2 - z$
10. Calculate the following
    (a) 3^{-1}
    (b) $2^{-1} \times 2^{-3}$
    (c) $\left(\frac{1}{2}\right)^{-2}$
    (d) $\left(2\frac{1}{2}\right)^{-2}$
11. In each of the following find a replacement for n which makes the statement true.
    (a) $1 = 100^n$
    (b) $\frac{3}{4} = \left(\frac{4}{3}\right)^n$
    (c) $\left(\frac{2}{3}\right)^n = \frac{9}{4}$
    (d) $10 = 10^n$
12. Multiply out the brackets and simplify
    (a) $2(x - y) - 2(x + y)$
    (b) $pqr - pq(r + 1)$
    (c) $a^2(a + b^2) + b^2(a^2 + b)$
    (d) $x^2y^2 - (x^2y^2 - 1) + x^2(y^2 - 1) - y^2(x^2 + 1)$
13. List the common factors in each of the following expressions
    (a) $a^2bc^2 + a^3b^2c$
    (b) $12x^2y^3 - 9x^4y^2 + 18x^3y$
14. Factorise completely
    (a) $5xyz + 10x^2y$
    (b) $14x^2y - 21xy^2$
    (c) $3a^2 - 6a^3$
    (d) $2x^2 - x$

Expanding Products of Algebraic Factors

You have met examples of multiplying a bracketed expression by a number and letters. An example of a bracketed expression is $(2x + 5y)$.

We now have to look at multiplying two bracketed expressions together and for simplicity we shall from now on say 'brackets' rather than 'bracketed expression'.

Take a simple example:

$$(2 + x) \times (3 + y)$$

This could be separated into:

$$2 \times (3 + y) + x \times (3 + y)$$
$$= 6 + 2y + 3x + xy$$

Note how every term in the first pair of brackets multiplies every term in the second pair of brackets.

It is normal to write the brackets without the multiplication sign using algebraic shorthand.

$$(2 + x)(3 + y) = 6 + 2y + 3x + xy$$

You will get used to doing this multiplication quickly if you are systematic.

Start with the 2 from the first pair of brackets and multiply it first by the 3 and then by the $+y$ from the second pair of brackets, writing down both results. Now move on to the $+x$ and multiply it by first the 3 and then the $+y$, again writing down both results.

NOTE: If you really find this difficult you can draw a table like this:

×	3	+y
2		
+x		

and fill it in. However, this is fine to start with but is too complicated and lengthy for permanent use, so it is much better to practise working in the systematic way described above until it becomes easy for you.

Once again we have to take careful notice of any minus signs, as being careless with these is one of the commonest ways to get a wrong answer and lose marks.

Usually you will find that there is a second line of working after you have multiplied out the brackets where you can simplify by collecting like terms, as you will see in the next example. Multiplying out the brackets is sometimes called **expanding** the brackets.

Example 1
Multiply out the brackets and simplify where possible.
(a) $(x + 2)^2$ NOTE: Be careful! It is the brackets that are squared, not just the x and the $+2$!
(b) $(x + 2)(x - 3)$ (c) $(3a - 4)(2a + 5)$ (d) $(x + 1)(x - 1)$
(e) $(a + b)(a - b)$ (f) $(2 - s)(5 + s)$ (g) $(x^2 + 1)(x + 2)$
(h) $(a + b)(a + b + 1)$ NOTE: As usual, just be systematic.

Answer 1
(a) $(x + 2)^2 = (x + 2)(x + 2)$
$= x^2 + 2x + 2x + 4 = x^2 + 4x + 4$

(b) $(x + 2)(x - 3) = x^2 - 3x + 2x - 6$
$= x^2 - x - 6$

(c) $(3a - 4)(2a + 5) = 6a^2 + 15a - 8a - 20$
$= 6a^2 + 7a - 20$

(d) $(x + 1)(x - 1) = x^2 - x + x - 1$
$= x^2 - 1$

(e) $(a + b)(a - b) = a^2 - ab + ab - b^2$
$= a^2 - b^2$

(f) $(2 - s)(5 + s) = 10 + 2s - 5s - s^2$
$= 10 - 3s - s^2$

(g) $(x^2 + 1)(x + 2) = x^3 + 2x^2 + x + 2$

(h) $(a + b)(a + b + 1) = a^2 + ab + a + ba + b^2 + b$
$= a^2 + 2ab + a + b^2 + b$ NOTE: Remember that ba is the same as ab.

Exercise 2.1
Multiply out the brackets and simplify where possible
1. $(a + 1)(a + 1)$
2. $(x + 4)(x + 5)$
3. $(x + 4)(x - 5)$
4. $(x - 4)(x + 5)$
5. $(x - 4)(x - 5)$
6. $(2b + 1)(b + 1)$

7. $(5c - 2)(c - 2)$
8. $(6x + 5)(2x + 3)$
9. $(x + y)(x + y)$
10. $(x - y)(x - y)$
11. $(x - y)(x + y)$
12. $(2d + 3e)(2d - 3e)$
13. $(7z + 1)(2z - 1)$
14. $(2 + x)(4 + x)$
15. $(2 + x)(2 - x)$
16. $(a^2 + b)(a + b)$
17. $(x^2 - 1)(x + 1)$
18. $(b + c)(2b + 2c + 1)$
19. $(x - 1)(x^2 + x + 1)$
20. $(x^2 + 1)(x^2 - 1)$
21. $(2x + 3)^2$
22. $(2x + 3)(2x - 3)$
23. $(2x - 3)^2$
24. $(3x - 4)(3x + 4)$
25. $(2b - 1)(2b + 1)$
26. $(2a + 3b)(c + d)$

Factorising Quadratic Expressions

A quadratic expression contains a term in, say, x^2, and can contain a term in x and a constant term, but no other terms. For example, $x^2 + 2x + 1$ is a quadratic expression.
Now that we know how to multiply out two pairs of brackets and simplify the result we need to answer the question 'how do we get back to our original two pairs of brackets?'
This needs to be treated like a puzzle which can be solved by trial and error. Try a possible solution, and then multiply out again to see if you get back the original expression.
You will often find that you will not get the correct answer straight away, but this does not mean that you are doing anything particularly wrong, so just try again, with a different combination of letters and numbers. After a while you will begin to see the patterns and will find that you get the correct answer with fewer attempts. It is worth practising these until it becomes easier.
We will start with easy examples, and work on towards the more difficult. First we will look at an example which has been multiplied out, so we can see where the terms come from.

$$(x + 2)(x + 5)$$
$$= x^2 + 5x + 2x + 10$$
$$= x^2 + 7x + 10$$

If you are asked to factorise this expression you can start by drawing two empty pairs of brackets:

$$= x^2 + 7x + 10$$
$$= (\quad)(\quad)$$

The term at the beginning of each pair of brackets comes from factorising x^2, so this is easy:

$$= x^2 + 7x + 10$$
$$= (x\quad)(x\quad)$$

The term at the end of each pair of brackets comes from factorising the constant term, +10. Beside your working it is a good idea to write down all the factors of +10, as usual being systematic about it.
The factors of 10 are {1, 2, 5, 10}.
Start with the smallest and write it down multiplied by the other factor which gives the product 10. (In this case 1×10.) Then move on to the next factor. (In this case 2×5.) This is as far as you can go with 10 because you will start repeating pairs of factors. By working in this way you will not be in danger of leaving out any factors when the examples become more difficult.

$x^2 + 7x + 10$ 1×10
$= (x\quad)(x\quad)$ 2×5

The first part of the puzzle is to decide which of these two pairs of factors you need to choose. If you look back to the multiplying out you will see that the two factors have to add up to +7, so in this example there is not a lot of choice, and we pick 2 and 5, both being positive.

$= x^2 + 7x + 10$ $2 + 5$
$= (x + 2)(x + 5)$ $= 7$

Things start to get more complicated when the constant term has more pairs of factors, or when there are negative signs as well. We will come to these later.

As usual, you should work through the examples, and then work through the whole exercise.

NOTE: It is worth repeating that you should always check your answer by multiplying out the brackets again.

Example 2
Factorise the following quadratic expressions
(a) $x^2 + 11x + 10$
(b) $x^2 + 10x + 24$
(c) $x^2 + 11x + 24$

Answer 2
(a) $x^2 + 11x + 10$
$= (x\quad)(x\quad)$
$= (x + 1)(x + 10)$

1×10 $1 + 10 = 11$
2×5

(b) $x^2 + 10x + 24$
$= (x\quad)(x\quad)$
$= (x + 4)(x + 6)$

1×24
2×12
3×8
4×6 $4 + 6 = 10$

(c) Using the factors we have already set out in part (b) you should see that $3 + 8 = 11$, so:
$x^2 + 11x + 24$
$= (x + 3)(x + 8)$

NOTE: Remember to be systematic, as you can see in part (b) of the answer, and work through all the factors in order.

Exercise 2.2

Factorise
1. $x^2 + 4x + 3$
2. $x^2 + 6x + 5$
3. $x^2 + 13x + 12$
4. $x^2 + 8x + 12$
5. $x^2 + 7x + 12$
6. $x^2 + 8x + 16$
7. $x^2 + 17x + 16$
8. $x^2 + 10x + 16$
9. $x^2 + 18x + 17$
10. $x^2 + 2x + 1$

We will now look at quadratic expressions containing minus signs.

Algebra I 25

Example 3
Factorise
(a) $x^2 - 5x - 6$ (b) $x^2 + 5x - 6$ (c) $x^2 - 5x + 6$ (d) $x^2 - 7x + 6$

Answer 3
In each one of these we have to factorise 6: 1×6
                                                               2×3

(a) $x^2 - 5x - 6$ $+1 - 6 = -5$, and $+1 \times -6 = -6$
    $= (x + 1)(x - 6)$

(b) $x^2 + 5x - 6$ $-1 + 6 = +5$, and $-1 \times +6 = -6$
    $= (x - 1)(x + 6)$

(c) $x^2 - 5x + 6$ $-2 - 3 = -5$, and $-2 \times -3 = +6$
    $= (x - 2)(x - 3)$

(d) $x^2 - 7x + 6$ $-1 - 6 = -7$, and $-1 \times -6 = +6$
    $= (x - 1)(x - 6)$

Exercise 2.3
Factorise
1. $x^2 + 6x + 5$
2. $x^2 + 4x - 5$
3. $x^2 - 4x - 5$
4. $x^2 - 6x + 5$
5. $x^2 - 7x - 8$
6. $x^2 - 6x + 8$
7. $x^2 + 2x - 8$
8. $x^2 + 35x - 36$
9. $x^2 - 16x - 36$
10. $x^2 - 13x + 36$

Example 4
Factorise
(a) $2x^2 + 3x + 1$ (b) $2x^2 + x - 3$ (c) $2x^2 - 5x + 3$ (d) $6x^2 + 17x + 5$
(e) $6x^2 + 13x + 6$ (f) $6x^2 + 5x - 6$ (g) $6x^2 + 5xy - 6y^2$

Answer 4
(a) $2x^2 + 3x + 1$
    $= (2x + 1)(x + 1)$

NOTE: In this case, 1 can only be factorised to 1×1 so there is no choice of combination and the answer is simple.
$2 = 1 \times 2,$ $1 = 1 \times 1$

(b) $2x^2 + 5x + 3$
    $= (2x + 3)(x + 1)$

NOTE: There is now an element of choice because the 2 and the 3 will both factorise.
$2 = 1 \times 2$ and $3 = 1 \times 3$
Try both combinations to find the right one.
       $(2x + 1)(x + 3) = 2x^2 + 7x + 3$
or $(2x + 3)(x + 1) = 2x^2 + 5x + 3$
The second combination is the one we want.

(c) $2x^2 - 5x + 3$
$= (2x - 3)(x - 1)$

NOTE: This is like (b), but with minus signs in both pairs of brackets.

(d) $6x^2 + 17x + 5$
$= (2x + 5)(3x + 1)$

NOTE: The 6 will factorise in two different ways, making more choices.

$6 = 1 \times 6$ $\quad\quad 5 = 1 \times 5$
or $\quad 6 = 2 \times 3$

Again, try each in turn with the factors of 5 until you find the correct combination.

$\quad\quad (x + 1)(6x + 5) = 6x^2 + 11x + 5$
or $\quad (x + 5)(6x + 1) = 6x^2 + 31x + 5$
or $\quad (2x + 1)(3x + 5) = 6x^2 + 13x + 5$
or $\quad (2x + 5)(3x + 1) = 6x^2 + 17x + 5$

We see that the last combination gives the correct value (17) for the coefficient of the x term, so this is the solution we want.

(e) $6x^2 + 13x + 6$
$= (2x + 3)(3x + 2)$

NOTE: This is more difficult because there are two sixes, both of which will factorise in two different ways.

$6 = 1 \times 6$ $\quad\quad 6 = 1 \times 6$
or $\quad 6 = 2 \times 3$ $\quad\quad 6 = 2 \times 3$

Systematic trials with all the possible combinations will get you the right answer.

(f) $6x^2 + 5x - 6$
$(2x + 3)(3x - 2)$

NOTE: This is probably the most difficult one you would have to tackle, because it is like (e), but also with a minus sign.

(g) $6x^2 + 5xy - 6y^2$
$= (2x + 3y)(3x - 2y)$

NOTE: This is the same as (f), but with a term in y at the end of each pair of brackets.

As you can see, you must be prepared for some trial and error when factorising the more difficult quadratics. Practise as much as you can and you will begin to see the solutions without having to write down every possible combination. It does get easier. Always remember to multiply out again to check that you do have the correct solution.

Exercise 2.4

Factorise
1. $3x^2 + 4x + 1$
2. $6x^2 + 5x + 1$
3. $3x^2 + 5x + 2$
4. $3x^2 + 7x + 2$
5. $3x^2 - 2x - 1$
6. $3x^2 + 2x - 1$
7. $3x^2 - 4x + 1$
8. $6x^2 + 7x + 1$
9. $4x^2 - 7x - 2$
10. $4x^2 + 8x + 3$
11. $4x^2 + 13x + 3$
12. $4x^2 - 4x - 3$
13. $4x^2 + 4x - 3$
14. $4x^2 - 8x + 3$
15. $8x^2 + 14x + 3$
16. $8x^2 + 10x - 3$
17. $8x^2 + 10x + 3$
18. $8x^2 + 2x - 3$
19. $8x^2 + 26x + 15$
20. $9x^2 + 30x + 25$
21. $6x^2 - 25x + 25$
22. $8x^2 + 26xy + 15y^2$
23. $9x^2 + 30xy + 25y^2$
24. $6x^2y^2 - 25xy + 25$

NOTE: Look again at question 24. It is the same as question 21, but in $(xy)^2$ and xy instead of x^2 and x.

Algebra I 27

Factorising a Difference of Squares

Do you remember what happened in some of the questions on multiplying out the brackets in Exercise 2.1, for example, $(x + y)(x - y)$?
To take another example:
$$(c + 4d)(c - 4d)$$
$$= c^2 - 4cd + 4cd - 16d^2$$
$$= c^2 - 16d^2$$

This result is called a **difference of squares**. The minus sign means to find the *difference* between c^2 and $16d^2$, and c^2 and $16d^2$ are both terms that are perfect *squares*.
There is no easy way of working out how to factorise a difference of squares, so it is very important that you learn the following result:
$$x^2 - y^2 = (x + y)(x - y)$$

This *only works for a difference of squares*, so do not try to find something similar for a sum of squares, which will not factorise in this way.
One example to look out for is:
$$x^2 - 1 = (x - 1)(x + 1)$$
This is not obvious until you remember that 1 is the same as 1^2.

Example 5
Factorise where possible
(a) $1 - 25x^2$
(b) $9y^2 - 16x^2$
(c) $y^2 + x^2$

Answer 5
(a) $1 - 25x^2$
$= (1 - 5x)(1 + 5x)$
(b) $9y^2 - 16x^2$
$= (3y - 4x)(3y + 4x)$
(c) $y^2 + x^2$
Nothing can be done with this because it is a sum of squares, not a difference

Exercise 2.5

Factorise
1. $x^2 - y^2$
2. $a^2 - 1$
3. $x^2 - 9$
4. $4y^2 - 9$
5. $25 - a^2$
6. $36a^2 - 49b^2$
7. $a^2b^2 - x^2y^2$
8. $1 - 4c^2$

Factorising by Pairing

Looking back at Exercise 2.1 question 26:
$$(2a + 3b)(c + d)$$
$$= 2ac + 2ad + 3bc + 3bd$$

We see that it is not always possible to finish off by simplifying two of the terms. These expressions can often be factorised by first putting them into two pairs.

28 Extended Mathematics for Cambridge IGCSE

For example,
$$ab + xy + ay + bx$$
This can be grouped into two pairs in which each of the pairs has a common factor. We rewrite
$$ab + xy + ay + bx$$
$$= ab + ay + xy + bx$$
$$= a(b + y) + x(y + b)$$
Remembering that $b + y$ is the same as $y + b$, we can take $(b + y)$ out as a common factor.
$$= a(b + y) + x(b + y)$$
$$= (b + y)(a + x)$$
Check that this is correct by multiplying out again.
The same result would have been obtained if the terms had been grouped the other possible way:
$$ab + bx + xy + ay$$
$$= b(a + x) + y(x + a)$$
and so on.

Factorising Systematically

When you are asked to factorise an expression it is worth looking at the question systematically.
Always look for any common factors first, and take them outside a pair of brackets. Then think about difference of squares, then a quadratic and then pairing.

Factorising systematically

1. *Are there any common factors?*
   If so, factorise them out first.
2. *Is there a difference of squares?*
   If so, factorise.
3. *Is the expression quadratic?*
   If so, factorise.
4. *Lastly think about pairing.*

Example 6
Factorise completely
(a) $2x^2 - 8$
(b) $x^2y^2 - 4z^2$
(c) $8ab + 4ay + 4bx + 2xy$
(d) $2a^2 + 6ab + 4b^2$

Answer 6
(a) $2x^2 - 8$
   $= 2(x^2 - 4)$
   $= 2(x - 2)(x + 2)$

NOTE: You would not be able to carry on if you did not spot the common factor first.

(b) $x^2y^2 - 4z^2$
    $= (xy - 2z)(xy + 2z)$
(c) $8ab + 4ay + 4bx + 2xy = 2(4ab + 2ay + 2bx + xy)$
    $= 2[2a(2b + y) + x(2b + y)]$
    $= 2(2b + y)(2a + x)$
(d) $2a^2 + 6ab + 4b^2$
    $= 2(a^2 + 3ab + 2b^2)$
    $= 2(a + 2b)(a + b)$

NOTE: If you had factorised this without first looking for common factors you might have ended up with $(2a + 4b)(a + b)$, or $(a + 2b)(2a + 2b)$, which is fine, but not completely factorised, and as it stands is only worth part of the marks in an examination. In each case, if you go on to factorise out the common factor of 2 you would have completed the factorisation. A similar problem would have arisen if the common factor in (c) had not been taken out first.

Exercise 2.6

Factorise completely
1. $9a + 15b$
2. $1 - x^2$
3. $18 - 2x^2$
4. $2x^2 - x - 1$
5. $6x^2 - 18x$
6. $2x^2 - 12x + 18$
7. $20x^3 - 5x$
8. $3x^2 + 3x - 6$
9. $2y^3 - 4y^2 + 2y$
10. $16xy^2 - 4x^2y$
11. $16xy^3 - 4x^3y$
12. $12ab + 6ay + 3xy + 6bx$

More about Indices

Earlier in this chapter, we looked at indices that were taken from the set of integers (positive and negative whole numbers and zero). Now we must look at fractional indices.

You should remember that finding a square root is the inverse of squaring, and so squaring a number's square root gives back the number itself.

Look at the following statements:

$$\sqrt{2} \times \sqrt{2} = 2$$

$$2^{\frac{1}{2}} \times 2^{\frac{1}{2}} = 2^{\frac{1}{2} + \frac{1}{2}} = 2^1 = 2$$

It makes sense to say that $\sqrt{2}$ is the same as $2^{\frac{1}{2}}$.

In the same way,

$$\sqrt[3]{2} \times \sqrt[3]{2} \times \sqrt[3]{2} = 2^{\frac{1}{3}} \times 2^{\frac{1}{3}} \times 2^{\frac{1}{3}} = 2^{\frac{1}{3} + \frac{1}{3} + \frac{1}{3}} = 2^1 = 2$$

You can see that the root is denoted by the denominator in a fractional index.

Thus a cube root is written as a power of $\frac{1}{3}$, a fourth root is written as a power of $\frac{1}{4}$ and so on.

Example 7
Calculate

(a) $25^{\frac{1}{2}}$

(b) $32^{\frac{1}{5}}$ NOTE: This means the fifth root of 32. In other words, which number raised to the power 5 would equal 32?

When you are trying to find the root of a number, start investigating small numbers first. You will not be given anything too difficult, the most likely roots are 2, 3, 4 or 5.

Answer 7

(a) $25^{\frac{1}{2}}$ means the square root of 25, which is 5

$25^{\frac{1}{2}} = 5$

(b) $32^{\frac{1}{5}}$ trying 2 first, $2 \times 2 \times 2 \times 2 \times 2 = 32$

$32^{\frac{1}{5}} = 2$

The normal rules for working with indices apply to fractional indices as well. In particular it is useful to understand the following:

$$32^{\frac{3}{5}} = (32^3)^{\frac{1}{5}} = (32^{\frac{1}{5}})^3 = 2^3 = 8$$

$$32^{\frac{3}{5}} = (32^3)^{\frac{1}{5}} = 32768^{\frac{1}{5}} = 8$$

Hence, you can multiply the indices in any order but it is much simpler to take the root first so that you are dealing with smaller numbers.

Example 8
(a) Calculate

  (i) $(100)^{\frac{3}{2}}$ (ii) $(27)^{-\frac{2}{3}}$

(b) Simplify

  (i) $(x^{\frac{1}{2}} \times y^{\frac{3}{2}})^2$ (ii) $x^{-\frac{1}{2}} \times x^{\frac{5}{2}}$ (iii) $(x^{-\frac{3}{2}})^{-4}$

(c) Find replacements for x which will make the following statements true

  (i) $25^x = 5$ (ii) $27^x = 3$ (iii) $256^x = 4$

Answer 8

(a) (i) $(100)^{\frac{3}{2}} = (100^{\frac{1}{2}})^3 = 10^3 = 1000$

  (ii) $(27)^{-\frac{2}{3}} = \dfrac{1}{27^{\frac{2}{3}}} = \dfrac{1}{\left(27^{\frac{1}{3}}\right)^2} = \dfrac{1}{3^2} = \dfrac{1}{9}$

(b) (i) $(x^{\frac{1}{2}} \times y^{\frac{3}{2}})^2 = x^{(2 \times \frac{1}{2})} \times y^{(2 \times \frac{3}{2})} = x \times y^3 = xy^3$

(ii) $x^{-\frac{1}{2}} \times x^{\frac{5}{2}} = x^{\left(-\frac{1}{2}+\frac{5}{2}\right)} = x^2$ NOTE: $-\frac{1}{2} + \frac{5}{2} = \frac{4}{2} = 2$

(iii) $(x^{-\frac{3}{2}})^{-4} = x^{(-\frac{3}{2} \times -4)} = x^6$ NOTE: $-\frac{3}{2} \times -4 = +6$

NOTE: Ask yourself these questions the other way round:

(c) (i) $25^x = 5$ NOTE: Which power of 5 would make 25?

$5^2 = 25$, so $25^{\frac{1}{2}} = 5$ NOTE: 5 squared equals 25, so the square root of $25 = 5$

$x = \frac{1}{2}$

(ii) $27^x = 3$

$3^3 = 27$, so $27^{\frac{1}{3}} = 3$ NOTE: Check your answer! Remember that while $27^{\frac{1}{3}} = 3$, $27^3 \neq 3$, so the

$x = \frac{1}{3}$ answer is $x = \frac{1}{3}$, not $x = 3$!

(iii) $256^x = 4$

$4^4 = 256$, so $256^{\frac{1}{4}} = 4$ NOTE: 4 to the power 4 equals 256, so the fourth root of $256 = 4$

$x = \frac{1}{4}$

Exercise 2.7

Calculate

1. $81^{\frac{1}{4}}$
2. $81^{-\frac{1}{4}}$
3. $81^{-\frac{3}{4}}$

Simplify, giving your answer in a form with positive powers.

4. $2x^{\frac{2}{5}} + 5x^{\frac{2}{5}}$
5. $2x^{\frac{2}{5}} \times 5x^{\frac{2}{5}}$
6. $2x^{\frac{2}{5}} \div 5x^{\frac{2}{5}}$

7. $2x^{\frac{2}{5}} \times 5x^{-\frac{2}{5}}$
8. $2x^{-\frac{2}{5}} \times 5x^{-\frac{2}{5}}$
9. $2x^{\frac{2}{5}} \div 5x^{-\frac{2}{5}}$

10. $2y^{\frac{3}{4}} \times y^{\frac{1}{2}}$
11. $2y^{\frac{3}{4}} \div y$
12. $(x^{\frac{1}{2}} + 1)(x^{\frac{1}{2}} - 1)$

13. $(xy^2)^{-\frac{1}{2}}$
14. $2x^{-\frac{1}{2}}(x^{\frac{1}{2}} + x)$
15. $(2x)^{\frac{2}{5}} \times (16x)^{\frac{2}{5}}$

16. $(x^{\frac{1}{2}} - 1)^2$
17. $(x^{\frac{1}{2}} - x^{-\frac{1}{2}})^2$

18. Find replacements for x which would make each of the following statements true.
    (a) $32^x = 2$
    (b) $81^x = 3$
    (c) $125^x = 5$

Algebraic Fractions

In this section we are going to see how we can simplify algebraic fractions, and also how to add, subtract, multiply and divide algebraic fractions. The same rules apply for arithmetic, with a few changes to allow for the variables (letters).

Simplifying algebraic fractions

Look at the expressions below.

$$\frac{x}{2}, \frac{y^2}{y+1}, \frac{3}{a}, \frac{\frac{1}{a}+\frac{1}{b}}{\frac{a}{b}}$$

These are all examples of algebraic fractions.
Now we must see how to simplify some typical examples following the familiar rules of arithmetic.

The fraction $\frac{10}{15}$ can be simplified by dividing the denominator and numerator by the highest common factor. In this case it is 5.

$$\frac{10}{15} = \frac{5 \times 2}{5 \times 3} = \frac{2}{3}$$

In a similar way $\frac{abc^2}{a^2c}$ can be simplified by dividing the numerator and denominator by the HCF of abc^2 and a^2c, which is ac.

$$\frac{abc^2}{a^2c} = \frac{ac \times bc}{ac \times a} = \frac{bc}{a}$$

Remember:

- You *must* check that whatever you are dividing by can in fact divide the *whole* of the numerator and the *whole* of the denominator.
- It **really** is best to factorise first to avoid one of the most common mistakes in algebra. Two further examples should demonstrate this.

$$\frac{xy+zy}{y^2} = \frac{y(x+z)}{y \times y} = \frac{x+z}{y}$$

But,

$$\frac{xy+yz}{2y+z} = \frac{y(x+z)}{2y+z}$$

The y cannot divide the whole of the denominator, so no further simplifying can be done.
Complicated fractions can often be simplified by *multiplying* the denominator and numerator by the same number or letter as in this example:

$$\frac{\frac{1}{2}+\frac{2}{3}}{\frac{2}{3}}$$

Algebra I   33

The lowest common multiple of the denominators 2 and 3 is 6, so if the top and bottom of this fraction are multiplied by 6 these denominators can be cancelled out.

It is a good idea to multiply by $\dfrac{6}{1}$ to avoid mistakes.

Remember that $\dfrac{1}{2} \times \dfrac{6}{1} = 3$ and $\dfrac{2}{3} \times \dfrac{6}{1} = 4$.

$$\dfrac{\dfrac{1}{2} + \dfrac{2}{3}}{\dfrac{2}{3}} = \dfrac{\left(\dfrac{1}{2} + \dfrac{2}{3}\right) \times \dfrac{6}{1}}{\dfrac{2}{3} \times \dfrac{6}{1}} = \dfrac{(3+4)}{4} = \dfrac{7}{4} = 1\dfrac{3}{4}$$

Work through this carefully to make sure you understand what is happening.

To avoid some of the common errors in working with complex algebraic fractions it is best, if the numerator or denominator (or both) contain sums or differences of terms, to draw pairs of brackets round them before you start any work.

For example,

$$\dfrac{x+y}{x-y}$$

should be written as $\dfrac{(x+y)}{(x-y)}$ before any work is done.

Remember:

- Before starting work with algebraic fractions draw pairs of brackets round the numerator and denominator if necessary.

Example 9

(a) Calculate $\dfrac{1 + \dfrac{2}{3}}{\dfrac{1}{12} + \dfrac{3}{4}}$ **NOTE: Multiply top and bottom by 12.**

(b) Simplify

(i) $\dfrac{\dfrac{1}{a} + \dfrac{1}{b}}{\dfrac{1}{ab}}$ **NOTE: Multiply top and bottom by ab.**

(ii) $\dfrac{x+y}{3x+3y}$ **NOTE: Factorise the denominator.**

(iii) $\dfrac{x^2 - y^2}{x+y}$ **NOTE: Factorise the numerator (a difference of squares).**

(iv) $\dfrac{x-1}{x^2+x-2}$

(v) $\dfrac{x-2y}{x^2-xy-2y^2}$

(vi) $\dfrac{a^2+3a+2}{a^2+a-2}$ NOTE: Factorise both the numerator and the denominator.

Answer 9

(a) $\dfrac{1+\dfrac{2}{3}}{\dfrac{1}{12}+\dfrac{3}{4}} = \dfrac{\left(1+\dfrac{2}{3}\right)}{\left(\dfrac{1}{12}+\dfrac{3}{4}\right)} = \dfrac{\left(1+\dfrac{2}{3}\right)\times 12}{\left(\dfrac{1}{12}+\dfrac{3}{4}\right)\times 12} = \dfrac{12+8}{1+9} = \dfrac{20}{10} = 2$

(b) Simplify

(i) $\dfrac{\dfrac{1}{a}+\dfrac{1}{b}}{\dfrac{1}{ab}} = \dfrac{\left(\dfrac{1}{a}+\dfrac{1}{b}\right)\times\dfrac{ab}{1}}{\dfrac{1}{ab}\times\dfrac{ab}{1}} = \dfrac{b+a}{1} = b+a$

(ii) $\dfrac{x+y}{3x+3y} = \dfrac{(x+y)}{(3x+3y)} = \dfrac{(x+y)}{3(x+y)} = \dfrac{1}{3}$

(iii) $\dfrac{x^2-y^2}{x+y} = \dfrac{(x+y)(x-y)}{(x+y)} = x-y$

(iv) $\dfrac{x-1}{x^2+x-2} = \dfrac{(x-1)}{(x+2)(x-1)} = \dfrac{1}{x+2}$

(v) $\dfrac{x-2y}{x^2-xy-2y^2} = \dfrac{(x-2y)}{(x-2y)(x+y)} = \dfrac{1}{x+y}$

(vi) $\dfrac{a^2+3a+2}{a^2+a-2} = \dfrac{(a+2)(a+1)}{(a+2)(a-1)} = \dfrac{a+1}{a-1}$

Exercise 2.8

Calculate

1. $\dfrac{\dfrac{1}{2}-\dfrac{1}{3}}{1+\dfrac{1}{6}}$

2. $\dfrac{2-\dfrac{2}{5}}{\dfrac{2}{25}+\dfrac{4}{5}}$

3. $\dfrac{\dfrac{2}{3}+\dfrac{3}{4}}{\dfrac{1}{6}+\dfrac{5}{2}}$

Algebra I 35

Simplify

4. $\dfrac{4x-8}{2}$

5. $\dfrac{3}{6x+9}$

6. $\dfrac{xy+xz}{x}$

7. $\dfrac{xy+xz}{2y+2z}$

8. $\dfrac{3x}{9xy+15xz}$

9. $\dfrac{6x^2+4x}{10x^2-8x}$

10. $\dfrac{3xyz}{xy-4xz}$

11. $\dfrac{x+y}{x^2-y^2}$

12. $\dfrac{4x^2-9}{2x-3}$

13. $\dfrac{y+2}{y^2+3y+2}$

14. $\dfrac{x^2-1}{x^2-2x+1}$

15. $\dfrac{x^2-1}{x^2+2x+1}$

16. $\dfrac{\frac{1}{x}\times\frac{1}{y}}{\frac{3}{x}\times\frac{3}{y}}$

17. $\dfrac{\frac{1}{x}+\frac{x}{y}}{\frac{1}{y}+\frac{y}{x}}$

18. $\dfrac{\frac{a}{b}+\frac{b}{a}}{\frac{1}{ab}}$

Multiplying and dividing algebraic fractions

Let us revise how to multiply and divide fractions involving numbers only:

$$\frac{2}{5}\times\frac{3}{7}=\frac{2\times 3}{5\times 7}=\frac{6}{35} \quad \text{and} \quad \frac{2}{5}\div\frac{3}{7}=\frac{2}{5}\times\frac{7}{3}=\frac{14}{15}$$

You can save work by cancelling before multiplying where possible.
For example,

$$\frac{5}{12}\div\frac{15}{4}=\frac{5}{12}\times\frac{4}{15}=\frac{1\times 4}{12\times 3}=\frac{1\times 1}{3\times 3}=\frac{1}{9}$$

The numerator and denominator were first divided by 5 and then by 4, but these two divisions can be done at the same time.
The alternative is:

$$\frac{5}{12}\div\frac{15}{4}=\frac{5}{12}\times\frac{4}{15}=\frac{20}{180}$$

which then has to be simplified.
The same applies to multiplying and dividing algebraic fractions.

NOTE: In all work with algebraic fractions, if the denominator of the simplified answer is factorised it is better to leave it that way.

For example, $\dfrac{x}{(x+1)(x+2)}$ is preferable to $\dfrac{x}{x^2+3x+2}$.

Example 10
(a) Calculate

(i) $\dfrac{3}{8} \times \dfrac{4}{9}$

(ii) $\dfrac{3}{8} \div \dfrac{1}{6}$

(b) Simplify

(i) $\dfrac{x}{y+1} \times \dfrac{y}{x+1}$

(ii) $\dfrac{x}{y+1} \div \dfrac{x^2}{(y+1)^2}$

(iii) $\dfrac{x^2+2x-3}{x} \div \dfrac{x+3}{x-1}$

Answer 10

(a) (i) $\dfrac{3}{8} \times \dfrac{4}{9} = \dfrac{3 \times 4}{8 \times 9} = \dfrac{1 \times 1}{2 \times 3} = \dfrac{1}{6}$

(ii) $\dfrac{3}{8} \div \dfrac{1}{6} = \dfrac{3}{8} \times \dfrac{6}{1} = \dfrac{3 \times 6}{8 \times 1} = \dfrac{3 \times 3}{4} = \dfrac{9}{4}$

(b) (i) $\dfrac{x}{y+1} \times \dfrac{y}{x+1}$

$= \dfrac{x}{(y+1)} \times \dfrac{y}{(x+1)}$

$= \dfrac{xy}{(y+1)(x+1)}$ NOTE: In this case no further simplification is possible.

(ii) $\dfrac{x}{y+1} \div \dfrac{x^2}{(y+1)^2} = \dfrac{x}{(y+1)} \div \dfrac{x^2}{(y+1)^2}$

$= \dfrac{x}{(y+1)} \times \dfrac{(y+1)^2}{x^2}$

$= \dfrac{y+1}{x}$ NOTE: Divide top and bottom by $(y+1)$ and x.

(iii) $\dfrac{x^2+2x-3}{x} \div \dfrac{x+3}{x-1} = \dfrac{(x^2+2x-3)}{x} \div \dfrac{(x+3)}{(x-1)}$

$= \dfrac{(x+3)(x-1)}{x} \times \dfrac{(x-1)}{(x+3)}$ NOTE: Divide top and bottom by $(x+3)$.

$= \dfrac{(x-1)^2}{x}$

Exercise 2.9

Calculate

1. $\dfrac{4}{9} \times \dfrac{2}{3}$

2. $\dfrac{4}{9} \div \dfrac{2}{3}$

Algebra I 37

Simplify

3. $\dfrac{1}{xy} \times \dfrac{x^2 y^2}{x^2 + y^2}$

4. $\dfrac{x}{x-1} \times \dfrac{x+1}{xy}$

5. $\dfrac{x}{x-1} \div \dfrac{x+1}{xy}$

6. $\dfrac{x+y}{x^2+y^2} \times \dfrac{x-y}{x^2-y^2}$

7. $\dfrac{xyz}{x-y} \div \dfrac{x^2 y^2 z^2}{x^2 - y^2}$

8. $\dfrac{x^2 + 2x + 1}{x} \times \dfrac{1}{(x+1)^2}$

Adding and subtracting algebraic fractions

As with numerical fractions, the common denominator has to be found first. Also, finding the lowest common denominator saves time, as it avoids extra cancelling at the end.

$$\frac{2}{3} + \frac{3}{4} = \frac{2 \times 4}{3 \times 4} + \frac{3 \times 3}{4 \times 3} = \frac{8}{12} + \frac{9}{12} = \frac{17}{12}$$

Remember:

- You *cannot* cancel across the addition sign!

If, as in the last case, the lowest common denominator is merely the product of the two denominators, you might find that a mental picture helps you to work more quickly:

$$= \frac{2 \times 4 + 3 \times 3}{3 \times 4}$$

Example 11

(a) Calculate $\dfrac{2}{5} + \dfrac{3}{8}$

NOTE: The common denominator is 40, so multiply the first fraction top and bottom by 8 and the second one by 5.

(b) Write as a single fraction in its simplest form

(i) $\dfrac{3}{x} - \dfrac{x}{y}$

NOTE: The common denominator is xy.

(ii) $\dfrac{3}{x-1} - \dfrac{4}{2x+1}$

NOTE: The common denominator is $(x - 1)(2x + 1)$. Start by putting brackets round both denominators, then multiply the first fraction top and bottom by $(2x + 1)$ and the second by $(x - 1)$. Beware of the Minus Sign.

(iii) $\dfrac{2x-1}{x-1} + \dfrac{x+1}{2x+1}$

(iv) $\dfrac{3}{xy} - \dfrac{x}{y}$

NOTE: The common denominator is xy, so you only have to multiply the second fraction top and bottom by x.

38 Extended Mathematics for Cambridge IGCSE

(v) $\dfrac{1}{x(x-1)} - \dfrac{1}{(x-1)(x-2)}$ NOTE: The common denominator is $x(x-1)(x-2)$, so multiply the first fraction top and bottom by $(x-2)$ and the second by x.

(vi) $\dfrac{1}{x-1} - \dfrac{1}{x^2-1}$ NOTE: Factorise the second denominator first to find the lowest common denominator.

Answer 11

(a) $\dfrac{2}{5} + \dfrac{3}{8} = \dfrac{2\times 8}{5\times 8} + \dfrac{3\times 5}{8\times 5} = \dfrac{16}{40} + \dfrac{15}{40} = \dfrac{31}{40}$

(b) (i) $\dfrac{3}{x} - \dfrac{x}{y} = \dfrac{3y}{xy} - \dfrac{x^2}{xy}$

$= \dfrac{3y - x^2}{xy}$

(ii) $\dfrac{3}{x-1} - \dfrac{4}{2x+1} = \dfrac{3}{(x-1)} - \dfrac{4}{(2x+1)}$

$= \dfrac{3(2x+1)}{(x-1)(2x+1)} - \dfrac{4(x-1)}{(2x+1)(x-1)}$

$= \dfrac{3(2x+1) - 4(x-1)}{(x-1)(2x+1)}$

$= \dfrac{6x+3-4x+4}{(x-1)(2x+1)}$ NOTE: Here you can see the importance of the brackets, because in this case the last term in the numerator is the product of two negative numbers.

$= \dfrac{2x+7}{(x-1)(2x+1)}$ NOTE: It is usual to leave the denominator in factorised form unless the question says otherwise.

(iii) $\dfrac{2x-1}{x-1} + \dfrac{x+1}{2x+1} = \dfrac{(2x-1)}{(x-1)} + \dfrac{(x+1)}{(2x+1)}$

$= \dfrac{(2x-1)(2x+1)}{(x-1)(2x+1)} + \dfrac{(x-1)(x+1)}{(x-1)(2x+1)} = \dfrac{(4x^2-1)+(x^2-1)}{(x-1)(2x+1)}$

$= \dfrac{5x^2-2}{(x-1)(2x+1)}$

(iv) $\dfrac{3}{xy} - \dfrac{x}{y} = \dfrac{3}{xy} - \dfrac{x^2}{xy} = \dfrac{3-x^2}{xy}$

(v) $\dfrac{1}{x(x-1)} - \dfrac{1}{(x-1)(x-2)} = \dfrac{(x-2)}{x(x-1)(x-2)} - \dfrac{x}{x(x-1)(x-2)}$

$= \dfrac{x-2-x}{x(x-1)(x-2)} = -\dfrac{2}{x(x-1)(x-2)}$

(vi) $\dfrac{1}{x-1} - \dfrac{1}{x^2-1} = \dfrac{1}{(x-1)} - \dfrac{1}{(x-1)(x+1)}$

$= \dfrac{(x+1)}{(x-1)(x+1)} - \dfrac{1}{(x-1)(x+1)}$

$= \dfrac{x+1-1}{(x-1)(x+1)}$

$= \dfrac{x}{(x-1)(x+1)}$

Exercise 2.10

Write in the simplest form

1. $\dfrac{2}{3} + \dfrac{3}{5}$
2. $1 + \dfrac{1}{x}$
3. $\dfrac{x}{3} + \dfrac{x}{4}$
4. $\dfrac{x}{y} - \dfrac{y}{x}$
5. $\dfrac{1}{a-b} + \dfrac{1}{a+b}$
6. $\dfrac{1}{a-b} - \dfrac{1}{a+b}$
7. $\dfrac{3}{x-1} + \dfrac{3}{x}$
8. $\dfrac{3}{x-1} - \dfrac{3}{x}$
9. $\dfrac{2x-3}{3} + \dfrac{2-3x}{4}$
10. $\dfrac{2x-3}{3} - \dfrac{2-3x}{4}$
11. $\dfrac{4}{x+y} - \dfrac{3}{x+2y}$
12. $\dfrac{x}{x-y} - \dfrac{y}{x+y}$
13. $\dfrac{x}{x-y} + \dfrac{y}{x+y}$
14. $\dfrac{a+b}{a-b} + \dfrac{a-b}{a+b}$
15. $\dfrac{a+b}{a-b} - \dfrac{a-b}{a+b}$
16. $\dfrac{x}{2x-y} - \dfrac{2x-y}{x+y}$
17. $\dfrac{2x-1}{xy} + \dfrac{x-2}{y}$
18. $\dfrac{1}{x(x+1)} + \dfrac{1}{x^2-1}$

Exercise 2.11

Mixed Exercise

1. Calculate

   (a) $\dfrac{15}{27} \times \dfrac{9}{15}$

   (b) $\dfrac{\frac{1}{2} + \frac{1}{6}}{\frac{3}{4} - \frac{5}{12}}$

   (c) $(27)^{-\frac{4}{3}}$

   (d) $\left(36^{-2}\right)^{-\frac{1}{4}}$

   (e) $8^2 + 8^0$

   (f) $\dfrac{1}{64^{-\frac{2}{3}}}$

(g) $\left(\dfrac{6}{7}\right)^{-1}$ (h) $\dfrac{6^{-1}}{7}$ (i) $\dfrac{6}{7^{-1}}$

2. Simplify

(a) $\left(x^0 y^3\right)^{-\frac{2}{3}}$ (b) $y^{\frac{1}{2}} \div y^{\frac{5}{6}}$ (c) $\dfrac{(2x^2)^2}{(x^{\frac{1}{2}})^3}$

(d) $3a^{\frac{3}{2}} \times 4b^{\frac{1}{3}} \times (4b)^{\frac{1}{2}} \times (ab)^2$ (e) $\left(\dfrac{x^2}{y^{-3}}\right)^{-1} \times \dfrac{x^{\frac{5}{2}}}{y^{-1}}$

3. Expand the brackets and simplify
   (a) $(3x + 7)(2x - 7)$ (b) $(x - 1)^2$ (c) $(x^2 - y^2)^2$
   (d) $(xy + 1)(xy - 1)$ (e) $(x + a)(y + b)$ (f) $(3a - 5b)(2c + 3d)$

4. Factorise completely
   (a) $3x^2 + x - 2$ (b) $x^2 - 5x - 50$ (c) $2x^2 - 21x + 49$
   (d) $50x^2 - 18$ (e) $x^3y - xy$ (f) $6x^2 - 15x - 36$
   (g) $2ac + ad + 2bc + bd$ (h) $2ac - 2bc + bd - ad$ (i) $8a^3b - 18ab^3$

5. Simplify

(a) $\dfrac{ax}{a^2x^2 - ax}$ (b) $\dfrac{x^2 - 1}{x^2 + 5x - 6}$ (c) $\dfrac{20x}{10x^2}$

(d) $\dfrac{4xy + 2x^2y}{4xy - 2x^2y}$ (e) $\dfrac{2x^2y}{5x} \times \dfrac{15xy^2}{4y^3}$ (f) $\dfrac{\frac{1}{x} + \frac{x}{y}}{1 + \frac{x}{y}}$

6. Simplify

(a) $\dfrac{a}{a+b} + \dfrac{a}{a-b}$ (b) $\dfrac{x+1}{x-1} - \dfrac{x-1}{x+1}$ (c) $\dfrac{x+1}{x-1} \div \dfrac{x-1}{x+1}$ (d) $\dfrac{x+1}{x-1} \times \dfrac{x-1}{x+1}$

7. Write as a single fraction in its lowest terms

(a) $\dfrac{c}{c-d} - \dfrac{1}{c^2 - d^2}$ (b) $\dfrac{c}{c-d} + \dfrac{c}{c^2 - d^2}$

Examination Questions

8. Simplify $\dfrac{2}{3}p^{12} \times \dfrac{3}{4}p^8$.   (0580/02 May/June 2004 q 4)

9. Work out the value of $\dfrac{\frac{1}{2} - \frac{3}{8}}{-\frac{1}{2} + \frac{3}{8}}$ (0580/02 May/June 2004 q 7)

10. (a) Factorise completely $12x^2 - 3y^2$.
    (b) (i) Expand $(x - 3)^2$.
        (ii) $x^2 - 6x + 10$ is to be written in the form $(x - p)^2 + q$.
             Find the values of p and q.

    (0580/02 May/June 2004 q 20)

11. (a) $3^x = \dfrac{1}{3}$. Write down the value of x.
    (b) $5^y = k$. Find 5^{y+1}, in terms of k.

    (0580/02 Oct/Nov 2003 q 8)

12. Write $\dfrac{3}{x} - \dfrac{2}{x+1}$ as a single fraction in its simplest form.

    (0580/02 Oct/Nov 2003 q 14 a)

13. Work out as a single fraction
    $\dfrac{2}{x-3} - \dfrac{1}{x+4}$.

    (0580/02 May/June 2003 q 10)

14. (a) Factorise $(a - 2b) - 3c(a - 2b)$
    (b) Simplify $5t(t + 3) - 3(5t - 2)$

    (4024/01 Oct/Nov 2004 q 22 a and b)

15. Express as a single fraction in its simplest form $\dfrac{2}{x-3} - \dfrac{1}{x+2}$.

    (4024/01 May/June 2004 q 8)

16. Find a, b and c when
    (a) $3^a \div 3^5 = 27$,   (b) $125^b = 5$,   (c) $10^c = 0.001$.

    (4024/01 Oct/Nov 2003 q 12)

17. (a) It is given that $5^{-2} \times 5^k = 1$.
        Write down the value of k.
    (b) It is given that $\sqrt[3]{7} = 7^m$.
        Write down the value of m.

    (4024/01 May/June 2003 q 7)

18. (a) Expand and simplify $(x - 1)(x^2 + x + 1)$
    (b) Factorise $ax - bx - 3ay + 3by$

    (4024/01 May/June 2003 q 20)

19. Simplify
$$\frac{x+2}{x} - \frac{x}{x+2}.$$
Write your answer as a fraction in its simplest form.

(0580/02 May/June 2005 q 16)

20. Simplify

(a) $\left(\dfrac{x^{27}}{27}\right)^{\frac{2}{3}}$ (b) $\left(\dfrac{x^{-2}}{4}\right)^{-\frac{1}{2}}$

(0580/02 Oct/Nov 2005 q 18)

21. Factorise
    (a) $4x^2 - 9$,   (b) $4x^2 - 9x$,   (c) $4x^2 - 9x + 2$.

(0580/02 May/June 2006 q 19)

22. (a) Evaluate $5^2 + 5^0$.
    (b) Simplify

    (i) $\left(\dfrac{1}{x}\right)^{-2}$ (ii) $\left(x^6\right)^{\frac{1}{2}}$

(4024/01 May/June 2006 q 13)

23. (a) Simplify

    (i) $x(3x + 2) - (2x + 4)$ (ii) $\dfrac{ax^2 - x^2}{ax - x}$

    (b) Factorise completely $7x^2 - 63$.

(4024/01 May/June 2006 q 23)

24. Write as a fraction in its simplest form $\dfrac{x-3}{4} + \dfrac{4}{x-3}$.

(0580/02 May/June 2007 q 10)

25. (a) $\sqrt{32} = 2^p$. Find the value of p.

    (b) $\sqrt[3]{\dfrac{1}{8}} = 2^q$. Find the value of q.

(0580/02 May/June 2007 q 17)

26. Simplify $\dfrac{x}{3} + \dfrac{5x}{9} - \dfrac{5x}{18}$

(0580/21 May/June 2008 q 2)

27. Simplify $(27x^3)^{\frac{2}{3}}$

(0580/21 May/June 2008 q 8)

28. Write as a single fraction in its simplest form
$$\frac{4}{2x+3} - \frac{2}{x-3}$$

(0580/21 Oct/Nov 2008 q 11)

29. Find the value of n in each of the following statements
   (a) $32^n = 1$ (b) $32^n = 2$ (c) $32^n = 8$

(0580/02 Oct/Nov 2006 q 7)

30. (a) Simplify $(27x^6)^{\frac{1}{3}}$.

   (b) $(512)^{-\frac{2}{3}} = 2^p$. Find p.

(0580/02 Oct/Nov 2007 q 21)

31. Write the following in order of size, smallest first.

$$\sqrt{\frac{9}{17}} \qquad \frac{5}{7} \qquad 72\% \qquad \left(\frac{4}{3}\right)^{-1}$$

(0580/21 May/June 2009 q 2)

32. Write as a single fraction in its simplest form.
$$\frac{1}{c} + \frac{1}{d} - \frac{c-d}{cd}$$

(0580/21 May/June 2009 q 10)

Chapter 3

Working with Numbers

This is a short chapter, but together with Chapter 4 in *Core Mathematics for Cambridge IGCSE* it covers much of the mathematics you need in everyday life. As before, it is essential that you work carefully through the Core Skills exercise before proceeding with the rest of the chapter, and refer back to the Core Book for a reminder if necessary. The Mixed Exercise at the end of the chapter will give you a good idea of the variety of questions you might meet in your examination.

Core Skills

1. Calculate (a) the area, and (b) the perimeter of a rectangle measuring 15 metres by 7 centimetres, stating the units in your answer.
2. Round the following to the degree of accuracy stated.
   (a) 4.6749 to 2 decimal places
   (b) 500.612 to 3 significant figures
   (c) 0.0093 to 2 decimal places
   (d) 0.01056 to 3 significant figures
   (e) 516.2 centimetres to the nearest centimetre
   (f) 99.8 kilograms to the nearest kilogram
   (g) 9197 to the nearest 10
   (h) 999 to the nearest hundred
3. Copy and complete the following to show the limits of accuracy in each case.
   (a) ..... $\leqslant 439 <$ ..... given that 439 is correct to the nearest whole number.
   (b) ..... $\leqslant 5670 <$ ..... given that 5670 is correct to three significant figures.
4. Calculate the volume of a cuboid measuring 100 centimetres by 2 metres by 10 metres, giving your answers in standard form
   (a) in mm^3 (b) in cm^3 (c) in m^3
5. Change
   (a) 50 mm^2 to cm^2 (b) 500 ml to litres   (c) 12 kilograms to grams

6. By writing each number correct to 1 significant figure estimate the answers to each of these calculations, giving your answers to 1 significant figure.

   (a) 659×0.712
   (b) $\dfrac{76}{81} \div \dfrac{218}{389}$
   (c) $\dfrac{1112.5}{501.9} + \dfrac{359}{31.6}$

7. Calculate, giving your answers in standard form

   (a) $(7.5 \times 10^{-5}) \times (1.9 \times 10^2)$
   (b) $(9.35 \times 10^3) \div (3.76 \times 10^{-9})$
   (c) $(1.23 \times 10^3) - (1.23 \times 10^2)$
   (d) $(5.49 \times 10^{-5}) + (5.12 \times 10^{-6})$

8. Calculate, giving your answers correct to 3 significant figures

   (a) $\dfrac{5.76 + 7.93}{4.1 + 2.98}$
   (b) $\dfrac{5.76}{4.1} + \dfrac{7.93}{2.98}$
   (c) $\dfrac{5.76 + 7.93}{4.1} + 2.98$

   (d) $\sqrt{\dfrac{5.76}{4.1}} + \sqrt{\dfrac{7.93}{2.98}}$
   (e) $\sqrt{\dfrac{5.76}{4.1} + \dfrac{7.93}{2.98}}$

9. Simplify the following ratios.

   (a) $50:625$
   (b) 7 centimetres : 28 metres
   (c) $\dfrac{2}{3} : \dfrac{1}{6}$
   (d) $1\dfrac{3}{5} : 2\dfrac{5}{6}$
   (e) $70:35:350$
   (f) $1.85:25$

10. Write the ratio 900 litres : 180 millilitres in the form
    (a) $1:n$ (b) $n:1$

11. Ana and Bette share €357 in the ratio $5:2$. How much does each receive?

12. 240 blocks are needed to build a wall 1.5 metres high. How many blocks would be needed to make the same wall only 70 centimetres high?

13. It takes 5 builders 9 days to build a large shed. How long would it take 3 builders working at the same rate?

14. A car travels 66 kilometres in 1 hour 12 minutes. Calculate its average speed in kilometres per hour.

15. Change 75 km/h to m/s.

16. Change 8.15 pm to the 24 hour clock.

17. A television program lasts 1 hour and 45 minutes. It finishes at 1.10 pm. At what time did it start?

Upper and Lower Bounds in Calculations

You have seen how to find the upper and lower limits or bounds of measurements that have been approximated to a given significant figure, decimal place or other unit. You also need to be able to use these values in calculations, and find the upper and lower bounds of the results of the calculations.

Let us assume that x and y are two measurements that have been given to a stated approximation.

The *largest* possible value of $x + y$ would obviously be obtained by adding the *upper* bounds of the two measurements together. Conversely, the smallest value would come from the two lower bounds.

The *largest* possible value of $x - y$ would come from subtracting the *lower* bound of y from the *upper* bound of x to give the greatest possible difference.

But how would you get the largest possible value of $x \times y$?

Also how would you get the largest possible value of $\dfrac{x}{y}$?

In the following example we are using just numbers, without putting them in any context in order to make the idea clear.

Example 1

Given that $x = 35$ correct to the nearest whole number and that $y = 25$ correct to the nearest whole number, copy and complete the following table.

		Value	Lower bound	Upper bound
(a)	x	35	34.5	35.5
(b)	y	25		
(c)	$x + y$	60		
(d)	$x - y$	10		
(e)	$x \times y$	875		
(f)	$x \div y$	1.4		
(g)	$2x + 3y$			
(h)	x^2			

Answer 1

		Value	Lower bound	Upper bound
(a)	x	35	34.5	35.5
(b)	y	25	24.5	25.5
(c)	$x + y$	60	34.5 + 24.5 = 59	35.5 + 25.5 = 61
(d)	$x - y$	10	34.5 − 25.5 = 9.0	35.5 − 24.5 = 11
(e)	$x \times y$	875	34.5 × 24.5 = 845.2	35.5 × 25.5 = 905.25
(f)	$x \div y$	1.4	34.5 ÷ 25.5 = 1.35 (3 s.f.)	35.5 ÷ 24.5 = 1.45 (3 s.f.)
(g)	$2x + 3y$	120	2 × 34.5 + 3 × 24.5 = 142.5	2 × 35.5 + 3 × 25.5 = 147.5
(h)	x^2	1225	1190.25	1260.25

Exercise 3.1

1. Given that $x = 14$ and $y = 12$, both to the nearest whole number, work out (i) the lower bound, and (ii) the upper bound of the following.
   (a) $x - y$ (b) $x^2 - y^2$ (c) xy (d) $\dfrac{x}{y}$

2. Given that $a = 16.9$ cm, $b = 7.3$ cm and $c = 5.8$ cm, all to 1 decimal place, work out (i) the lower bound, and (ii) the upper bound of the following.
   (a) abc (b) $\dfrac{ab}{c}$ (c) $\dfrac{(a+b)}{2} \times c$

   Give your answers correct to 4 significant figures if not exact.

Increase and Decrease by a Given Ratio

An example should be sufficient to explain how values may be increased or decreased in a given ratio.

Example 2
(a) Increase 45 cm in the ratio 2 : 3.   (b) Decrease 45 cm in the ratio 2 : 3.

Answer 2
NOTE: As usual, stop to ask yourself whether the answer should be larger or smaller than the original. Use the ratio as a multiplier as shown.

(a) Increase, so 45 needs to be made larger. Multiplying by a number greater than 1 will make the answer larger.

$$45 \times \frac{3}{2} = 67.5$$

45 cm *increased* in the ratio 2:3 = 67.5 cm

(b) Decrease, so 45 needs to be made smaller. Multiplying by a number less than 1 will make the answer smaller.

$$45 \times \frac{2}{3} = 30$$

45 cm *decreased* in the ratio 2:3 = 30 cm

Exercise 3.2

1. Increase 4.9 in the ratio 7:12.
2. Decrease 3.65 in the ratio 1:5.
3. Increase 5 cm in the ratio 1:10.
4. Increase 1000 kg in the ratio 5:6.
5. Decrease 120 in the ratio 3:4.
6. Decrease 81 in the ratio 9:1.

Reverse Percentages

Calculating a reverse percentage is a method for finding an original value if you have only been given the value *after* a percentage change has been made.

To understand how reverse percentages work we will first look at some simple percentage examples.

A shop offers a 20% reduction in a sale.

An item costing Rs 60 will be reduced by 20% of 60.

Reduction = Rs $(\frac{20}{100} \times 60)$ = Rs 12

Sale price = Rs (60 − 12) = Rs 48

There is another, quicker method for calculating the sale price.

The original price is 100%, so taking off 20% leaves 80% to pay in the sale.

Sale price = 80% of Rs 60

= Rs $(\frac{80}{100} \times 60)$ = Rs 48

So if you want to calculate the *reduction* you use 20%, but if you want to calculate the *new price* you calculate 80%. Either way the original price is 100%.

Setting this out another way:

original price   −   reduction   =   new price
   100%   −   20%   =   80%
              × 0.6   × 0.6   × 0.6   (100 ÷ 60 = 0.6)
   Rs 60   −   Rs 12   =   Rs 48

The same method is used for an increase in price.
If a price is increased by 20% then the new amount is (100 + 20)% = 120%.

$$\text{Rs 60 increased by 20\%} = \text{Rs } \left(\frac{120}{100} \times 60\right) = \text{Rs 72}$$

As before, if you want to calculate the *increase* use 20%, but if you want to calculate the *new price* use 120%.

Setting this out the other way:

     original price   +   increase   =   new price
        100%   +   20%   =   120%
                 × 0.6   × 0.6   × 0.6
        Rs 60   +   Rs 12   =   Rs 72

This leads us easily into calculating reverse percentages.
An example should show you how this is done.

Example 3
(a) An item in a sale costs $36, after a 10% reduction. Calculate the original price.
(b) A shop buys its goods at **wholesale** prices, it then adds a 40% **mark-up** (this is the extra they charge in order to make a profit). The **retail price** is the price the shop charges its customers.
An item in the shop is marked at $70. (This is the retail price.)
Calculate the price the shop paid for the item. (This is the wholesale price.)

Answer 3
(a) Original price   −   reduction   =   sale price
     100%   −   **10%**   =   **90%**
              × 0.4   × 0.4   × 0.4   (36 ÷ 90 = 0.4)
      ?   − $4   =   **$36**
   Original price = 100 × 0.4 = $40
(b) Wholesale price   + mark up   =   retail price
     100%   +   **40%**   =   **140%**
              × 0.5   × 0.5   × 0.5   (70 ÷ 140 = 0.5)
      ?   + $20   =   **$70**
The wholesale price was 100 × 0.5 = $50

Exercise 3.3

1. A toy was reduced by 20% in a sale. The sale price was $72. Calculate the price before the reduction.
2. Niko is given an increase of 6% on his hourly rate. He now earns $13.25 per hour. Calculate his previous hourly rate.
3. In 2010 a new car costs $10063. This is an increase of 16% on its new price in 2009. Calculate the increase in dollars.
4. An item in a sale is reduced by 15%. The actual reduction is Rs 72. Calculate
   (a) the original price,
   (b) the sale price.

Exercise 3.4

Mixed Exercise

1. A rectangle measures 6.5 cm by 3.5 cm, each measurement correct to 1 decimal place. Calculate the lower bound of
   (a) the perimeter, (b) the area of the rectangle.
   Give your exact answer.
2. Yash is cutting short pieces of wood from a longer piece. The shorter pieces of wood are 10 cm long, correct to the nearest centimetre, and the whole piece is 1 metre, correct to the nearest centimetre.
   (a) Show that it may not be possible for him to cut 10 shorter pieces of wood.
   (b) What is the maximum amount that might be left over?
3. Increase the following in the ratios given.
   (a) 66 in the ratio $1:4$ (b) 105 in the ratio $5:7$ (c) 98 in the ratio $8:7$
4. Decrease the following in the ratios given.
   (a) 72 in the ratio $5:3$ (b) 11.5 in the ratio $5:2$ (c) 0.719 in the ratio $1:10$
5. In 2006 a car is valued at $8607 when it is 1 year old. The rate of depreciation in the first year is 58%.
   (a) What did the car cost in 2005 when it was new?
   (b) Calculate its value in 2008 if the depreciation rate is 20% per annum for these two years.
   Give your answers to the nearest dollar.
6. (a) A shop has a mark up of 46% on an item selling at €49.64. Calculate the wholesale price.
   (b) The shop has a sale. An item is marked at €160 after a reduction of 20%. Calculate the amount **saved** if the item is purchased at the sale price rather than the normal price.

Examination Questions

7. The population of Newtown is 45 000.
   The population of Villeneuve is 39 000.
   (a) Calculate the ratio of these populations in its simplest form.
   (b) In Newtown, 28% of the population are below the age of twenty.
   Calculate how many people in Newtown are below the age of twenty.
   (c) In Villeneuve, 16 000 people are below the age of twenty.
   Calculate the percentage of people in Villeneuve below the age of twenty.
   (d) The population of Newtown is 125% greater than it was fifty years ago.
   Calculate the population of Newtown fifty years ago.
   (e) The two towns are combined and made into one city called Monocity.
   In Monocity the ratio of men : women : children is 12 : 13 : 5.
   Calculate the number of children in Monocity.

   (0580/04 Oct/Nov 2004 q 1)

8. Fatima and Mohammed each buy a bike.
   (a) Fatima buys a city-bike which has a price of $120.
   She pays 60% of this price and then pays $10 per month for 6 months.
   (i) How much does Fatima pay altogether?
   (ii) Work out your answer to **part (a) (i)** as a percentage of the original price of $120.
   (b) Mohammed pays $159.10 for a mountain-bike in a sale.
   The original price has been reduced by 14%.
   Calculate the original price of the mountain-bike.
   (c) Mohammed's height is 169 cm and Fatima's height is 156 cm.
   The frame sizes of their bikes are in the same ratio as their heights.
   The frame size of Mohammed's bike is 52 cm.
   Calculate the frame size of Fatima's bike.
   (d) Fatima and Mohammed are members of a school team which takes part in a bike ride for charity.
   (i) Fatima and Mohammed ride a total distance of 36 km.
   The ratio distance Fatima rides : distance Mohammed rides is 11 : 9.
   Work out the distance Fatima rides.
   (ii) The distance of 36 km is only $\frac{2}{23}$ of the total distance the team rides.
   Calculate this total distance.

   (0580/04 May/June 2004 q 1)

9. In 1997 the population of China was 1.24×10^9.
   In 2002 the population of China was 1.28×10^9.
   Calculate the percentage increase from 1997 to 2002.

   (0580/02 Oct/Nov 2004 q 8)

10. A square has sides of length d metres.
    This length is 120 metres, correct to the nearest 10 metres.
    (a) Copy and complete this statement.
    .... $\leqslant d <$ ....
    (b) Calculate the difference between the largest and the smallest possible areas of the square.

    (0580/02 Oct/Nov 2004 q 13)

11. A train left Sydney at 23 20 on December 18th and arrived in Brisbane at 02 04 on December 19th.
    How long, in hours and minutes, was the journey?

    (0580/02 May/June 2004 q 1)

12. The population of Europe is 580 000 000 people.
    The land area of Europe is 5 900 000 square kilometres.
    (a) Write 580 000 000 in standard form.
    (b) Calculate the number of people per square kilometre, to the nearest whole number.
    (c) Calculate the number of square **metres** per person.

    (0580/02 Oct/Nov 2003 q 18)

13. A rectangle has sides of length 6.1 cm and 8.1 cm correct to 1 decimal place.
    Calculate the upper bound for the area of the rectangle as accurately as possible.

    (0580/21 Oct/Nov 2008 q 7)

14. A rectangular field is 18 metres long and 12 metres wide. Both measurements are correct to the nearest metre. Work out exactly the smallest possible area of the field.

    (0580/02 May/June 2003 q 6)

15. The ratios of teachers : male students : female students in a school are 2 : 17 : 18.
    The total number of **students** is 665. Find the number of **teachers**.

    (0580/02 May/June 2003 q 5)

16. A holiday in Europe was advertised at a cost of €245.
    The exchange rate was $1 = €1.06.
    Calculate the cost of the holiday in dollars, giving your answer correct to the nearest cent.

    (0580/21 May/June 2008 q 5)

17. Write the number 1045.2781 correct to
    (a) 2 decimal places   (b) 2 significant figures

    (0580/21 May/June 2008 q 7)

18. Marcus receives $800 from his grandmother.
    (a) He decides to spend $150 and to divide the remaining $650 in the ratio
        savings : holiday = 9 : 4
        Calculate the amount of his savings.
    (b) (i) He uses 80% of the $150 to buy some clothes.
            Calculate the cost of the clothes.
        (ii) The money remaining from the $150 is $37\frac{1}{2}\%$ of the cost of a day trip to Cairo. Calculate the cost of the trip.
    (c) (i) Marcus invests $400 of his savings for 2 years at 5% per year **compound** interest. Calculate the amount he has at the end of the 2 years.
        (ii) Marcus's sister also invests $400, at r% per year **simple** interest. At the end of 2 years she has exactly the same amount as Marcus. Calculate the value of r.

        (0580/04 May/June 2009 q 1)

19. A student played a computer game 500 times and won 370 of these games.
    He then won the next x games and lost none.
    He has now won 75% of the games he has played.
    Find the value of x.

    (0580/21 May/June 2008 q 17)

20. In January Sunanda changed £25 000 into dollars when the exchange rate was $1.96 = £1.
    In June she changed the dollars back into pounds when the exchange rate was $1.75 = £1.
    Calculate the profit she made, giving your answer in pounds (£).

    (0580/21 May/June 2009 q 11)

21. In 2005, there were 9 million bicycles in Beijing, correct to the nearest million.
    The average distance travelled by each bicycle in one day was 6.5 km correct to 1 decimal place.
    Work out the upper bound for the **total** distance travelled by all the bicycles in one day.

    (0580/21 May/June 2009 q 6)

22. A light on a computer comes on for 26 700 microseconds.
    One microsecond is 10^{-6} seconds.
    Work out the length of time, in seconds, that the light is on
    (a) in standard form,   (b) as a decimal.

    (0580/21 Oct/Nov 2008 q 4)

23. A rectangle has sides of length 6.1 cm and 8.1 cm correct to 1 decimal place.
    Calculate the upper bound for the area of the rectangle as accurately as possible.

    (0580/21 Oct/Nov 2008 q 7)

24. Beatrice has an income of $40 000 in one year.
   (a) She pays
           no tax on the first $10 000 of her income;
           10% tax on the next $10 000 of her income;
           25% tax on the rest of her income.
   Calculate
     (i) the total amount of tax Beatrice pays,
    (ii) the total amount of tax as a percentage of the $40 000.
   (b) Beatrice pays a yearly rent of $10 800.
   After she has paid her tax, rent and bills, she has $12 000.
   Calculate how much Beatrice spends on bills.
   (c) Beatrice divides the $12 000 between shopping and saving in the ratio
   shopping : saving = 5 : 3.
     (i) Calculate how much Beatrice spends on shopping in one year.
    (ii) What fraction of the original $40 000 does Beatrice **save**?
        Give your answer in its lowest terms.
   (d) The rent of $10 800 is an increase of 25% on her previous rent.
   Calculate her previous rent.
                                                                                        (0580/04 Oct/Nov 2008 q 1)

25. At 05 06 Mr Ho bought 850 fish at a fish market for $2.62 each.
   95 minutes later he sold them all to a supermarket for $2.86 each.
   (a) What was the time when he sold the fish?
   (b) Calculate his total profit.
                                                                                         (0580/21 May/June 2009 q 3)

26. Vreni took part in a charity walk.
   She walked a distance of 20 kilometres.
   (a) She raised money at a rate of $12.50 for each kilometre.
      (i) How much money did she raise by walking the 20 kilometre?
     (ii) The money she raised in **part (a) (i)** was $\dfrac{5}{52}$ of the total money raised.
        Work out the total money raised.
    (iii) In the previous year the total money raised was $2450.
        Calculate the percentage increase on the previous year's total.
   (b) Part of the 20 kilometres was on a road and the rest was on a footpath.
   The ratio road distance : footpath distance was 3 : 2.
      (i) Work out the road distance.
     (ii) Vreni walked along the road at 3 km/h and along the footpath at
        2.5 km/h. How long, in hours and minutes, did Vreni take to walk the
        20 kilometres?
    (iii) Work out Vreni's average speed.
    (iv) Vreni started at 08 55. At what time did she finish?

(c) On a map, the distance of 20 kilometres was represented by a length of 80 centimetres.
The scale of the map was $1:n$.
Calculate the value of n.

(0580/04 May/June 2008 q 1)

27. Angharad sleeps for 8 hours each night, correct to the nearest 10 minutes.
The total time she sleeps in the month of November (30 nights) is T hours.
Between what limits does T lie?

(0580/02 Oct/Nov 2006 q 4)

28. $p = \dfrac{0.002751 \times 3400}{(9.8923 + 24.7777)^2}$

(a) Write each number in this calculation correct to 1 significant figure.
(b) Use your answer to **part (a)** to **estimate** the value of p.

(0580/02 Oct/Nov 2007 q 6)

29. (a) In October the cost of a car in euros was €20 000.
The cost of this car in pounds was £14 020.
Calculate the **exact** value of the exchange rate in October, writing your answer in the form €1 = £....
(b) In November the car still cost €20 000 and the exchange rate was €1 = £0.6915.
Calculate the difference, in pounds, between the cost in October and November.

(0580/02 Oct/Nov 2007 q 8)

30. Carmen spends 5 minutes, correct to the nearest minute, preparing one meal.
She spends a total time of T minutes preparing 30 meals. Between what limits does T lie?

(0580/02 May/June 2007 q 6)

Chapter 4

Algebra II

In this chapter we start to develop algebra as a tool for solving problems.
The chapter follows from Chapter 5 in *Core Mathematics for Cambridge IGCSE*, and if you have any difficulty with the Core Skills exercise you should refer back to the Core Book.

Core Skills

1. Solve the following equations
   (a) $4x - 1 = 9$
   (b) $x + 5 = 17 - x$
   (c) $-5x = x - 12$
   (d) $3(x + 4) = 2(x - 1) + 5x$
   (e) $3 - x = -1$
   (f) $6y - (y + 1) = 5y + 2(3 - y)$
   (g) $4(a - 1) - 3(2a - 1) = 5(1 + a) - 6$

2. Rearrange the formulae to make x the subject
   (a) $s = \dfrac{x}{h}$
   (b) $A = lx$
   (c) $V = xy + c$
   (d) $v^2 = u^2 - 2ax$
   (e) $A = \dfrac{1}{2}(a + b)x$
   (f) $c = \dfrac{b}{x}$

3. Find the nth term in the following sequences.
   (a) 2, 7, 12, 17, …
   (b) −10, −5, 0, 5, …
   (c) 2, 5, 10, 17, 26, …
   (d) 10, 8, 6, 4, 2, …
   (e) 1, 4, 9, c16, 25, …

4. Use each of the following formulae to find (i) the first term, and (ii) the 100th term in the sequence.
   (a) nth term $= \dfrac{(n+2)^2}{2}$
   (b) nth term $= n^2 - 10n$
   (c) nth term $= -8n - 1$

5. Solve these pairs of simultaneous equations
   (a) $x + y = 7$
       $x + 2y = 3$
   (b) $2x - y = 1$
       $y + x = 5$
   (c) $3x + y = 2$
       $x + 2y = 4$
   (d) $2x - y = 6$
       $2y - x = 3$

Algebra II 57

More Equations

You will be expected to solve more difficult equations, putting into practise the techniques you have already learned in Chapter 5 of the Core Book and in Chapter 2 of this book.

Example 1
Solve the following equations.

(a) $\dfrac{x+2}{3} - \dfrac{x-1}{4} = \dfrac{3}{2}$

(b) $1.3x + 2.5 = 4x$

Answer 1

(a) $\dfrac{x+2}{3} - \dfrac{x-1}{4} = \dfrac{3}{2}$

$12 \times \left(\dfrac{x+2}{3} - \dfrac{x-1}{4} = \dfrac{3}{2} \right)$

$4(x + 2) - 3(x - 1) = 6 \times 3$
$4x + 8 - 3x + 3 = 18$
$x = 18 - 11$
$x = 7$

NOTE: Remember that you can do the same thing to both sides of an equation, so to get rid of the fractions multiply the whole equation by the lowest common multiple of 3, 4 and 2. This is called the *lowest common denominator*.

Now multiply each term by 12 and 3, 4 and 2 will all divide into the twelve, removing the denominators of each fraction.

In the first fraction $12 \div 3 = 4$, and so on.
Be careful with the minus sign in front of the second bracket!!

(b) $1.3x + 2.5 = 4x$

$13x + 25 = 40x$
$13x - 40x = -25$
$-27x = -25$
$x = \dfrac{-25}{-27}$
$x = \dfrac{25}{27}$

NOTE: This time the equation would look more friendly multiplied by 10.

Exercise 4.1

Solve the following equations

1. $2.5x - 1.5 = 6$

2. $\dfrac{x}{2} + \dfrac{x}{3} = 5$

3. $\dfrac{x}{2} - \dfrac{x}{3} = \dfrac{7}{3}$

4. $\dfrac{x+1}{5} - \dfrac{x-1}{2} = \dfrac{1}{10}$

5. $\dfrac{1}{4}(x-1) - \dfrac{3}{8}(x+1) = 2$

6. $\dfrac{1}{x} - \dfrac{3}{8x} = 1$

NOTE: Multiply all through by $8x$.

7. $\dfrac{x}{x+1} + 1 = 0$ NOTE: Multiply all through by $(x + 1)$.

8. $x(x + 2) + 2 = x^2$ NOTE: Multiply out the brackets then subtract x^2 from both sides.

Quadratic Equations

Quadratic equations have an x squared term, and also possibly an x term and a number term. For example, $y = x^2 + 2x - 3$ is a quadratic equation. It has two variables, x and y. There are an infinite number of solutions to this equation.

Two possible solutions are: $x = 0$, $y = -3$, or $x = 2.6$, $y = 8.96$ because both these sets of values 'satisfy' the equation. 'Satisfy' in this case means 'make the equation true'.

Later in the course you will plot equations like this on an xy graph, and, by joining up the points you plot will be able to see where all the solutions lie, and the shape of curves produced by quadratic equations.

For the moment we will be finding solutions algebraically.

Of particular interest are those quadratic equations where $y = 0$.

For example, $x^2 + 2x - 3 = 0$.

This is an equation in a single variable, x, and we may be able to find values of x which satisfy it.

What happens when we replace x by 1?

$$x^2 + 2x - 3$$
$$= 1^2 + 2 \times 1 - 3$$
$$= 0$$

so $x = 1$ is a solution to this equation.

There is another solution ($x = -3$) and we could try different values for x until we find it, but this could take some time, and would be impossible in many cases, so we must learn the algebraic methods.

Solving quadratic equations by factorising

Look at the equation $2x^2 + 5x + 2 = 0$.

The left hand side of the equation can be factorised to give:

$$(2x + 1)(x + 2) = 0$$

The left hand side of the equation is now written as two brackets multiplied together. If the x in this equation is replaced by a number, for example 3, the equation would become:

$$(2 \times 3 + 1)(3 + 1) = 0$$
$$7 \times 4 = 0$$
$$28 = 0$$

This is clearly nonsense since 28 is not equal to zero!

The only way that two different numbers or terms can be multiplied to give zero is if one of them is already zero. So either $(2x + 1) = 0$ or $(x + 2) = 0$, which gives us two possible answers to the quadratic equation.

Setting out the answer to a question asking you to solve a quadratic equation:
$$2x^2 + 5x + 2 = 0$$
$$(2x + 1)(x + 2) = 0$$
either $\qquad 2x + 1 = 0 \qquad$ or $\quad x + 2 = 0$
$$x = -\frac{1}{2} \qquad\qquad x = -2$$

The two solutions are $x = -\frac{1}{2}$ or $x = -2$.

Finally, to test that this has worked correctly we can substitute each solution into the original equation:

when $\qquad\qquad\qquad\qquad x = -\frac{1}{2}$

$$2x^2 + 5x + 2$$
$$= 2 \times \left(-\frac{1}{2}\right)^2 + 5 \times \left(-\frac{1}{2}\right) + 2$$
$$= 2 \times \frac{1}{4} - \frac{5}{2} + 2$$
$$= \frac{1}{2} - \frac{5}{2} + 2$$
$$= 0$$

So $x = -\frac{1}{2}$ is a solution to the quadratic equation $2x^2 + 5x + 2 = 0$.

You should check for yourself that the other solution is also correct.

Warning! This method *only* works for two terms multiplied together to give *zero*. It *cannot* be used unless one side is zero because the method depends on the special fact that the only way two (or more) different numbers can be multiplied together to give zero is if one of them *is* zero. This of course does not apply to two numbers multiplied together to give, for example, 2.

Example 2
Solve the quadratic equations
(a) $x^2 + 2x - 3 = 0$ $\qquad\qquad$ (b) $3x^2 - 11x + 8 = 0$

Answer 2
(a) $x^2 + 2x - 3 = 0$ $\qquad\qquad$ (b) $3x^2 - 11x + 8 = 0$
$\quad\;\;(x + 3)(x - 1) = 0$ $\qquad\qquad\quad\;\;(3x - 8)(x - 1) = 0$
$\quad\;\;$either $x + 3 = 0\;$ or $\;x - 1 = 0$ $\qquad\;\;$either $3x - 8 = 0 \quad$ or $\quad x - 1 = 0$
$\qquad\qquad\;\;x = -3 \qquad\quad\; x = 1$ $\qquad\qquad\qquad\quad 3x = 8 \qquad\qquad\;\; x = 1$
$\qquad\qquad\qquad\qquad\qquad\qquad\qquad\qquad\qquad\qquad\;\; x = \frac{8}{3}$

$\qquad\qquad\qquad\qquad\qquad\qquad\qquad\qquad\qquad x = \frac{8}{3}$ or $\quad x = 1$

Exercise 4.2

Solve the following equations
1. $x^2 + 5x + 6 = 0$
2. $x^2 + x - 2 = 0$
3. $x^2 - 6x + 5 = 0$
4. $2x^2 - x - 1 = 0$
5. $x^2 - 1 = 0$

NOTE: Remember how to factorise a difference of squares?

6. $2x^2 - 7x + 5 = 0$
7. $3x^2 - x - 4 = 0$
8. $4x^2 - 8x + 3 = 0$
9. $4x^2 - 9 = 0$
10. $x^2 - 8x + 15 = 0$
11. $4x^2 - 16 = 0$

NOTE: Divide the equation by 4 first, remembering that $\frac{0}{4} = 0$

12. $x^2 + 2x + 1 = 0$

NOTE: The two solutions to this equation are the same.

13. $3x^2 - 6x + 3 = 0$

NOTE: Remember to check for common factors first!

14. $6x^2 - 22x + 16 = 0$

Using the quadratic formula to solve quadratic equations

Some quadratic equations may not factorise easily, because their solutions are not whole numbers or simple fractions.

For these we may use a formula, often called the **quadratic formula**. We will derive a simplified form of it in the next section, but for the moment we will get used to using it. The formula refers to the general equation:

$$ax^2 + bx + c = 0$$

and states that:

$$x = \frac{-b \pm \sqrt{b^2 - 4ac}}{2a}$$

To see how it works we will use it to solve the equation $x^2 + 2x - 3 = 0$, which we have already solved by factorising in Example 2(a).

Comparing $\qquad x^2 + 2x - 3 = 0$
with $\qquad ax^2 + bx + c = 0$,
we see that $a = 1$, $b = 2$ and $c = -3$.

Using the quadratic formula:

$$x = \frac{-b \pm \sqrt{b^2 - 4ac}}{2a}$$

$$x = \frac{-2 \pm \sqrt{2^2 - 4 \times 1 \times (-3)}}{2 \times 1}$$

$$x = \frac{-2 \pm \sqrt{4 + 12}}{2}$$

$$x = \frac{-2 \pm \sqrt{16}}{2}$$

$$x = \frac{-2 \pm 4}{2}$$

NOTE: Notice that the whole numerator is divided by $2a$. It is a very common mistake to only divide the square root by $2a$, and this will earn you no marks!

Algebra II 61

The ± sign means 'plus or minus', and gives us the two solutions.

either $\qquad x = \dfrac{-2-4}{2} \qquad$ or $\qquad x = \dfrac{-2+4}{2}$

$\qquad\qquad\qquad x = -3 \qquad$ or $\qquad x = 1$

This is more time-consuming than the method used in Example 2(a), so you should use the method of factorising if it is possible. However, questions often ask you to solve the given quadratic equation giving the answers correct to two decimal places, or correct to three significant figures. This is a big hint that it is not going to factorise and you should then use the formula without spending time attempting a factorisation.

Example 3
Solve the following equations, giving your answers to 3 significant figures.
(a) $5x^2 + 2x - 1 = 0$ (b) $2x^2 - x - 1 = 0$

Answer 3
(a) $5x^2 + 2x - 1 = 0$
Comparing with $\quad ax^2 + bx + c = 0$
then $a = 5$, $b = 2$ and $c = -1$
Using $x = \dfrac{-b \pm \sqrt{b^2 - 4ac}}{2a}$

$x = \dfrac{-2 \pm \sqrt{2^2 - 4 \times 5 \times -1}}{2 \times 5} \qquad x = \dfrac{-2 \pm \sqrt{4 + 20}}{10}$

$x = \dfrac{-2 + \sqrt{24}}{10} \quad$ or $\quad x = \dfrac{-2 - \sqrt{24}}{10}$

$x = 0.290 \qquad$ or $\quad x = -0.690$ to 3 significant figures

(b) $2x^2 - x - 1 = 0$
In this equation $a = 2$, $b = -1$ and $c = -1$
$x = \dfrac{-b \pm \sqrt{b^2 - 4ac}}{2a}$

$x = \dfrac{1 \pm \sqrt{(-1)^2 - 4 \times 2 \times (-1)}}{2 \times 2}$

$x = \dfrac{1 \pm \sqrt{1 + 8}}{4}$

$x = \dfrac{1 + 3}{4} \quad$ or $\quad x = \dfrac{1 - 3}{4}$

$x = 1 \qquad\qquad x = -\dfrac{1}{2}$

These answers are exact, so do not need to be given to 3 significant figures. In fact the equation could have easily been solved by factorising.
You must take great care with the plus and minus signs in this formula.

For example, in the answer to part 1(b) you will see that:
$-b = -(-1) = +1$,
$b^2 = (-1)^2 = +1$ and
$-4ac = -4 \times 2 \times (-1) = +8$

Exercise 4.3
Solve the following, giving your answers to 3 significant figures.
1. $x^2 + 5x - 7 = 0$
2. $x^2 - 5x - 7 = 0$
3. $3x^2 + 10x + 2 = 0$
4. $x^2 - x - 1 = 0$
5. $-x^2 + x + 4 = 0$
6. $2x^2 - 7x + 2 = 0$

Completing the square to solve quadratic equations

You may sometimes be asked to rearrange a quadratic expression or solve a quadratic equation by *completing the square*. This method sometimes seems a bit daunting at first, but is relatively easy if you follow a routine. The easiest way to show you this routine is by one or two examples.

$$x^2 + 2x - 3 = 0$$

To solve the above equation by completing the square the equation first needs to be written in another form. We begin by rewriting the first two terms of the equation:

- Take the first two terms:

$$x^2 + 2x$$

- Open a bracket, insert x, $+$, and *half* the coefficient of the x term, close the bracket and square it:

$$(x + 1)^2$$

- Squaring this bracket gives:

$$(x + 1)(x + 1)$$
$$= x^2 + 2x + 1$$

- Comparing this with the first two terms of our original equation we can see that we have acquired an extra $+1$, so this must now be subtracted, and the rest of the original equation put in place:

$$(x + 1)^2 - 1 - 3 = 0$$

- Collecting the last two terms finishes the rearrangement and completes the square:

$$(x + 1)^2 - 4 = 0$$

You should multiply out the brackets again to check that this equation is still the same as the original equation.

Before going on to use this new form to solve the equation it would be beneficial to practise some rearrangements.

Example 4
(a) $x^2 - 4x + 3$
(b) $x^2 + 5x - 4$
(c) $2x^2 + 4x - 1$

Write these expressions in the form $(x+b)^2 + c$ or $a(x+b)^2 + c$

Algebra II   63

Answer 4
(a) $x^2 - 4x + 3$ $(x - 2)^2 = x^2 - 4x + 4$ so subtract 4
   $= (x - 2)^2 - 4 + 3$ and replace the +3
   $= (x - 2)^2 - 1$

(b) $x^2 + 5x - 4$ $\left(x + \dfrac{5}{2}\right)^2 = x^2 + 5x + \dfrac{25}{4}$ so subtract $\dfrac{25}{4}$

   $= \left(x + \dfrac{5}{2}\right)^2 - \dfrac{25}{4} - 4$ and replace the -4

   $= \left(x + \dfrac{5}{2}\right)^2 - \dfrac{41}{4}$

(c) $2x^2 + 4x - 1$ NOTE: Divide the expression by 2 to get the x^2 term with a coefficient of 1, keeping the 2 outside the brackets as a factor.

   $= 2\left[x^2 + 2x - \dfrac{1}{2}\right]$

   NOTE: Using square brackets helps to show that the 2 has divided the whole expression, and avoids confusion with the other set of brackets.

   $= 2\left[(x+1)^2 - 1 - \dfrac{1}{2}\right]$

   $= 2\left[(x+1)^2 - \dfrac{3}{2}\right]$

   $= 2(x + 1)^2 - 3$ NOTE: The factor 2 outside the square brackets now multiplies the terms inside these brackets to finish the working.

Exercise 4.4
Write these expressions in the form $(x + b)^2 + c$ or $a(x + b)^2 + c$
1. $x^2 - 6x + 1$ 2. $x^2 + 5x + 2$ 3. $x^2 - 3x - 3$ 4. $2x^2 + 4x - 5$
5. $3x^2 - 6x - 4$ 6. $2x^2 - 3x + 2$ 7. $2x^2 + 5x - 5$

Solving the equation after completing the square
This is now quite straightforward.
Looking again at our original quadratic equation:
$$x^2 + 2x - 3 = 0$$
- Rewrite in the completed square form:
$$(x + 1)^2 - 4 = 0$$
- Add 4 to both sides:
$$(x + 1)^2 = 4$$
- Square root both sides:
$$x + 1 = \pm \sqrt{4}$$
$$x + 1 = \pm 2$$

64 Extended Mathematics for Cambridge IGCSE

- Subtract 1 from both sides:

$$x = +2 - 1 \quad \text{or} \quad x = -2 - 1$$
$$x = 1 \qquad\qquad\qquad x = -3$$

We have now solved our original equation in three different ways. The choice of method depends on the type of equation, or on the question you have been set. There is a fourth method, which you will see when you work on Chapter 6, Algebra and Graphs.

Example 5
Solve these quadratic equations by completing the square. Leave your answers in surd form where appropriate. (A *surd* is the square root of a number which itself is not a perfect square. So leaving in *surd* form means leaving in square root form.)
(a) $x^2 + 3x - 5 = 0$ (b) $2x^2 + x - 3 = 0$ (c) $x^2 + bx + c = 0$

Answer 5
(a) $x^2 + 3x - 5 = 0$

$$\left(x + \frac{3}{2}\right)^2 - \frac{9}{4} - 5 = 0$$

$$\left(x + \frac{3}{2}\right)^2 - \frac{29}{4} = 0$$

$$\left(x + \frac{3}{2}\right)^2 = \frac{29}{4}$$

$$x + \frac{3}{2} = \pm\sqrt{\frac{29}{4}}$$

$$x = -\frac{3}{2} \pm \frac{\sqrt{29}}{2} \qquad \left(\text{or } \frac{-3 \pm \sqrt{29}}{2}\right)$$

(b) $2x^2 + x - 3 = 0$ We can now divide the equation all through by 2.

$$x^2 + \frac{x}{2} - \frac{3}{2} = 0 \qquad\qquad \text{Remember that } 0 \div 2 = 0!$$

$$\left(x + \frac{1}{4}\right)^2 - \frac{1}{16} - \frac{3}{2} = 0 \qquad \left(x + \frac{1}{4}\right)^2 - \frac{25}{16} = 0$$

$$x + \frac{1}{4} = \pm\sqrt{\frac{25}{16}}$$

$$x = -\frac{1}{4} \pm \frac{5}{4}$$

either $x = -\frac{3}{2}$ or $x = 1$

(c) $x^2 + bx + c = 0$

$$\left(x + \frac{b}{2}\right)^2 - \frac{b^2}{4} + c = 0$$

$$\left(x+\frac{b}{2}\right)^2 = \frac{b^2}{4} - c$$

$$\left(x+\frac{b}{2}\right)^2 = \frac{b^2 - 4c}{4}$$

$$x + \frac{b}{2} = \pm\sqrt{\frac{b^2 - 4c}{4}}$$

$$x = -\frac{b}{2} \pm \frac{\sqrt{b^2 - 4c}}{2}$$

$$x = \frac{-b \pm \sqrt{b^2 - 4c}}{2}$$

This is the proof of the simplified version of the quadratic formula, with $a = 1$.
You could be asked for this proof, so make sure you follow the example thoroughly and understand it.

Exercise 4.5

Solve these equations by completing the square and leaving in surd form (square root form) if the answers are not exact.

1. $x^2 + x - 1 = 0$
2. $x^2 - \frac{1}{2}x - 2 = 0$
3. $x^2 - 4x - 5 = 0$
4. $4x^2 + 8x - 1 = 0$
5. $x^2 - 7x + 2 = 0$
6. $2x^2 - 3x + 1 = 0$

Exercise 4.6

Solve the following equations by any appropriate method, giving your answers to 3 significant figures if they are not exact.

1. $x^2 - x - 1 = 0$
2. $x^2 + 8x + 1 = 0$
3. $x^2 - 2x + 1 = 0$
4. $2x^2 + x - 6 = 0$
5. $5x^2 - 5x - 2 = 0$
6. $x^2 - 3x + 1 = 0$
7. $x^2 - 3x - 1 = 0$
8. $3x^2 + 4x + 1 = 0$
9. $3x^2 - 4x - 1 = 0$
10. $-x^2 + 2x - 1 = 0$
11. $8x^2 + 342x + 35 = 0$
12. $6x^2 - 13x - 15 = 0$

Quadratic equations may appear in disguise, so you may need to do a little algebraic work before they are ready to solve.
The next example should make this clear.

Example 6

Solve for x

(a) $-\frac{1}{x} = 3x - 4$

(b) $\frac{3x-1}{x} + \frac{x+1}{x-1} = \frac{1}{x(x-1)}$ $x \neq 0$ and $x \neq 1$ (because we cannot divide by zero)

Answer 6

(a) $-\dfrac{1}{x} = 3x - 4$

Multiply both sides by x:
$-1 = x(3x - 4)$
$-1 = 3x^2 - 4x$
$0 = 3x^2 - 4x + 1$
$0 = (3x - 1)(x - 1)$
either $(3x - 1) = 0$ or $(x - 1) = 0$
$x = \dfrac{1}{3}$ $\qquad$ $x = 1$

(b) $\dfrac{3x-1}{x} + \dfrac{x+1}{x-1} = \dfrac{1}{x(x-1)}$ $\qquad$ NOTE: The common denominator is $x(x - 1)$, so multiply all through by $x(x - 1)$

$\dfrac{(3x-1)}{\cancel{x}} \times \dfrac{\cancel{x}(x-1)}{1} + \dfrac{(x+1)}{\cancel{(x-1)}} \times \dfrac{x\cancel{(x-1)}}{1} = \dfrac{1}{\cancel{x(x-1)}} \times \dfrac{\cancel{x(x-1)}}{1}$

(Cancel out common factors in numerator and denominator of each fraction)
$(3x - 1)(x - 1) + x(x + 1) = 1$
$3x^2 - 4x + 1 + x^2 + x - 1 = 0$
$4x^2 - 3x = 0$
$x(4x - 3) = 0$
either $x = 0$ or $(4x - 3) = 0$

$\qquad\qquad\qquad x = \dfrac{3}{4}$ $\qquad$ NOTE: The question states that $x \neq 0$ so $x = \dfrac{3}{4}$ is the answer.

Alternative method for (b):

Multiplying all through by the lowest common denominator can look very cumbersome when the denominators are algebraic as in the above example.

An alternative method is to multiply each fraction top and bottom by the factor or factors that are needed to make all the denominators the same.

In the above example, the first fraction $\dfrac{3x-1}{x}$ needs to be multiplied top and bottom by $(x - 1)$ and the second needs to be multiplied top and bottom by x. The right hand side already has the common denominator.

Thinking of this in the simplest way: you decide what the lowest common denominator is, and then, taking each fraction in turn, look for what is 'missing' in its denominator, and correct it.

Your first line of working would become:

$$\frac{(3x-1)(x-1)}{x(x-1)} + \frac{x(x+1)}{x(x-1)} = \frac{1}{x(x-1)}$$

Now the denominators are all the same we can see that multiplying the whole equation by $x(x-1)$ would cancel the denominators, so the next line of working will be:
$(3x-1)(x-1) + x(x+1) = 1$
Then proceed as before.

Example 7
Solve the following equation

$$\frac{1}{x} - \frac{1}{x+2} = \frac{2}{2x+1}$$

Answer 7

$$\frac{1}{x} - \frac{1}{x+2} = \frac{2}{2x+1}$$

The lowest common denominator is $x(x+2)(2x+1)$, so multiply the first fraction top and bottom by the 'missing' factors $(x+2)$ and $(2x+1)$, multiply the second top and bottom by x and $(2x+1)$ and so on.

$$\frac{(x+2)(2x+1)}{x(x+2)(2x+1)} - \frac{x(2x+1)}{x(x+2)(2x+1)} = \frac{2x(x+2)}{x(x+2)(2x+1)}$$

Now the whole equation has a common denominator, so multiplying through by that common denominator will cancel it out of every term, leaving:
$(x+2)(2x+1) - x(2x+1) = 2x(x+2)$.

This can now be multiplied out and simplified (taking care with the minus sign!)

$(2x^2 + 5x + 2) - (2x^2 + x) = (2x^2 + 4x)$

$2x^2 + 5x + 2 - 2x^2 - x - 2x^2 - 4x = 0$

$-2x^2 + 2 = 0$

Dividing through by -2:
$x^2 - 1 = 0$
$\quad x = 1 \quad \text{or} \quad x = -1$

Exercise 4.7
Solve the following

1. $x + 3 + \dfrac{2}{x} = 0$

2. $2 - \dfrac{5}{x} + \dfrac{2}{x^2} = 0$

3. $\dfrac{6x}{x-1} + (x-2) = 0$

68   Extended Mathematics for Cambridge IGCSE

4. $x^3 - x(x^2 - x) = 6 - x$
5. $\dfrac{1}{x(x+1)} = \dfrac{1}{2}$
6. $\dfrac{x-4}{x+1} + \dfrac{1}{x(x+1)} = -1$
7. $\dfrac{x^2}{(5+x)(6-x)} + \dfrac{1}{5+x} = \dfrac{1}{6-x}$
8. $\dfrac{1}{x} + \dfrac{1}{x+1} = \dfrac{3}{3x-1}$
9. $\dfrac{1}{x-1} - \dfrac{1}{x} = \dfrac{2}{x+2}$
10. $\dfrac{x}{2x+5} - \dfrac{1}{x} = \dfrac{3x-9}{2x^2+5x}$

NOTE: Factorise $2x^2 + 5x$ before proceeding as normal.

11. Rectangle with length $x - 4$ cm and width $x - 5$ cm.

NOTE: Remember that the area of a rectangle is obtained by multiplying the length by the width.

The area of the rectangle is 12 square centimetres.
   (a) Form an equation in x.
   (b) Solve your equation for x.
   (c) Write down the dimensions of the rectangle.

12. The sum of the squares of two consecutive numbers is 145. Calculate the numbers.
    NOTE: Let the two numbers be x and $x + 1$.
13. The product of two consecutive **even** numbers is 168. Calculate the numbers.
14. The length of a rectangle is 3 cm more than its breadth. The area of the rectangle is 40 cm². Let the breadth of the rectangle be x cm.
    (a) Show that this information leads to the equation $x^2 + 3x - 40 = 0$.
    (b) Solve this equation.
    (c) Write down the dimensions of the rectangle.

More Sequences

You will be expected to be able to answer slightly more difficult questions on sequences.

Example 8
(a) Find the nth term of the following sequence
   $\dfrac{1}{3}$ $\dfrac{3}{5}$ $\dfrac{5}{7}$ $\dfrac{7}{9}$ ...
(b) The nth term of a sequence is given by $n^2 + n + 1$.
   (i) Calculate the 50th term.
   (ii) Find n when the nth term $= 111$.
(c) Find the nth term of the following sequences
   (i) 16   25   36   49   ...
   (ii) 2   6   12   20   30   ...

Answer 8
(a) The numerators and denominators form separate sequences.
   The numerators are   1   3   5   7   ...
   The nth term is $2n - 1$

The denominators are   3   5   7   9   ...
The nth term is $2n + 1$

So the nth term for the sequence is $= \dfrac{2n-1}{2n+1}$

(b) (i) nth term $= n^2 + n + 1$
50th term $= 50^2 + 50 + 1$
$= 2551$
(ii) nth term $= 111$
$n^2 + n + 1 = 111$
$n^2 + n - 110 = 0$
$(n + 11)(n - 10) = 0$
$n = -11$ or $n = 10$
So the 10th term $= 111$

(c) (i) 16   25   36   49 ...
This sequence is clearly based on the sequence of square numbers, but the first term is 16 (4^2), not 1^2, and the second term is 25 (5^2) not 2^2.
So the nth term is not n^2, but $(n + 3)^2$.
nth term $= (n + 3)^2$

(ii) 2   6   12   20   30 ...
The *first* set of differences is
+4   +6   +8   +10 ...
and the *second* set is
+2   +2   +2 ...
so the sequence is based on n^2.
For the first term $1^2 + 1 = 2$, for the second term $2^2 + 2 = 6$, and for the third term $3^2 + 3 = 12$.
So the formula for the nth term is $n^2 + n$.
nth term $= n^2 + n$

Exercise 4.8

1. Find the first four terms of the following sequences
   (a) nth term $= n^2$
   (b) nth term $= n^2 - 2$
   (c) nth term $= n^2 - 2n$
   (d) nth term $= n^2 - 2n - 2$
   (e) nth term $= \dfrac{n}{n+1}$
   (f) nth term $= \dfrac{1}{n^2+1}$
   (g) nth term $= (n - 1)^2$
   (h) nth term $= n^3$
   (i) nth term $= n^3 + 2$
   (j) nth term $= n^3 - n$
   (k) nth term $= 3n$

2. Find the nth term of the following sequences
   (a) 1×2 2×3 3×4 4×5 ...
   (b) 0   2   6   12   20   30   ...
   (c) 4   9   16   25   ...
   (d) 2   8   18   32   50   ...
   (e) $\dfrac{1}{2}$ $\dfrac{4}{3}$ $\dfrac{9}{4}$ $\dfrac{16}{5}$ ...
   (f) $\dfrac{3}{4}$ $\dfrac{6}{5}$ $\dfrac{9}{6}$ $\dfrac{12}{7}$ $\dfrac{15}{8}$ ...

3. The nth term of a sequence is given by $2n^2 + 1$.
   (a) Find the 3rd, 7th and 10th terms.
   (b) Find n when the nth term is
      (i) 33   (ii) 243
4. The nth term of a sequence is given by $n^2 + 2n$.
   (a) Find the 4th, 7th and 100th terms.
   (b) Find n when the nth term is
      (i) 3   (ii) 99   (iii) 120   (iv) 288

More Simultaneous Equations

You will be expected to solve simultaneous equations which might require rearranging or simplifying first or where *both* equations need to be multiplied by constants in order to equalise coefficients.

The method is called **elimination** because one of the variables is eliminated to give an equation in the other.

For example, solve this pair of simultaneous equations:
$$-2x + 3y = 4$$
$$5x = 8 + 6y$$

It is convenient to number the two equations so that you can explain your method.

$$-2x + 3y = 4 \quad \ldots(i)$$
$$5x = 8 + 6y \quad \ldots(ii)$$

Rearrange (ii)
$$5x - 6y = 8 \quad \ldots(iii)$$

Multiply (i) by 5 and (iii) by 2
$$-10x + 15y = 20$$
$$10x - 12y = 16$$

Add these two equations to eliminate the x terms.
$$0 + 3y = 36$$
$$y = 12$$

Substitute for y in (ii)
$$5x = 8 + 6 \times 12$$
$$5x = 8 + 72$$
$$5x = 80$$
$$x = \frac{80}{5}$$
$$x = 16$$

So $x = 16$ and $y = 12$

This example shows the method when both equations are multiplied by constants, and the x terms have been eliminated. But in this particular case it would have been simpler to multiply (i) by 2, and then solve simultaneously with (iii), eliminating the y terms first. Try this to check the answer.

Shown below is a convenient and clear method for showing your working.

Algebra II

Example 9
Solve these two equations by the method of elimination
$3x - 2y = 4$
$5x - 3y = 7$

Answer 9

(i) ... $3x - 2y = 4$ $\xrightarrow{\times 5}$ $15x - 10y = 20$ $\longrightarrow$ $15x - 10y = 20$

(ii) ... $5x - 3y = 7$ $\xrightarrow{\times 3}$ $15x - 9y = 21$ $\xrightarrow{\times -1}$ $-15x + 9y = -21$ add

$-y = -1$
$y = 1$

Substitute $y = 1$ in (i) $3x - 2 \times 1 = 4$
$3x = 6$
$x = 2$

$x = 2$ and $y = 1$ **NOTE:** Substitute $y = 1$ in equation (ii) to check the working.

In the example above we have equalised the x coefficients to eliminate the x terms. Try the same question yourself, equalising the y coefficients by multiplying equation (i) by 3 and equation (ii) by 2, to eliminate the y terms.

Exercise 4.9

Solve these simultaneous equations by the method of elimination

1. $3x + 4y = 5$ **NOTE:** Look carefully!   2. $7y = 5 - 3x$
   $3y + 2x = 4$ $4y = 3 - 2x$
3. $3x + 3y = 2$ 4. $6x - 5y = 2$
   $2x - 8y = 3$ $5x + 3y = 16$

In some cases an alternative method, called solving by **substitution**, may be simpler. For example, look at this pair of simultaneous equations:

$$3x + 5y = 15$$
$$y = 2x - 1$$

In this case putting the right hand side of the second equation in brackets gives:

$$y = (2x - 1)$$

which can be substituted for y in the first equation.

$$3x + 5(2x - 1) = 15$$

We now have a linear equation in x only, so we can solve it in the normal way.

$$3x + 10x - 5 = 15$$
$$13x = 20$$
$$x = \frac{20}{13}$$

Now substitute for x in one of the equations (the second one is simpler).

$$y = \left(2 \times \frac{20}{13} - 1\right)$$

$$y = \frac{40}{13} - 1$$

$$y = \frac{40-13}{13}$$

$$y = \frac{27}{13}$$

Your questions will probably have simpler calculations than this, but there is no reason why fractions cannot be involved.

> **Example 10**
> Solve the following pair of equations
> $$\frac{1}{2}y + \frac{1}{5}x = 2$$
> $$y - x = 5$$
>
> **Answer 10**
> Number the two equations so that you can explain your method:
> $$\frac{1}{2}y + \frac{1}{5}x = 2 \quad \ldots\text{(i)}$$
> $$y - x = 5 \quad \ldots\text{(ii)}$$
> Multiply equation (i) by 10 to simplify.
> $$5y + 2x = 20 \quad \ldots\text{(iii)}$$
> Rearrange equation (ii)
> $$y = 5 + x \quad \ldots\text{(iv)}$$
> Substitute $(5 + x)$ for y in (iii)   NOTE: Remember to put $5 + x$ in a bracket!
> $$5(5 + x) + 2x = 20$$
> $$25 + 5x + 2x = 20$$
> $$7x = 20 - 25$$
> $$7x = -5$$
> $$x = -\frac{5}{7}$$
> Substitute for x in (iv)
> $$y = 5 + (-\frac{5}{7})$$
> $$y = \frac{35 - 5}{7}$$
> $$y = \frac{30}{7}$$
> so $x = -\frac{5}{7}$ and $y = \frac{30}{7}$
>
> As you should see, equations (ii) and (iii) can also be solved by elimination, either by multiplying (ii) by 5 to eliminate y, or by multiplying (ii) by 2 to eliminate x. You could try these for practice.

Exercise 4.10

Solve these pairs of equations by the method of substitution.
1. $y + 2x = 5$
   $3y - 5x = 4$
2. $y = 5x + 4$
   $3y - 2x + 1 = 0$
3. $x - 3 = 4y$
   $7y + 2x = 31$
4. $3x + y = 4$
   $2(y - 5) = -5x$
5. $x + 3y = 3$
   $5x = 6 - 5y$
6. $\frac{3}{2}x - y = \frac{1}{4}$
   $x + \frac{1}{2}y = 1$

The method you choose to use to solve your simultaneous equations is your choice, but by practising both methods you will have a choice and be able to choose the simplest for each particular pair of equations.

Exercise 4.11

Solve the following pairs of simultaneous equations
1. $3y + 2x = 12$
   $y + x = 4$
2. $5y = x - 15$
   $x + y = 9$
3. $2x + 3y = 15$
   $3x + 2y = 15$
4. $4y = x - 10$
   $3y - 2x = 5$
5. $0.1x - 0.2y = 2$
   $x + y = 17$
6. $1.4x + 3.9y = 6.4$
   $0.2x - 1.3y = 1.2$
7. $x = 7 - 2y$
   $y = 5 + 2x$
8. $\frac{1}{2}x - \frac{3}{4}y = 5$
   $x + y = 5$
9. $-x + 3y = 10$
   $2x + 5y = 2$
10. $\frac{2}{3}x + \frac{1}{4}y = 1$
    $5x + y = 1$

Inequalities

You have already seen the inequality signs: $>$ $\geqslant$ $<$ $\leqslant$.
It is useful to be able to represent inequalities on a number line, using a circle and an arrow. If the inequality is strict ($<$ or $>$) use an open circle, otherwise ($\leqslant$ or $\geqslant$) use a filled circle to show that the inequality includes the end number. If the inequality has no end use an arrow to show it continuing in that direction.
This is best made clear by some examples.

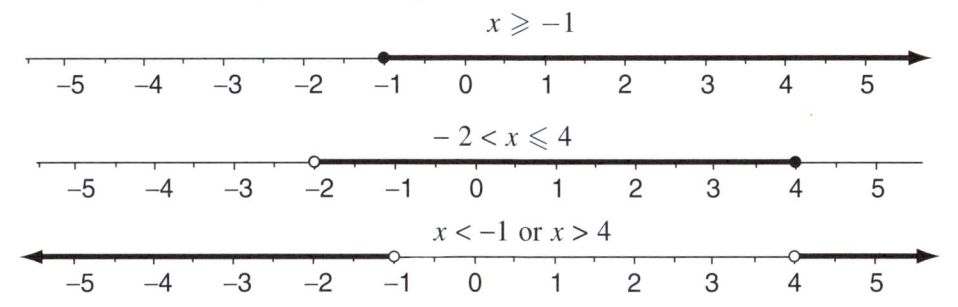

$x \geqslant -1$

$-2 < x \leqslant 4$

$x < -1$ or $x > 4$

74 Extended Mathematics for Cambridge IGCSE

Inequalities are similar to equations, but the solutions are generally a range of values rather than individual values. The methods used to solve inequalities are the same as those for equations with one very important exception which you must remember.
For example, consider the following:
$$x - 4 < 3x + 6$$
$$x - 3x < 6 + 4 \quad \text{(taking } 3x \text{ from both sides and adding 4 to both sides)}$$
$$-2x < 10$$

The next step is to divide both sides by -2, which we will do, but *at the same time* the inequality sign must be changed from 'less than' to 'greater than'.
$$x > \frac{10}{-2}$$
$$x > -5$$

The rule is: **When you multiply or divide an inequality by a negative number you must change the direction of the inequality sign.**

There are several ways to explain why it is necessary to turn the sign around.

Look at the number line below, and remember that numbers on the right are larger than numbers on the left.

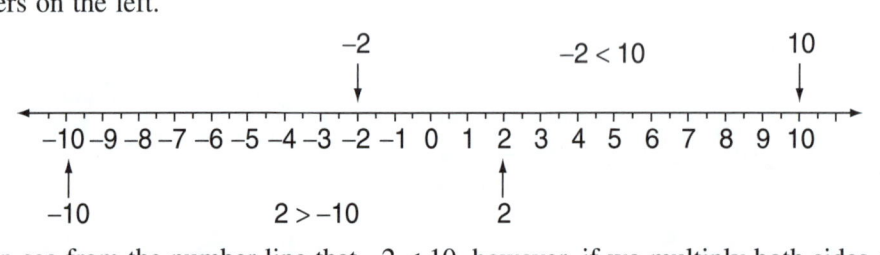

We can see from the number line that $-2 < 10$, however, if we multiply both sides of this inequality by -1 is it still true that $2 < -10$? From the number line you can see that it is not, and in fact $2 > -10$.

When you are working with inequalities try to picture (or sketch) a number line and think of its symmetry around zero to help you remember to alter the sign.

Another approach is to avoid multiplying or dividing by negative numbers.

Going back to the inequality above, when we reached the stage where we were about to divide by -2, we could do the following instead:
$$-2x < 10$$
Now add $-2x$ to both sides and take 10 from both sides:
$$-10 < 2x$$
Divide both sides by 2:
$$-5 < x$$
Now, since -5 is less than x it follows that x is greater than -5, and the answer may be given as:
$$x > -5$$
Note, however, that without this last step your solution is incomplete.

NOTE: Remember that it is only when multiplying or dividing by a negative number that the sign has to be changed, not when adding or subtracting.

Algebra II   75

Example 11
(a) Solve the following inequalities
   (i) $5x - 3 \geqslant 2x - 1$ (ii) $2 \leqslant 3x - 5 < 7$
(b) List the integers which satisfy the following inequality
   $3.5 < -x < 10.5$

Answer 11
(a) (i) $5x - 3 \geqslant 2x - 1$
   $5x - 2x \geqslant -1 + 3$ (taking $2x$ from both sides and adding 3 to both sides)
   $3x \geqslant +2$
   $x \geqslant \dfrac{2}{3}$

   (ii) $2 \leqslant 3x - 5 < 7$ (add 5 to each part of the inequality)
   $2 + 5 \leqslant 3x < 7 + 5$
   $7 \leqslant 3x < 12$ (divide all through by 3)
   $\dfrac{7}{3} \leqslant x < 4$
   or $2\dfrac{1}{3} \leqslant x < 4$

   so x is greater than or equal to $2\dfrac{1}{3}$ and less than 4, which could be shown on the number line as:

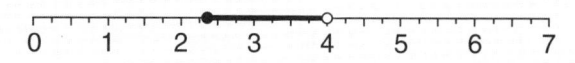

   0 1 2 3 4 5 6 7

(b) $3.5 < -x < 10.5$ (multiply all through by -1, and remember to change the inequality signs)
   $-3.5 > x > -10.5$ or $-10.5 < x < -3.5$
   This is shown on the number line below:

   −12 −11 −10 −9 −8 −7 −6 −5 −4 −3 −2 −1 0

   so the solution is $\{-10, -9, -8, -7, -6, -5, -4\}$

(Drawing the number line makes it easier to see which integers are included. You will note that -10 is included, but -3 is outside the range of the inequality)

Exercise 4.12

1. Solve the following inequalities, showing the solutions on a number line
   (a) $-2x < 1$
   (b) $6x > 7 - x$
   (c) $3(x + 1) \leqslant x - 2$
   (d) $2(3x - 2) \leqslant 7(x - 1)$
   (e) $\dfrac{1}{2}x - 3 < x + \dfrac{1}{5}$
   (f) $-7 < 2x < 4$
   (g) $1 \leqslant x - 1 < 6$
   (h) $3 < -x < 5$

2. List the integers which satisfy the following inequalities
   (a) $7x < -14$
   (b) $-3x \geqslant 6$
   (c) $2x - 1 < 2$
   (d) $-5 < 2x + 1 < 2$
   (e) $-2 \leqslant x - 1 < 3$
   (f) $1 \leqslant -x \leqslant 4$
   (g) $-1 < -x < 3$

Variation

We have already looked at the ideas of direct and inverse proportion, and now we must think about them algebraically, or in more general terms.

An example of *direct* proportion would be the number of similar items you buy (N) and the total cost (T) of those items, because as the number *increases* the total cost will *increase* in proportion.

An example of *inverse* proportion would be the number of items you could buy (N) for a given amount of money and the individual cost (c) of those items, because as the cost of each item *increases* the number you could buy *decreases*.

We will now look at how you are able to express these variations algebraically.

To demonstrate direct proportion or variation we can think of the first example above. If N = number of items, and T = the total cost then we can say that T is directly proportional to N and write:

$$T \propto N$$

where $\propto$ is the sign meaning 'is proportional to'. We read this as 'T is proportional to N', or 'T varies as (or with) N'.

For inverse proportion, as in the second example, if N is the number of items you can buy for the amount of money you have and c is the cost of each item, then we can say that N is inversely proportional to c, and write:

$$N \propto \frac{1}{c}$$

This is read 'N is inversely proportional to c' or 'N varies inversely as (or with) c'

Having written down the statement showing the variation we then replace the $\propto$ sign by '$= k \times$', where k is a constant, known as the constant of variation.

$$T \propto N \qquad \text{and} \qquad N \propto \frac{1}{c}$$
$$T = k \times N \qquad \qquad N = k \times \frac{1}{c}$$

Before we can make use of these equations it is necessary in each case to find the value of the constant k. Questions on variation will give you a pair of values so that you can find k. Work through the example to see how this is done.

The constant of variation may of course be any letter, but k is often used to represent a constant.

For example, we may be told that when the total cost, T, is Rs 980, the number of items, N, is 20.

Substituting in the equation:

$$T = k \times N$$
$$980 = k \times 20$$
$$k = 49$$

We now have a formula:
$$T = 49N$$
This formula can now be used to find other values of T or N.

A quantity may be directly or inversely proportional to a simple function of another quantity, such as the cube, or the square root as shown in the example below.

Example 12
(a) Given that N varies inversely as c, and that $N = 550$ when $c = 0.20$, find a formula connecting N and c.
(b) Given that x is inversely proportional to y^3, and that $x = 1$ when $y = 2$, find a formula connecting x and y.
(c) Given that h is proportional to the square root of A, and that $h = 3$ when $A = 81$,
  (i) find a formula connecting h and A,
  (ii) use your formula to find h when $A = 16$.

Answer 12

(a) $N \propto \dfrac{1}{c}$

$N = k \times \dfrac{1}{c}$

Given $N = 550$ when $c = 0.20$

$550 = k \times \dfrac{1}{0.20}$

$550 \times 0.20 = k$

$k = 110$

So the formula is $N = \dfrac{110}{c}$

(b) $x \propto \dfrac{1}{y^3}$

$x = \dfrac{k}{y^3}$

given that $x = 1$ when $y = 2$

$1 = \dfrac{k}{2^3}$

$k = 8$

The formula is $x = \dfrac{8}{y^3}$

(c) (i) $h \propto \sqrt{A}$

$h = k\sqrt{A}$

Given that $h = 3$ when $A = 81$

$3 = k \times \sqrt{81}$

$3 = k \times 9$

$k = \dfrac{1}{3}$

The formula is $h = \dfrac{1}{3}\sqrt{A}$

(ii) When $A = 16$

$h = \dfrac{1}{3}\sqrt{16}$

$h = \dfrac{4}{3}$

Exercise 4.13

1. The total cost (T) of a number of similar items (N) is directly proportional to the number of items.
   (a) Given that the total cost for 30 items is Rs 990, find an equation connecting T and N.
   (b) What does the constant of variation represent in this example?
2. Given that y is proportional to x^2, and that $y = 4$ when $x = 3$, find an equation in x and y.
3. (a) W is proportional to V and the constant of variation is d.
   Given that when $W = 10$ kg, $V = 7$ m^3 find the numerical value of d.
   (b) D is inversely proportional to V.
   When $D = 6$ mg/m^3, $V = 3.5$ m^3.
   Find a formula connecting D and V.
4. Given that a varies inversely as the square root of b, and that when $b = 9$, $a = 4$, find
   (a) the formula connecting a and b,
   (b) the value of a when $b = 16$,
   (c) the value of b when $a = 2.5$, giving your answer to 3 significant figures.
5. Gordon is making a set of cubic glass paperweights of various sizes.
   The mass, m g, of each paperweight varies as the cube of the length of one of its sides, l cm.
   A paperweight of side 5 cm has a mass of 375 g.
   (a) Find a formula to calculate the mass of each paperweight, given the length of one side.
   (b) Calculate the mass of a paperweight that measures 6.1 cm on each side.
6. The force of attraction (F) between two planets is inversely proportional to the square of the distance (d) between the planets.
   The units to measure force are newtons.
   When $F = 10^{20}$ newtons, $d = 10^9$ metres.
   (a) Find a formula connecting F and d.
   (b) Find the force between the two planets when they are 10^{15} metres apart.

Rearranging Formulae

For Extended Level you will need to be able to rearrange more complicated formulae. Remember that you are, in a sense, working backwards to rearrange the formulae, undoing each operation, so inverses need to be used, and BoDMAS may have to be used backwards. As before, it is helpful to underline the variable we want to make the subject of the formula. For example, a formula connecting speed (u), distance (d), time (t) and acceleration (a) is:

$$s = ut + \frac{1}{2}at^2$$

The formula is written in a way that makes it easy to find s, the distance. Rearrange this formula to make *a* the subject.

- Turn the formula round so that *a* is on the left hand side, and underline *a*.

$$ut + \frac{1}{2}\underline{a}t^2 = s$$

- Subtract *ut* from both sides.

$$\frac{1}{2}\underline{a}t^2 = s - ut$$

- Multiply both sides by 2.

$$\underline{a}t^2 = 2(s - ut)$$

- Divide both sides by t^2.

$$\underline{a} = \frac{2(s - ut)}{t^2}$$

The formula is now written in a way that makes it easy to find *a*, the acceleration, and you may have noticed that, contrary to BoDMAS subtraction came before multiplication and division.

Example 13

(a) $A = P\left(1 + \dfrac{r}{100}\right)^2$

Rearrange the formula to make *r* the subject.

(b) $ax - by = cx - d$

Rearrange the formula to make *x* the subject.

Answer 13

(a) $A = P\left(1 + \dfrac{r}{100}\right)^2$

$$P\left(1 + \frac{r}{100}\right)^2 = A$$

$$\left(1 + \frac{r}{100}\right)^2 = \frac{A}{P}$$

$$1 + \frac{r}{100} = \sqrt{\frac{A}{P}}$$

$$\frac{r}{100} = \sqrt{\frac{A}{P}} - 1$$

$$r = 100\left(\sqrt{\frac{A}{P}} - 1\right)$$

> (b) $ax - by = cx - d$
> $ax - cx = -d + by$
> $x(a - c) = by - d$
> $$x = \frac{by - d}{a - c}$$
>
> NOTE: x appears on both sides of the formula so the first step is to collect the x terms on one side.

There are two methods available to help you if you are not able to rearrange a formula. One is to compare it with a numerical version, so in part (a) of the example above we could write $20 = 5\left(1 + \frac{r}{100}\right)^2$, calculate r, and note the steps which had to be taken to find r. The numbers used are not too important as they are only there to help you understand the method.

The alternative is to use a number machine.
Always input the letter you need to make the new subject, and then work backwards.

input r → $\div 100$ → $\frac{r}{100}$ → $+1$ → $1 + \frac{r}{100}$ → square → $\left(1 + \frac{r}{100}\right)^2$ → $\times P$ → output $A = P\left(1 + \frac{r}{100}\right)^2$

output r ← $\times 100$ ← $\sqrt{\frac{A}{P}} - 1$ ← -1 ← $\sqrt{\frac{A}{P}}$ ← square rt ← $\frac{A}{P}$ ← $\div P$ ← input A

$$r = 100\left(\sqrt{\frac{A}{P}} - 1\right)$$

NOTE: The number machine does not work so well when the subject appears twice, as in Example 12 (b).

Exercise 4.14

1. $V = \frac{1}{12}(b - a)^2$ make b the subject

2. $S = \frac{a}{1 - r}$ make r the subject

3. $C = \frac{x^2 + y^2 - z^2}{2xy}$ make z the subject

4. $y = 4ax$ make a the subject

5. $S = \frac{n}{2}[2a + (n - 1)d]$ make a the subject

6. $E = \frac{1}{2}m(v^2 - u^2)$ make v the subject

7. $e = \dfrac{W}{W + w}$

   (a) make w the subject,

   (b) make W the subject.

8. $v^2 - u^2 = 2as$ make u the subject

9. $t = 2\pi \sqrt{\dfrac{l}{g}}$ make l the subject

10. $a^2 = b^2 + c^2 - 2bcA$ make A the subject

11. $ax + by = ac$ make a the subject

12. $2x - 3y = ax + by$

    (a) make x the subject,   (b) make y the subject,   (c) make b the subject.

Exercise 4.15

Mixed Exercise

1. Solve the following equations

   (a) $\dfrac{4x + 1}{3} = \dfrac{x + 5}{5}$ (b) $\dfrac{5x - 1}{x + 1} = 6$

2. Solve the following equations, giving your answers correct to 3 significant figures if they are not exact.

   (a) $9x^2 - 25 = 0$ (b) $2x^2 + 3x - 27 = 0$ (c) $3x^2 + 8x - 4 = 0$

3. Write $x^2 - x - 5$ in the form $(x + a)^2 + b$.

4. Find the nth term of the following sequences

   (a) $\dfrac{1}{3}$, $\dfrac{3}{6}$, $\dfrac{5}{9}$, $\dfrac{7}{12}$ ...

   (b) 0, $\dfrac{1}{4}$, $\dfrac{4}{9}$, $\dfrac{9}{16}$ ...   NOTE: $\dfrac{0}{1} = 0$

5. Given that y is inversely proportional to $\sqrt{x}$, and that $y = 10$ when $x = 9$, find

   (a) a formula connecting x and y.

   (b) (i) y when $x = 40$

        (ii) x when $y = 20$

6. Rearrange the following formulae

   (a) $\dfrac{A - B}{c} = D$ make B the subject

   (b) $\dfrac{A}{a} = \dfrac{B}{b}$ make b the subject

7. Solve the following inequality.

   $3x - 5 < 5x \leqslant 3x + 2$

8. Solve the following pairs of simultaneous equations
   (a) $2y + 3x = 10$
   $3y - 2x = -11$
   (b) $7y - x = 17$
   $5x + y = -73$

9. The nth term of a sequence is $\dfrac{(n-2)^2}{(n+2)^2}$. Write down the first 5 terms.

Examination Questions

10. (a) Solve
    (i) $9 - k < 7$
    (ii) $\dfrac{5}{2t} = \dfrac{1}{12}$
    (b) Solve the simultaneous equations $x + y = 29$,
    $4x = 95 - 2y.$
    (4024/01 May/June 2007 q 22)

11. Solve the simultaneous equations $\dfrac{1}{2}x + 2y = 16$,
    $2x + \dfrac{1}{2}y = 19.$
    (0580/02 May/June 2005 q 8)

12. Factorise
    (a) $4x^2 - 9$
    (b) $4x^2 - 9x$
    (c) $4x^2 - 9x + 2$
    (0580/02 May/June 2006 q 19)

13. Solve the equation
    $\dfrac{x-2}{4} = \dfrac{2x+5}{3}.$
    (0580/02 May/June 2006 q 13)

14. (a) (i) The cost of a book is $\$x$.
    Write down an expression in terms of x for the number of these books which are bought for $\$40$.
    (ii) The cost of each book is increased by $\$2$.
    The number of books which are bought for $\$40$ is now one less than before.
    Write down an equation in x and show that it simplifies to $x^2 + 2x - 80 = 0$.
    (iii) Solve the equation $x^2 + 2x - 80 = 0$.
    (iv) Find the original cost of one book.
    (b) Magazines cost $\$m$ each and newspapers cost $\$n$ each. One magazine costs $\$2.55$ more than one newspaper. The cost of two magazines is the same as the cost of five newspapers.
    (i) Write down two equations in m and n to show this information.
    (ii) Find the values of m and n.
    (0580/04 Oct/Nov 2005 q 8)

15. The wavelength, w, of a radio signal is inversely proportional to its frequency, f.
    When $f = 200$, $w = 1500$.
    (a) Find an equation connecting f and w.
    (b) Find the value of f when $w = 600$.

    (0580/02 May/June 2005 q 9)

16.

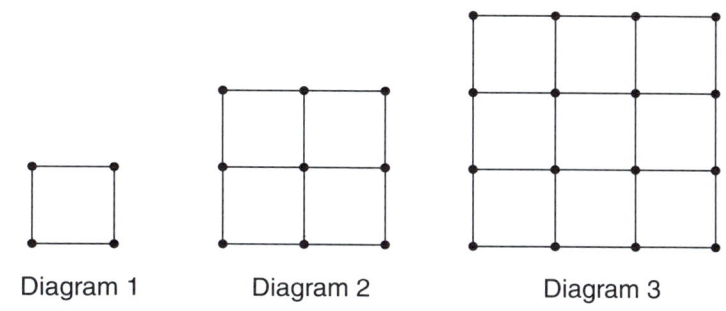

Diagram 1   Diagram 2   Diagram 3

The first three diagrams in a sequence are shown above.
The diagrams are made up of dots and lines. Each line is one centimetre long.
(a) Make a sketch of the next diagram in the sequence.
(b) The table below shows some information about the diagrams.

Diagram	1	2	3	4	....	n
Area	1	4	9	16	....	x
Number of dots	4	9	16	p	....	y
Number of one centimetre lines	4	12	24	q	....	z

(i) Write down the values of p and q.
(ii) Write down each of x, y and z in terms of n.

(c) The **total** number of one centimetre lines in the first n diagrams is given by the expression
$$\frac{2}{3}n^3 + fn^2 + gn$$
(i) Use $n = 1$ in this expression to show that $f + g = \frac{10}{3}$.
(ii) Use $n = 2$ in this expression to show that $4f + 2g = \frac{32}{3}$.
(iii) Find the values of f and g.
(iv) Find the total number of one centimetre lines in the first 10 diagrams.

(0580/04 May/June 2007 q 9)

17. Showing all your working, solve
    (a) $\frac{5x}{2} - 9 = 0$,
    (b) $x^2 + 12x + 3 = 0$, giving your answers correct to 1 decimal place.

    (0580/02 Oct/Nov 2005 q 23)

18. Solve the inequality
    $4 - 5x < 2(x + 4)$.

    (0580/02 Oct/Nov 2005 q 16)

19. Solve
    (a) $0.2x + 3.6 = 1.2$,
    (b) $\dfrac{2-3x}{5} < x+2$.

    (0580/02 May/June 2005 q 19)

20. The air resistance (R) to a car is proportional to the square of its speed (v).
    When $R = 1800$, $v = 30$.
    Calculate R when $v = 40$.

    (0580/02 Oct/Nov 2004 q 7)

21. The force of attraction (F) between two objects is inversely proportional to the square of the distance (d) between them.
    When $d = 4$, $F = 30$.
    Calculate F when $d = 8$.

    (0580/02 Oct/Nov 2005 q 13)

22. The length, y, of a solid is inversely proportional to the square of its height, x.
    (a) Write down a general equation for x and y.
        Show that when $x = 5$ and $y = 4.8$ the equation becomes $x^2y = 120$.
    (b) Find y when $x = 2$.
    (c) Find x when $y = 10$.
    (d) Find x when $y = x$.
    (e) Describe exactly what happens to y when x is doubled.
    (f) Describe exactly what happens to x when y is decreased by 36%.
    (g) Make x the subject of the formula $x^2y = 120$.

    (0580/04 May/June 2006 q 5)

23. Make c the subject of the formula
    $\sqrt{3c-5} = b$.

    (0580/02 Oct/Nov 2004 q 12)

24.

    The number of tennis balls (T) in the diagram is given by the formula

    $T = \dfrac{1}{2}n(n+1)$,

    where n is the number of rows.
    The diagram above has 4 rows.
    How many tennis balls will there be in a diagram with 20 rows?

    (0580/02 Oct/Nov 2005 q 1)

25. Make d the subject of the formula
$$c = \frac{d^3}{2} + 5.$$
(0580/02 Oct/Nov 2005 q 12)

26. It is given that $p = \frac{12}{\sqrt{q}}$.
    (a) Describe the relationship between p and q in words.
    (b) Calculate q when $p = 4$.
(4024/01 May/June 2006 q 6)

27. Use the formula, $P = \frac{V^2}{R}$ to calculate the value of P when $V = 6 \times 10^6$ and $R = 7.2 \times 10^8$.
(0580/02 Oct/Nov 2007 q 2)

28. The table shows some terms of several sequences.

Term	1	2	3	4	8
Sequence **P**	7	5	3	1	p
Sequence **Q**	1	8	27	64	q
Sequence **R**	$\frac{1}{2}$	$\frac{2}{3}$	$\frac{3}{4}$	$\frac{4}{5}$	r
Sequence **S**	4	9	16	25	s
Sequence **T**	1	3	9	27	t
Sequence **U**	3	6	7	−2	u

(a) Find the values of p, q, r, s, t and u.
(b) Find the nth term of sequence
   (i) **P**   (ii) **Q**   (iii) **R**
   (iv) **S**   (v) **T**   (vi) **U**.
(c) Which term in sequence **P** is equal to −777?
(d) Which term in sequence **T** is equal to 177147?
(0580/04 Oct/Nov 2007 q 9)

29. Solve the equations
    (a) $\frac{2x}{3} - 9 = 0$ (b) $x^2 - 3x - 4 = 0$
(0580/02 Oct/Nov 2007 q 14)

30. (a) The formula for the nth term of the sequence 1, 5, 14, 30, 55, 91, ... is
$$\frac{n(n+1)(2n+1)}{6}.$$
Find the 20th term.
(b) The nth term of the sequence 10, 17, 26, 37, 50, ... is $(n+2)^2 + 1$
Write down the formula for the nth term of the sequence 17, 26, 37, 50, 65, ...
(0580/21 May/June 2008 q 4)

31. Solve the inequality
$$\frac{2x-5}{8} > \frac{x+4}{3}$$

(0580/21 May/June 2008 q 13)

32. (a) There are 10^9 nanoseconds in 1 second.
Find the number of nanoseconds in 5 minutes, giving your answer in standard form.
(b) Solve the equation
$5(x + 3 \times 10^6) = 4 \times 10^7$

(0580/21 May/June 2009 q 14)

33. (a) (i) Factorise $x^2 - x - 20$.
(ii) Solve the equation $x^2 - x - 20 = 0$.
(b) Solve the equation: $3x^2 - 2x - 2 = 0$.
Show all your working and give your answers correct to 2 decimal places.
(c) $y = m^2 - 4n^2$.
(i) Factorise $m^2 - 4n^2$.
(ii) Find the value of y when $m = 4.4$ and $n = 2.8$.
(iii) $m = 2x + 3$ and $n = x - 1$.
Find y in terms of x in its simplest form.
(iv) Make n the subject of the formula $y = m^2 - 4n^2$.
(d) (i) $m^4 - 16n^4$ can be written as $(m^2 - kn^2)(m^2 + kn^2)$.
Write down the value of k.
(ii) Factorise completely $m^4n - 16n^5$.

(0580/04 May/June 2008 q 2)

34. Solve the simultaneous equations $2y + 3x = 6$,
$x = 4y + 16$.

(0580/21 May/June 2009 q 12)

35. A spray can is used to paint a wall.
The thickness of the paint on the wall is t. The distance of the spray can from the wall is d.
t is inversely proportional to the square of d.
$t = 0.2$ when $d = 8$.
Find t when $d = 10$.

(0580/21 May/June 2009 q 13)

36. Solve the inequality $\frac{2-5x}{7} < \frac{2}{5}$

(0580/21 Oct/Nov 2008 q 12)

37. The length of time, T seconds, that the pendulum in a clock takes to swing is given by the formula
$$T = \frac{6}{\sqrt{(1+g^2)}}$$

Rearrange the formula to make g the subject.

(0580/02 Oct/Nov 2007 q 17)

38. (a) Factorise $ax^2 + bx^2$.
    (b) Make x the subject of the formula $ax^2 + bx^2 - d^2 = p^2$.
    (0580/21 Oct/Nov 2008 q 8)

39. The quantity p varies inversely as the square of $(q + 2)$.
    $p = 5$ when $q = 3$.
    Find p when $q = 8$.
    (0580/21 Oct/Nov 2008 q 13)

40. Rearrange the formula to make y the subject.
    $$x + \frac{\sqrt{y}}{9} = 1.$$
    (0580/21 May/June 2009 q 9)

41. Solve the simultaneous equations $2x + \frac{1}{2}y = 1$,
    $$6x - \frac{3}{2}y = 21.$$
    (0580/02 Oct/Nov 2007 q 7)

42. $\frac{4c}{5} - \frac{3c}{35} = \frac{10}{7}$. Find c.
    c(0580/02 Oct/Nov 2007 q 5)

43. (a) Solve the equation $\frac{m-3}{4} + \frac{m+4}{3} = -7$

    (b) (i) $y = \frac{3}{x-1} - \frac{2}{x+3}$
        Find the value of y when $x = 5$.

        (ii) Write $\frac{3}{x-1} - \frac{2}{x+3}$ as a single fraction.

        (iii) Solve the equation $\frac{3}{x-1} - \frac{2}{x+3} = \frac{1}{x}$

    (c) $p = \frac{t}{q-1}$
        Find q in terms of p and t.
        (0580/04 Oct/Nov 2009 q 9)

Chapter 5

Geometry and Shape

This chapter develops the ideas that began in Chapter 6 of *Core Mathematics for Cambridge IGCSE*. It is essential that you fully understand that work. The Core Skills exercise below can only touch on those ideas, and you should use it as a reminder. Work through it, and go back to the Core Book for anything that you have forgotten before continuing with the rest of this chapter.

Core Skills

1. Find x in the following diagrams

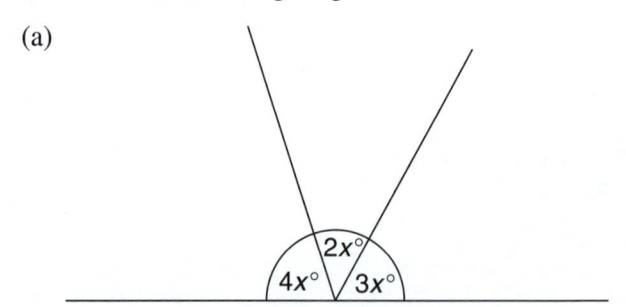

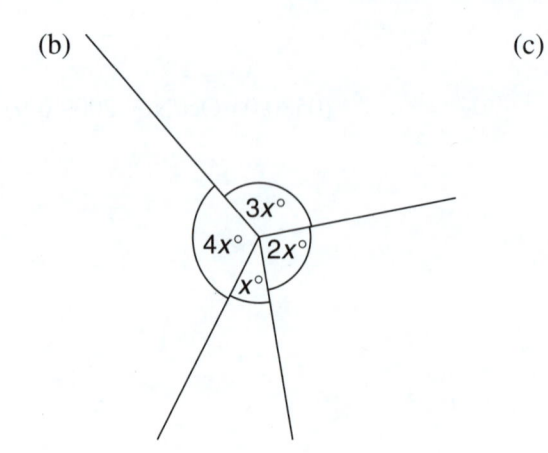

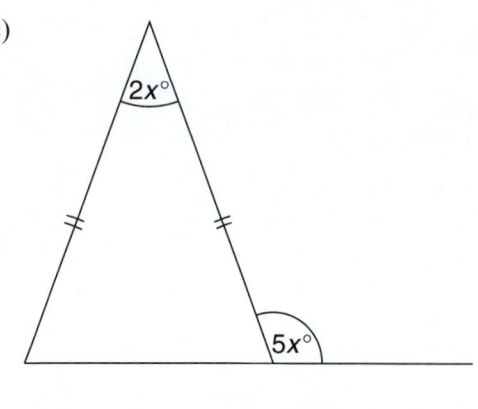

   (a)

   (b)   (c)

2. Construct accurately a triangle with sides 6 cm, 4 cm and 5 cm. Measure the angles of the triangle.
3. Calculate the interior angle of a regular nonagon.
4. Calculate the total interior angle of a 13 sided polygon.
5. Name the quadrilaterals with the following properties
   (a) The diagonals bisect each other but not at right angles.
   (b) One diagonal is the perpendicular bisector of the other.
   (c) One pair of opposite sides is parallel.
   (d) Each diagonal is the perpendicular bisector of the other. The diagonals are different lengths.
6.

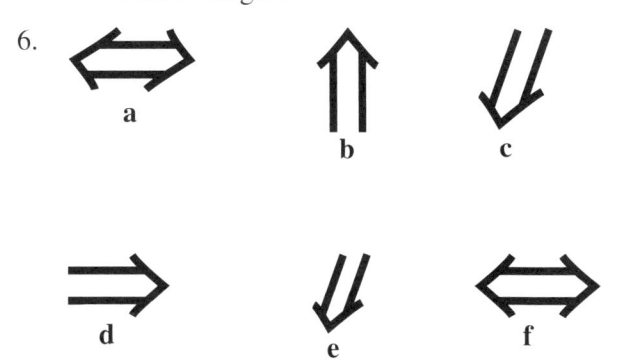

   In the figures above
   (a) which two figures are congruent,
   (b) which two figures are similar?
7. Describe the symmetry of each figure shown below 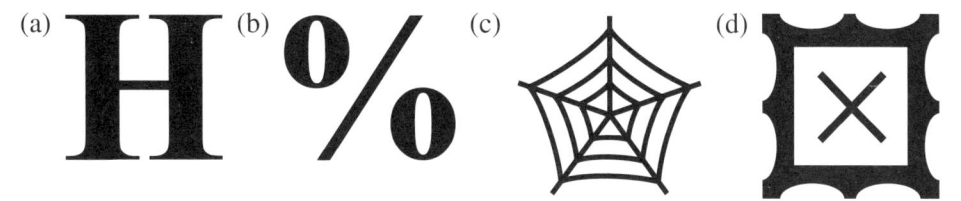
8. Describe the locus of points that are 5 centimetres from a given point, A.
9. The figure below is the net for which solid shape?

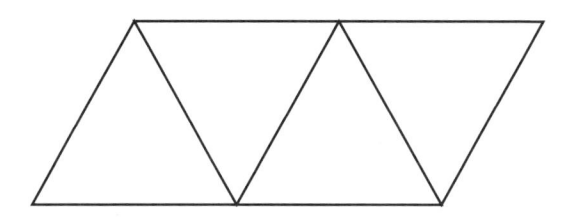

10.

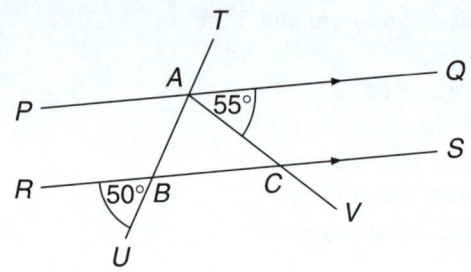

PQ and *RS* are parallel straight lines.
Angle *QAC* = 55° and angle *RBU* = 50°.
   (a) Name a pair of corresponding angles.
   (b) Name a pair of alternate angles.
   (c) Name a pair of vertically opposite angles.
   (d) Calculate the size of angles *BAC*, giving reasons for your answer.

Symmetry in Three-Dimensional Shapes

We have seen that two-dimensional shapes can have lines of symmetry, with one side of the line of symmetry perfectly matching the other when the shape is folded along the line. The line of symmetry is a mirror line, and one side of the line of symmetry is the mirror image of the other. We will look at mirror images in more depth in a later chapter.

In a similar way, three-dimensional shapes can have **planes of symmetry**. One side of the plane of symmetry perfectly matches the other, and is in fact the mirror image of the other side.

A simple example is a cuboid as is shown in the diagrams below. The cuboid has 3 planes of symmetry, one in a horizontal direction as is shown in the first diagram, and two in vertical directions, shown in the second and third diagrams.

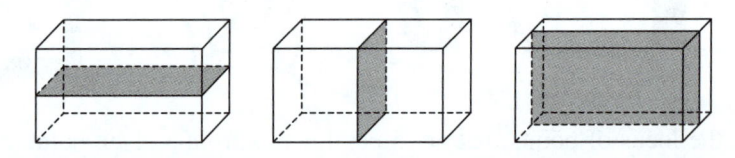

Example 1

(a) Draw a plane of symmetry in these shapes
   (i) a square-based pyramid,
   (ii) a regular tetrahedron (triangular-based pyramid),
   (iii) a cylinder.

(b) How many planes of symmetry are there in
   (i) the square-based pyramid, (ii) the regular tetrahedron, (iii) the cylinder?

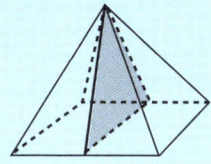

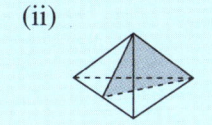

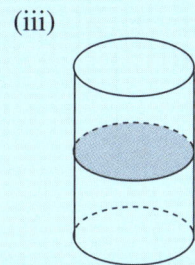

Answer 1
(a) (i)   (ii)   (iii)

(b) (i) 4   **NOTE: Remember the diagonals.**

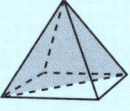

(ii) 6   (iii) 1 horizontal plane of symmetry as shown in part (a), but an infinite number of vertical planes of symmetry.

We have also seen that a two-dimensional shape may have rotational symmetry about a point, or centre of symmetry. The order of rotational symmetry is given by the number of different ways that the shape can be rotated and fitted into its own outline until it returns to the first position.

A three-dimensional shape may also have rotational symmetry, but it will be about an axis, rather than a centre. An example is shown in the diagrams below of a square-based pyramid which has a vertical axis of rotational symmetry, and the symmetry is of order 4 because the pyramid will fit its own outline in four different ways. One vertex of the base of the pyramid is marked with a cross to show the four different positions.

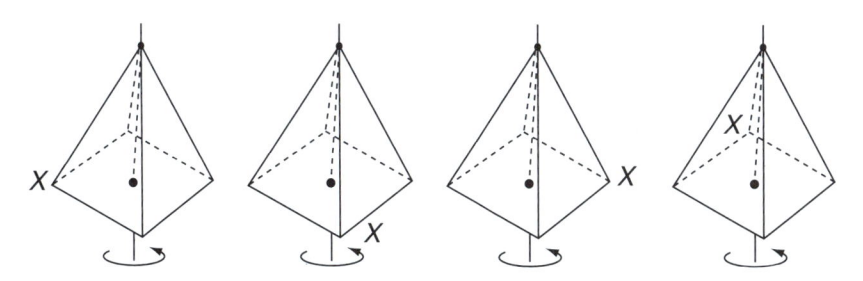

Example 2
The cuboid shown below has two square faces and four rectangular faces.
(a) Describe the symmetry of the cuboid about the axis of symmetry XY.
(b) How many other axes of symmetry does the cuboid have?

(c) Copy the diagram, draw in the other axes of symmetry, and beside each axis write down the order of symmetry about that axis.

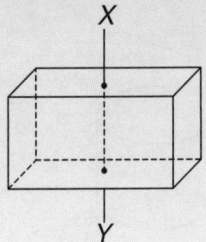

Answer 2
(a) The cuboid has rotational symmetry of order 2 about the axis of symmetry *XY*
(b) The cuboid has two more axes of symmetry
(c)

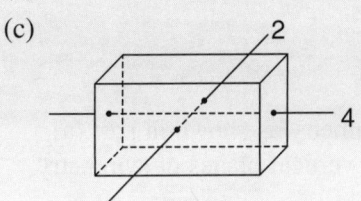

Example 3
(a) Sketch a cone.
(b) Draw an axis of symmetry of the cone.
(c) What is the order of rotational symmetry of the cone about this axis?

Answer 3
(a), (b)

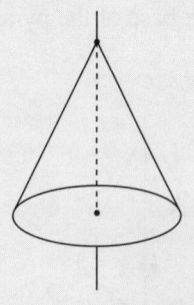

(c) The order of rotational symmetry is infinite

Exercise 5.1

1. (a) Sketch an equilateral triangular prism.
   (b) How many axes of symmetry does the prism have?
   (c) Draw two of the axes of symmetry on your sketch, and beside each one write down the order of rotational symmetry about that axis.

Geometry and Shape

(d) How many planes of symmetry does the prism have?

2. (a) Sketch a regular tetrahedron.
   (b) Draw one of its axes of symmetry.
   (c) Write down the order of rotational symmetry beside this axis.
3. Describe the symmetry of the shapes shown below.
   (a) circular base
   (b) square base
   (c) prism with a regular hexagonal cross-section

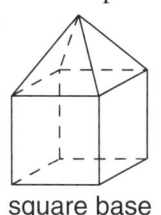

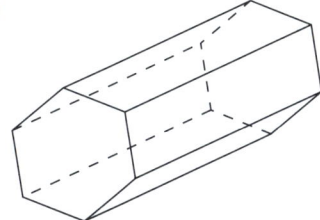

Further Circle Facts

We have seen how the two-dimensional symmetry of the circle means that the tangent at a point is at right angles to the radius at that point. We now look at three more properties of the circle which are due to the symmetry of the circle. In each of the following diagrams a line of symmetry is drawn as a dotted line. In each case folding the circle along that line will result in one side fitting exactly over the other.

- **Equal chords are equidistant from the centre of the circle:**
  $AB = CD$
  OP is perpendicular to AB
  OQ is perpendicular to CD
  $OP = OQ$

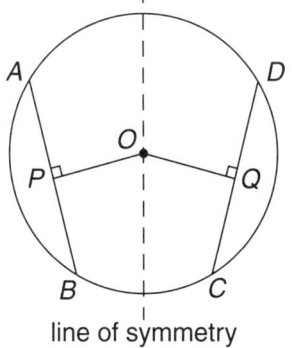

  line of symmetry

- **The perpendicular bisector of a chord passes through the centre of the circle:**

  line of symmetry

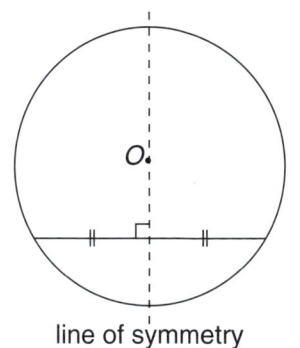

94　*Extended Mathematics for Cambridge IGCSE*

- **Tangents from a point outside the circle are equal in length:**

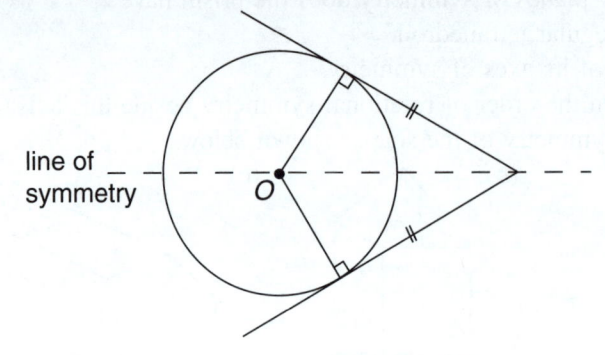

You need to be able to recall these facts and use them in solving circle problems.

Example 4

1.

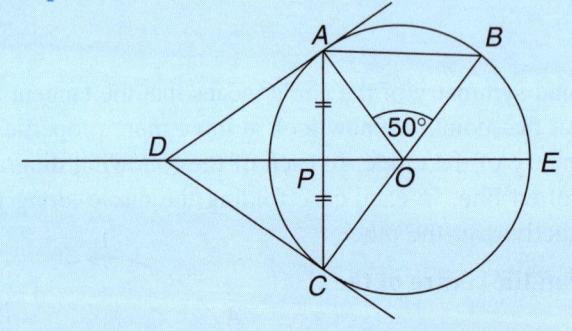

In the diagram, O is the centre of circle, angle $AOB = 50°$, and DA and DC are tangents to the circle. BOC is a straight line. $AP = PC$.
(a) Show that AB is parallel to DE.
(b) Find angle ADC.

Answer 4
(a) $\angle BAC = 90°$ (angle in a semicircle)
　　$\angle APD = 90°$ (perpendicular bisector of a chord
　　　　　　　　　　passes through the centre of the circle)
　　So AB is parallel to DE (alternate angles)
(b) $\angle AOC = 180 - 50 = 130°$ (angles on a straight line)
　　$\angle DAO = \angle DCO = 90°$ (angle between a tangent and a radius)
　　$\angle ADC = 360 - 130 - 2 \times 90$ (angle sum of a quadrilateral)
　　$\angle ADC = 50°$

NOTE: This is just one way to arrive at the answer. There are other ways also.

Exercise 5.2

1.

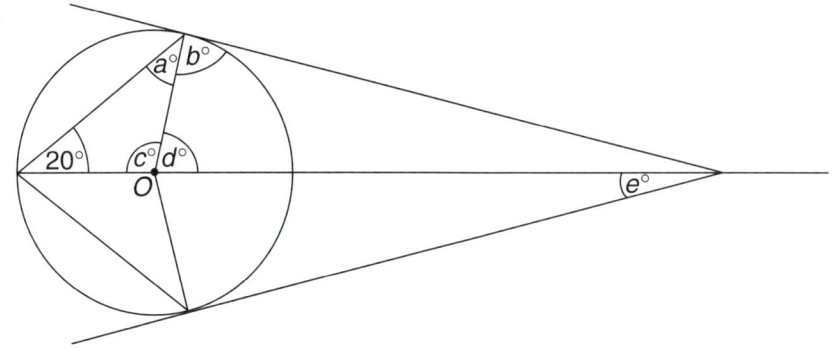

In the diagram, O is the centre of the circle. Find a, b, c, d and e.

2.

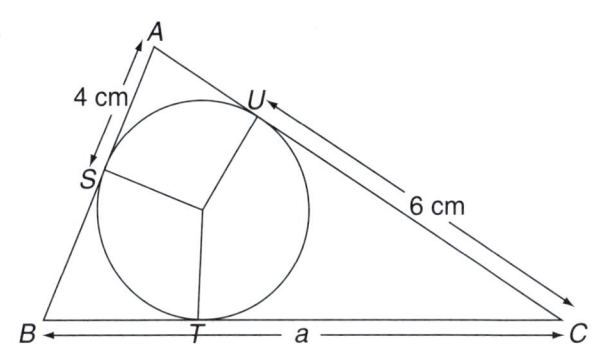

The perimeter of the triangle is 29 cm. Find a, giving reasons.

3.

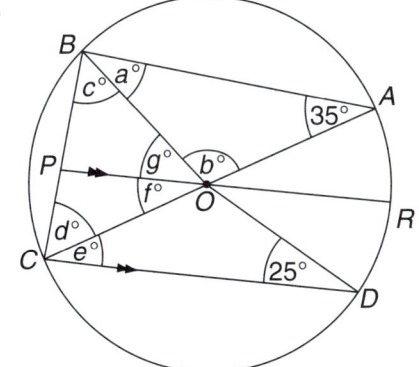

In the diagram, O is the centre of the circle and PR is parallel to CD. AOC is a straight line.
Find, a, b, c, d, e, f and g, giving reasons.

96 Extended Mathematics for Cambridge IGCSE

Investigations

In each of these investigations draw the diagrams and measure the angles accurately. Record the measured angles for comparison.

1. Draw a circle, centre O and radius 6 centimetres. Mark the centre with a dot as soon as you remove your compass point so that you can find it accurately.
   Choose any three points on the circumference, and label them A, B and C as in the diagram below.
   Join AB, AO, BC and OC as shown.
   Measure angles ABC and AOC.

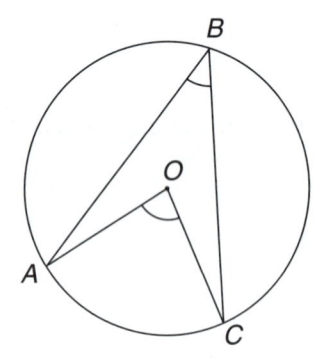

   Repeat with circles of different radii and different positions on the circumference.
   You should find that:

   - **the angle at the centre of the circle is twice the angle at the circumference.**

2. Draw another circle, centre O and radius 6 centimetres.
   Mark four points on the circumference and label them A, B, C and D as shown in the digram below.
   Join AB, BD, AC, CD and AD as shown.
   Measure angles ABD and ACD.

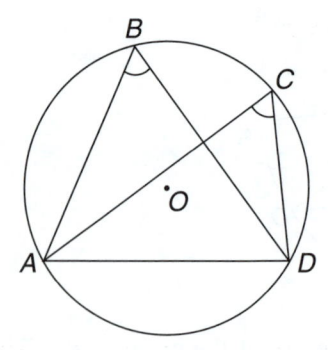

   Repeat with other circles and points on the circumference.
   Draw some with C above the chord AD and some with C below the chord AD as shown in the next diagram.

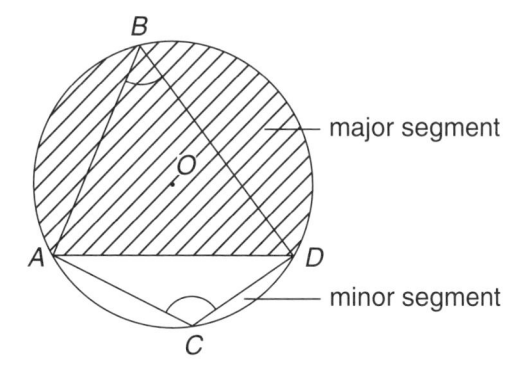

Remember that the chord *AD* divides the circle into two segments. When *B* and *C* are both above the chord they are in the same segment, when one is below the chord they are in opposite segments.

You should have found that:
- **angles in the same segment standing on the same chord are equal,**
- **angles in opposite segments standing on the same chord add up to 180°.**

The term for pairs of angles which add up to 180° is **supplementary**, so we can reword the second of the above facts:
- **angles in opposite segments are supplementary.**

3. Draw one final circle and mark any four points on the circumference. Join the points to form a quadrilateral, as shown in the diagram below. Measure all the angles.

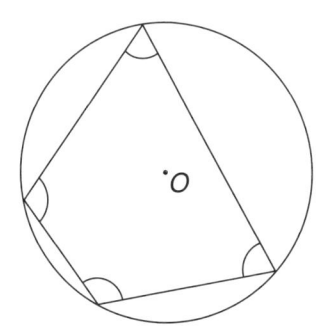

You should find that the opposite angles of the quadrilateral add up to 180°.

This also follows from the fact that we have just found, namely that angles in opposite segments of the circle are supplementary, because the diagonals of the quadrilateral form chords which divide the circle into segments.

A quadrilateral that has its vertices on a circle is called a **cyclic quadrilateral**, so we have a new fact:

- **opposite angles of a cyclic quadrilateral are supplementary.**

A cyclic quadrilateral can have a circle drawn through all its vertices, but this does not apply to all quadrilaterals. We can use the fact that if opposite angles of a quadrilateral are supplementary then the quadrilateral is a cyclic quadrilateral.

- **If the opposite angles of a quadrilateral are supplementary then the quadrilateral is cyclic.**

It follows that all squares and rectangles are always cyclic quadrilaterals. Why? Which of the other special quadrilaterals *can* be cyclic, and what conditions would apply?

To test your answers to the above try to draw each type of quadrilateral in a circle, and see if you are right.

Finding the Centre of a Circle

We can use the fact that the perpendicular bisector of a chord passes through the centre of the circle to find the centre of any circle.
Try the following:
Draw carefully round any circular object, such as a tin lid or a cylinder.
Draw a chord, and construct the perpendicular bisector of the chord.
Draw another chord, and construct the perpendicular bisector of the second chord. The two perpendicular bisectors will intersect at the centre of the circle.

Example 5

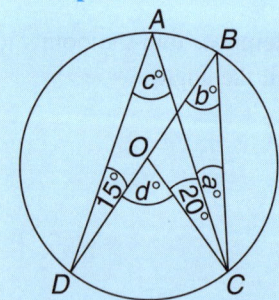

Find a, b, c and d, giving reasons.

Answer 5

$\angle ACB = \angle ADB = 15°$ (angles in the same segment standing on the same chord or arc, AB)

$a = 15°$

NOTE: The chords AB and DC are not drawn in the diagram, so we could say 'standing on the same arc AB' instead. It makes no difference.

$\angle OBC = \angle OCB = 20 + 15 = 35°$ (isosceles triangle)

$b = 35°$

$\angle DAC = \angle DBC = 35°$ (angles in the same segment standing on the same chord or arc, DC)

$c = 35°$

$d = 70°$ (angle at the centre is twice the angle at the circumference.)

Example 6

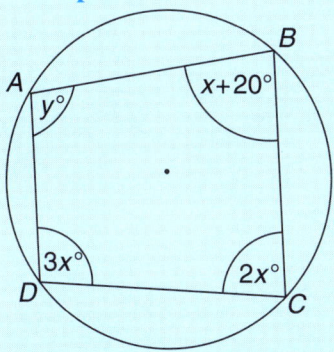

Calculate x and y.

Answer 6

$3x + (x + 20) = 180°$ (opposite angles of a cyclic quadrilateral)
$\quad 4x + 20 = 180°$
$\quad\quad\quad 4x = 160°$
$\quad\quad\quad\ x = 40°$
$y + 2x = 180°$ (opposite angles of a cyclic quadrilateral)
$y + 2 \times 40 = 180°$
$\quad\quad\quad y = 100°$

Exercise 5.3

1. Find the angles marked x in the diagrams below. Give reasons for your answers.

(a)

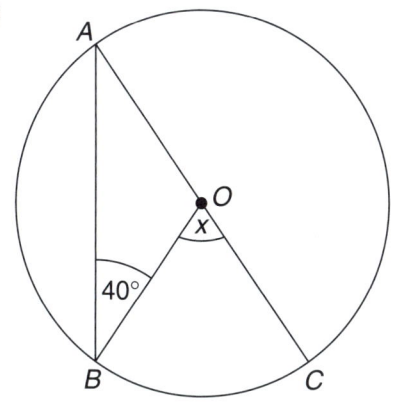

(b)

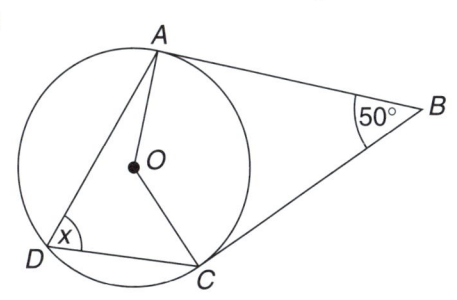

AB and CB are tangents to the circle, centre O

In questions 2, 3 and 4, use the information in the diagrams to find the angles marked with letters. Give reasons.

2.

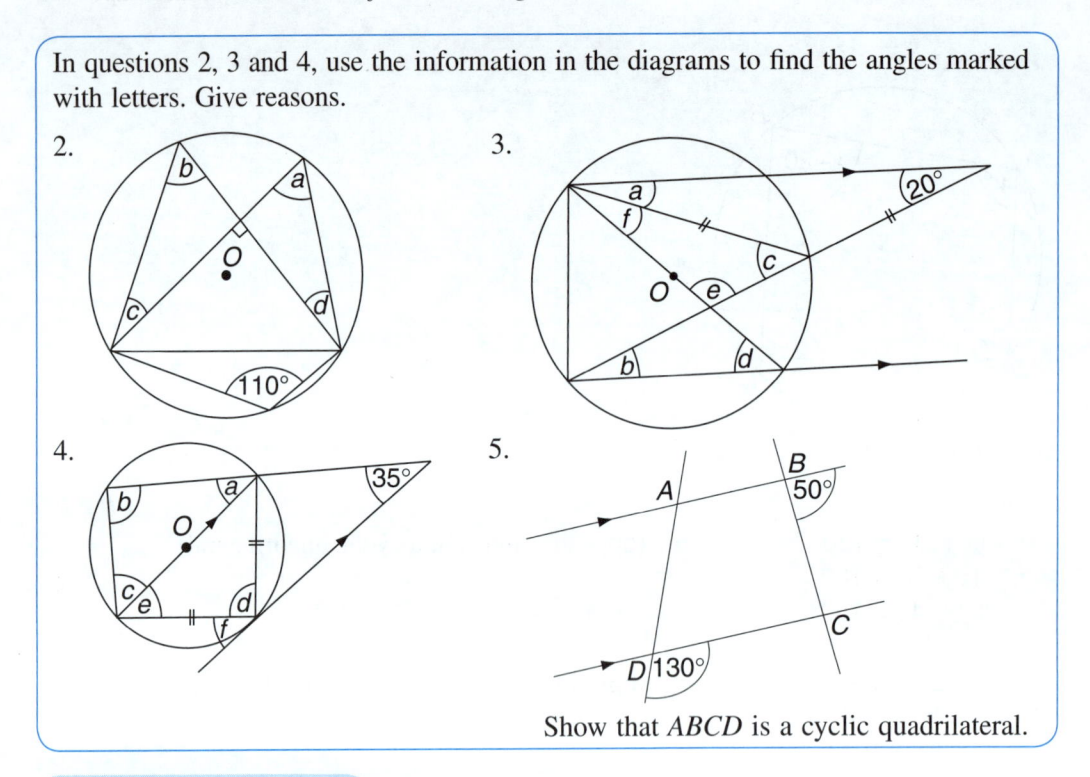

3.

4.

5.

Show that *ABCD* is a cyclic quadrilateral.

Irregular Polygons

Regular polygons have all angles and all sides equal, but this does not apply to irregular polygons. However, the angle sum of irregular polygons can be calculated in the same way as before. The diagram shows an irregular hexagon divided into 4 triangles. The angle sum is $4 \times 180° = 720°$.

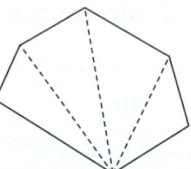

The exterior angles of each irregular polygon add up to 360° as before.

Example 7
Calculate x in the following polygons.

(a)

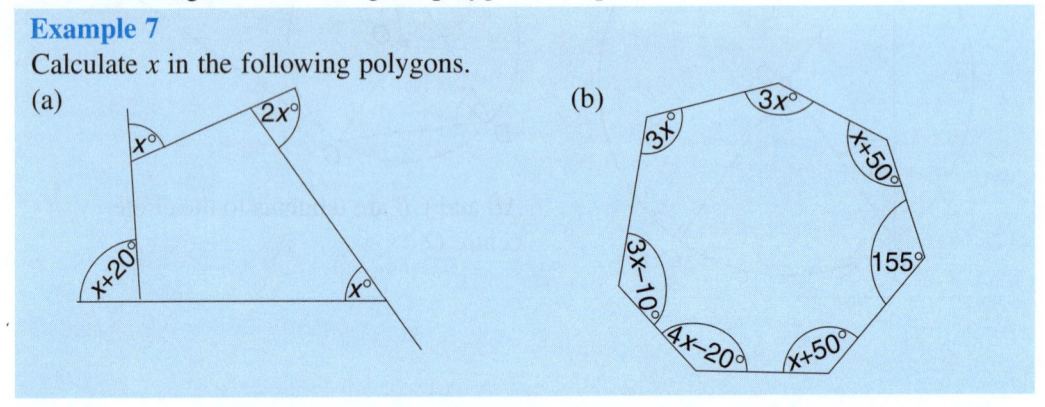

(b)

Answer 7
(a) Using the fact that the exterior angles add up to 360°.
$$x + 2x + (180 - x) + (x + 20) = 360$$
$$3x + 200 = 360$$
$$3x = 160$$
$$x = 53\frac{1}{3}$$

(b) The angle sum of the interior angles = $(7 - 2) \times 180 = 900°$
$$3x + 3x + (x + 50) + 155 + (x + 50) + (4x - 20) + (3x - 10) = 900$$
$$15x + 225 = 900$$
$$15x = 675$$
$$x = 45$$

Exercise 5.4

1. An irregular polygon has an angle sum of 1980°. How many sides does the polygon have?
2. An irregular nonagon (9 sides) has interior angles $2x°$, $2x + 1°$, $x - 1°$, $x°$, $3x°$, $2x - 1°$, $3x°$, $x + 1°$ and $2(x + 1)°$. Calculate x.
3. An irregular pentagon has angles $3a°$, $a + b°$, $2b°$, $3b°$ and $40°$.
   A triangle has angles $a°$, $b°$, and $a + b°$.
   Form two equations in a and b and solve them simultaneously to find a and b.

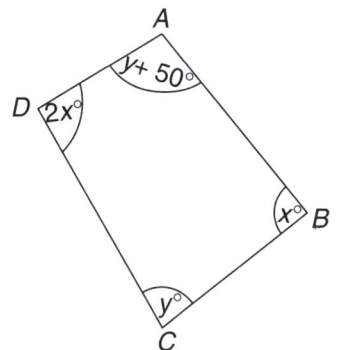

Exercise 5.5

Mixed Exercise

1. $ABCD$ is a cyclic quadrilateral. Calculate x and y.

2.

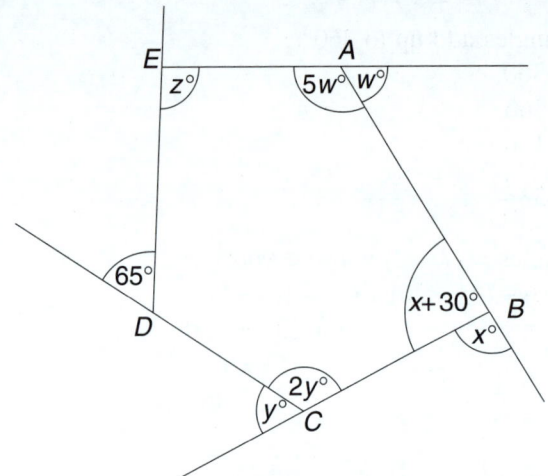

ABCDE is an irregular pentagon. Calculate w, x, y and z.

3.

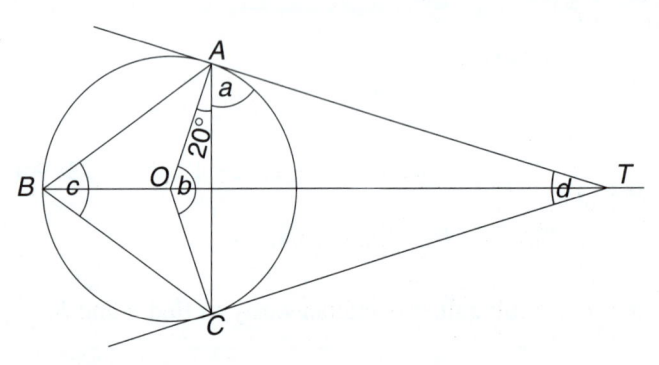

(a) Find the angles marked a, b, c and d.
(b) Show that AOCT is a cyclic quadrilateral.
(c) Describe where the centre of the circle AOCT would be located.

4. Describe the symmetry of a pyramid whose base is a regular hexagon.

5. What can you say about   (a) AE and BD,   (b) AF, FE, BC and CD in this diagram? Give reasons for your answers.

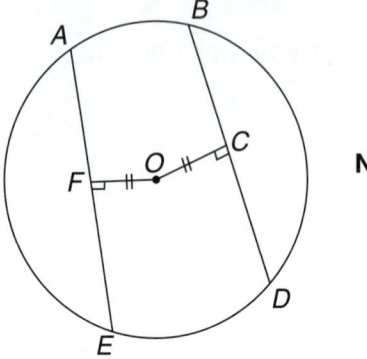

NOT TO SCALE

6.

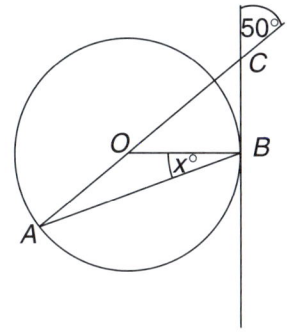

Find x, giving reasons for your answer.

7.

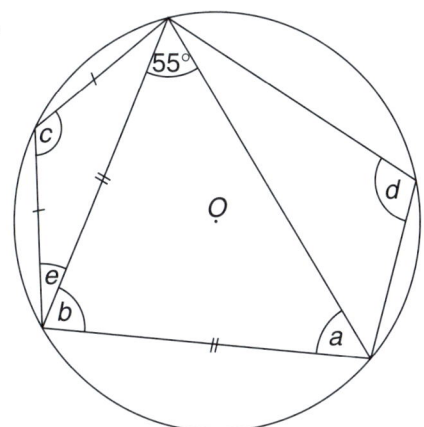

Find the angles marked a, b, c, d and e, giving reasons for your answers.

Examination Questions

8.

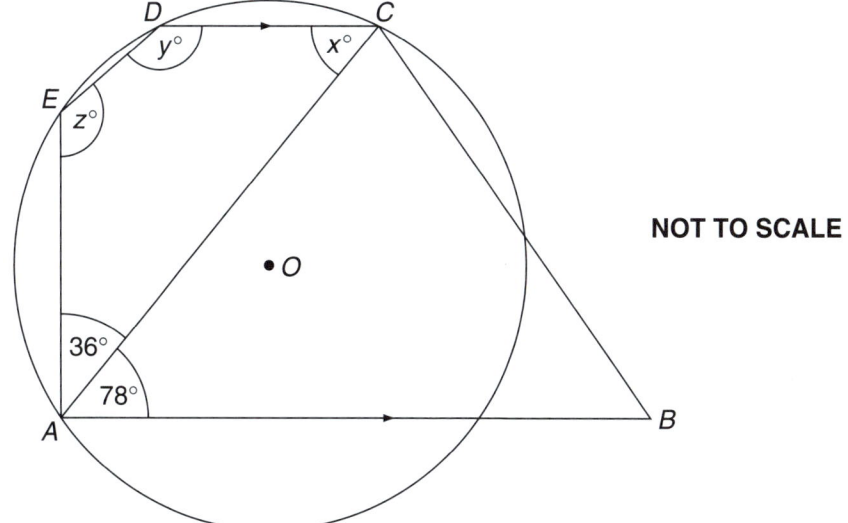

NOT TO SCALE

ABCDE is a pentagon.

A circle, centre *O*, passes through the points *A*, *C*, *D* and *E*.

Angle *EAC* = 36°, angle *CAB* = 78° and *AB* is parallel to *DC*.

(a) Find the values of *x*, *y* and *z*, giving a reason for each.

(b) Explain why *ED* is **not** parallel to *AC*.

(c) Find the value of angle *EOC*.

(d) *AB* = *AC*.

Find the value of angle *ABC*.

(0580/04 Oct/Nov 2008 q 8)

9. In triangle *PLQ*, *PL* = 14 cm, *PQ* = 10 cm and *LQ* = 7 cm.
   (a) Using ruler and compasses only, complete triangle *PLQ* where *Q* is above *PL*.
   (b) Measure and write down *PQ̂L*.
   (c) Draw a semicircle with *PL* as diameter.
      The line *LQ* produced meets the semicircle at *M*.
      Measure and write down the length of *QM*.
   (d) Explain why *PM* is perpendicular to *LM*.

(4024/01 May/June 2005 q 24 (part))

10.

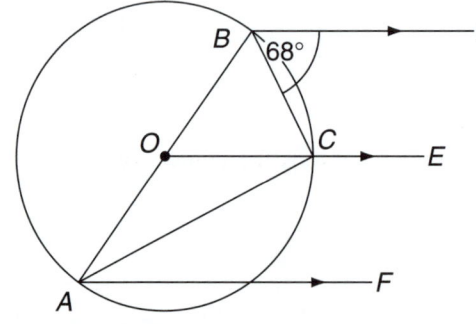

Points *A*, *B* and *C* lie on a circle, centre *O*, with diameter *AB*.
BD, *OCE* and *AF* are parallel lines.
Angle *CBD* = 68°
Calculate
(a) angle *BOC*,
(b) angle *ACE*.

(0580/21 Oct/Nov 2009 q 19)

11. A map is drawn using a scale of 1 cm to 5 m.
    (a) The point *B* is 70 m due East of *A*.
       Draw the line representing *AB*.
    (b) The point *C* is North of *AB* and equidistant from *A* and *B*.
       Angle *BAC* = 40°.

(i) By drawing appropriate lines, find and label the point C.
(ii) Find the actual distance AC.
(iii) State the size of the reflex angle BAC.

(4024/01 May/June 2006 q 22)

12. ABCD is a cyclic quadrilateral. The tangents at C and D meet at E. Calculate the values of p, q and r.

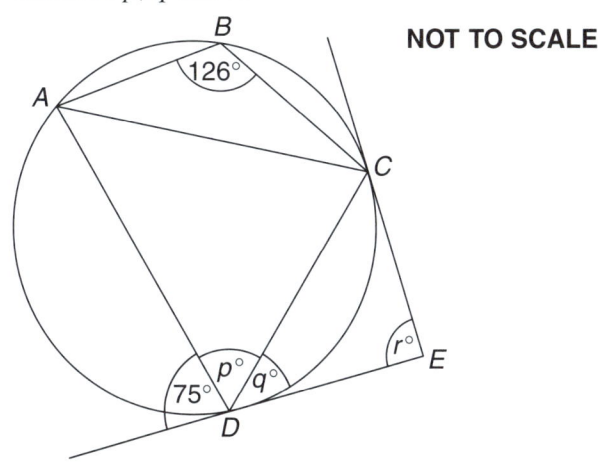

NOT TO SCALE

(0580/02 Oct/ Nov 2004 q 16)

13. AD is a diameter of the circle ABCDE.
Angle BAC = 22° and angle ADC = 60°
AB and ED are parallel lines.
Find the values of w, x, y and z.

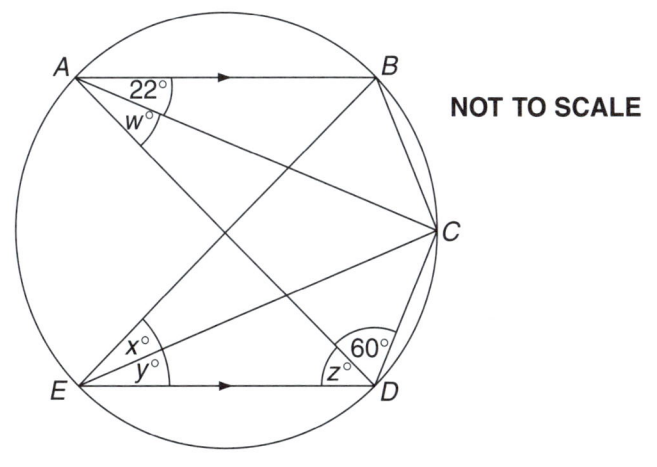

NOT TO SCALE

(0580/02 May/June 2006 q 18)

14. In the diagram, A, B, C and D lie on a circle with centre O.
The tangent to the circle at C meets the diameter AD produced at T.
$D\hat{O}C = 66°$

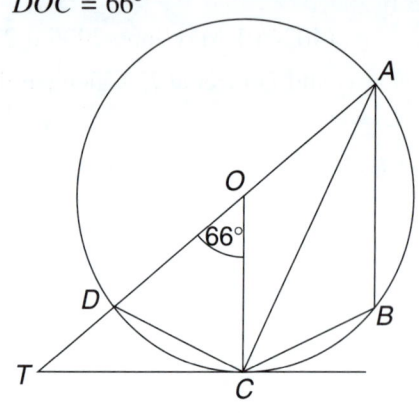

Calculate
(i) $D\hat{A}C$ (ii) $D\hat{T}C$
(iii) $A\hat{D}C$ (iv) $A\hat{B}C$

(4024/02 May/June 2007 q 4 b)

15.

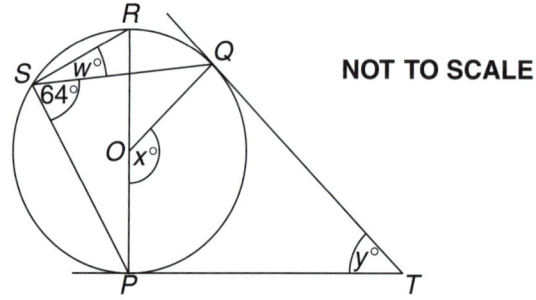

P, Q, R and S lie on a circle, centre O.
TP and TQ are tangents to the circle.
PR is a diameter and angle $PSQ = 64°$.
(a) Work out the values of w and x.
(b) **Showing all your working**, find the value of y.

(0580/02 May/June 2007 q 19)

16. A, B, C and D lie on a circle with centre O. AC is a diameter of the circle. AD, BE and CF are parallel lines. Angle $ABE = 48°$ and angle $ACF = 126°$.

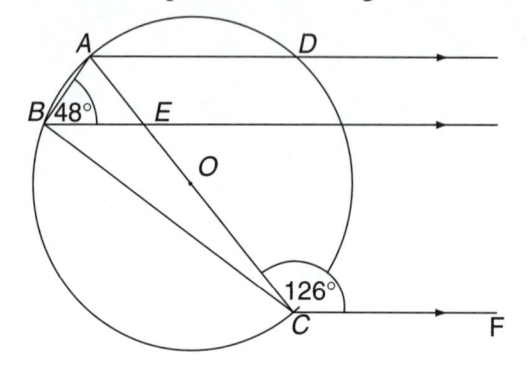

Find
(a) angle DAE,
(b) angle EBC,
(c) angle BAE.

(0580/02 Oct/Nov 2005 q 15)

17. (a) In the diagram, *ABCD* is a parallelogram.
    ADE and *BFE* are straight lines.
    $AF = BF$.
    $A\hat{B}F = 54°$ and $C\hat{B}F = 57°$
    Find the value of
       (i) *t*,
       (ii) *u*,
       (iii) *x*,
       (iv) *y*.

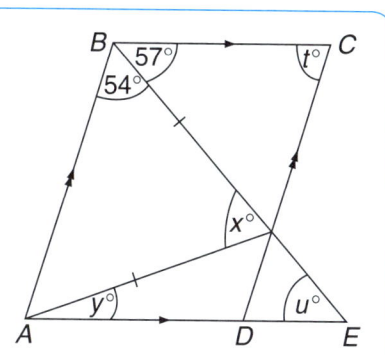

    (b) This hexagon has rotational symmetry of order 3. Calculate the value of *z*.

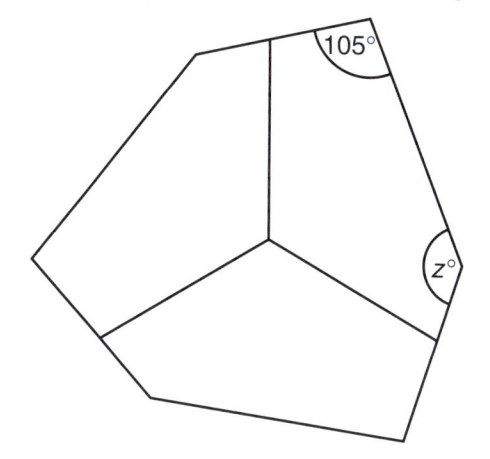

(4024/02 May/June 2005 q 3 a b)

18.

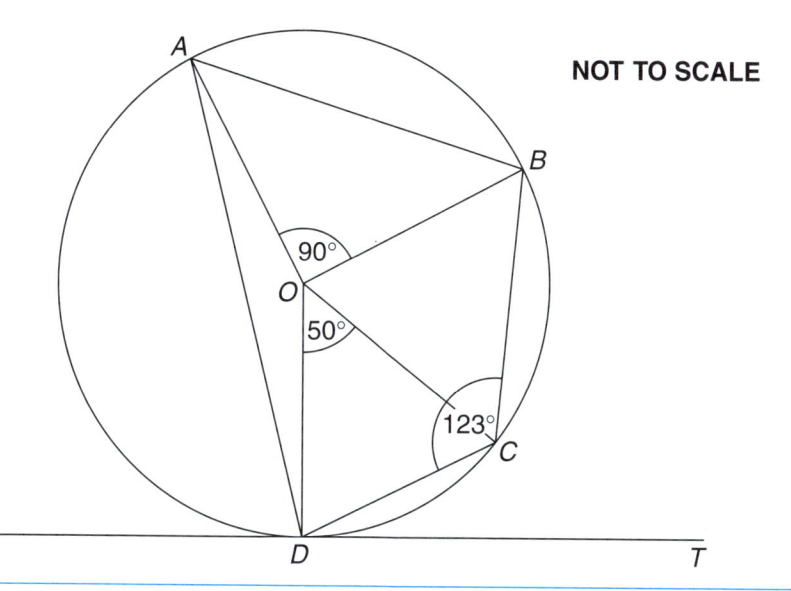

The points A, B, C and D lie on a circle centre O.
Angle AOB = 90°, angle COD = 50° and angle BCD = 123°.
The line DT is a tangent to the circle at D.
Find
(a) angle OCD,
(b) angle TDC,
(c) angle ABC,
(d) reflex angle AOC.

(0580/02 Oct/Nov 2007 q 20)

19.

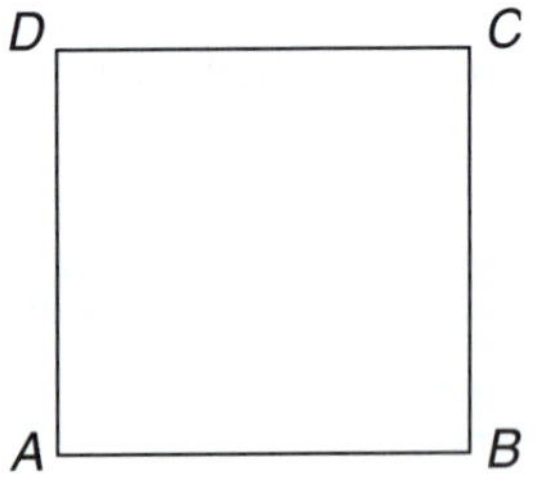

ABCD is a square.
It is rotated through 90° clockwise about B.
Copy the square and draw accurately the locus of the point D.

(0580/21 Oct/Nov 2008 q 5)

20.

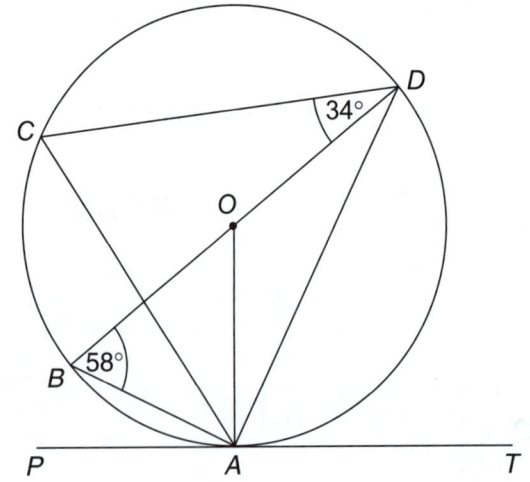

A, B, C and D lie on a circle, centre O.
BD is a diameter and PAT is the tangent at A.
Angle ABD = 58° and angle CDB = 34°.
Find
    (a) angle ACD,   (b) angle ADB,
    (c) angle DAT,   (d) angle CAO.

(0580/21 May/June 2009 q 22)

Chapter 6

Algebra and Graphs

This chapter continues the work in *Core Mathematics for Cambridge IGCSE* Chapter 7. Travel graphs, straight line graphs, and curves are studied in more depth.

Core Skills

1. Draw the following graphs for $-3 \leqslant x \leqslant 3$.
   (a) $y = x + 1$
   (b) $y = -x - 3$
   (c) $y = \dfrac{1}{2}x$
   (d) $y = 5$
   (e) $y = -2x + 1$
   (f) $2y + 3x = 4$

2. By writing in the form $y = mx + c$ find the y-intercept and the gradient of the line $2x = -5y + 3$.

3. Write down (i) the gradient and (ii) the y-intercept of the following.
   (a) 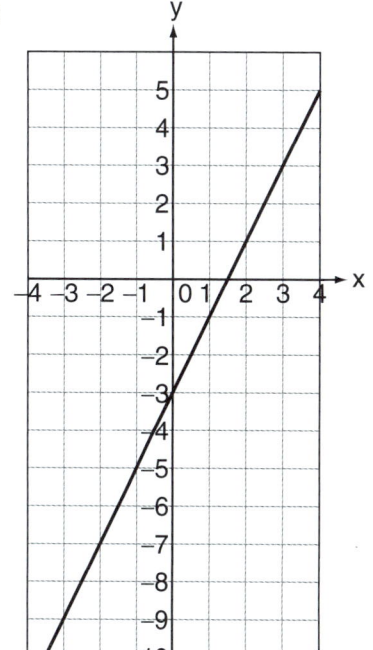
   (b)

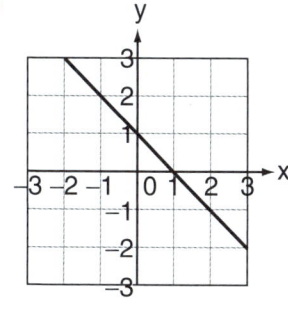

110 *Extended Mathematics for Cambridge IGCSE*

4. Write down the equations of the following

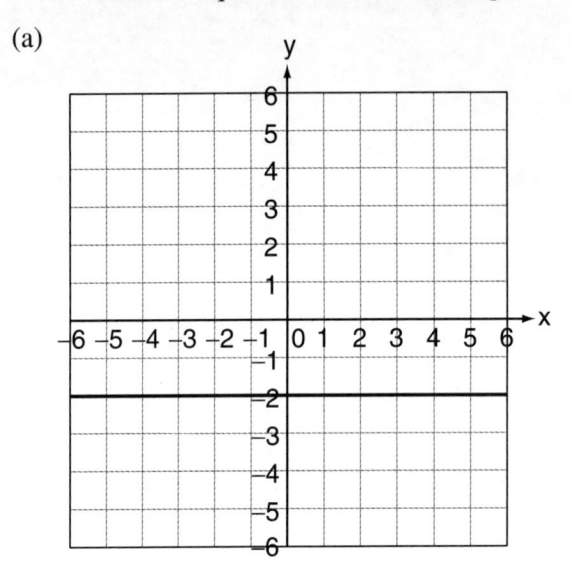

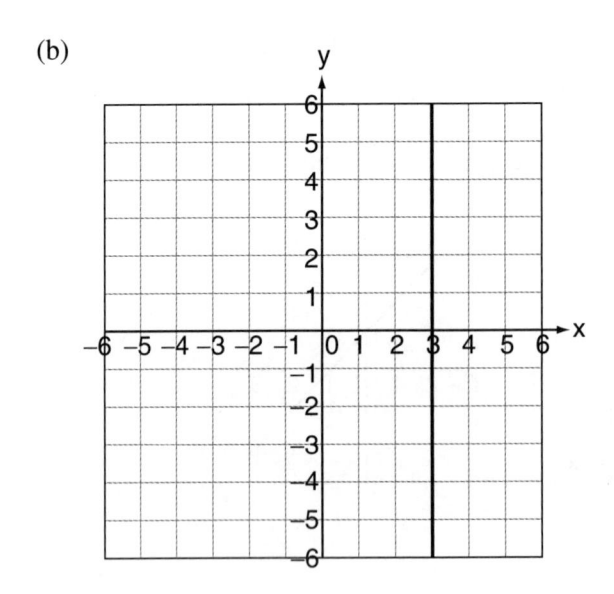

(a)

(b)

(c)

(d)

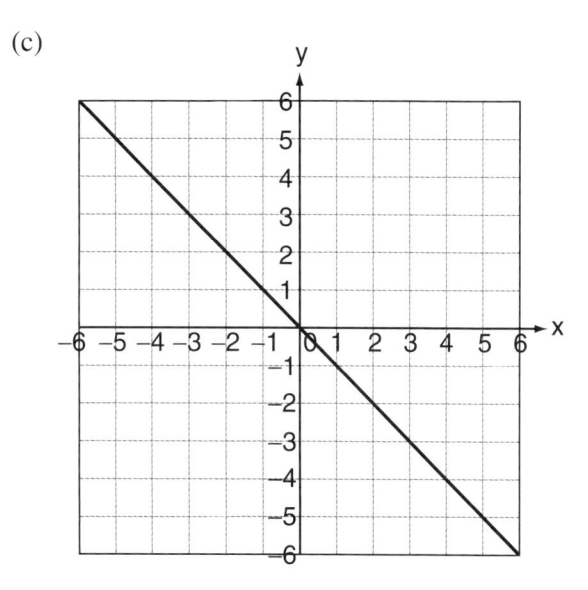

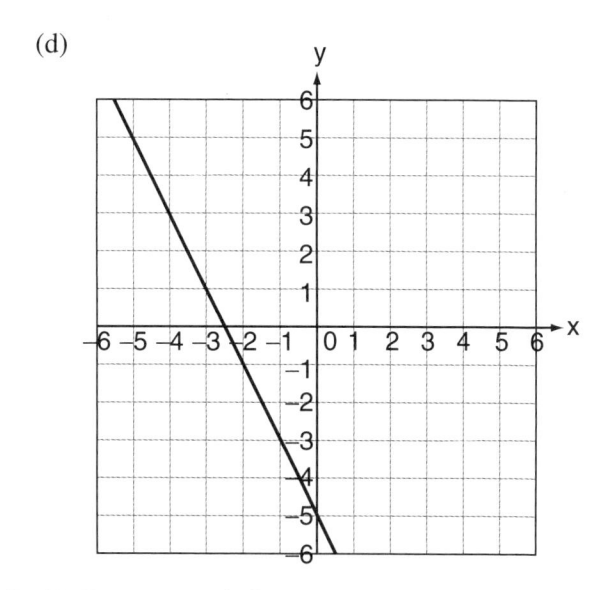

5. (a) Draw a graph for converting degrees Centigrade (°C) to degrees Fahrenheit (°F), given that 0°C = 32°F and that 100°C = 212°F. (The graph should be a straight line connecting these two points.)
   (b) Use your graph to convert (i) 25°C to °F and (ii) 125°F to °C.
6. (a) Complete the table of values for $y = x^2 - x - 2$.

x	−3	−2	−1	0	1	2	3	4
y	10	4		−2			4	10

(b) Draw the graph of $y = x^2 - x - 2$ for $-3 \leqslant x \leqslant 4$.
    Draw the line of symmetry on the graph.
(c) What is the equation of the line of symmetry?

7. Use a graphical method to solve the pair of simultaneous equations
$$y = -2x - 1$$
$$y = 2x + 2$$

Distance/Time Graphs

For your Extended course you will need to know some more about travel graphs.
In the Core Book you studied distance/time graphs, and saw that the gradient of the line gave the speed of movement because speed = $\dfrac{\text{distance gone}}{\text{time taken}}$, which is the gradient of the line (gradient = $\dfrac{\text{up}}{\text{along}}$) when the distance travelled is on the vertical (y) axis and the time taken is on the horizontal (x) axis.

You may, for example, need to use a distance/time graph to work out when and where one object overtakes another. The graph below shows the journeys of two cyclists, Brendan and Amit, leaving school and travelling along the same route. Brendan leaves school later than Amit, but travels faster. Brendan overtakes Amit at the precise moment when they are both in the same place at the same time, that is at point P on the graph.

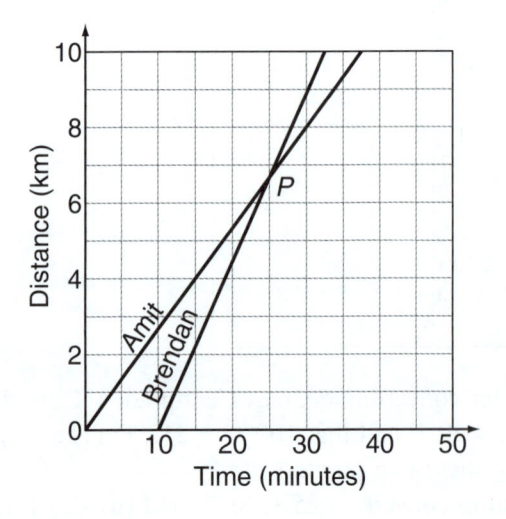

If the graph shows straight lines, as in the above case, it means that the gradient is the same at any point on each line, so the speed is the same at any point on the journey. The cyclists are travelling at a constant speed.

Algebra and Graphs 113

However, if we think more about a typical journey we realise that the object starts from rest and gradually increases speed until it is travelling at a constant speed. It may then gradually reduce speed until it stops. Such a journey could be shown by the graph below, showing the short journey of a car.

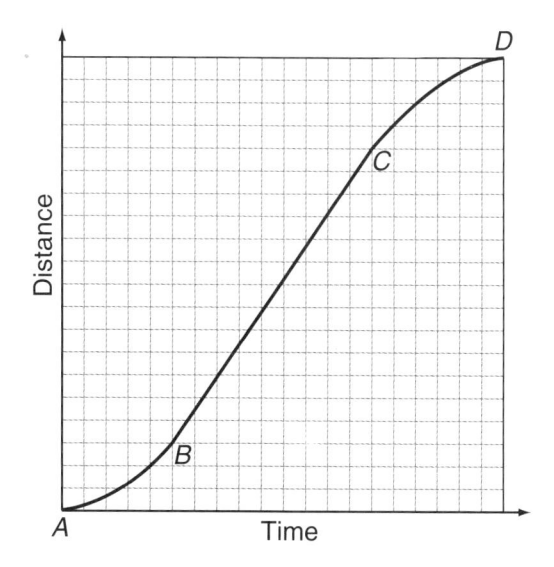

From *A* to *B* on the graph the car is accelerating (getting faster), from *B* to *C* it is at constant speed, and from *C* to *D* it is decelerating or retarding (getting slower).

How could we work out its speed at any precise moment on a curved section, for example, when it is accelerating?

Imagine that you could zoom in to look very closely, in great detail, at a curved section. You could imagine it broken into a series of short line segments as in the first diagram below. The gradient at any point on the curve would be the gradient of the line segment at that point.

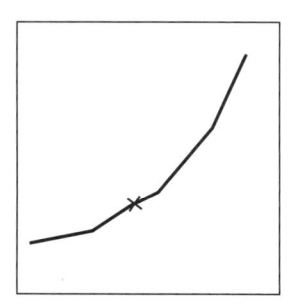

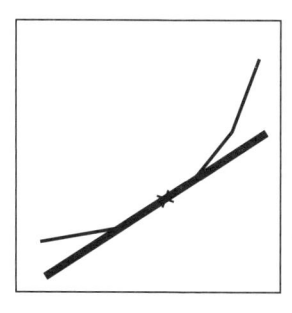

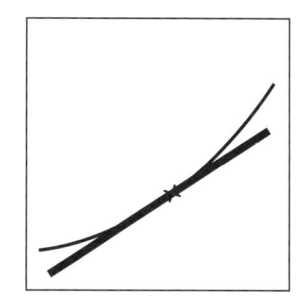

We could work out the gradient of the line at that point by extending it as in the second diagram, and then finding $\dfrac{\text{the change in } y}{\text{the change in } x}$ by constructing a right-angled triangle at convenient points.

114   Extended Mathematics for Cambridge IGCSE

Now imagine that we can zoom out so that the series of line segments become a smooth curve again, as in the third diagram. The line we have drawn to work out the gradient is now the *tangent* to the curve.

We have shown that the gradient at any point on a curve is the same as the gradient of the tangent at that point. You have met tangents to circles in Chapter 6 of the Core Book and you should know that a tangent is a line that just touches a curve at *one* point. Drawing a tangent to a curve is not something that can be done exactly, but at least it gives an estimate of the speed at any particular time. You will see more about tangents later in this section.

To draw the tangent as accurately as possible it is best to approach the point slowly by sliding your ruler along the paper. If you do this from the inside of the curve, you can make sure that you are cutting off equal arcs on each side of the point until you reach the point where the ruler nearly leaves the curve altogether. Then draw the tangent.

The diagrams below show this.

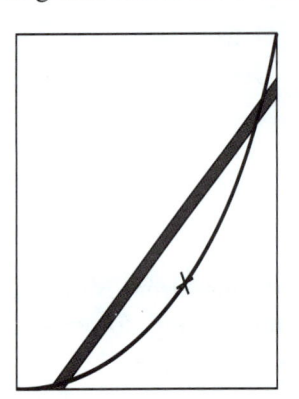

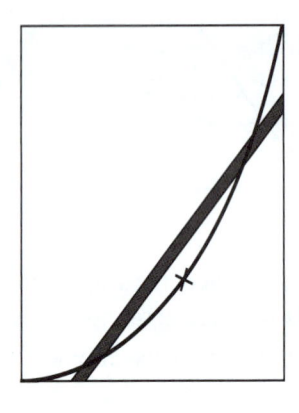

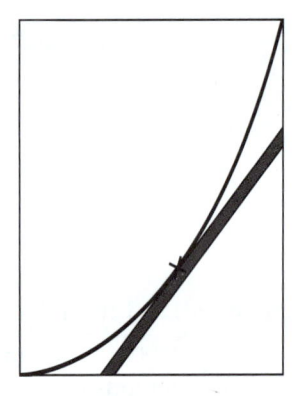

The gradient found by drawing a tangent will only give an *estimate* of the speed because it would be impossible to be sure of drawing the tangent at precisely the right place.

Speed/Time Graphs

Finding the acceleration

Another type of travel graph is a *speed/time* graph. With speed on the vertical (y) axis and time on the horizontal (x) axis the gradient gives $\frac{\text{change in speed}}{\text{time taken}}$ which is, as we know, the acceleration of the object. Constant acceleration will give a straight line graph because the gradient (acceleration) stays the same. An example of a speed/time graph is shown below.

At A the object is stationary. From A to B the object is accelerating with a constant acceleration. From B to C it is travelling at constant speed so the gradient is zero, as is the acceleration. From C to D it is decelerating, until at D it is once again stationary.

The acceleration of the object in the diagram below, from A to B is $\frac{45}{20} = 2.25$ m/s².

Algebra and Graphs 115

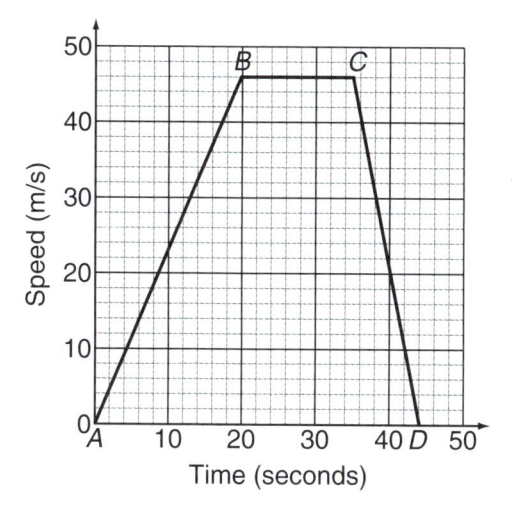

The deceleration from *C* to *D* is $\frac{45}{7.5} = 6$ m/s².

You will notice that the gradient while the object is decelerating is negative. We can either say that the acceleration is −6 m/s² or that the deceleration (retardation) is 6 m/s².
In other words the word deceleration takes account of the negative gradient.

Finding the distance travelled

The area under the graph from the time at *A* to the time at *D* in the above diagram gives the distance travelled in that time.

NOTE: It might help you to remember this if you think of area as involving the multiplication of two lengths in two perpendicular directions. Taking these directions to be the *x* and *y* directions, we will be multiplying the change in speed by the time taken. Remembering that speed = $\frac{\text{distance gone}}{\text{time taken}}$, then distance gone = speed × time taken.

In the above diagram the distance travelled over the whole journey may be calculated by using the formula for the area of a trapezium, or it can be split into triangles and a rectangle.

Using the trapezium formula: Area = $\frac{1}{2}$ (*AD* + *BC*) × distance between them

$$= \frac{1}{2} (42.5 + 15) \times 45 = 1293.75 \text{ metres}$$

So distance gone during the journey = 1.29 km.

Remember:

Travel graphs
- on travel graphs the time is shown on the horizontal axis.

Distance/time graphs
- on a distance/time graph the gradient of the line gives the speed,
- a straight line indicates constant speed,
- a curve indicates varying speed, and an estimate of the speed at any time can be found from the gradient of the tangent,
- a horizontal line indicates that the object is stationary,
- a negative gradient indicates that the object has turned round and is heading back to the start. (This means that the distance from the starting point is getting less, not that the object is going back to the origin on the graph). The gradient still indicates speed, but the sign of the gradient indicates that the object is going in the opposite direction.

Speed/time graphs
- on a speed/time graph the gradient of the line gives the acceleration,
- a straight sloping line indicates constant acceleration,
- a straight horizontal line indicates constant speed,
- a negative gradient indicates that the object is decelerating, but, unlike on a distance/time graph, it does not necessarily mean that the object has turned around,
- the area under the line gives the distance travelled.

Example 1

A car starts from rest at a set of traffic lights at A, shown on the distance/time graph below. It gradually increases speed until it is travelling at a constant speed, shown from B to C on the graph. It then slows to stop at the next set of lights at E.

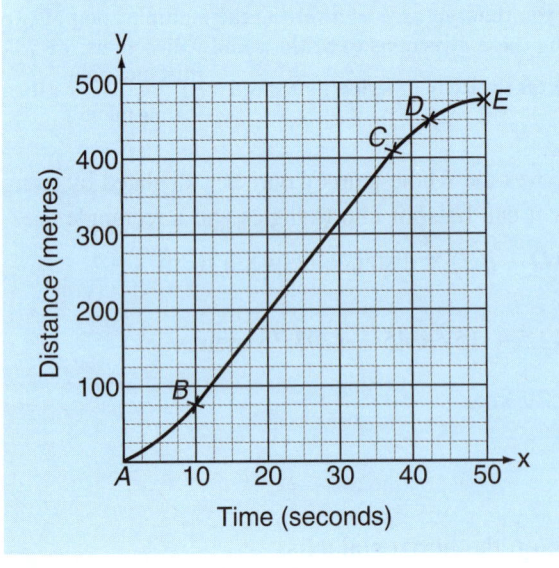

(a) Calculate the constant speed of the car from *B* to *C*.
(b) Calculate the average speed for the whole time between the traffic lights. Copy the graph.
(c) By drawing a tangent, estimate the speed at *D*.

Answer 1
(a) The speed of the car from *B* to *C* is:
$$\frac{\text{distance gone}}{\text{time taken}} = \frac{410-75}{37.5-10} = \frac{335}{27.5} = 12.1\dot{8}$$
$$= 12.2 \text{ m/s to 3 significant figures}$$
(b) The average speed of the car over the total distance is:
$$\frac{\text{total distance gone}}{\text{total time taken}} = \frac{475}{50} = 9.5 \text{ m/s}$$

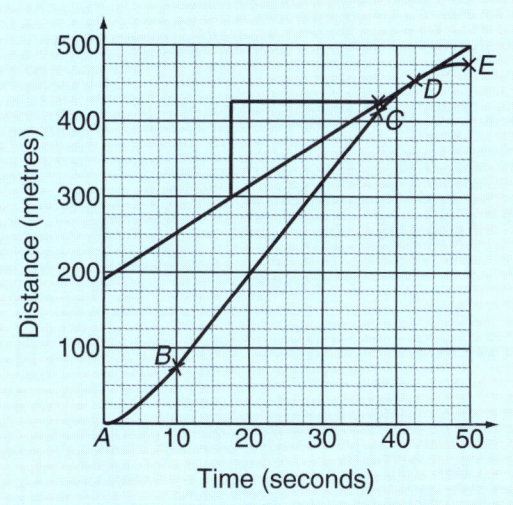

(c) From the triangle drawn the gradient of the tangent is:
$$\frac{\text{up}}{\text{along}} = \frac{425-300}{37.5-17.5} = \frac{125}{20}$$
Speed at *D* = 6.25 m/s

Example 2
A train leaves station *A* and steadily increases speed for 60 seconds.
It then travels at a constant speed of 30 metres per second for 2 minutes, finally steadily reducing speed until it comes to rest at station *B* 5 minutes after leaving station *A*.
(a) Draw a speed/time graph to show this journey.
(b) Calculate the total distance between stations *A* and *B*.
(c) Calculate the acceleration from station *A* to the maximum speed.

Answer 2

(a)

[Speed-time graph: speed (m/s) on y-axis from 0 to 30, time (seconds) on x-axis from 0 to 320. Line rises from (0,0) to (60,30), horizontal to (240,30), then down to (300,0).]

(b) Total distance = area under graph.

Area of a trapezium = $\frac{1}{2}$ sum of the parallel sides × distance between them

$= \frac{1}{2} (120 + 300) \times 30 = 6300$ m

Total distance = 6.3 km

(c) Acceleration = gradient of line = $\frac{30}{60}$ = 0.5 m/s^2

Exercise 6.1

1. Calculate the areas under the following

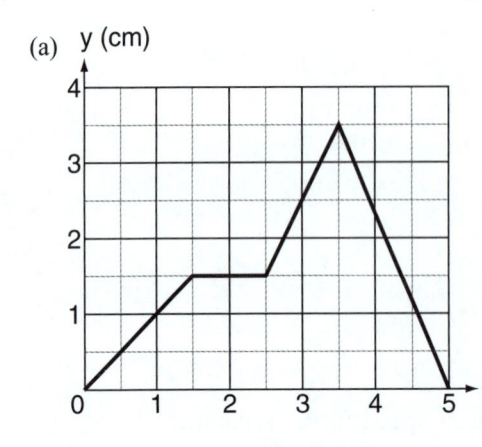

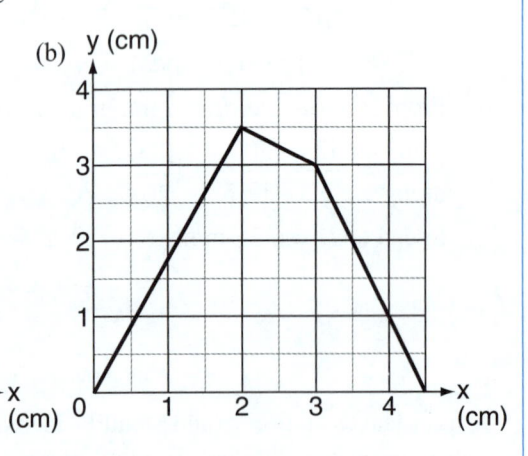

   (a) [Graph with y (cm) vs x (cm): line from (0,0) to (1,1.5), horizontal to (2,1.5), up to (3,4), down to (5,0).]

   (b) [Graph with y (cm) vs x (cm): line from (0,0) to (2,3.5), down to (3,3), down to (4,0).]

2. Sister and brother, Svetlana and Igor, go to the same school 2 kilometres from home. Svetlana walks to school, and Igor cycles.
   Igor stops on the way at the shops, but Svetlana goes straight to school.
   The graph shows their journeys.

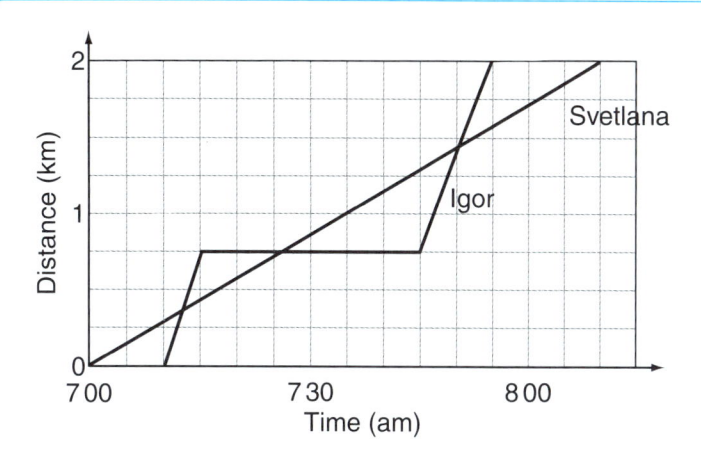

Give your answers to parts (b), (c) and (e) in kilometres per hour.
(a) How long does it take Igor to get to school?
(b) What is his average speed?
(c) Calculate his speed before and after his visit to the shops.
(d) At what time does Svetlana overtake Igor?
(e) Calculate Svetlana's speed.
(f) When does Igor overtake Svetlana?

3. A small plane leaves an airfield at 12 00, and takes 20 minutes to accelerate to its steady cruising speed of 350 km/h. At 14 35 it starts to descend for landing. The graph shows this information. (The zig-zag line indicates a break in the scale.)

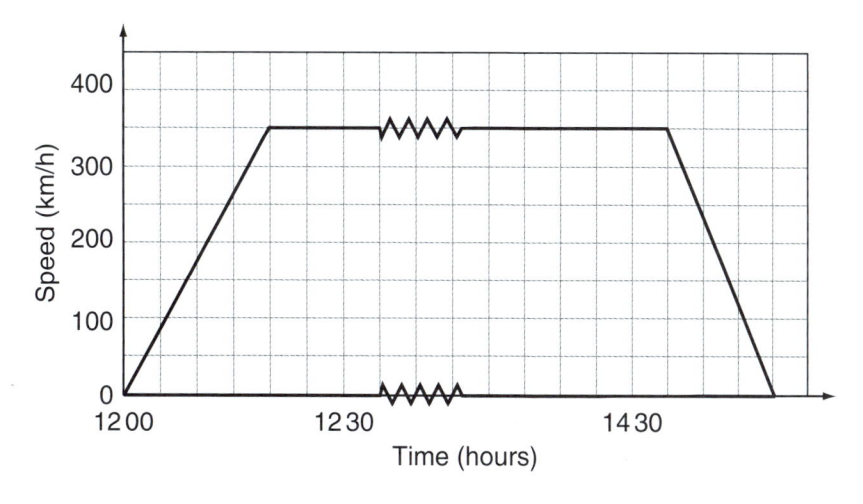

(a) Calculate the acceleration of the plane during take-off.
(b) Calculate the total distance the plane travelled.

More Graphs of Curves

Plotting graphs of curves

You need to be able to plot and recognise more curves.
If you do all the questions in the following exercise, you will build up a reference of some of the more frequently used curves together with their equations and special features. You will then be able to recognise curves more easily.

Example 3

Draw the following graphs for $-3 \leqslant x \leqslant 3$.

(a) $y = 2^x$
(b) $y = 3 \times 2^x + x^2$

Answer 3

(a)

x	−3	−2	−1	0	1	2	3
y	0.13	0.25	0.5	1	2	4	8

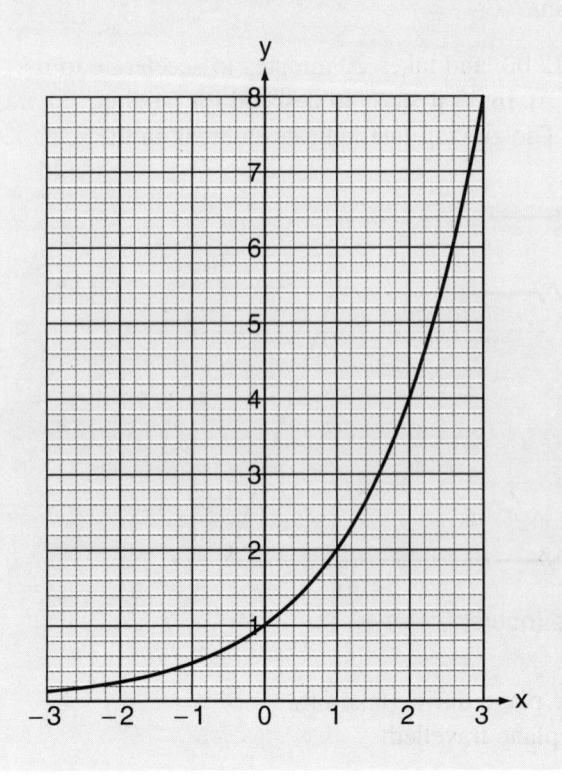

(b)

x	−3	−2	−1	0	1	2	3
y	9.4	4.8	2.5	3	7	16	33

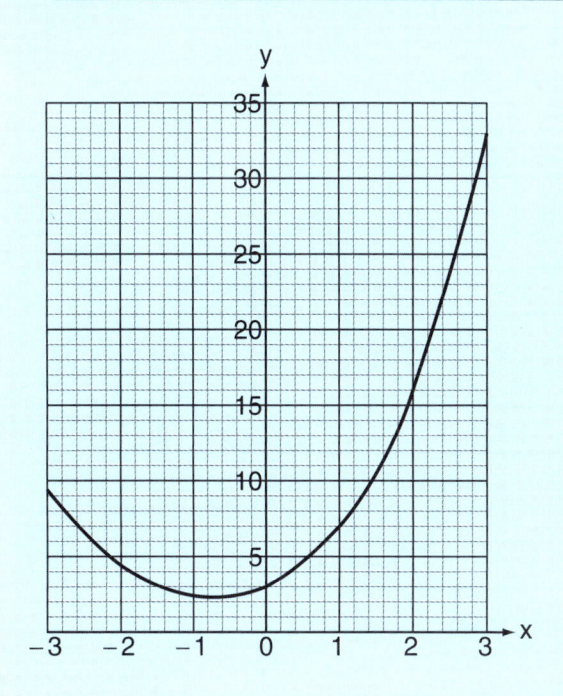

Exercise 6.2

For each of the functions shown below

(i) copy and complete the table, giving your answers correct to 1 decimal place,
(ii) draw and label the graph.

1.

Question	Function	x	−3	−2	−1	0	1	2	3
(a)	$y = x^2$	y	9		1	0			9
(b)	$y = -x^2$	y			−1	0	−1		−9
(c)	$y = (x + 1)^2$	y	4	1	0	1	4		
(d)	$y = x^2 - 4$	y		0	−3			0	
(e)	$y = x^2 - 2x - 8$	y	7	0		−8	−9		

2.

Question	Function	x	−3	−2	−1	0	1	2	3
(a)	$y = x^3$	y	−27		−1	0	1		
(b)	$y = -x^3$	y			1	0			−27
(c)	$y = (x+1)^3$	y		−1	0		8		
(d)	$y = x^3 - 4$	y	−31		−5	−4		4	23

3. (a)

x	−4.5	−4	−3	−2.5	−2	−1.5	−1	−0.5	0	0.5	1
$y = x^3 + 4x^2 + 1$	−9.1	1	10		9		4		1	2.1	6

(b)

x	−2.5	−2	−1.5	−1	−0.5	0	0.5	1	2
$y = x^3 + x^2 - 2x$	−4.4		1.9	2		0		0	8

(c)

x	−2.5	−2	−1.5	−1	−0.5	0	0.5	1
$y = x^2(x+2)$	−3.1	0		1	0.4	0		3

(d)

x	−2.5	−2	−1.5	−1	−0.5	0	0.5	1
$y = -x^2(x+2)$		0		−1	−0.4	0		−3

4.

Question	Function	x	−3	−2.5	−2	−1.5	−1	−0.5	0.5	1	1.5	2	2.5	3
(a)	$y = \dfrac{3}{x}$	y	−1	−1.2		−2		−6	6	3	2			1
(b)	$y = \dfrac{9}{x^2}$	y	1		2.3	4	9		36		4		1.4	1
(c)	$y = x^2 + \dfrac{3}{x}$	y	8		2.5	0.3	−2		6.3	4		5.5		10

5.

Function	x	−0.5	−0.25	0	0.25	0.5	0.75	1
$y = 10^x$	y	0.3		1			5.6	10

Algebra and Graphs

Using graphs of curves

Having drawn your curve you will probably be asked in an examination to use it to solve an equation, or to estimate the gradient at any point.

Example 4

The diagram shows the graphs of the curve $y = x^2 + 2x - 1$, and the straight line $y = x + 5$.
(a) Use the graph to find the solutions to the equation $x^2 + 2x - 1 = 0$.
    Give your answers correct to 1 decimal place.
(b) Use the curve and the straight line to find the solutions to the equation $x^2 + x - 6 = 0$.
(c) What is the minimum value of $y = x^2 + 2x - 1$?
(d) Give the coordinates of the point where the gradient of the curve = 0.
(e) Find an estimate of the gradient of the curve at the point $(-2, -1)$.

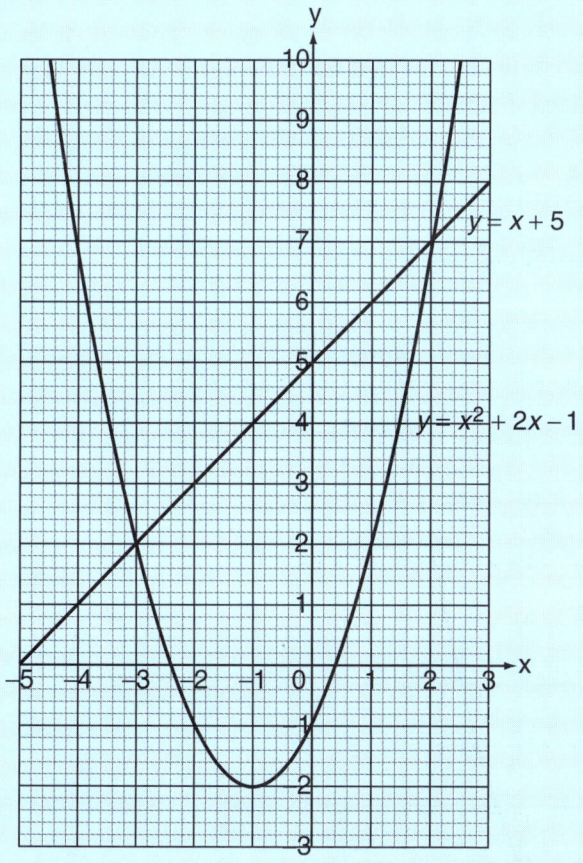

Answer 4

(a) The solutions to $x^2 + 2x - 1 = 0$ are the points where $y = 0$ (that is, on the x-axis).
    So from the graph, the solutions are $x = -2.4$ or $x = 0.4$

(b) The curve and the line meet when the x and y values of the two equations are the same, that is, when they are solved simultaneously.
$y = x^2 + 2x - 1$
$y = x + 5$
Substituting $y = x + 5$ in the first equation
$x + 5 = x^2 + 2x - 1$
$x^2 + x - 6 = 0$ $(-x - 5)$ This is the equation you were asked to solve, so its solutions can be found where the graphs intersect and are $x = -3$ and $x = 2$.

NOTE: The equation you are asked to solve is in x only, so do *not* give the y values of the points of intersection!

(c) The minimum value of $y = x^2 + 2x - 1$ is the minimum value of y as you see it on the graph. The minimum value is $y = -2$.

(d) The gradient of the curve is zero where the curve is horizontal, so the coordinates are $(-1, -2)$.

NOTE: Although you do not need to know this for your course, it might interest you to know that this point is called the *turning point*, because it is the point at which the curve turns around and the gradient changes from negative to positive.

(e) The gradient at the point $(-2, -1)$ is estimated by drawing the tangent at that point. This is shown in the next diagram.

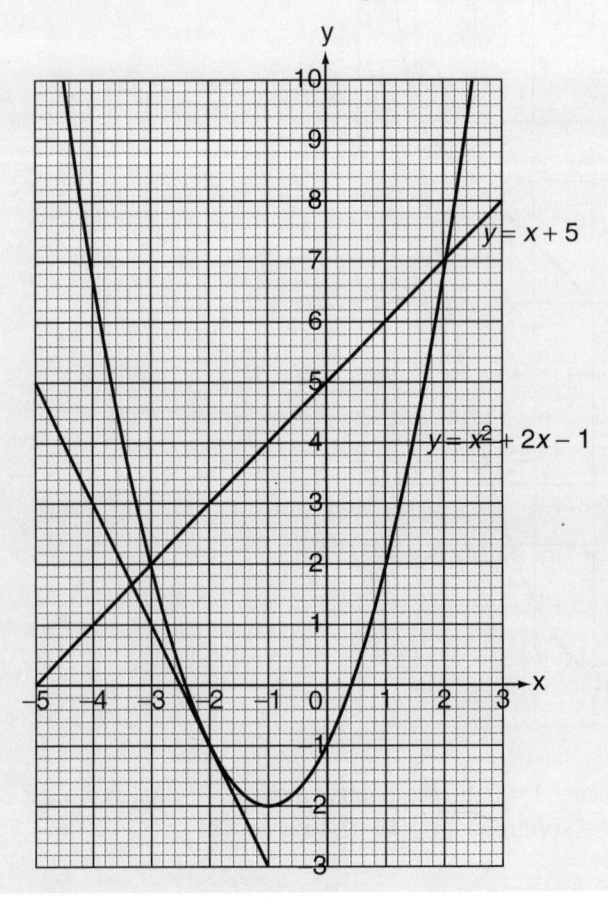

Algebra and Graphs

The gradient is $\dfrac{\text{change in } y}{\text{change in } x}$.

From the tangent drawn on the graph an estimate of the gradient is given by $\dfrac{5--3}{-5--1}$. Gradient $= \dfrac{8}{-4} = -2$

Sketching graphs of curves

You should have noticed features in the graphs you plotted in Exercise 6.2.

For example, the constant term gives the *y*-intercept just as it does in the equation of a straight line.

The sign in front of the x^2 term in the graph of a parabola also tells us about the parabola. If the sign is negative the graph is 'n' shaped. If the sign is positive the graph is 'u' shaped. If in doubt about the orientation of any curve try putting in a high positive value for *x*, such as 10, or 100, and see whether the corresponding *y* value is positive or negative.

When you are asked to 'sketch' a curve you do not have to plot points, but just draw a sketch showing key features such as where the curve crosses the axes and its general shape.

For a parabola the line of symmetry will always be a vertical line through the midpoint of the two points where the curve crosses the *x*-axis as shown in the example below.

Example 5

(a) For the curve $y = x^2 + x - 2$,
  (i) factorise the right hand side.
  (ii) Hence find the two values of *x* which make $y = 0$.
  (iii) Write down the *y*-intercept.
  (iv) Sketch the curve.
  (v) Write down the equation of the line of symmetry.
(b) Repeat the parts (i) to (v) for the curve $y = x^2 - 2x + 1$.

Answer 5

(a) (i) $y = (x + 2)(x - 1)$
  (ii) $y = 0$ when $x = -2$ or $x = 1$
  (iii) The *y*-intercept is $y = -2$
  (iv)

  (v) The line of symmetry is $x = -0.5$

(b) (i) $y = (x-1)(x-1)$
   (ii) $y = 0$ when $x = 1$
   (iii) The y-intercept is $y = +1$
   (iv)

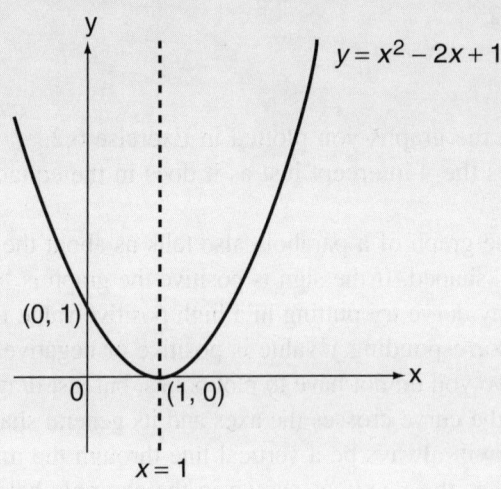

[Graph showing parabola $y = x^2 - 2x + 1$ with vertex at (1, 0), y-intercept at (0, 1), and line of symmetry $x = 1$]

   (v) The line of symmetry is $x = 1$

Using function notation

You will be working with function notation in Chapter 8. Meanwhile, you may meet it briefly in questions about graphs. The following example should show you that you do not have to worry about it at this stage. The $f(x)$ or $g(x)$ just replace y in the equation as you will see in the example.

Example 6

$$f(x) = \frac{x^2}{x^2 + 1}$$

(a) Copy and complete the table of values for $0 \leqslant x \leqslant 5$.

x	0	1	2	3	4	5
$f(x)$	0	0.5				0.96

(b) Using 2 cm to 1 unit on the x-axis, and 10 cm to one unit on the y-axis, draw the graph of $y = f(x)$.

(c) Why can $f(x) = \dfrac{x^2}{x^2 + 1}$ never be $\geqslant 1$?

Algebra and Graphs

(d) Why can $f(x) = \dfrac{x^2}{x^2+1}$ never be < 0?

$g(x) = \dfrac{5-x}{5}$

(e) On the same axes draw the graph of $y = g(x)$.

(f) Use your graphs to solve the equation: $\dfrac{x^2}{x^2+1} = \dfrac{5-x}{5}$, giving your answer to 1 decimal place.

Answer 6

(a)

x	0	1	2	3	4	5
$f(x)$	0	0.5	0.8	0.9	0.94	0.96

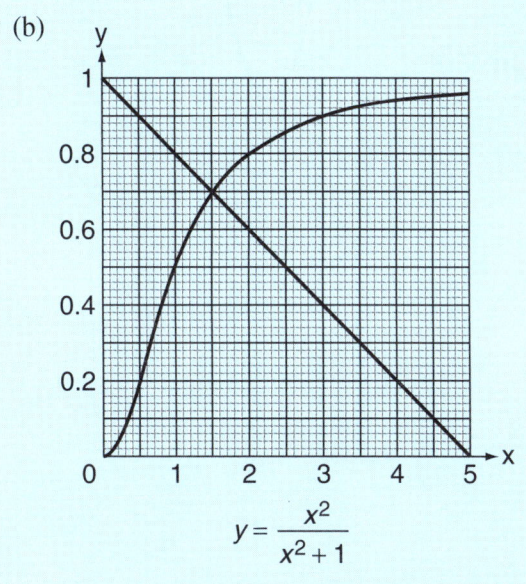

(b)

$y = \dfrac{x^2}{x^2+1}$

(c) $f(x)$ can never be greater than or equal to 1 because $x^2 + 1$ is always 1 more than x^2, so that the numerator is always less than the denominator

(d) $f(x)$ can never be negative because both the numerator and the denominator will be positive or zero because x is squared in both, and then 1 is added in the denominator

(e) This is a straight line graph

Using $y = \dfrac{5-x}{5}$, when $x = 0$, $y = 1$ and when $y = 0$, $x = 5$ so the line goes through (0, 1) and (5, 0) as shown on the graph

(f) $x = 1.5$ to 1 decimal place

Exercise 6.3

1. Draw sketch graphs of the following curves.
   On each parabola show the line of symmetry together with its equation.
   Show the coordinates of the points where the curves cut the axes.
   (a) $y = 2x^2 + 3x - 5$
   (b) $y = x(x - 2)^2$
   (c) $y = -x^2 + 3x - 2$
   (d) $y = (x - 1)(x + 2)(x - 3)$

2. Match the following graphs to their possible equations.

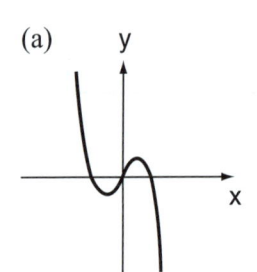

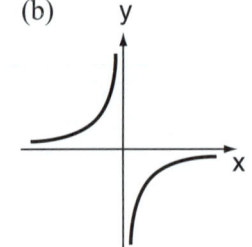

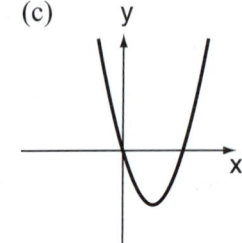

   (a)   (b)   (c)

   (i) $y = \dfrac{1}{x}$

   (ii) $y = x^3 - x$

   (iii) $y = x - x^3$

   (iv) $y = 2x - x^2$

   (v) $y = -\dfrac{1}{x}$

   (vi) $y = x^2 - 2x$

3. $y = \dfrac{10x}{(x+1)^2}$

   (a) Copy and complete the following table, giving the values of y to 1 decimal place.

x	0	1	2	3	3.5	4
y	0		2.2			

   (b) Plot these points, using a scale of 2 cm represents 1 unit on each axis.
   (c) By drawing a tangent estimate the gradient when $x = 0.5$.

Algebra and Graphs

Straight Line Segments

Line segments are sections, or segments, of lines.

Two points on a coordinate grid may be joined by a line segment, for example, the line segment AB joining the points (1, 2) and (9, 8) in the diagram below.

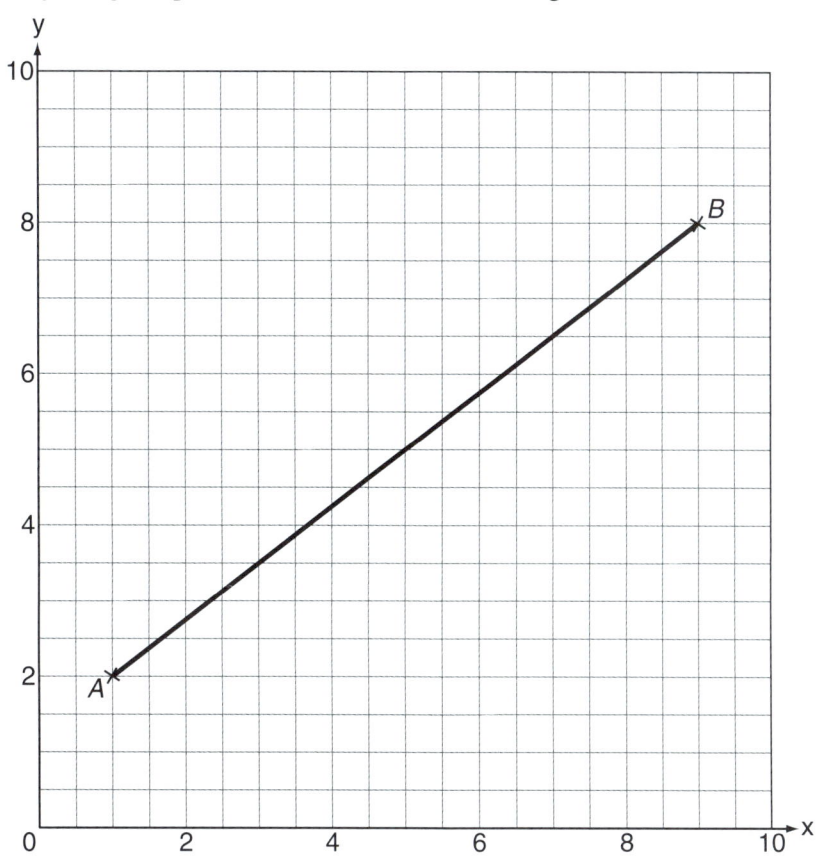

Finding the gradient of a line segment

Since the gradient of a line is $\dfrac{\text{change in } y}{\text{change in } x}$, we can easily see from the diagram that the gradient of the line AB is $\dfrac{6}{8} = \dfrac{3}{4}$.

If you do not have a diagram there are two choices.

Either you can use a formula, or you can write the answer down almost 'by inspection'.

The formula is: $$\text{gradient} = \frac{y_2 - y_1}{x_2 - x_1},$$

where the line segment joins the points (x_1, y_1) and (x_2, y_2).

NOTE: $\dfrac{y_2 - y_1}{x_2 - x_1} = \dfrac{y_1 - y_2}{x_1 - x_2}$ since both the numerator and the denominator have been multiplied by −1. You may choose which points to call (x_1, y_1) and (x_2, y_2), but do not change in mid calculation!

In the above diagram the gradient would be calculated using
$$(x_1, y_1) = (1, 2), \quad \text{and} \quad (x_2, y_2) = (9, 8).$$

$$\text{gradient} = \frac{8-2}{9-1} = \frac{6}{8} = \frac{3}{4}$$

The alternative is to say to yourself "how do I get from 2 to 8?", and write down +6, followed by "how do I get from 1 to 9?", and write the +8 under the +6: $\frac{+6}{+8} = \frac{3}{4}$.

The method you use depends on whether you like learning formulae or not! Either way is acceptable.

The danger with the formula method lies in the misuse of minus signs.

Both methods will be shown in the example.

Finding the length of a line segment

The length of the segment is easily found using Pythagoras' Theorem.

This important Theorem is introduced in Chapter 9 of the Core Book. The theorem states that "the square on the hypotenuse of a right-angled triangle is equal to the sum of the squares on the other two sides." The hypotenuse is the longest side of a right-angled triangle and is opposite the right angle.

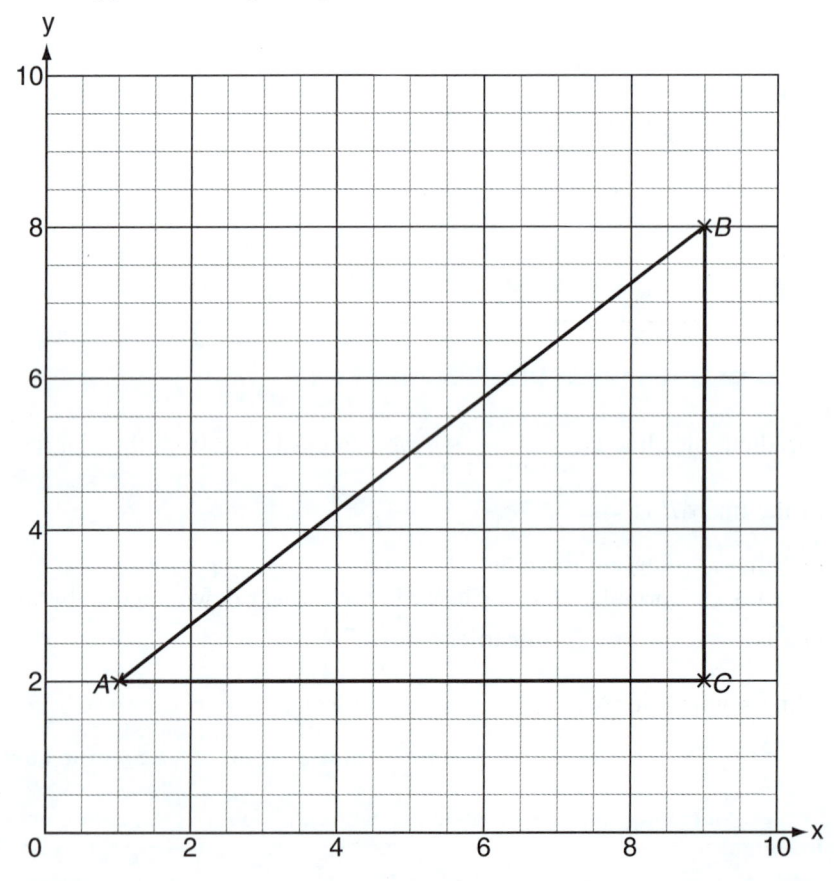

Algebra and Graphs 131

The diagram shows a right-angled triangle *ABC* drawn on the segment.
The length of the line *AC* is 9 − 1 = 8, and the length of the side *BC* is 8 − 2 = 6.
By Pythagoras Theorem $AB^2 = AC^2 + BC^2$ so

$$AB = \sqrt{8^2 + 6^2} = 10 \text{ units.}$$

The formula would be $AB = \sqrt{(x_1 - x_2)^2 + (y_1 - y_2)^2}$.

Notice that it does not matter which way round the subtraction is performed (either $x_1 - x_2$ or $x_2 - x_1$), because the result will be squared, thus losing any minus sign.
The alternative would be to think "how do I get from ...?" as before, and straight away write down:

$$AB = \sqrt{8^2 + 6^2} = 10 \text{ units.}$$

Again, the choice is yours, as either method is acceptable.

Finding the midpoint of a line segment

Finding the midpoint could hardly be easier as it is found by taking the *mean* of the *x* values and the *mean* of the *y* values. (The mean is dealt with in more detail in Chapter 11, but it is obtained by adding the two values together and then dividing by 2).
In the example above, the midpoint is given by:

$$\left(\frac{1+9}{2}, \frac{2+8}{2} \right) = (5, 5)$$

Check on the diagram and you will see this is correct.

For formula fans, the midpoint is given by $\left(\dfrac{x_1 + x_2}{2}, \dfrac{y_1 + y_2}{2} \right)$.

Example 7
Find
(a) the length,
(b) the gradient,
(c) the midpoint of the line joining $A(5, -2)$ and $B(-7, -8)$.

Answer 7
For this example we will draw a diagram so that you can see that the methods are correct, but normally it is not necessary. However, it does no harm to draw a diagram if that gives you more confidence.

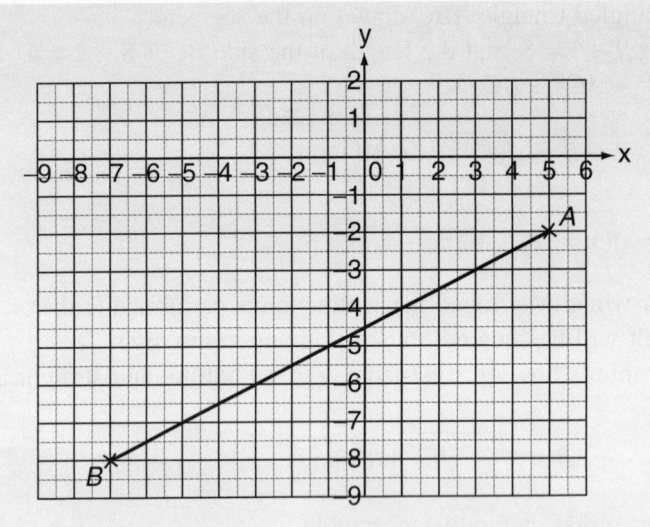

(a) Let $A(5, -2)$ be (x_1, y_1), and $B(-7, -8)$ be (x_2, y_2).
Then AB is given by

$$\sqrt{(5--7)^2 + (-2--8)^2}$$
$$= \sqrt{12^2 + 6^2}$$
$$= \sqrt{180}$$
$$= 6\sqrt{5}$$

NOTE: Watch those minus signs!

NOTE: Leaving the answer as a surd leaves it in its most accurate form. Give to 3 significant figures if requested.

(b) The gradient of AB is $\dfrac{\text{change in } y}{\text{change in } x}$,

$$= \frac{-8--2}{-7-5}$$
$$= \frac{-6}{-12} = \frac{1}{2}$$

Or
To get from A to B:
-2 to -8 is subtract 6
5 to -7 is subtract 12,
so gradient $= \dfrac{-6}{-12} = \dfrac{1}{2}$

NOTE: Remember to go in the same direction (from A to B) each time.

(c) Midpoint of AB is the mean of the two points.

$$x = \frac{5 + -7}{2} = \frac{-2}{2} = -1$$

$$y = \frac{-2 + -8}{2} = \frac{-10}{2} = -5$$

Midpoint is (−1, −5)

NOTE: Notice that the two values are added and then divided by 2 to get the mean.

Using the formula gives exactly the same result; you just have to decide how you are going to remember how to get the midpoint.

Exercise 6.4

Find (a) the length (giving your answer to 3 significant figures if necessary),
(b) the gradient (as simplified fractions if not whole numbers),
(c) the midpoint
for the line segments joining each of the following pairs of points.

1. $A = (0, 2)$, $B = (1, 6)$
2. $A = (-1, -1)$, $B = (-2, -2)$
3. $A = (5, 10)$, $B = (-5, -10)$
4. $A = (-3, -4)$, $B = (13, 14)$
5. $A = (4, -7)$, $B = (-3, 1)$
6. $A = (-1, 3)$, $B = (3, -1)$
7. $A = (-3, 5)$, $B = (-1, -4)$
8. $A = (5, 3)$, $B = (1, -2)$

Exercise 6.5

Mixed Exercise

1. A ball is thrown vertically up into the air from ground level.
   Its height varies with the time of flight according to the following equation.
   $h = 15t - 5t^2$

   (a) Copy and complete the table, giving the values for h correct to 1 decimal place.

t (seconds)	0	0.25	0.5	0.75	1	1.25	1.5	1.75	2	2.25	2.5	2.75	3
h (metres)	0	3.4			10				10				

   (b) Using scales of 1 cm to represent 1 metre on the vertical axis and 4 cm to represent 1 second on the horizontal axis, draw the distance(height)/time graph for the ball.
   (c) What is the maximum height of the ball?
   (d) Draw a tangent to estimate the speed of the ball after 1.25 seconds.

(e) What happens to the speed of the ball at its maximum height?
(f) With approximately what speed does the ball return to the ground?

2. The speed of the same ball as in the previous question varies with the time according to the following equation.

   speed = 15 − 10*t*.

   (a) Copy and complete the table showing how the speed varies for the first part of the ball's flight.

t (seconds)	0	0.25	0.5	0.75	1	1.25	1.5
speed (m/s)			10	5			

   (b) Using scales of 1 cm represents 2 metres per second on the vertical axis, and 1 cm represents 0.25 seconds on the horizontal axis, draw the speed/time graph for the first part of the flight of the ball.
   (c) (i) What would happen to the graph if you were to carry on and plot the whole of the flight?
   (ii) Why is this?
   (d) Using the graph calculate the height after 1 second.
   (e) Using the graph calculate the acceleration of the ball.

3. The line $3y + 2x - 12 = 0$ cuts the y-axis at A and the x-axis at B. Find the length of AB.

4. $y = x^2 + \dfrac{1}{x}$

   (a) Copy and complete the table of values for x and y. Give values of y correct to 1 decimal place.

x	0.1	0.2	0.4	0.6	0.8	1	1.5	2	2.5	3
y	10		2.7				2.9			9.3

   (b) Using a scale of 1 cm to 1 unit on the y-axis, and 1 cm to 0.2 units on the x-axis, plot the points in the table and draw a smooth curve.
   (c) On your graph draw the line $y = x$.

   NOTE: Be careful! The scales are not the same on each axis!

   (d) By drawing a tangent parallel to the line $y = x$, estimate the coordinates of the point where the gradient of the curve = 1. Give the coordinates to the nearest whole number.
   (e) Write down the equation of the tangent at this point in the form $y = mx + c$, giving c to the nearest whole number.

5. $y = x^2 - x - 6$
   $y = -x^2 - x + 6$
   Using a graphical method, solve these two equations simultaneously.
   Use integer values of x: $-3 \leqslant x \leqslant 4$.

Examination Questions

6. A cyclist is training for a competition and the graph shows one part of his training.

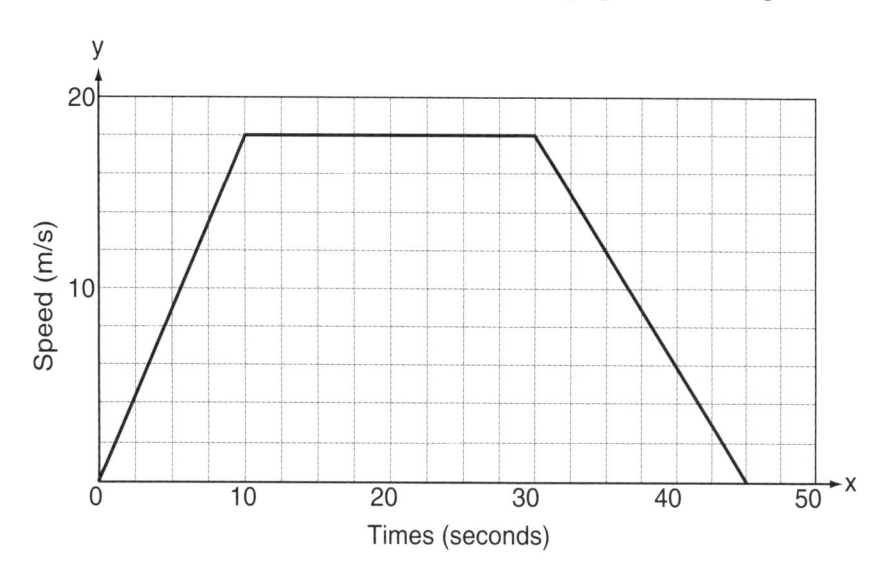

   (a) Calculate the acceleration during the first 10 seconds.
   (b) Calculate the distance travelled in the first 30 seconds.
   (c) Calculate the average speed for the entire 45 seconds.

   (0580/02 May/June 2004 q 21)

7. The straight line graph of $y = 3x - 6$ cuts the x-axis at A, and the y-axis at B.
   (a) Find the coordinates of A and the coordinates of B.
   (b) Calculate the length of AB.
   (c) M is the midpoint of AB.
       Find the coordinates of M.

   (0580/02 Oct/Nov 2003 q 16)

8. (a) Copy and complete the table of values for $y = 3^x$.

x	-2	-1.5	-1	-0.5	0	0.5	1	1.5	2
y		0.2						5.2	9

(b) Using your table copy and complete the graph of $y = 3^x$ for $-2 \leqslant x \leqslant 2$.

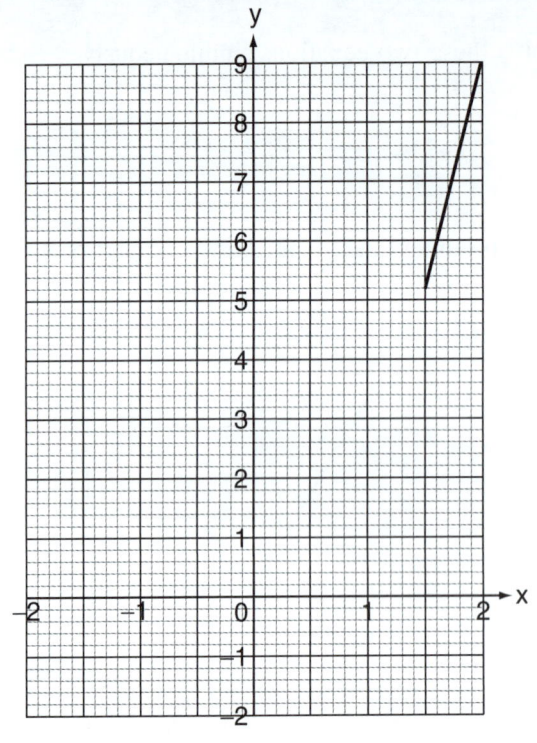

(c) Use the graph to find the solution of the equation
$3^x = 6$.

(0580/02 May/June 2003 q 20)

9. (a) P is the point $(-3, 3)$ and Q is the point $(13, -2)$.
Find the coordinates of the midpoint of PQ.

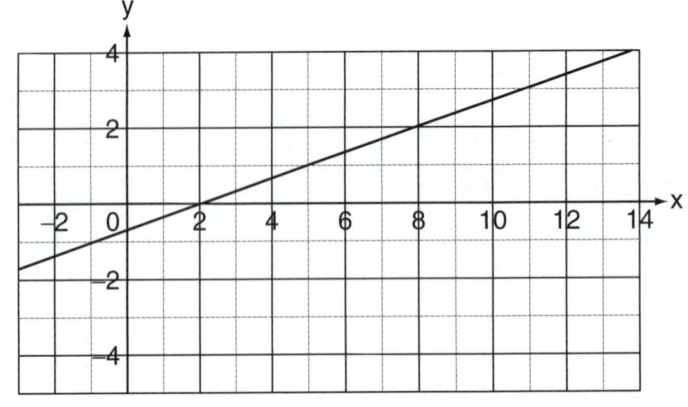

(b) The line $x - 3y = 2$ is shown on the diagram.
The line $x - 3y = k$ cuts the y-axis at the point $(0, -4)$.
   (i) Copy the diagram and draw the line $x - 3y = k$ on your diagram.
   (ii) Calculate the value of k.

(4024/01 May/June 2005 q 13)

10.

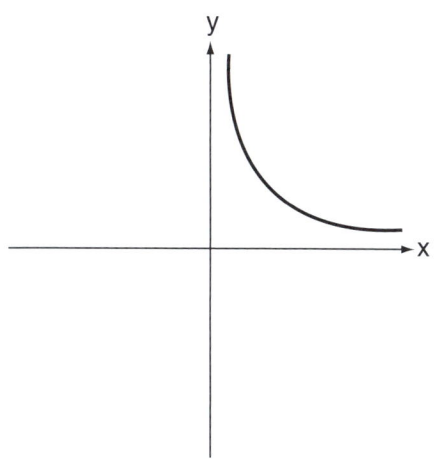

(a) The diagram is a sketch of the graph of $y = \dfrac{3}{x}$ for $x > 0$.

Copy the diagram and complete the sketch for $x < 0$.

(b) Sketch the graph of $y = x$ on the same diagram.

(c) The graphs of $y = \dfrac{3}{x}$ and $y = x$ meet at $x = k$.

Find the values of k.

(4024/01 May/June 2005 q 22)

11. The table below gives some values of x and the corresponding values of y, correct to one decimal place, where

$$y = \dfrac{x^2}{8} + \dfrac{18}{x} - 5$$

x	1	1.5	2	2.5	3	4	5	6	7	8
y	13.2	7.3	4.5	3.0	2.1	1.5	1.7	p	3.7	5.3

(a) Find the value of p.

(b) Using a sheet of graph paper and a scale of 2 cm to 1 unit, draw a horizontal x-axis for $0 \leqslant x \leqslant 8$.

Using a scale of 1 cm to 1 unit, draw a vertical y-axis for $0 \leqslant y \leqslant 14$.

On your axes plot the points given in the table, and join them with a smooth curve.

(c) Use your graph to find
 (i) the value of x when $y = 8$,
 (ii) the least value of $\dfrac{x^2}{8} + \dfrac{18}{x}$ for values of x in the range $0 \leqslant x \leqslant 8$.

(d) By drawing a tangent, find the gradient of the curve at the point where $x = 2.5$.

(e) On the axes used in part (b) draw the graph of $y = 12 - x$.

(f) The x-coordinates of the points where the two graphs intersect are solutions of the equation
$x^3 + Ax_2 + Bx + 144 = 0$.

Find the value of A and the value of B.

(4024/02 May/June 2006 q 8)

12. (a) $f(x) = \dfrac{12}{x+1}$

x	0	1	2	3	4	5	6	7	8	9	10	11
f(x)	p	6	4	3	2.4	2	1.71	q	1.33	r	1.09	1

(i) Calculate the values of p, q, and r.
(ii) On a sheet of graph paper draw the graph of $y = f(x)$ for $0 \leqslant x \leqslant 11$. Use a scale of 1 cm to 1 unit on each axis.
(iii) By drawing a suitable line, find an estimate of the gradient of the graph at the point (3, 3).

(b) On the same grid draw the graph of $y = 8 - x$ for $0 \leqslant x \leqslant 8$.
(c) (i) Show that the equation $f(x) = 8 - x$ simplifies to $x^2 - 7x + 4 = 0$.
(ii) **Use your graph** to solve this equation, giving your answers correct to 1 decimal place.

(0580/04 Oct/Nov 2004 q 2)

13. A sketch of the graph of the quadratic function $y = px^2 + qx + r$ is shown in the diagram.

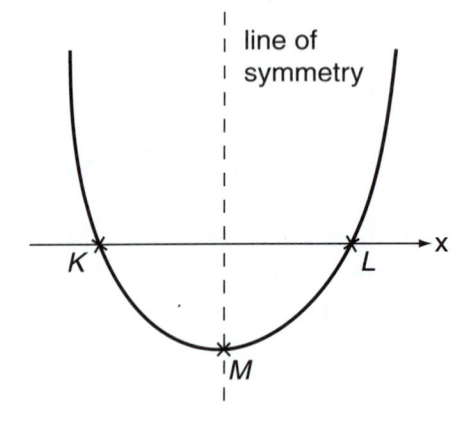

The graph cuts the *x*-axis at *K* and *L*.
The point *M* lies on the graph and on the lines of symmetry.
(a) When $p = 1$, $q = 2$, $r = -3$, find
   (i) the *y*-coordinate of the point where $x = 4$,
   (ii) the coordinates of *K* and *L*,
   (iii) the coordinates of *M*.
(b) Describe how the above sketch of the graph would change in each of the following cases.
   (i) *p* is negative,   (ii) $p = 1$, $q = r = 0$.
(c) Another quadratic function is $y = ax^2 + bx + c$.
   (i) Its graph passes through the origin.
       Write down the value of *c*.
   (ii) The graph also passes through the points (3, 0) and (4, 8).
        Find the values of *a* and *b*.

(0580/04 Oct/Nov 2004 q 7)

14. The points $A\left(0, \frac{1}{2}\right)$ and $B\left(2, 4\frac{1}{2}\right)$, lie on the curve as shown in the diagram.

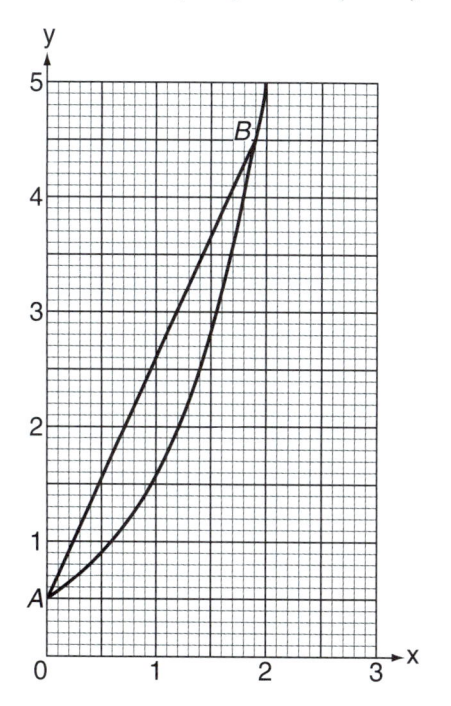

(i) Calculate the gradient of the straight line *AB*.
(ii) Using the diagram, estimate the value of *x* at which the gradient of the curve is equal to the gradient of the straight line *AB*.

(4024/01 May/June 2007 13b)

15.

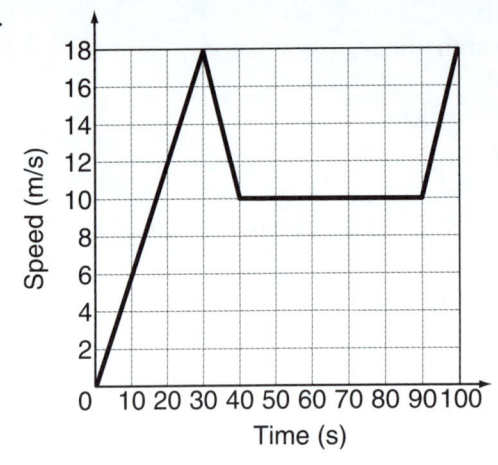

The diagram shows part of a journey by a truck.
(a) The truck accelerates from rest to 18 m/s in 30 seconds.
    Calculate the acceleration of the truck.
(b) The truck then slows down in 10 seconds for some road works and travels through the road works at 12 m/s.
    At the end of the road works it accelerates back to a speed of 18 m/s in 10 seconds.
    Find the **total** distance travelled by the truck in the 100 seconds.
                                                    (0580/02 May/June 2007 q 21)

16.

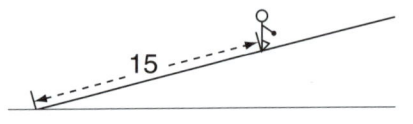

Adam stood on a slope, 15 m from the bottom.
He rolled a heavy ball directly up the slope.
After t seconds the ball was y metres from the bottom of the slope.
The table below gives some values of t and the corresponding values of y.

t	0	1	2	2.5	3	3.5	4	4.5	5	5.5
y	15	22	25	25	24	22	19	15	10	4

(a) Using a scale of 2 cm to represent 1 unit, draw a horizontal t-axis for $0 \leqslant t \leqslant 6$.
    Using a scale of 2 cm to represent 5 units, draw a vertical y-axis for $0 \leqslant y \leqslant 30$.
    On your axes, plot the points given in the table and join them with a smooth curve.
(b) Extend the curve to find the value of t when the ball reached the bottom of the slope.

(c) (i) By drawing a tangent, find the gradient of the curve when $t = 3.5$.
   (ii) State briefly what this gradient represents.
(d) Immediately after he rolled the ball, Adam ran down the slope at a constant speed of 1.5 m/s.
   (i) Write down the distance of Adam from the bottom of the slope when
      (a) $t = 0$,
      (b) $t = 4$.
   (ii) On the same axes, draw the graph that represents the distance of Adam from the bottom of the slope for $0 \leqslant t \leqslant 6$.
   (iii) Hence find the distance of Adam from the bottom of the slope when the ball passed him.

(4024/02 May/June 2007 q 8)

17. (a) Find the values of k, m and n in each of the following equations, where $a > 0$.
    (i) $a^0 = k$
    (ii) $a^m = \dfrac{1}{a}$
    (iii) $a^n = \sqrt{a^3}$.

(b) The table shows some values of the function $f(x) = 2^x$.

x	-2	-1	-0.5	0	0.5	1	1.5	2	3
$f(x)$	r	0.5	0.71	s	1.41	2	2.83	4	t

   (i) Write down the values of r, s, and t.
   (ii) Using a scale of 2 cm to represent 1 unit on each axis, draw an x-axis from -2 to 3 and a y-axis from 0 to 10.
   (iii) On your grid, draw the graph of $y = f(x)$ for $-2 \leqslant x \leqslant 3$.
(c) The function g is given by $g(x) = 6 - 2x$.
   (i) On the same grid as **part (b)**, draw the graph of $y = g(x)$ for $-2 \leqslant x \leqslant 3$.
   (ii) Use your graphs to solve the equation $2^x = 6 - 2x$.
   (iii) Write down the value of x for which $2^x < 6 - 2x$ for $x \in \{\text{positive integers}\}$.

(0580/04 May/June 2006 q 3)

18. $f(x) = 1 - \dfrac{1}{x^2}$, $x \neq 0$.

(a)

x	-3	-2	-1	-0.5	-0.4	-0.3	0.3	0.4	0.5	1	2	3
$f(x)$	p	0.75	0	-3	-5.25	q	q	-5.25	-3	0	0.75	p

Find the values of p and q.

142 *Extended Mathematics for Cambridge IGCSE*

(b) (i) Draw an *x*-axis for $-3 \leqslant x \leqslant 3$ using 2 cm to represent 1 unit and a *y*-axis for $-11 \leqslant y \leqslant 2$ using 1 cm to represent 1 unit.
  (ii) Draw the graph of $y = f(x)$ for $-3 \leqslant x \leqslant -0.3$ and for $0.3 \leqslant x \leqslant 3$.
(c) Write down an integer *k* such that $f(x) = k$ has no solutions.
(d) **On the same grid**, draw the graph of $y = 2x - 5$, for $-3 \leqslant x \leqslant 3$.
(e) (i) Use your graphs to find solutions of the equation $1 - \frac{1}{x^2} = 2x - 5$.
  (ii) Rearrange $1 - \frac{1}{x^2} = 2x - 5$ into the form $ax^3 + bx^2 + c = 0$, where *a*, *b* and *c* are integers.
(f) (i) Draw a tangent to the graph of $y = f(x)$ which is parallel to the line $y = 2x - 5$.
  (ii) Write down the equation of this tangent.

  (0580/04 Oct/Nov 2005 q 5)

19. During one day, at a point *P* in a small harbour, the height of the surface of the sea above the seabed was noted.
The results are shown in the table.

Time (*t* hours) after 8 am	0	1	2	3	4	5	6	7	8	9
Height (*y* metres) above the seabed	3.8	3.3	2.5	1.8	1.2	1.0	1.2	1.8	2.5	3.3

(a) Using a scale of 1 cm to represent 1 hour, draw a horizontal *t*-axis for $0 \leqslant t \leqslant 9$.
Using a scale of 2 cm to represent 1 metre, draw a vertical *y*-axis for $0 \leqslant y \leqslant 4$.
On your axes, plot the points given in the table and join them with a smooth curve.
(b) (i) By drawing a tangent, find the gradient of the curve at the point where $t = 4$.
  (ii) Explain the meaning of this gradient.
(c) On the same day, a straight pole was driven vertically into the seabed at the point *P*.
Work started at 8 am.
The pole was driven in at a constant rate.
The height, *y* metres, of the top of the pole above the seabed, *t* hours after 8 am, is given by the equation
$y = 4 - \frac{1}{2}t.$
  (i) Write down the length of the pole.
  (ii) On the same axes as the curve, draw the graph of $y = 4 - \frac{1}{2}t$.
  (iii) How many **centimetres** was the top of the pole above the surface of the sea at noon?
  (iv) Find the value of *t* when the top of the pole was level with the surface of the sea.

  (4024/02 May/June 2005 q 8)

20. The diagram shows the points $A(1, 2)$, $B(4, 6)$, and $D(-5, 2)$.

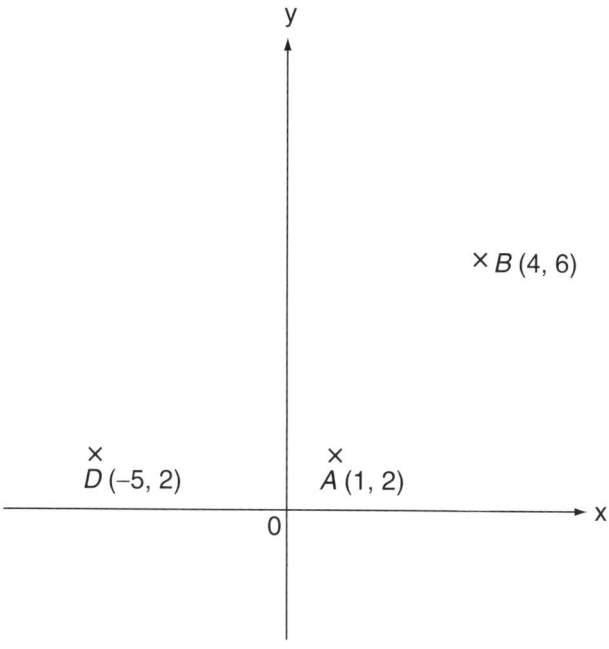

(a) Find the coordinates of the midpoint of AB.
(b) Calculate the length of AB.
(c) Calculate the gradient of the line AB.
(d) Find the equation of the line AB.
(e) The triangle ABC has line of symmetry $x = 4$.
   Find the coordinates of C.

(4024/01 May/June 2006 q 25 a to e)

21.

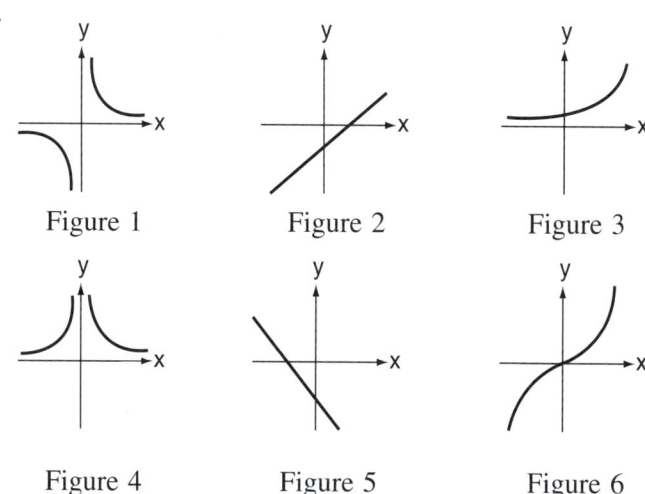

Figure 1   Figure 2   Figure 3

Figure 4   Figure 5   Figure 6

Which of the graphs shown above could be the graph of
(a) $y = x^3$,

(b) $y = \dfrac{1}{x^2}$,

(c) $y = x - 1$? (4024/01 Oct/Nov 2005 q 9)

22. A straight line passes through two points with coordinates (6, 8) and (0.5). Work out the equation of the line.
(0580/21 May/June 2008 q 9)

23. Alaric invests $100 at 4% per year **compound interest**.
   (a) How many dollars will Alaric have after 2 years?
   (b) After x years, Alaric will have y dollars.
       He knows a formula to calculate y.
       The formula is $y = 100 \times 1.04^x$

x (years)	0	10	20	30	40
y (dollars)	100	p	219	q	480

   Use this formula to calculate the values of p and q in the table.

   (c) Using a scale of 2 cm to represent 5 years on the x-axis and 2 cm to represent $50 on the y-axis draw an x-axis for $0 \leqslant x \leqslant 40$ and a y-axis for $0 \leqslant y \leqslant 500$.
       Plot the five points on the table and draw a smooth curve through them.

   (d) Use your graph to estimate
       (i) how many dollars Alaric will have after 5 years,
       (ii) how many years, to the nearest year, it takes for Alaric to have $200.

   (e) Beatrice invests $100 at 7% per year **simple interest**.
       (i) Show that after 20 years Beatrice has $240.
       (ii) How many dollars will Beatrice have after 40 years?
       (iii) On the **same grid**, draw a graph to show how the $100 which Beatrice invests will increase during the 40 years.

   (f) Alaric first has more than Beatrice after n years.
       Use your graphs to find the value of n.
(0580/04 May/June 2008 q 8)

24. Find the coordinates of the midpoint of the line joining the points $A(2, -5)$ and $B(6, 9)$.
(0580/21 May/June 2009 q 7)

25. (a) The table shows some values for the equation

$$y = \frac{x}{2} - \frac{2}{x} \text{ for } -4 \leq x \leq -0.5 \text{ and } 0.5 \leq x \leq 4.$$

x	-4	-3	-2	-1.5	-1	-0.5	0.5	1	1.5	2	3	4
y	-1.5	-0.83	0	0.58			-3.75		-0.58	0	0.83	1.5

 (i) Copy the table and write the missing values of y in the empty spaces.
 (ii) On a sheet of 2 mm graph paper draw the graph of

$$y = \frac{x}{2} - \frac{2}{x} \text{ for } -4 \leq x \leq -0.5 \text{ and } 0.5 \leq x \leq 4,$$ using a scale of 2 cm to represent 1 unit on each axis.

(b) Use your graph to solve the equation $\frac{x}{2} - \frac{2}{x} = 1$.

(c) (i) By drawing a tangent, work out the gradient of the graph where $x = 2$.
 (ii) Write down the gradient of the graph where $x = -2$.

(d) (i) On the grid, draw the line $y = -x$ for $-4 \leq x \leq 4$.
 (ii) Use your graphs to solve the equation $\frac{x}{2} - \frac{2}{x} = -x$.

(e) Write down the equation of a straight line which passes through the origin and does **not** intersect the graph of $y = \frac{x}{2} - \frac{2}{x}$.

(0580/04 May/June 2009 q 5)

Chapter 7

Length, Area and Volume

This chapter continues the work from Chapter 8 of the Core course. The Core Skills exercise will remind you of the work already covered.

Core Skills

1. Convert
   (a) 250 mm to metres,
   (b) 20 cm³ to cubic metres,
   (c) 15 m² to square centimetres,
   (d) 1.7 litres to millilitres.
2. Calculate
   (a) the circumference,
   (b) the area
   of a circle with diameter 18 cm.
3. Calculate
   (a) the perimeter,
   (b) the area
   of each of the following shapes.

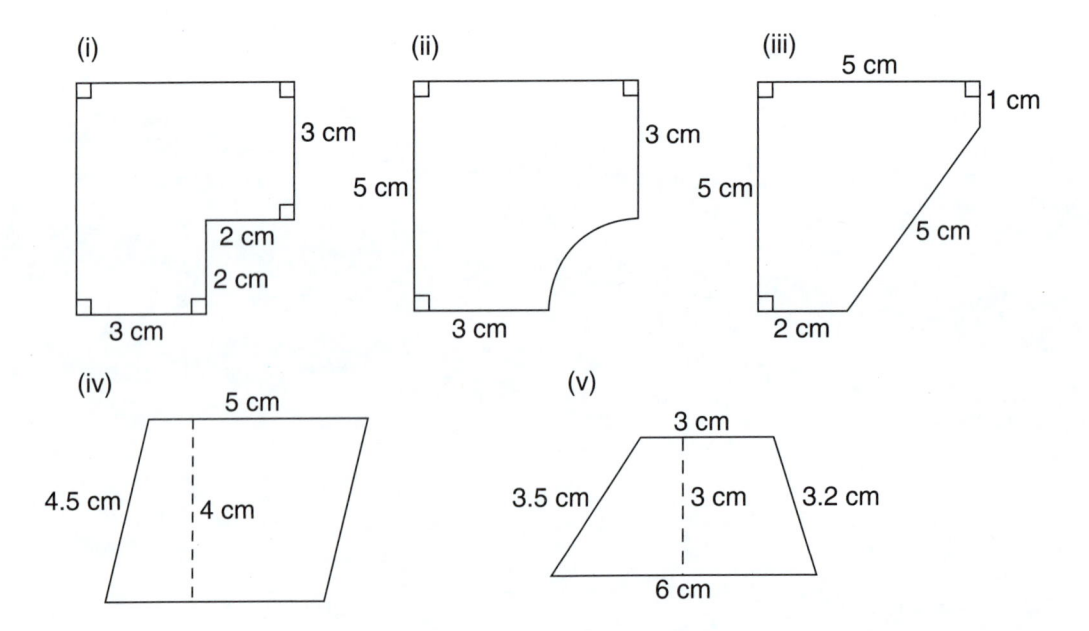

(i), (ii), (iii), (iv), (v)

4. Calculate
   (a) the total surface area,   (b) the volume
   of a cuboid measuring 12 cm by 5 cm by 10 cm.
5. Calculate
   (a) the total surface area,   (b) the volume
   of a solid cylinder with a radius of 5 cm and a height of 10 cm.
6. Calculate the capacity in millilitres of a cylinder measuring 6 cm in diameter and 25 cm in height.

Arc Lengths in Circles

As you saw in Chapter 6 of the Core Book, an arc is a section of the circumference of a circle.

The diagram below is of a circle, centre O and a radius shown as an arrow. If the arrow is rotated clockwise, to the position shown as a dotted line, through an angle of 60° at the centre, it traces out an *arc* shown as the heavy curved line on the circle.

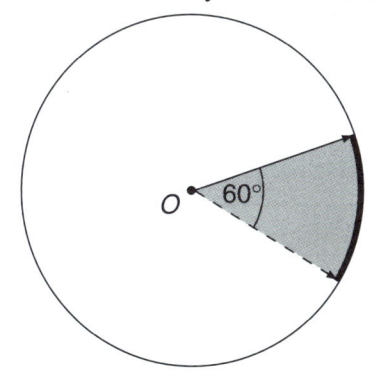

The length of the arc depends on the angle at the centre of the circle. The larger the angle, the longer the arc. The angle at the centre is called the angle *subtended* at the centre by the arc. The angle can be expressed as a fraction of the complete turn (360°). In the above diagram, the arc length will be $\frac{60}{360} = \frac{1}{6}$ of the whole circumference.

Example 1

Calculate the length of an arc subtended by the angle 60° at the centre of a circle, radius 4 cm.

Answer 1

Circumference = $2\pi r = 2 \times \pi \times 4$ cm

Arc length = $\frac{60}{360} \times 2 \times \pi \times 4 = 4.18879...$

Arc length = 4.19 cm

Sector Areas in Circles

In the diagram on page 147, the region between the two arrows (shaded) is a *sector* of the circle. The area of the sector is a fraction of the area of the whole circle, in this case $\frac{60}{360}$ of the total area of the circle.

Example 2

Calculate the area of the sector subtended by the angle 100° at the centre of the circle with radius 4 cm.

Answer 2

Area of the whole circle $= \pi r^2 = \pi \times 4^2$ cm²

Area of the sector $= \frac{100}{360} \times \pi \times 4^2 = 13.96263 \ldots$

Area of sector $= 14.0$ cm²

Exercise 7.1

Calculate the quantities represented by letters in the following table.

Angle at the centre	Radius	Length of arc	Area of sector
30°	10 cm	a	b
140°	8 cm	c	d
200°	5 cm	e	f
g	3 cm	10 cm	h
60°	i	12 cm	j
k	15 cm	l	30 cm²
20°	m	n	40 cm²
p	9 cm	q	25 cm²
100°	r	s	60 cm²

More Volumes and Surface Areas of Solids

You may be asked to calculate the volume or surface area of spheres, cones, or pyramids, but you will be given any necessary formulae. (See page 366 for the formulae you are expected to learn).

Length, Area and Volume

Example 3
Calculate the total surface area of a square based pyramid as shown in the diagram below.

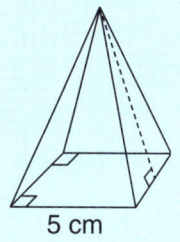

5 cm

The base is a square with side = 5 cm. The four isosceles triangles have a height of 6 cm, shown by the dotted line.

Answer 3
The diagram below shows the net of the pyramid.

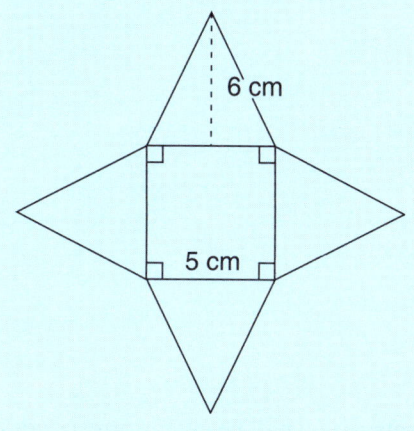

6 cm

5 cm

The total area of the net = area of square + 4 × area of triangle.
$$= 5^2 + 4 \times \frac{1}{2} \times 5 \times 6$$
$$= 85 \text{ cm}^2$$

Curved Surface Areas

Cylinders, cones and spheres all have curved surfaces.
- If you are asked to calculate the curved surface of a cone or a sphere you will be given the corresponding formula.
- If you are asked to calculate the *total* surface area of a cone you will need to add the area of the circular base.
- The net (Core Book Chapter 6) of a cylinder is a rectangle and two circles. The rectangle makes the *curved* surface of the cylinder. The length of the rectangle is the circumference of the the circular ends. The *total* surface area includes the two circles. You will be expected to answer questions on the surface area of the cylinder without being given a formula.

Example 4

Calculate the total surface area of a cone with a base radius = 10 cm and a slant height = 30 cm.

(The curved surface area of a cone = $\pi r l$, where l is the slant height.)

Answer 4

The slant height of a cone is the height measured along the surface as in the diagram.

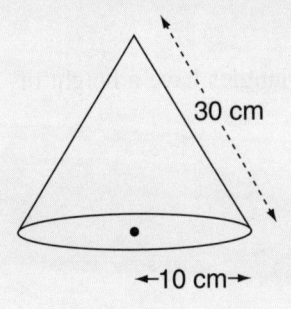

Curved surface area = $\pi \times 10 \times 30 = 300\pi$ cm^2
Circular base = $\pi \times 10^2 = 100\pi$ cm^2
Total surface area = $100\pi + 300\pi = 1256.637 \ldots$ cm^2
Total surface area = 1260 cm^2 to 3 significant figures

Volumes

If you are asked to calculate the volume of a cone or a sphere you will be given corresponding formula.

Example 5

Calculate the volume of a child's toy which is in the shape of a cone on a hemisphere.

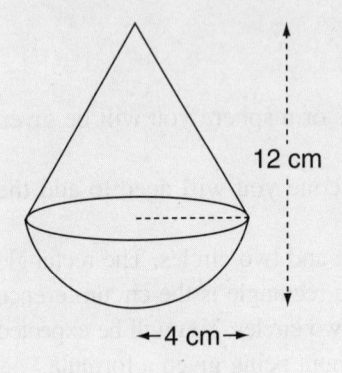

The total height of the toy is 12 cm, and the radius is 4 cm.

(The volume of a cone = $\frac{1}{3}\pi r^2 h$, where h is the perpendicular height, and the volume of a sphere = $\frac{4}{3}\pi r^3$)

Answer 5

The height of the cone = 12 − 4 = 8 cm

NOTE: The height of the cone = total height − radius of hemisphere.

Volume of cone = $\frac{1}{3}\pi r^2 h = \frac{1}{3} \times \pi \times 4^2 \times 8$ cm³

Volume of hemisphere = $\frac{1}{2} \times \frac{4}{3}\pi r^3 = \frac{1}{2} \times \frac{4}{3} \times \pi \times 4^3$ cm³

Total volume = $(\frac{1}{3} \times \pi \times 4^2 \times 8) + (\frac{1}{2} \times \frac{4}{3} \times \pi \times 4^3)$

= 268.08257... cm³

Total volume of toy = 268 cm³ to 3 significant figures.

Exercise 7.2

You can use the following formulae in this exercise.

Surface area of a sphere = $4\pi r^2$

Volume of a sphere = $\frac{4}{3}\pi r^3$

Curved surface area of a cone = $\pi r l$, where l is the slant height of the cone

Volume of a cone = $\frac{1}{3}\pi r^2 h$, where h is the perpendicular height of the cone.

1. Calculate the quantities represented by letters in the following table.

radius	Sphere		Hemisphere	
	surface area	volume	total surface area	volume
3 cm	a	b	c	d
7 cm	e	f	g	h
i	20 cm²	j	k	l
m	n	30 cm³	p	q
r	s	t	u	50 cm³

2. Calculate the quantities represented by letters in the following table.

Cone				
radius	slant height	perpendicular height	curved surface area	volume
3 cm	5 cm		a	
6 cm		7 cm		b
c	10 cm	8 cm	60 cm²	d
e	13 cm	10 cm		f	15 cm³

3. Calculate
   (a) the curved surface area,
   (b) the total surface area,
   (c) the volume
   of a cylinder with radius 4.5 cm and height 12 cm.

4. A cylinder has a height of 10 cm and a volume 283 cm³. Calculate
   (a) the radius of the cylinder,
   (b) its total surface area.

Similar Shapes

You have already met similar shapes in Chapter 6 of the Core Book, but this is a reminder. Similar shapes have the same shape but are different sizes. This means that they have corresponding angles equal and corresponding sides in proportion.

For example, all squares are similar because they all have 4 angles = 90°, and 4 equal sides.

All cubes are similar because they all have 6 equal faces at right angles to each other.

The diagrams below show two lines (A and B) with lengths in the ratio 1 : 2, and two squares (C and D) drawn from sides with these lengths, followed by two cubes (E and F).

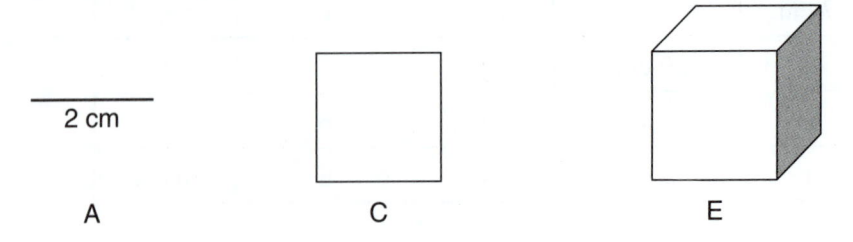

2 cm

A   C   E

Length, Area and Volume   153

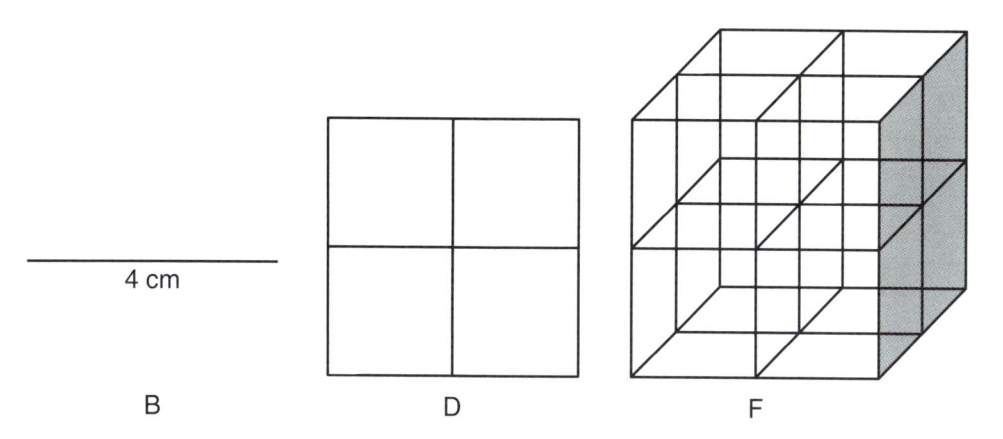

    4 cm

    B   D   F

The areas of the two squares are 2 × 2 = 4 cm² and 4 × 4 = 16 cm², so the ratios of the areas of the squares are 4 : 16 = 1 : 4. The area of D is four times the area of C.
The volumes of the two cubes are 2 × 2 × 2 = 8 cm³ and 4 × 4 × 4 = 64 cm³, so the ratios of the volumes are 8 : 64 = 1 : 8. The volume of F is eight times the volume of E. This is an important fact to understand.

If the sides or lengths of two *similar* shapes are in the ratio 1 : 2, then the areas are in the ratio 1 : 4 (1² : 2²), and the volumes are in the ratio 1 : 8 (1³ : 2³).

In general, for *similar* shapes:

- if lengths are in the ratio $a : b$
- then areas are in the ratio $a^2 : b^2$
- and volumes are in the ratio $a^3 : b^3$

Notice the importance of the word *similar*. These ratios only apply to similar shapes.

Example 6

The heights of two similar cylinders are 4 cm and 5 cm respectively.
The total surface area of the smaller cylinder is 30 cm².
Calculate the total surface area of the larger cylinder.

Answer 6

Length ratio = 4 : 5
Area ratio = 16 : 25
Total surface area of small cylinder = 30 cm²
Total surface area of large cylinder = $30 \times \dfrac{25}{16}$ = 46.875 cm²

NOTE: *Always* check that your answer is sensible. In this case, the surface area *is* larger for the larger cylinder. Also, if the area of the smaller cylinder had been 32 cm², then the ratio 16 : 25 would have made the larger cylinder 50 cm², so the answer does seem reasonable.

Example 7
The surface areas of two spheres are in the ratio 4 : 9.
The volume of the smaller sphere is 20 cm³.
Calculate the volume of the larger sphere.

Answer 7
Area ratio = 4 : 9
Length ratio = $\sqrt{4} : \sqrt{9}$ = 2 : 3
Volume ratio = $2^3 : 3^3$ = 8 : 27
Volume of the smaller sphere = 20 cm³
Volume of the larger sphere = $20 \times \dfrac{27}{8}$ = 67.5 cm³

Example 8
The heights of two similar bottles are in the ratio 3 : 5.
The capacity of the larger bottle is 1.5 litres.
Calculate the capacity of the smaller bottle in millilitres.

Answer 8
Capacity is the same as volume.
Length ratio = 3 : 5
Volume ratio = 27 : 125
Capacity of larger bottle = 1.5 litres.
Capacity of smaller bottle = $1.5 \times \dfrac{27}{125}$ = 0.324 litres.
Capacity of smaller bottle = 324 millilitres.

NOTE: This seems reasonable because it is smaller. Also 27 is approximately $\dfrac{1}{5}$ of 125, and $\dfrac{1}{5}$ of 1.5 litres is 300 millilitres.

Exercise 7.3
1. The volumes of two similar cuboids are in the ratio 1 : 125.
   (a) Find the ratio of
      (i) the surface areas
      (ii) the lengths of the cuboids
   (b) Copy and complete the following table for these two cuboids.

	length	height	total surface area	volume
smaller cuboid	20 cm	(i)	(ii)	80 cm³
larger cuboid	(iii)	2 cm	10600 cm²	(iv)

2. The areas of two 'smiley faces' are in the ratio 10 : 1.

   NOT TO SCALE

   The height of the larger face is 4 cm. Calculate the height of the smaller face.
3. The volumes of two similar cones are in the raio 8 : 27.
   (a) Find the ratio of the heights of the two cones.
   (b) Find the ratio of the areas of the bases of the two cones.
   (c) The curved surface area of the smaller cone is 20 cm². Calculate the curved surface area of the larger cone.
4. A soft drink is sold in two different sizes of bottles. The bottles are similar in shape. The height of the smaller bottle is 25 cm and the height of the larger bottle is 35 cm.
   (a) Calculate the ratio of the volumes.
   (b) The volume of the smaller bottle is 730 ml.
      Calculate the volume of the larger bottle. Give your answer correct to 3 significant figures.
5. A map is in the scale 1 : 25000.
   A lake on the map has an area of 5 cm².
   Calculate the area of the actual lake. Give your answer in square metres in standard form.

Similar Triangles

Similar triangles have the same shape, which means they have equal angles.
The corresponding sides of similar triangles are in the same ratio. *Corresponding* sides are the sides opposite equal angles, as shown in the diagram.

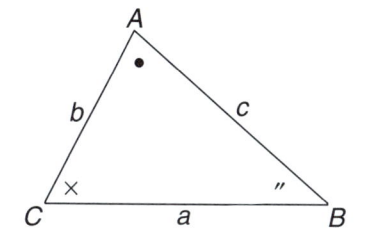

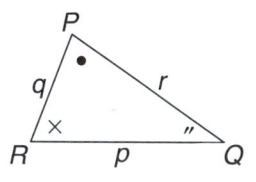

156 Extended Mathematics for Cambridge IGCSE

We will use the notation of a single capital letter for each angle where the angle is unambiguous, and the lower case letter for the length of the opposite side.
It is convenient to mark the angles that are equal with the same symbols as you will see in the diagram.
The corresponding sides are as follows:
a corresponds to p, b corresponds to q and c corresponds to r.
So we can write:

$$\frac{a}{p} = \frac{b}{q} = \frac{c}{r}$$

In order to keep everything in the correct order, as you will see later in the less obvious examples, it is wise to put the triangles in as well.

$$\frac{\text{large triangle}}{\text{small triangle}} : \frac{a}{p} = \frac{b}{q} = \frac{c}{r}$$

However, you must remember that this refers to the lengths of the sides, *not* to the areas of the triangles!

Example 9

In the diagram *PT* is parallel to *QS*. Find the lengths of *QS*, and *PR*.

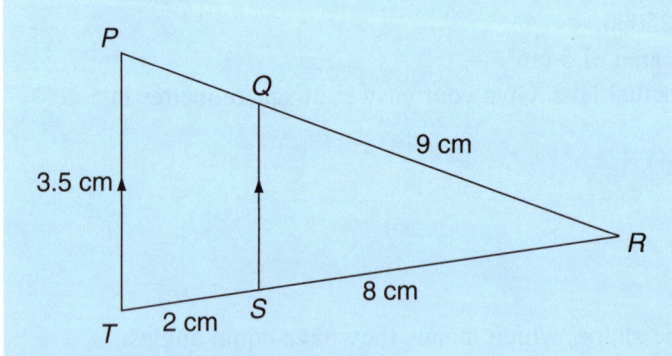

Answer 9

First of all we have to be sure the triangles *PRT* and *QRS* are similar.

$\angle P = \angle SQR$ (corresponding angles in parallel lines)

$\angle T = \angle QSR$ (corresponding angles in parallel lines)

$\angle R$ is common to both triangles.

So the angles are equiangular, and therefore the triangles are similar.

Copying the diagram and marking in the equal angles:

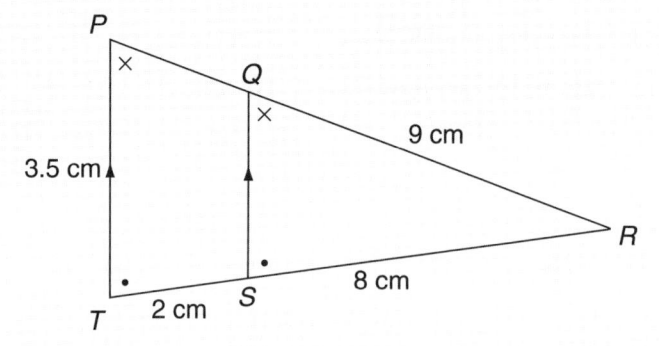

$$\frac{\text{large triangle}}{\text{small triangle}} : \frac{TR}{SR} = \frac{PR}{QR} = \frac{PT}{QS}$$

$$\frac{2+8}{8} = \frac{PR}{9} = \frac{3.5}{QS} \qquad \frac{10}{8} = \frac{PR}{9}$$

$$PR = 9 \times \frac{10}{8} = 11.25 \text{ cm}$$

Also, $\dfrac{10}{8} = \dfrac{3.5}{QS}$

$QS \times 10 = 3.5 \times 8$ ($\times QS$ and $\times 8$)

$QS = 3.5 \times \dfrac{8}{10}$ ($\div 10$)

$QS = 2.8$ cm $PR = 11.25$ cm and $QS = 2.8$ cm.

NOTE: A quick check shows that these are reasonable because *PR* is in the large triangle, and is larger than *QR*, and *QS* is in the small triangle and is smaller than *PT*.

Example 10

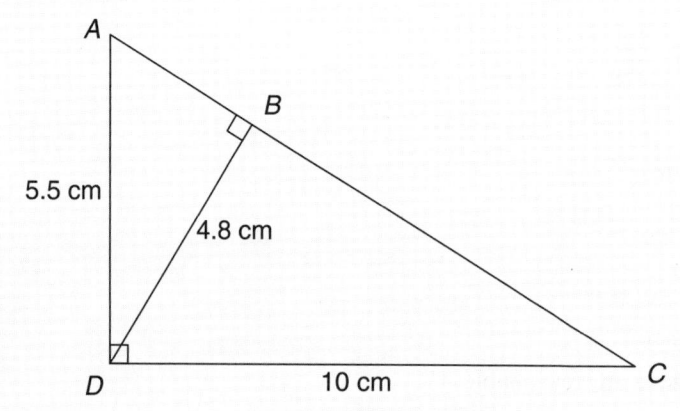

158 *Extended Mathematics for Cambridge IGCSE*

Using the diagram above

(a) prove that triangles *ACD* and *BCD* are similar.

(b) Use similar triangles to calculate the lengths of *AB* and *AC*. All the measurements are correct to 1 decimal place.

Answer 10

(a) In triangles *ACD* and *BCD*, $\angle ADC = \angle CBD = 90°$. (Given in the diagram).

$\angle A$ is common to both, so $\angle ACD = \angle ADB$. (Angle sum of a triangle.)

Triangles *ACD* and *BCD* are equiangular so they are similar.

(b) It is a good idea to split the diagram into the two triangles unless you can easily see which sides are corresponding. Mark the equal angles.

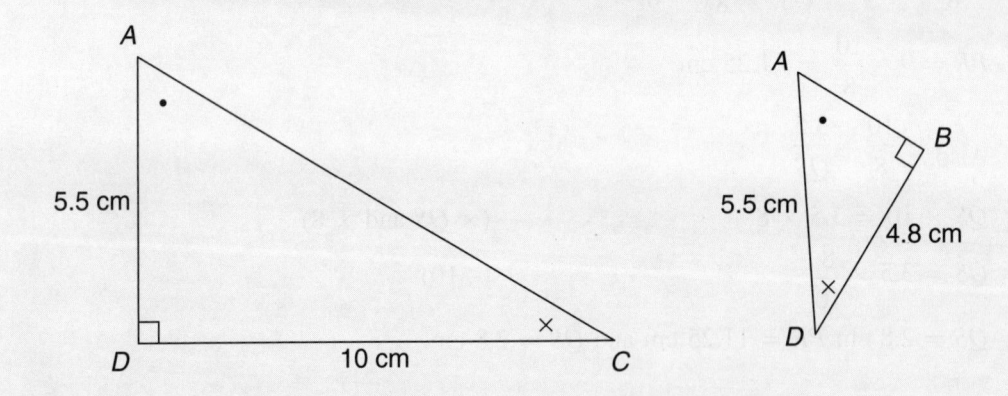

$$\frac{\text{large triangle}}{\text{small triangle}} : \frac{AC}{AD} = \frac{DC}{BD} = \frac{AD}{AB}$$

NOTE: Remember that corresponding sides are opposite the sides marked as equal by using the same signs. (× and . and so on.)

$$\frac{\text{large triangle}}{\text{small triangle}} : \frac{AC}{5.5} = \frac{10}{4.8} = \frac{5.5}{AB}$$

$$AC = 5.5 \times \frac{10}{4.8} = 11.5 \text{ cm}$$

$$AB = 5.5 \times \frac{4.8}{10} = 2.6 \text{ cm}$$

All correct to 1 decimal place.

Remember to check that these are reasonable.

Exercise 7.4

1. Triangles *ABC* and *PQR* are shown below. *AD* is the height of triangle *ABC* and *PS* is the height of *PQR*. *AD* = *DC*, and *PS* = *SR*.
   (a) Show that triangles *ABC* and *PQR* are similar.
   (b) Use similar triangles to
      (i) Calculate the length of *SR*.
      (ii) Calculate the area of triangle *PQR*, given that the area of triangle *ABC* is 14 cm².

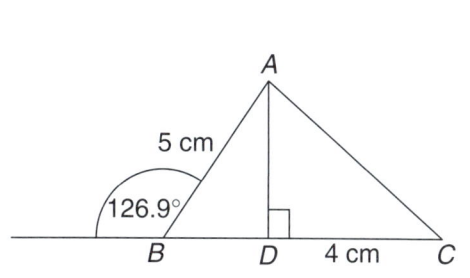

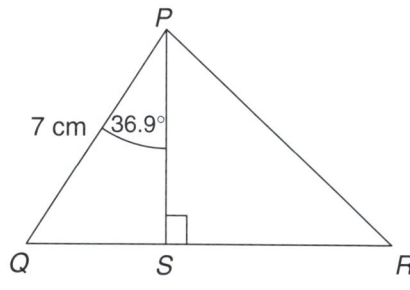

2. In the diagram below, *AB* and *DE* are parallel.
   (a) Show that triangles *ABC* and *CDE* are similar.
   (b) Calculate the length of *AB*.
   (c) The area of triangle *ABC* = 6 cm². Calculate the area of triangle *CDE*.

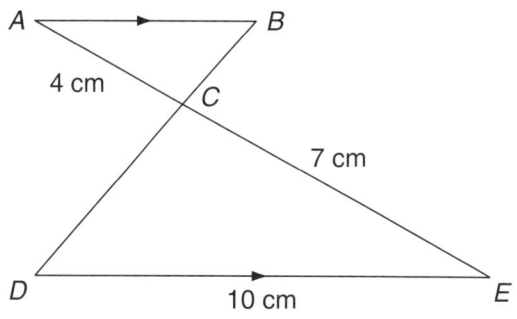

3. (a) Using your knowledge of cyclic quadrilaterals (Chapter 5); show that triangles *PRT* and *QRS* are similar.

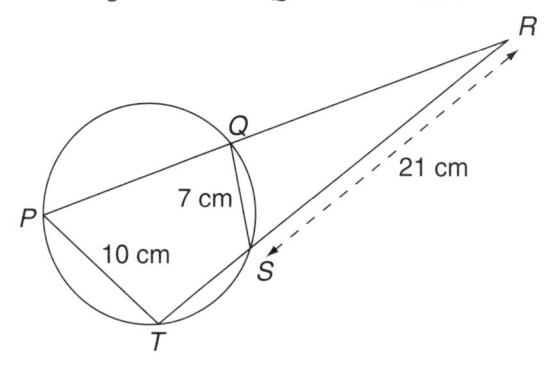

   (b) Calculate the length of *PR*.

160 Extended Mathematics for Cambridge IGCSE

Exercise 7.5

Mixed Exercise

1. A cone is to be made from a thin card. The cone will have a base radius of 4 cm, and a slant height of 10 cm.
   Raj draws a circle of radius 10 cm, and is going to cut out from that a sector to make the cone. Calculate the angle x.

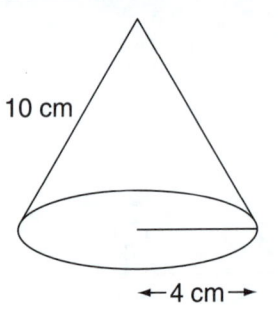

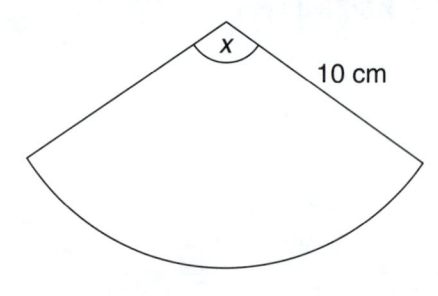

2. A map has a scale of 1 : 2500.
   On the map a reservoir has an area of 2 cm².
   What is the area of the reservoir? Give your answer in m².

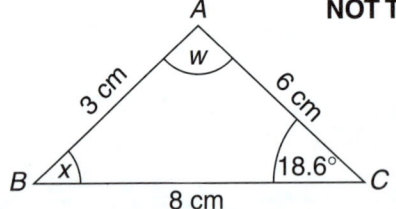

3. **NOT TO SCALE**

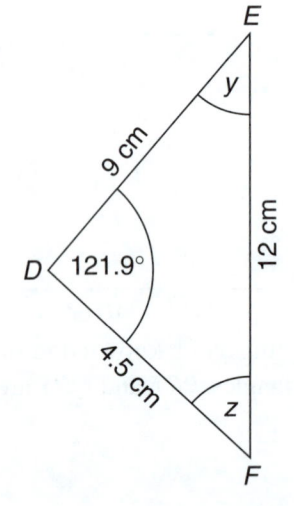

   (a) By showing that pairs of sides are in the same ratio prove that triangles ABC and DEF are similar.
   (b) Find the angles marked with letters in the two triangles.
   (c) Calculate the ratio of the areas of the two triangles.

4. A rectangle with length $(x + 6)$ and breadth $(x + 1)$ has the same area as a square with sides $= 2x$. (Measurements in centimetres.)
   Calculate the dimensions of the square and rectangle.

5. Two bowls are made from two identical wooden cylinders.
   One bowl is made by drilling out a hemisphere with the same radius as the cylinder. The other is made by drilling out a cone with the same radius as the cylinder, and perpendicular height equal to the height of the cylinder. The cylinders have height x cm, and radius x cm.
   (a) Calculate the volume of the wood which makes up the hemispherical bowl in terms of x.
   (b) Calculate the volume of the wood which makes up the conical bowl.

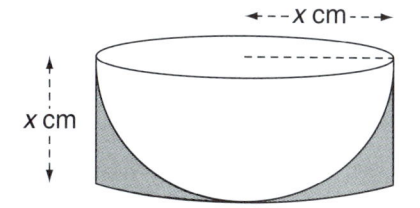

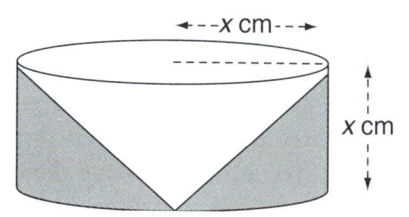

   (c) Comment on your answers to (a) and (b).

Examination Questions

6. In the diagram, $ABCD$ is a diameter of the circle, centre P.
   $AB = BC = CD = 2x$ centimetres.
   (a) Find an expression, in terms of x and π, for the circumference of this circle.
   (b) The perimeter of the shaded region consists of two semicircles whose diameters are AB and CD, and two semicircles whose diameters are AC and BD.
   Find an expression, in terms of x and π, for the area of the shaded region. (4024/01 May/June 2007 q 14)

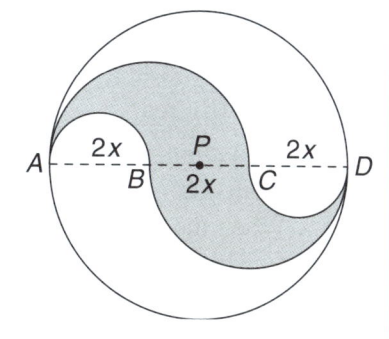

7. In the diagram, $BCDE$ is a trapezium, and the sides CD and BE are produced to meet at A. $CB = 12$ cm, $DE = 9$ cm, and the perpendicular distance from D to CB is 4 cm.
   Calculate
   (a) the area of $BCDE$,
   (b) the perpendicular distance from A to CB.

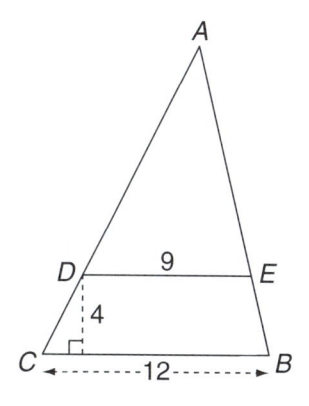

   (4024/01 May/June 2007 q 15)

8. (a) [The volume of a sphere is $\frac{4}{3}\pi r^3$.]

   [The surface area of a sphere is $4\pi r^2$.]

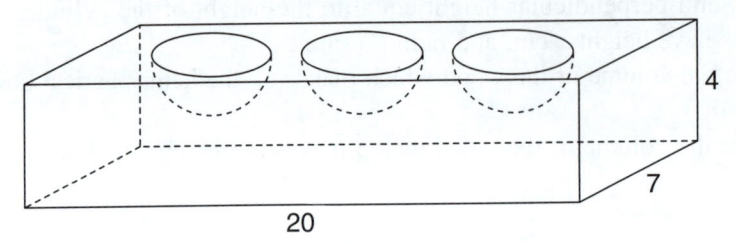

   A wooden cuboid has length 20 cm, width 7 cm, and height 4 cm.
   Three **hemispheres**, each of radius 2.5 cm, are hollowed out of the top of the cuboid, to leave the block as shown in the diagram.

   (i) Calculate the volume of wood in the block.
   (ii) The four vertical sides are painted blue.
       Calculate the total area that is painted blue.
   (iii) The inside of each **hemispherical** hollow is painted white.
        The flat part of the top of the block is painted red.
        Calculate the total area that is painted
        (a) white,
        (b) red.

(b) The volume of water in a container is directly proportional to the cube of its depth.
    When the depth is 12 cm, the volume is 576 cm³.
    Calculate
    (i) the volume when the depth is 6 cm,
    (ii) the depth when the volume is 1300 cm³.

   (4024/02 May/June 2007 q 7)

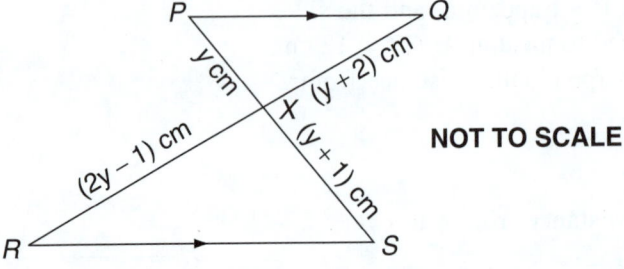

9. In the diagram, PQ is parallel to RS. PS and QR intersect at X.
   $PX = y$ cm, $QX = (y + 2)$ cm, $RX = (2y - 1)$ cm, and $SX = (y + 1)$ cm.
   (i) Show that $y^2 - 4y - 2 = 0$.
   (ii) Solve the equation $y^2 - 4y - 2 = 0$.

Show all your working and give your answers correct to two decimal places.
(iii) Write down the length of *RX*.

(0580/04 May/June 2007 q 3 b)

10.

NOT TO SCALE

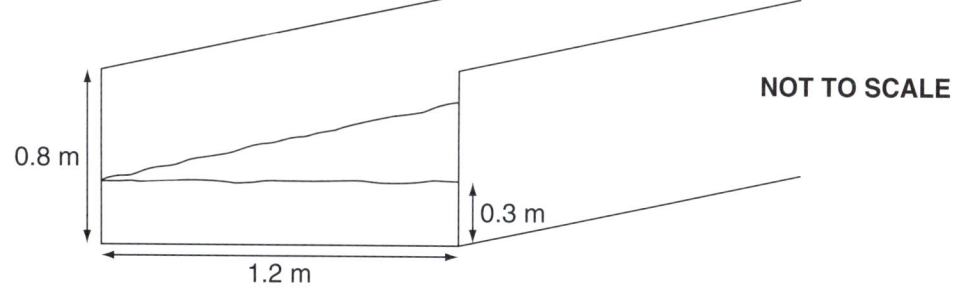

0.8 m

0.3 m

1.2 m

The diagram shows water in a channel.
This channel has a rectangular cross-section, 1.2 metres by 0.8 metres.

(a) When the depth of water is 0.3 metres, the water flows along the channel at 3 metres/**minute**.
Calculate the number of cubic metres which flows along the channel in one hour.

(b) When the depth of water in the channel increases to 0.8 metres, the water flows at 15 metres/minute.
Calculate the percentage increase in the number of cubic metres of water which flows along the channel in one hour.

(c) The water comes from a cylindrical tank.
When 2 cubic metres of water leave the tank, the level of water in the tank goes down by 1.3 **millimetres**.
Calculate the radius of the tank, in **metres**, correct to one decimal place.

(d) When the channel is empty, its **interior** surface is repaired.
This costs $0.12 per square metre. The total cost is $50.40.
Calculate the length, in metres, of the channel.

(0580/04 May/June 2007 q 7)

11.

NOT TO SCALE

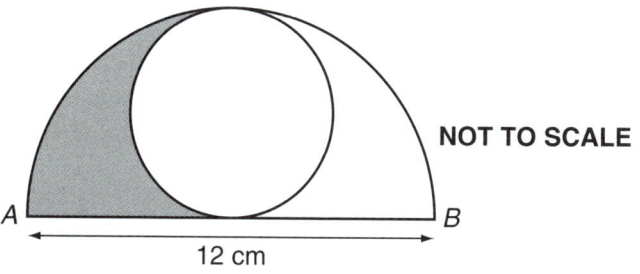

A   B
12 cm

The largest possible circle is drawn inside a semicircle, as shown in the diagram.

The distance AB is 12 centimetres.
(a) Find the shaded area.
(b) Find the perimeter of the shaded area.

(0580/02 May/June 2007 q 23)

12.

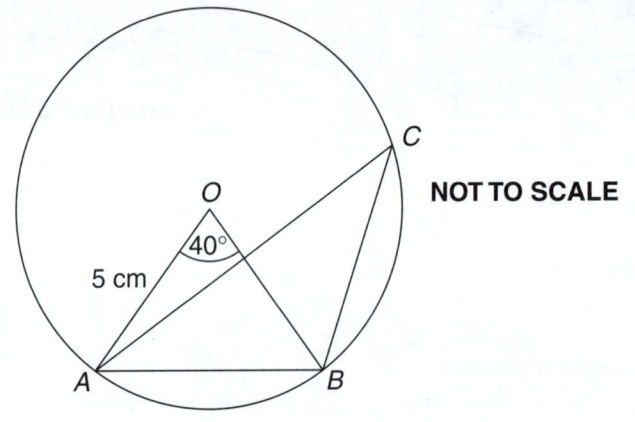

A, B and C are points on a circle, centre O.
Angle AOB = 40°
(a) (i) Write down the size of angle ACB.
 (ii) Find the size of angle OAB.
(b) The radius of the circle is 5 cm.
 (i) Calculate the length of the minor arc AB.
 (ii) Calculate the area of the minor sector OAB.

(0580/02 May/June 2005 q 21)

13.

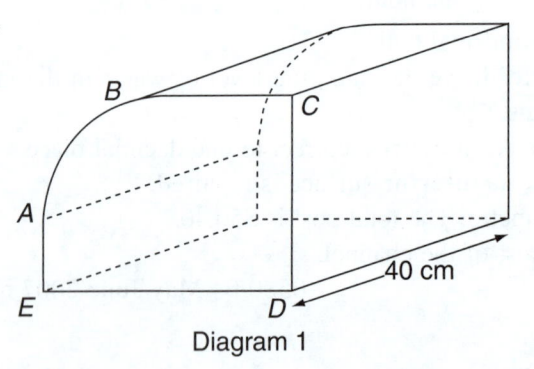

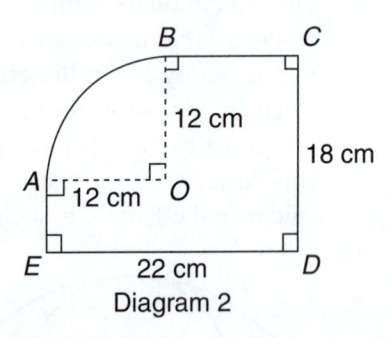

Diagram 1

Diagram 2

Diagram 1 shows a closed box. The box is a prism of length 40 cm.
The cross-section of the box is shown in Diagram 2, with all the right angles marked.
AB is an arc of a circle, centre O, radius 12 cm. ED = 22 cm and DC = 18 cm.
Calculate
(a) the perimeter of the cross-section,
(b) the area of the cross-section,

(c) the volume of the box,
(d) the **total** surface area of the box.

(0580/04 May/June 2006 q 2)

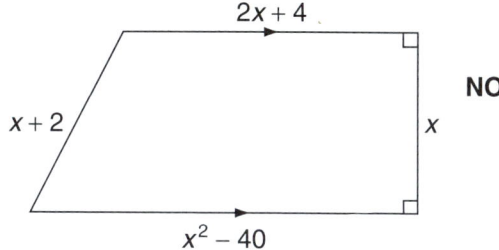

14. (a)

2x + 4

x + 2

x

x² – 40

NOT TO SCALE

The diagram shows a trapezium.
Two of its angles are 90°.
The lengths of the sides are given in terms of x.
The perimeter is 62 units.

    (i) Write down a quadratic equation in x to show this information. Simplify your equation.
    (ii) Solve your quadratic equation.
    (iii) Write down the only possible value of x.
    (iv) Calculate the area of the trapezium.

(0580/04 May/June 2006 q 8 a)

15.

35 m

24 m

1.1 m

D

C

2.5 m

B

A

NOT TO SCALE

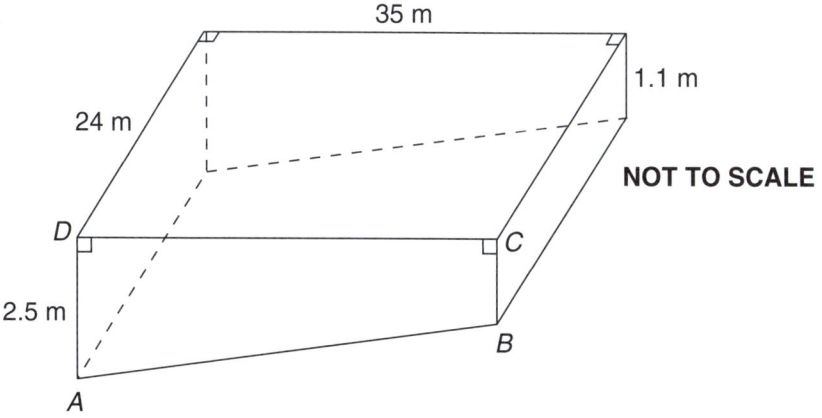

The diagram shows a swimming pool of length 35 m, and width 24 m.
A cross-section of the pool, $ABCD$, is a trapezium with $AD = 2.5$ m, and $BC = 1.1$ m.
(a) Calculate
    (i) the area of the trapezium $ABCD$,
    (ii) the volume of the pool,
    (iii) the number of litres of water in the pool, when it is full.
(b) $AB = 35.03$ m correct to 2 decimal places.

The sloping rectangular floor of the pool is painted. It costs $2.25 to paint one square metre.
    (i) Calculate the cost of painting the floor of the pool.
    (ii) Write your answer to **part (b)(i)** correct to nearest hundred dollars.
  (c) (i) Calculate the volume of a cylinder, radius 12.5 cm and height 4 cm.
    (ii) When the pool is emptied, the water flows through a cylindrical pipe of radius 12.5 cm. The water flows along this pipe at a rate of 14 centimetres per second.
Calculate the time taken to empty the pool.
Give your answer in days and hours, correct to the nearest hour.
                                                     (0580/04 Oct/Nov 2005 q 7)

16. A cylindrical glass has a radius of 3 centimetres and a height of 7 centimetres.
A large cylindrical jar full of water is a similar shape to the glass.
The glass can be filled with water from the jar exactly 216 times.
Work out the radius and height of the jar.
                                                     (0580/21 May/June 2008 q 10)

17. A spacecraft made 58376 orbits of the Earth and travelled a distance of 2.656×10^9 kilometres.
  (a) Calculate the distance travelled in 1 orbit correct to the nearest kilometre.
  (b) The orbit of the spacecraft is a circle. Calculate the radius of the orbit.
                                                     (0580/21 Oct/Nov 2008 q 14)

18. Two similar vases have heights which are in the ratio 3 : 2.
  (a) The volume of the larger vase is 1080 cm^3.
Calculate the volume of the smaller vase.
  (b) The surface area of the smaller vase is 252 cm^2.
Calculate the surface area of the larger vase.
                                                     (0580/21 May/June 2009 q 18)

19. A statue two metres high has a volume of five cubic metres.
A similar model of the statue has a height of four centimetres.
  (a) Calculate the volume of the model statue in cubic centimetres.
  (b) Write your answer to **part (a)** in cubic metres.
                                                     (0580/02 Oct/Nov 2006 q 13)

20. The surface area, A, of a cylinder, radius r and height h, is given by the formula $A = 2\pi rh + 2\pi r^2$.
  (a) Calculate the surface area of a cylinder of radius 5 cm and height 9 cm.
  (b) Make h the subject of the formula.
  (c) A cylinder has a radius of 6 cm and a surface area of 377 cm^2.
Calculate the height of this cylinder.
  (d) A cylinder has a surface area of 1200 cm^2 and its radius and height are equal.
Calculate the radius.
                                                     (0580/04 Oct/Nov 2006 q 8 (part))

21. (a) The scale of a map is 1 : 20 000 000.
On the map the distance between Cairo and Addis Ababa is 12 cm.
  (i) Calculate the distance, in kilometres, between Cairo and Addis Ababa.
  (ii) On the map the area of a desert region is 13 square centimetres.
  Calculate the actual area of this desert region, in square kilometres.
 (b) (i) The actual distance between Cairo and Khartoum is 1580 km.
  On a different map this distance is represented by 31.6 cm.
  Calculate, in the form 1 : n, the scale of this map.
  (ii) A plane flies the 1580 km from Cairo to Khartoum.
  It departs from Cairo at 11 55 and arrives in Khartoum at 14 03.
  Calculate the average speed of the plane, in kilometres per hour.

  (0580/04 May/June 2007 q 1)

22. A rectangle has sides of length 6.1 cm and 8.1 cm correct to 1 decimal place.
Copy and complete the statement about the perimeter of the rectangle.
..............cm ⩽ perimeter <...........cm

  (0580/02 Oct/Nov 2007 q 13)

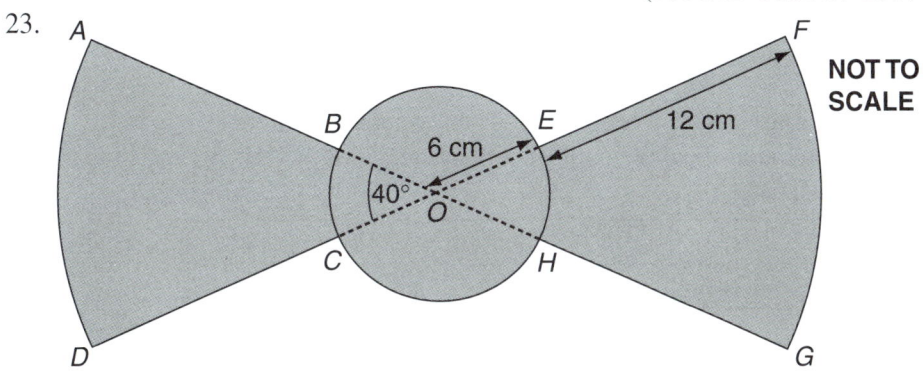

23.

The diagram shows part of a fan.
OFG and OAD are sectors, centre O, with radius 18 cm and sector angle 40°.
B, C, H and E lie on a circle, centre O and radius 6 cm.
Calculate the shaded area.

  (0580/21 May/June 2009 q 19)

Chapter 8

Further Algebra

This chapter completes the algebra part of your course.
You will learn about matrices and functions, and how to find solution sets for inequalities in two variables. The inequalities section leads to some simple programming work. You will see how matrices can be used in Chapter 10, in work on transformations.

Beginning Matrices

A **matrix** (plural **matrices**) is a rectangular arrangement of numbers.
For example, information about the numbers of boys and girls in years 10 and 11 at a school could be shown in a table.

	Year 10	Year 11
Boys	60	56
Girls	58	61

If this is presented as a matrix it becomes:

$$\begin{pmatrix} 60 & 56 \\ 58 & 61 \end{pmatrix}$$

The **order** of the matrix is given by the numbers of rows and columns. Rows are always given first. The example above is a square matrix with 2 rows and 2 columns, so it is a **2 × 2 matrix**. This is referred to as a 'two by two matrix' and written '2 × 2'. The sign does not imply multiplication.

Example 1
Write down the order of these matrices

(a) $\begin{pmatrix} a & b \\ c & d \end{pmatrix}$ (b) $\begin{pmatrix} 1 & 2 & 3 \\ 4 & 5 & 6 \end{pmatrix}$ (c) $\begin{pmatrix} 10 & 11 \\ 12 & 13 \\ 14 & 15 \end{pmatrix}$ (d) $\begin{pmatrix} x \\ y \\ z \end{pmatrix}$

Answer 1
(a) 2 × 2 (b) 2 × 3 (c) 3 × 2 (d) 3 × 1

Further Algebra 169

Operations on Matrices

Matrices of the same order can be added or subtracted by adding or subtracting the corresponding elements in each matrix.

For example, if the school shown above combines with another school we would have:

School A	Year 10	Year 11
Boys	60	56
Girls	58	61

School B	Year 10	Year 11
Boys	43	52
Girls	39	60

This can be shown in matrix form as:

$$\begin{pmatrix} 60 & 56 \\ 58 & 61 \end{pmatrix} + \begin{pmatrix} 43 & 52 \\ 39 & 60 \end{pmatrix} = \begin{pmatrix} 103 & 108 \\ 97 & 121 \end{pmatrix}$$

Example 2

Calculate

(a) $\begin{pmatrix} 5 & 6 \\ 7 & 3 \end{pmatrix} - \begin{pmatrix} 2 & 4 \\ 8 & 1 \end{pmatrix}$

(b) $\begin{pmatrix} 1 & 5 & 9 \\ 2 & 1 & 0 \end{pmatrix} + \begin{pmatrix} 0 & -4 & 8 \\ 3 & 6 & -2 \end{pmatrix}$

Answer 2

(a) $\begin{pmatrix} 5 & 6 \\ 7 & 3 \end{pmatrix} - \begin{pmatrix} 2 & 4 \\ 8 & 1 \end{pmatrix} = \begin{pmatrix} 3 & 2 \\ -1 & 2 \end{pmatrix}$

(b) $\begin{pmatrix} 1 & 5 & 9 \\ 2 & 1 & 0 \end{pmatrix} + \begin{pmatrix} 0 & -4 & 8 \\ 3 & 6 & -2 \end{pmatrix} = \begin{pmatrix} 1 & 1 & 17 \\ 5 & 7 & -2 \end{pmatrix}$

A matrix may be **transposed** by turning the columns into rows and the rows into columns. School A could be shown as:

	Boys	Girls
Year 10	60	58
Year 11	56	61

In matrix form this is written as:

$$\mathbf{A} = \begin{pmatrix} 60 & 56 \\ 58 & 61 \end{pmatrix}; \quad \mathbf{A} \text{ transpose} = \mathbf{A}' = \begin{pmatrix} 60 & 58 \\ 56 & 61 \end{pmatrix}$$

Matrices are normally denoted by capital letters.

Example 3

$C = \begin{pmatrix} 2 & 7 & 0 \\ 5 & 3 & 4 \end{pmatrix}$. Find C'.

Answer 3

$C' = \begin{pmatrix} 2 & 5 \\ 7 & 3 \\ 0 & 4 \end{pmatrix}$

A matrix may be multiplied by a number. For example,

$$3\begin{pmatrix} 1 & 6 \\ 0 & -1 \end{pmatrix} = \begin{pmatrix} 3 & 18 \\ 0 & -3 \end{pmatrix}$$

You will see that each element is multiplied by the number.

Example 4

Calculate $-\dfrac{1}{2}\begin{pmatrix} 5 & -1 \\ 8 & 0 \end{pmatrix}$.

Answer 4

$-\dfrac{1}{2}\begin{pmatrix} 5 & -1 \\ 8 & 0 \end{pmatrix} = \begin{pmatrix} \dfrac{-5}{2} & \dfrac{1}{2} \\ -4 & 0 \end{pmatrix}$

Exercise 8.1

1. Write the following information in a table and then in matrix form.
   (a) A manufacturer of computers makes laptops and desktops, both in either black or white cases. There are 10 black laptops, 20 white laptops, 4 black desktops, and 15 white desktops.
   (b) Four teams, the Reds, the Blues, the Greens, and the Yellows are in a friendly league. The Reds have 2 wins, 2 draws and lose 2 games. The Blues have 3 wins, 2 draws and lose 1 game. The Greens have 1 win, 5 draws and do not lose any games. The Yellows have 1 win, 1 draw and lose 4 games.
2. Work out the following matrix calculations

   (a) $\begin{pmatrix} 5 & 4 \\ 10 & 3 \end{pmatrix} + \begin{pmatrix} -2 & 6 \\ -18 & 0 \end{pmatrix}$ (b) $\begin{pmatrix} -5 & 0 \\ 1 & -3 \end{pmatrix} - \begin{pmatrix} -4 & -1 \\ 0 & 4 \end{pmatrix}$

(c) $2\begin{pmatrix} 1 & 0 \\ 3 & -8 \end{pmatrix}$ 	(d) $-1 \begin{pmatrix} -2 & 5 \\ 0 & 1 \end{pmatrix}$

3. $\mathbf{A} = \begin{pmatrix} 5 & 1 & -3 \\ 4 & 5 & 2 \end{pmatrix}$ 	$\mathbf{B} = \begin{pmatrix} 3 & 9 \\ 7 & 1 \end{pmatrix}$ 	$\mathbf{C} = \begin{pmatrix} 1 & 2 & 5 \\ 4 & 9 & 3 \end{pmatrix}$

$\mathbf{D} = \begin{pmatrix} 7 & 0 \\ 3 & 1 \end{pmatrix}$ 	$\mathbf{E} = \begin{pmatrix} 5 & 1 \\ 4 & 6 \\ 9 & 3 \end{pmatrix}$ 	$\mathbf{F} = \begin{pmatrix} -1 & 2 & 6 \end{pmatrix}$

(a) Write down the order of each matrix.
(b) Write down the transpose of each matrix.
(c) Write down the order of
   (i) $\mathbf{D'}$ (ii) $\mathbf{E'}$ (iii) $\mathbf{F'}$
(d) Calculate
   (i) $\mathbf{A} + \mathbf{C}$ (ii) $\mathbf{D} - \mathbf{B}$ (iii) $4 \times \mathbf{F}$

Multiplication of Matrices

We have seen that matrices can only be added or subtracted if they are of the same order, because corresponding elements from each matrix have to be added or subtracted. A similar restriction applies for multiplication of matrices, and we say that matrices have to be **conformable** for multiplication.

Take 4 matrices, for example,

$$\mathbf{A} = \begin{pmatrix} 2 & 5 & 6 \\ 7 & 1 & 3 \end{pmatrix} \quad \mathbf{B} = \begin{pmatrix} 1 & 2 \\ 3 & 4 \end{pmatrix}, \quad \mathbf{C} = \begin{pmatrix} 5 & 4 & 1 \\ 9 & 8 & 2 \end{pmatrix}, \quad \mathbf{D} = \begin{pmatrix} 1 & 2 \\ 5 & 6 \\ 9 & 1 \end{pmatrix}.$$

The orders of these matrices are:

$\mathbf{A} = 2 \times 3,$ 	$\mathbf{B} = 2 \times 2,$ 	$\mathbf{C} = 2 \times 3,$ 	$\mathbf{D} = 3 \times 2.$

Matrices are only conformable for multiplication if the number of columns in the first matrix is the same as the number of rows in the second. This means that, unlike in normal arithmetic when 5×10 is the same as 10×5, in matrix multiplication the order of multiplying does matter.

Looking at the matrices **A, B, C** and **D**, we see that $\mathbf{B} \times \mathbf{A}$ is possible because **B** has 2 columns and **A** has 2 rows:

```
          B                    A
   rows   columns       rows   columns
    2  ×    2            2   ×    3
                _____/
                 these two are
                     equal
```

However, **A × B** is not possible:

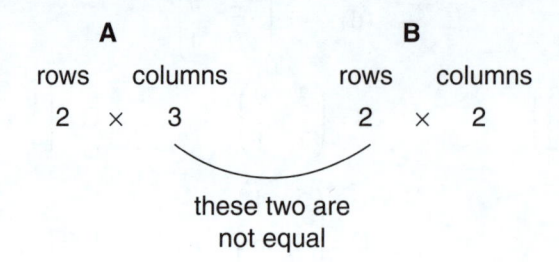

```
           A                           B
     rows    columns            rows    columns
      2   ×    3                 2   ×    2
              └──────────────────┘
                 these two are
                  not equal
```

When the two matrices are conformable for multiplication, the order of the resulting matrix is the rows from the first matrix and the columns from the second. A diagram should make this clearer.

matrix:	**B × A**	result: **BA**
order:	2 × ② ② × 3	2 × 3
	conformable	

Example 5

Using the matrices **A**, **B**, **C** and **D** above, state whether these matrices are conformable *in the order given*, and if they are, the order of the product.

(a) **AC**　　　(b) **CA**　　　(c) **DA**　　　(d) **DB**　　　(e) **BD**

Answer 5

(a)　　　A　　　　C
　　　2 × ③　　② × 3　　　not conformable

(b)　　　C　　　　A
　　　2 × ③　　② × 3　　　not conformable

(c)　　　D　　　　A　　　　　　　　　　　　　　DA
　　　3 × ②　　② × 3　　　conformable　　　　3 × 3

(d)　　　D　　　　B　　　　　　　　　　　　　　DB
　　　3 × ②　　② × 2　　　conformable　　　　3 × 2

(e)　　　B　　　　D
　　　2 × ②　　③ × 2　　　not conformable

Exercise 8.2

F, G, H, K, L are 5 matrices. Their orders are
F = 2 × 2,   **G** = 3 × 3,   **H** = 2 × 4,   **K** = 4 × 2,   **L** = 3 × 2.
For each of the following matrix products state whether they are conformable for multiplication, and if so what the order of the product will be.
1. **FG**   2. **GH**   3. **HK**   4. **KH**   5. **LG**   6. **GL**   7. **LF**   8. **FL**

We now need to find how to multiply two conformable matrices. The data in question 1(b) in Exercise 8.1, produces a table and matrix like this:

	Win	Draw	Lose
Reds	2	2	2
Blues	3	2	1
Greens	1	5	0
Yellows	1	1	4

$$A = \begin{pmatrix} 2 & 2 & 2 \\ 3 & 2 & 1 \\ 1 & 5 & 0 \\ 1 & 1 & 4 \end{pmatrix}$$

In order to decide which team wins the league it is decided to award 5 points for a win, 2 points for a draw, and no points for losing a game. This can be shown in a table and a matrix.

	Points
Win	5
Draw	2
Lose	0

$$B = \begin{pmatrix} 5 \\ 2 \\ 0 \end{pmatrix}$$

Now we can multiply the two matrices to find the overall winner.

$$AB = \begin{pmatrix} 2 & 2 & 2 \\ 3 & 2 & 1 \\ 1 & 5 & 0 \\ 1 & 1 & 4 \end{pmatrix} \times \begin{pmatrix} 5 \\ 2 \\ 0 \end{pmatrix}$$

The total points for the Reds are $2 \times 5 + 2 \times 2 + 2 \times 0 = 14$.
Check that you agree with this, and then find the points for the other three teams.
Check that the matrices are conformable for multiplication and find the order of the product.

$$\begin{array}{ccc} A & B & AB \\ 4 \times 3 & 3 \times 1 & 4 \times 1 \end{array}$$

$$AB = \begin{pmatrix} 2 & 2 & 2 \\ 3 & 2 & 1 \\ 1 & 5 & 0 \\ 1 & 1 & 4 \end{pmatrix} \begin{pmatrix} 5 \\ 2 \\ 0 \end{pmatrix} = \begin{pmatrix} 2\times5+2\times2+2\times0 \\ 3\times5+2\times2+1\times0 \\ 1\times5+5\times2+0\times0 \\ 1\times5+1\times2+4\times0 \end{pmatrix} = \begin{pmatrix} 14 \\ 19 \\ 15 \\ 7 \end{pmatrix}$$

Which team won the league?
Which team came second?
What would the result have been if 4 points were awarded for a win and 3 for a draw?
What about 3 for a win and 1 for a draw?

174 Extended Mathematics for Cambridge IGCSE

Example 6

Using the information above,

(a) write down **A′** and **B′**,   (b) write down the orders of **A′** and **B′**,

(c) multiply **A′** and **B′** together in the order in which they are conformable.

Answer 6

(a) $\mathbf{A'} = \begin{pmatrix} 2 & 3 & 1 & 1 \\ 2 & 2 & 5 & 1 \\ 2 & 1 & 0 & 4 \end{pmatrix}$ $\mathbf{B'} = \begin{pmatrix} 5 & 2 & 0 \end{pmatrix}$

(b) $\mathbf{A'} = 3 \times 4$ $\mathbf{B'} = 1 \times 3$

(c) $\mathbf{B'A'} = \begin{pmatrix} 5 & 2 & 0 \end{pmatrix} \times \begin{pmatrix} 2 & 3 & 1 & 1 \\ 2 & 2 & 5 & 1 \\ 2 & 1 & 0 & 4 \end{pmatrix}$ NOTE: $\mathbf{B'A'} = 1 \times 4$

$= \begin{pmatrix} 14 & 19 & 15 & 7 \end{pmatrix}$

We now need to find a general method for multiplying matrices.

You need to be particularly careful and systematic when multiplying matrices. It is strongly advised that you work out the order of the product and draw a matrix with dots in the positions where each element of the answer will go. We will now go step by step through the process.

For example, we will multiply these two matrices.

$$\mathbf{P} = \begin{pmatrix} 1 & 2 \\ 3 & 4 \\ 5 & 6 \end{pmatrix} \quad \mathbf{Q} = \begin{pmatrix} 7 & 8 \\ 9 & 10 \end{pmatrix} \quad \mathbf{PQ} = \begin{pmatrix} \cdot & \cdot \\ \cdot & \cdot \\ \cdot & \cdot \end{pmatrix}$$

$\mathbf{P}$ — 3×2 $\mathbf{Q}$ — 2×2 $\mathbf{PQ}$ — 3×2

PQ is conformable, but not **QP**. Check that you agree.

The rule is that you multiply the rows of the first matrix on to the columns of the second matrix, and add the results.

row 1 → $\begin{pmatrix} 1 & 2 \\ 3 & 4 \\ 5 & 6 \end{pmatrix} \begin{pmatrix} 7 & 8 \\ 9 & 10 \end{pmatrix} = \begin{pmatrix} 1\times 7 + 2\times 9 & \cdot \\ \cdot & \cdot \\ \cdot & \cdot \end{pmatrix} = $ row 1 → $\begin{pmatrix} 25 & \cdot \\ \cdot & \cdot \\ \cdot & \cdot \end{pmatrix}$

column 1 ↓   column 1 ↓

Then repeat for the remaining rows and columns.

Further Algebra

$$\text{row 1} \rightarrow \begin{pmatrix} \boxed{1 \; 2} \\ 3 \; 4 \\ 5 \; 6 \end{pmatrix} \begin{pmatrix} 7 & \boxed{8} \\ 9 & \boxed{10} \end{pmatrix} \overset{\text{column 2}}{\underset{\downarrow}{=}} \begin{pmatrix} 25 & 1 \times 8 + 2 \times 10 \\ \bullet & \bullet \\ \bullet & \bullet \end{pmatrix} = \text{row 1} \rightarrow \begin{pmatrix} 25 & \overset{\text{column 2}}{\underset{\downarrow}{28}} \\ \bullet & \bullet \\ \bullet & \bullet \end{pmatrix}$$

$$\text{row 2} \rightarrow \begin{pmatrix} 1 & 2 \\ \boxed{3 \; 4} \\ 5 & 6 \end{pmatrix} \begin{pmatrix} \boxed{7} & 8 \\ \boxed{9} & 10 \end{pmatrix} \overset{\text{column 1}}{\underset{\downarrow}{=}} \begin{pmatrix} 25 & 28 \\ 3 \times 7 + 4 \times 9 & \bullet \\ \bullet & \bullet \end{pmatrix} = \text{row 2} \rightarrow \begin{pmatrix} 25 & \overset{\text{column 1}}{\underset{\downarrow}{28}} \\ 57 & \bullet \\ \bullet & \bullet \end{pmatrix}$$

Check that you can follow this and complete the multiplication.

$$\begin{pmatrix} 1 & 2 \\ 3 & 4 \\ 5 & 6 \end{pmatrix} \begin{pmatrix} 7 & 8 \\ 9 & 10 \end{pmatrix} = \begin{pmatrix} 25 & 28 \\ 57 & 3 \times 8 + 4 \times 10 \\ 5 \times 7 + 6 \times 9 & 5 \times 8 + 6 \times 10 \end{pmatrix} = \begin{pmatrix} 25 & 28 \\ 57 & 64 \\ 89 & 100 \end{pmatrix}$$

NOTE: For Multiplying Matrices:
- Check that the two matrices are conformable.
- Write down the order of the product.
- Open brackets of a suitable size.
- Multiply rows on to columns and add.

Example 7

$$\mathbf{X} = \begin{pmatrix} 2 & 1 & 3 \\ 0 & 5 & 1 \end{pmatrix} \quad \mathbf{Y} = \begin{pmatrix} -1 & 0 \\ 2 & 2 \\ 1 & 4 \end{pmatrix}$$

Calculate
(a) **XY** (b) **YX**

Answer 7

(a) $\quad$ **X** $\qquad$ **Y** $\qquad$ **XY**
$\quad$ 2×3 $\qquad$ 3×2 $\qquad$ 2×2

$$\begin{pmatrix} 2 & 1 & 3 \\ 0 & 5 & 1 \end{pmatrix} \begin{pmatrix} -1 & 0 \\ 2 & 2 \\ 1 & 4 \end{pmatrix} = \begin{pmatrix} 2 \times {}^-1 + 1 \times 2 + 3 \times 1 & 2 \times 0 + 1 \times 2 + 3 \times 4 \\ 0 \times {}^-1 + 5 \times 2 + 1 \times 1 & 0 \times 0 + 5 \times 2 + 1 \times 4 \end{pmatrix} = \begin{pmatrix} 3 & 14 \\ 11 & 14 \end{pmatrix}$$

$$\mathbf{XY} = \begin{pmatrix} 3 & 14 \\ 11 & 14 \end{pmatrix}$$

(b) Y   X   YX
     3 × 2   2 × 3   3 × 3

$$\begin{pmatrix} -1 & 0 \\ 2 & 2 \\ 1 & 4 \end{pmatrix} \begin{pmatrix} 2 & 1 & 3 \\ 0 & 5 & 1 \end{pmatrix} = \begin{pmatrix} -1\times2+0\times0 & -1\times1+0\times5 & -1\times3+0\times1 \\ 2\times2+2\times0 & 2\times1+2\times5 & 2\times3+2\times1 \\ 1\times2+4\times0 & 1\times1+4\times5 & 1\times3+4\times1 \end{pmatrix} = \begin{pmatrix} -2 & -1 & -3 \\ 4 & 12 & 8 \\ 2 & 21 & 7 \end{pmatrix}$$

$$YX = \begin{pmatrix} -2 & -1 & -3 \\ 4 & 12 & 8 \\ 2 & 21 & 7 \end{pmatrix}$$

Exercise 8.3

1. $\begin{pmatrix} 2 & 5 \\ 0 & 1 \end{pmatrix} \begin{pmatrix} 6 & 7 \\ 3 & 4 \end{pmatrix} = \begin{pmatrix} 27 & a \\ 3 & 4 \end{pmatrix}$

   Find a.

2. $\begin{pmatrix} -1 & 6 \\ 4 & 0 \end{pmatrix} \begin{pmatrix} 5 & 2 \\ 3 & -1 \end{pmatrix} = \begin{pmatrix} 13 & b \\ 20 & c \end{pmatrix}$

   Find b and c.

3. $A = \begin{pmatrix} 5 & 1 & -1 \\ 0 & 1 & 3 \end{pmatrix}$ $B = \begin{pmatrix} 1 & 2 \\ 3 & 4 \end{pmatrix}$ $C = \begin{pmatrix} 3 & -1 & 2 \end{pmatrix}$ $D = \begin{pmatrix} 6 & 7 \\ 4 & 2 \\ -1 & 3 \end{pmatrix}$

   Where possible calculate

   (a) **AB**   (b) **BA**   (c) **CD**   (d) **DA**

4. $\begin{pmatrix} 2 & 3 \\ 1 & -1 \end{pmatrix} \begin{pmatrix} 1 & 1 & x \\ -1 & y & 4 \end{pmatrix} = \begin{pmatrix} -1 & 2 & 16 \\ z & 1 & -2 \end{pmatrix}$

   Find x, y and z.

5. $E = \begin{pmatrix} -1 & 3 \\ 0 & -2 \end{pmatrix}$

   Calculate (a) $3E$ (b) $-E$ NOTE: $-1 \times E$ (c) E^2 NOTE: $E \times E$
              (d) $3E + E^2$ (e) $E + E'$

The Zero Matrix

Since $a - a = 0$ and $b \times 0 = 0$, can similar results be obtained with matrices?

If $\quad A = \begin{pmatrix} a & b \\ c & d \end{pmatrix}$ and $X = \begin{pmatrix} w & x \\ y & z \end{pmatrix}$, then $$\mathbf{A} - \mathbf{A} = \begin{pmatrix} a & b \\ c & d \end{pmatrix} - \begin{pmatrix} a & b \\ c & d \end{pmatrix} = \begin{pmatrix} a-a & b-b \\ c-c & d-d \end{pmatrix} = \begin{pmatrix} 0 & 0 \\ 0 & 0 \end{pmatrix},$$

and $$\mathbf{X} \times \begin{pmatrix} 0 & 0 \\ 0 & 0 \end{pmatrix} = \begin{pmatrix} w & x \\ y & z \end{pmatrix} \begin{pmatrix} 0 & 0 \\ 0 & 0 \end{pmatrix} = \begin{pmatrix} 0 & 0 \\ 0 & 0 \end{pmatrix}.$$

So $\begin{pmatrix} 0 & 0 \\ 0 & 0 \end{pmatrix}$ is the **zero 2 × 2 matrix**.

From now on we will be dealing mainly with 2 × 2 matrices.

Example 8

$$\begin{pmatrix} 2 & -3 \\ -c & 10 \end{pmatrix} - \begin{pmatrix} a & -b \\ 5 & -d \end{pmatrix} = \begin{pmatrix} 0 & 0 \\ 0 & 0 \end{pmatrix}$$

Find a, b, c, and d.

Answer 8

$2 - a = 0 \qquad a = 2$

$-3 - {}^-b = 0 \qquad b = 3$

$-c - 5 = 0 \qquad c = -5$

$10 - {}^-d = 0 \qquad d = -10$

The Identity Matrix and Inverse Matrices

Remembering that $a \times 1 = a$, is there a matrix that has the same property?
We might think that $\begin{pmatrix} 1 & 1 \\ 1 & 1 \end{pmatrix}$ would multiply another matrix and leave it unchanged.
Investigating this we find

$$\begin{pmatrix} a & b \\ c & d \end{pmatrix} \times \begin{pmatrix} 1 & 1 \\ 1 & 1 \end{pmatrix} = \begin{pmatrix} a+b & a+b \\ c+d & c+d \end{pmatrix},$$

so that does not work.
The matrix we are looking for is the **identity matrix**, $\mathbf{I} = \begin{pmatrix} 1 & 0 \\ 0 & 1 \end{pmatrix}$.
Testing this:

$$\begin{pmatrix} 1 & 0 \\ 0 & 1 \end{pmatrix} \begin{pmatrix} a & b \\ c & d \end{pmatrix} = \begin{pmatrix} a+0 \times c & b+0 \times d \\ 0 \times a + c & 0 \times b + d \end{pmatrix} = \begin{pmatrix} a & b \\ c & d \end{pmatrix}$$

This is an important result because it helps us to find a method for dividing one matrix by another, which brings us on to **Inverse matrices**.

Suppose you need to work out $a \div b$? This can be written $\dfrac{a}{b}$ or $a \times \dfrac{1}{b}$. $\dfrac{1}{b}$ is the inverse of b so $b \times \dfrac{1}{b} = 1$.

The inverse of a matrix **A** is written as $\mathbf{A}^{-1}$, so $\mathbf{A} \times \mathbf{A}^{-1} = \mathbf{I}$, that is the matrix **A** multiplied by its inverse should give the identity matrix.

For example, let $\mathbf{A} = \begin{pmatrix} 2 & 3 \\ 4 & 5 \end{pmatrix}$ and $\mathbf{A}^{-1} = \begin{pmatrix} a & b \\ c & d \end{pmatrix}$.

If $\mathbf{A}^{-1}$ is the inverse of **A** then $\mathbf{AA}^{-1} = \mathbf{I}$.

$$\begin{pmatrix} 2 & 3 \\ 4 & 5 \end{pmatrix} \times \begin{pmatrix} a & b \\ c & d \end{pmatrix} = \begin{pmatrix} 1 & 0 \\ 0 & 1 \end{pmatrix}$$

$$\begin{pmatrix} 2a+3c & 2b+3d \\ 4a+5c & 4b+5d \end{pmatrix} = \begin{pmatrix} 1 & 0 \\ 0 & 1 \end{pmatrix}$$

Comparing each element of the two matrices:

$$2a + 3c = 1 \qquad 2b + 3d = 0$$
$$4a + 5c = 0 \qquad 4b + 5d = 1$$

Solve each pair of equations simultaneously and you will see that $a = \dfrac{-5}{2}$, $b = \dfrac{3}{2}$, $c = 2$, and $d = -1$.

$$\mathbf{A}^{-1} = \begin{pmatrix} \dfrac{-5}{2} & \dfrac{3}{2} \\ 2 & -1 \end{pmatrix}$$

This can also be written as:

$$\mathbf{A}^{-1} = \frac{1}{2}\begin{pmatrix} -5 & 3 \\ 4 & -2 \end{pmatrix} = -\frac{1}{2}\begin{pmatrix} 5 & -3 \\ -4 & 2 \end{pmatrix}$$

Compare with $\mathbf{A} = \begin{pmatrix} 2 & 3 \\ 4 & 5 \end{pmatrix}$.

We have shown one particular numerical form of the general rule for finding inverse matrices.

If $\mathbf{A} = \begin{pmatrix} a & b \\ c & d \end{pmatrix}$, then $\mathbf{A}^{-1} = \dfrac{1}{ad-bc}\begin{pmatrix} d & -b \\ -c & a \end{pmatrix}$.

$ad - bc$ is called the **determinant** of **A** and can be written as **det A** or $|\mathbf{A}|$. Determinants can only be evaluated for square matrices, and here only 2×2 determinants will be considered.

Apart from the determinant, we can see that the elements a and d have changed places, and the elements b and c have remained in the same place but have had their signs changed.

Example 9

Find the determinant of each of these matrices.

(a) $X = \begin{pmatrix} 1 & 3 \\ 5 & 2 \end{pmatrix}$ (b) $Y = \begin{pmatrix} -1 & 0 \\ 2 & 4 \end{pmatrix}$ (c) $Z = \begin{pmatrix} 10 & 3 \\ -7 & -2 \end{pmatrix}$

Answer 9

(a) det $X = 1 \times 2 - 5 \times 3 = -13$ (b) det $Y = -1 \times 4 - 2 \times 0 = -4$

(c) det $Z = 10 \times -2 - {}^-7 \times 3 = -20 + 21 = 1$

If the determinant comes to zero, then the matrix has no inverse because $\dfrac{1}{0}$ does not exist.

Exercise 8.4

Find the determinant of each of these matrices.

1. $P = \begin{pmatrix} 1 & 5 \\ 3 & 20 \end{pmatrix}$ 2. $Q = \begin{pmatrix} 2 & 6 \\ -1 & -2 \end{pmatrix}$ 3. $R = \begin{pmatrix} 9 & 1 \\ 0 & 3 \end{pmatrix}$ 4. $S = \begin{pmatrix} -3 & 1 \\ 4 & 5 \end{pmatrix}$

Having found the determinants we can find the inverse matrices.

Example 10

$X = \begin{pmatrix} 7 & 1 \\ 3 & 4 \end{pmatrix}$ $Y = \begin{pmatrix} -5 & 6 \\ 2 & -8 \end{pmatrix}$ $Z = \begin{pmatrix} 2 & 4 \\ 3 & 6 \end{pmatrix}$

Find (a) X^{-1} (b) Y^{-1} (c) Z^{-1}

Answer 10

(a) det $X = 7 \times 4 - 3 \times 1 = 28 - 3 = 25$

$X^{-1} = \dfrac{1}{25} \begin{pmatrix} 4 & -1 \\ -3 & 7 \end{pmatrix}$ NOTE: It is acceptable to leave the inverse in this form.

(b) det $Y = -5 \times {}^-8 - 2 \times 6 = +40 - 12 = 28$

$Y^{-1} = \dfrac{1}{28} \begin{pmatrix} -8 & -6 \\ -2 & -5 \end{pmatrix}$ NOTE: In this case the entries could be simplified by factorising out the −1.

$Y^{-1} = \dfrac{-1}{28} \begin{pmatrix} 8 & 6 \\ 2 & 5 \end{pmatrix}$

(c) det $Z = 2 \times 6 - 3 \times 4 = 12 - 12 = 0$

Z^{-1} does not exist

Exercise 8.5

$$A = \begin{pmatrix} 7 & 8 \\ 9 & 10 \end{pmatrix} \quad B = \begin{pmatrix} 2 & 1 \\ -1 & -2 \end{pmatrix} \quad C = \begin{pmatrix} -1 & -6 \\ 3 & 10 \end{pmatrix} \quad D = \begin{pmatrix} 0 & 1 \\ 1 & 0 \end{pmatrix}$$

Find
1. A^{-1} 2. B^{-1} 3. C^{-1} 4. D^{-1}

NOTE: D is *not* the identity matrix I, so make sure you can see the difference in the positions of the ones and zeroes.

5. $I = \begin{pmatrix} 1 & 0 \\ 0 & 1 \end{pmatrix}$. Find I^{-1}.

Example 11

(a) $A = \begin{pmatrix} 3 & 4 \\ 1 & 2 \end{pmatrix}$

Find A^{-1}.

(b) Given that $A \begin{pmatrix} x \\ y \end{pmatrix} = \begin{pmatrix} 10 \\ 6 \end{pmatrix}$,

solve $A^{-1} A \begin{pmatrix} x \\ y \end{pmatrix} = A^{-1} \begin{pmatrix} 10 \\ 6 \end{pmatrix}$

(c) Write down the values of x and y.

Answer 11

(a) det $A = 3 \times 2 - 1 \times 4 = 2$

$$A^{-1} = \frac{1}{2} \begin{pmatrix} 2 & -4 \\ -1 & 3 \end{pmatrix}$$

(b) $A \begin{pmatrix} x \\ y \end{pmatrix} = \begin{pmatrix} 10 \\ 6 \end{pmatrix}$

$$A^{-1} A \begin{pmatrix} x \\ y \end{pmatrix} = \frac{1}{2} \begin{pmatrix} 2 & -4 \\ -1 & 3 \end{pmatrix} \begin{pmatrix} 10 \\ 6 \end{pmatrix}$$

$$I \begin{pmatrix} x \\ y \end{pmatrix} = \frac{1}{2} \begin{pmatrix} 2 \times 10 - 4 \times 6 \\ -1 \times 10 + 3 \times 6 \end{pmatrix} = \frac{1}{2} \begin{pmatrix} -4 \\ 8 \end{pmatrix} = \begin{pmatrix} -2 \\ 4 \end{pmatrix}$$

NOTE: Remember that $A^{-1} A = I$

Since $I \begin{pmatrix} x \\ y \end{pmatrix} = \begin{pmatrix} x \\ y \end{pmatrix}$,

then $\begin{pmatrix} x \\ y \end{pmatrix} = \begin{pmatrix} -2 \\ 4 \end{pmatrix}$

(c) $x = -2$, $y = 4$

This example illustrates another use for matrices, which is for solving simultaneous equations.

To understand how this works, go back to part (b) of the example.

$$A\begin{pmatrix} x \\ y \end{pmatrix} = \begin{pmatrix} 10 \\ 6 \end{pmatrix}$$

$$\begin{pmatrix} 3 & 4 \\ 1 & 2 \end{pmatrix}\begin{pmatrix} x \\ y \end{pmatrix} = \begin{pmatrix} 10 \\ 6 \end{pmatrix}$$

Simplifying the left hand side of this equation we have:

$$\begin{pmatrix} 3x + 4y \\ x + 2y \end{pmatrix} = \begin{pmatrix} 10 \\ 6 \end{pmatrix}$$

These two matrices are equal, so the corresponding elements are equal:

$$3x + 4y = 10$$
$$x + 2y = 6$$

Solving these two equations simultaneously gives:

$$x = -2 \quad \text{and} \quad y = 4.$$

Using matrices may not be the quickest way to solve a pair of simultaneous equations, but it leads on to solving multiple simultaneous equations in multiple variables. The methods for these are complicated and not required for your course.

Exercise 8.6

$$E = \begin{pmatrix} -1 & 0 \\ 0 & 2 \end{pmatrix} \quad F = \begin{pmatrix} 7 & 1 & 6 \\ 3 & 5 & 2 \end{pmatrix} \quad G = \begin{pmatrix} 0 & 2 \\ 2 & 0 \end{pmatrix}$$

$$H = \begin{pmatrix} 1 \\ 2 \end{pmatrix} \quad J = \begin{pmatrix} 4 & 6 \end{pmatrix} \quad K = \begin{pmatrix} 1 & -1 \\ 2 & 3 \\ -1 & 0 \end{pmatrix}$$

1. Write down the order of each matrix.
2. Write down the orders of these products.
   (a) **FK**   (b) **KF**   (c) **HJ**   (d) **JH**   (e) **KH**
3. Write down **F′** (F transpose), and the order of **F′**.
4. Calculate
   (a) **EF**   (b) **GF**   (c) **HJ**   (d) **JH**
5. Write down the identity matrix, **I**.
6. Find
   (a) E^{-1} (the inverse of **E**)   (b) G^{-1} (c) EE^{-1} (d) **IE**   (e) **GI**

7. (a) Simplify
$$\begin{pmatrix} a \\ b \end{pmatrix} = \begin{pmatrix} -1 & 0 \\ 0 & -1 \end{pmatrix} \begin{pmatrix} -5 \\ 7 \end{pmatrix}.$$
   (b) Write down the values of a and b.

8. (a) Simplify
$$\begin{pmatrix} p & q \\ r & s \end{pmatrix} = \begin{pmatrix} 0 & -1 \\ 1 & 0 \end{pmatrix} \begin{pmatrix} -2 & 4 \\ 3 & -4 \end{pmatrix}$$
   (b) Write down the values of p, q, r and s.

9. Simplify the following

   (a) $2\begin{pmatrix} 7 \\ 8 \\ 9 \end{pmatrix} - 3\begin{pmatrix} -1 \\ 2 \\ 0 \end{pmatrix}$ (b) $x\begin{pmatrix} 1 \\ 2 \\ 3 \end{pmatrix} + 2\begin{pmatrix} 0 \\ 4x \\ x \end{pmatrix}$

10. Find the values of the letters in these statements

    (a) $\begin{pmatrix} 3 & 2 \\ -5 & 4 \end{pmatrix} = 2\begin{pmatrix} a & b \\ c & d \end{pmatrix}$ (b) $\begin{pmatrix} 4 & 16 \\ 24 & -8 \end{pmatrix} = x\begin{pmatrix} 1 & 4 \\ b & -2 \end{pmatrix}$

    (c) $\begin{pmatrix} 1 & -2 \\ 3 & -4 \end{pmatrix} = y\begin{pmatrix} 3 & -6 \\ 9 & -12 \end{pmatrix}$ (d) $\begin{pmatrix} \frac{1}{2} & \frac{3}{4} \\ r & \frac{-1}{2} \end{pmatrix} = p\begin{pmatrix} 2 & q \\ 28 & -2 \end{pmatrix}$

11. A car sales person sells two types of cars, called Reliable and Gofaster. She sells 10 blue, 5 red and 2 black Reliables, and 5 blue, 16 red and 3 black Gofasters. Reliables retail at $5000 and Gofasters at $6000.
    (a) Copy and complete the two tables below.

	Reliables	Gofasters
Retail value	$5000	-----

	Blue	Red	Black
Reliables	10	-----	-----
Gofasters	-----	16	3

    (b) Copy and complete the two matrices.

    $$\mathbf{R} = (5000 \quad \ldots) \quad \mathbf{N} = \begin{pmatrix} 10 & \ldots & \ldots \\ \ldots & 16 & 3 \end{pmatrix}$$

    Where **R** is the matrix of the retail values and **N** is the matrix of the number of cars sold.

(c) Evaluate **RN**.

(d) Evaluate $\mathbf{RN}\begin{pmatrix} 1 \\ 1 \\ 1 \end{pmatrix}$ (e) What does $\mathbf{RN}\begin{pmatrix} 1 \\ 1 \\ 1 \end{pmatrix}$ represent?

12. Why is it not possible to find the product of

$$\mathbf{A} = \begin{pmatrix} 3 & 5 & 0 \\ -1 & 6 & 7 \end{pmatrix} \quad \mathbf{B} = \begin{pmatrix} 1 & 2 & 3 & 4 \end{pmatrix}?$$

13. Why is it not possible to find the inverse of $\mathbf{C} = \begin{pmatrix} 2 & 6 \\ 3 & 9 \end{pmatrix}$?

Functions

What is a function?

A **function** is a **mapping** of one set of numbers on to another according to some rule. For example, the rule might be 'square and add 2'. If the first set of numbers is {0, 1, 2, 3} the second set would be {2, 3, 6, 11}.
In algebraic notation, this is written as:

$$\mathbf{f}: x \mapsto x^2 + 2,$$

which reads 'the function, f, maps x on to $x^2 + 2$'.
The first set of numbers is called the **domain** (think of 'home'), and the second set is called the **range** (think of going out on to the range).
This can also be shown in a diagram.

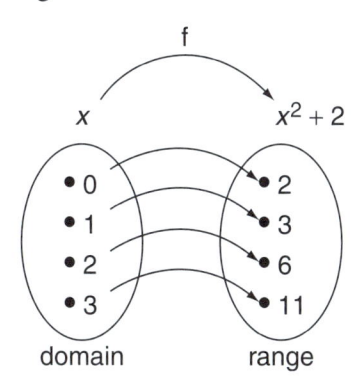

For this mapping diagram the domain is part of the set of integers, $\{x : 0 \leqslant x \leqslant 3, x \in \mathbf{Z}\}$, and there are only four pairs of values. However, if the domain is part of the set of Real numbers, $\{x : 0 \leqslant x \leqslant 3, x \in \mathbf{R}\}$, there are an infinite number of pairs of values in the domain and the function is shown by a line on a graph.

184 Extended Mathematics for Cambridge IGCSE

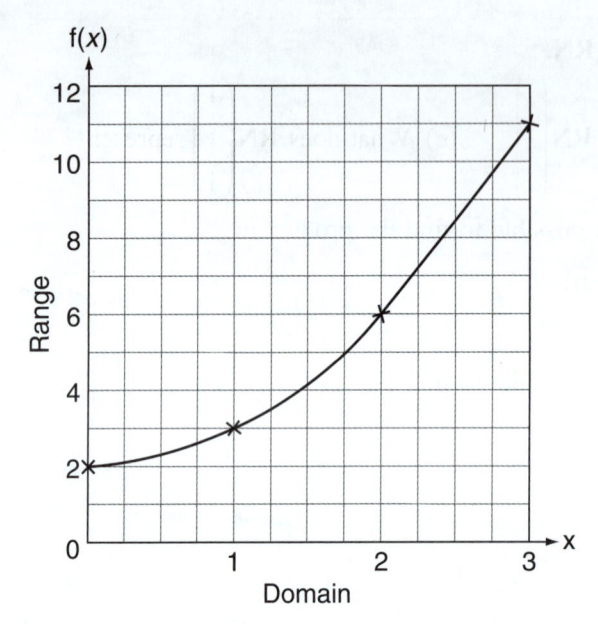

The graph illustrates the range of the function with the given domain.

As you can see the range is from 2 to 11, which is written $\{f(x) : 2 \leqslant f(x) \leqslant 11, f(x) \in \mathbf{R}\}$ and read as 'the set of values of $f(x)$ such that $f(x)$ is greater than or equal to 2 and less than or equal to 11, $f(x)$ is a member of the set of real numbers'.

There is another, very convenient notation for f. We can write $f(x) = x^2 + 2$, which is read 'f of x equals x squared plus two'. It is convenient because we can then write $f(0) = 0^2 + 2 = 2$,

$f(1) = 3$, and so on, as you will see in the next example.

Example 12

Using the domain $\{-2, -1, 0, 1, 2\}$, list the values of $f(x)$ in the range for

(a) $f(x) = x^2 - 2$ (b) $f(x) = (x+1)^2$ (c) $f(x) = (3x - 4)$

Answer 12

Using a mapping diagram

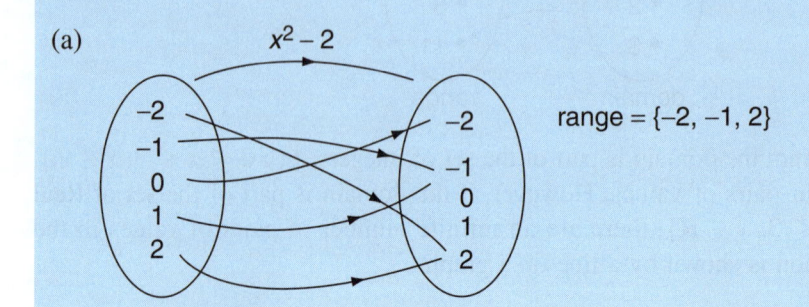

(a)

range = $\{-2, -1, 2\}$

(b)

$(x+1)^2$ mapping with domain $\{-2, -1, 0, 1, 2\}$ to $\{0, 1, 4, 9\}$

range = {0, 1, 4, 9}

(c)

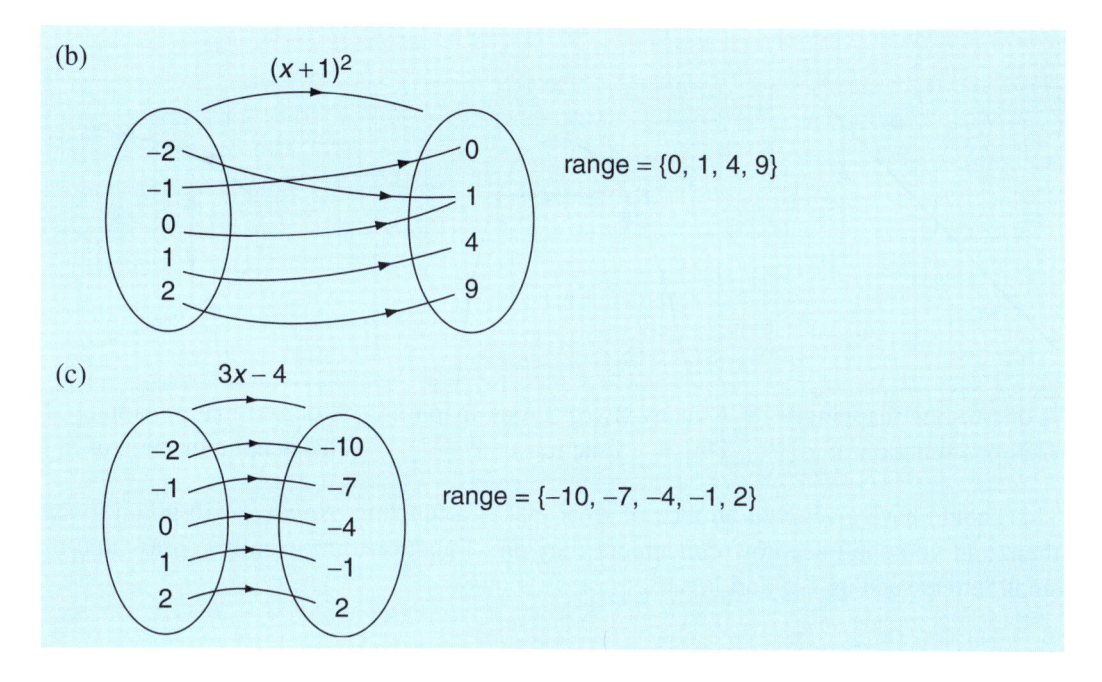

$3x - 4$ mapping with domain $\{-2, -1, 0, 1, 2\}$ to $\{-10, -7, -4, -1, 2\}$

range = {−10, −7, −4, −1, 2}

The last part (c) of the above example shows what is known as a 'one-to-one' mapping. Every member or element of the domain maps to exactly one member of the range.

Parts (a) and (b) show a 'many-to-one' mapping because two members of the domain sometimes map to one member of the range.

To be a proper function every member of the domain must map to one and only one member of the range.

Looking at the diagram below we can see two reasons why this does not represent a proper function.

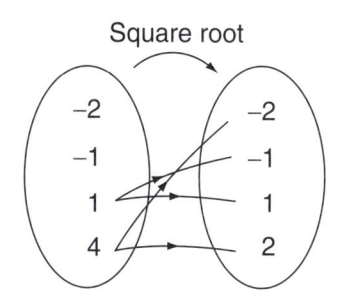

Square root mapping with domain $\{-2, -1, 1, 4\}$ to $\{-2, -1, 1, 2\}$

First of all there are two members of the domain that cannot be mapped to the range, and also there are two members of the domain that map to more than one element of the range.

It is easy to see the different types of mappings in these sketches of graphs.

186   *Extended Mathematics for Cambridge IGCSE*

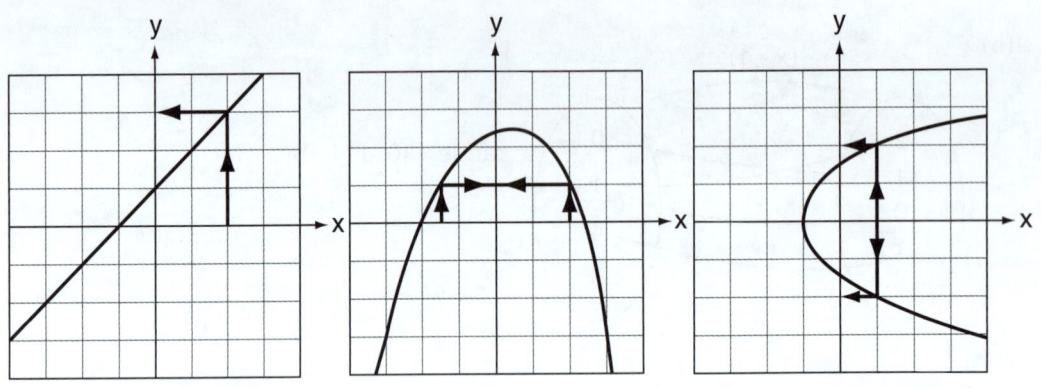

A one-to-one mapping.   A many (two)-to-one mapping.   A one-to-many mapping.
This is a function of *x*.   This is a function of *x*.   This is **not** a function of *x*.

This should have given you an idea of what makes a function. Now we will go on to use them and you shall see how convenient they are. Functions are usually represented by small letters such as f, g and h.

Example 13

If $f(x) = x^3 + x$ and $g(x) = +\sqrt{x^2 + 1}$, find

(a) f(1), f(4) and f(−5)   (b) g(1), g(0) and g(7).

Answer 13

(a) $f(1) = 1^3 + 1 = 2$

$f(4) = 4^3 + 4 = 68$

$f(-5) = -125 - 5 = -130$

(b) $g(1) = +\sqrt{1^2 + 1} = +\sqrt{2}$

$g(0) = +\sqrt{1} = 1$

$g(7) = +\sqrt{7^2 + 1} = +\sqrt{49 + 1} = +\sqrt{50} = +\sqrt{25 \times 2} = +\sqrt{25} \times \sqrt{2} = +5\sqrt{2}$

Example 14

If $f(x) = x^2 + 3x$, find *x* when $f(x) = 10$.

Answer 14

$x^2 + 3x = 10$

$x^2 + 3x - 10 = 0$

$(x + 5)(x - 2) = 0$

either $x = -5$ or $x = 2$

From now on the domains will all be from the set of Real Numbers unless otherwise stated. However, you might come across an extra restriction in the description of the domain. Take, for example,

$$f(x) = \frac{1}{x-2}, x \in \mathbf{R}, x \neq 2.$$

The value $x = 2$ has been specifically excluded from the domain because $f(2) = \frac{1}{2-2} = \frac{1}{0}$. As you know, you cannot divide by zero, so $f(2)$ does not exist and $x = 2$ must be excluded from the domain if $f(x)$ is a function. An alternative restriction might be to make $x > 2$, so that you eliminate 2 and all values less than 2.

Exercise 8.7

1. $f(x) = \dfrac{x}{x^2 + 2}$,

    Find   (a) f(1)   (b) f(−1)   (c) $f\left(\dfrac{1}{2}\right)$ (d) f(0)

2. $f(x) = 2x^2 + x - 1$.
   $g(x) = x^2 - 5x - 6$.
   Solve $f(x) = g(x)$.

3. $h(x) = \dfrac{x}{6-x}, x \neq 6$.

    Find   (a) h(1)   (b) h(1.2)   (c) $h\left(2\dfrac{3}{4}\right)$

4. $g(x) = x^3 + 5$
   Find x when $g(x) = -22$.

5.

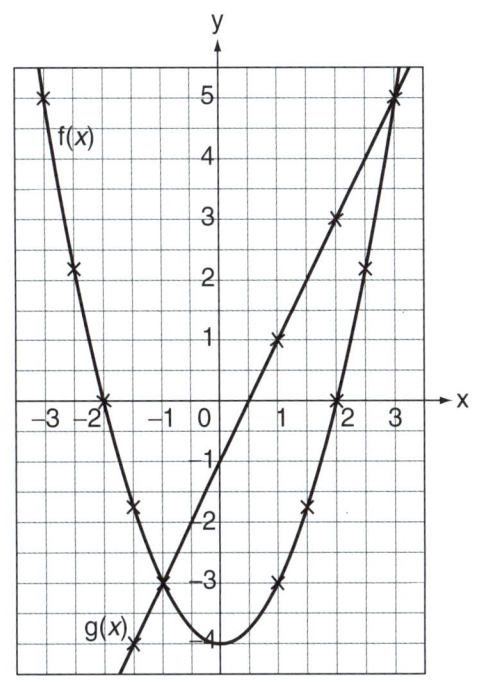

Using the graph evaluate the following
(a) f(2) (b) f(−2) (c) g(x) = 0   NOTE: Find x if g(x) = 0.
(d) g(x) = f(x) (e) f(0) (f) g(0)

Inverse Functions

As usual we have to be able to work backwards to undo anything we have just done. For example, the mapping of f(x) = 2x + 1 on the domain {2, 3, 4} can be shown as:

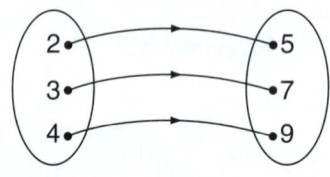

$$2 \to 5$$
$$3 \to 7$$
$$4 \to 9$$

But how do we get back from 5 to 2, from 7 to 3, and from 9 to 4?
The method is quite simple.
- Write y = 2x + 1.
- Rearrange to make x the subject.
- Change x to f⁻¹(x), and y to x.

The **inverse** of the function f(x) is written as f⁻¹(x). Going through the steps:
f(x) = 2x + 1
- y = 2x + 1
- $\dfrac{y-1}{2} = x$
- $x = \dfrac{y-1}{2}$
- $f^{-1}(x) = \dfrac{x-1}{2}$

Example 15

Find the inverse of (a) $f(x) = \dfrac{1}{x-2}, x > 2$ (b) $g(x) = \dfrac{x+5}{x}, x > 0$

Answer 15

(a) $y = \dfrac{1}{x-2}$ (b) $y = \dfrac{x+5}{x}$

y(x − 2) = 1   yx = x + 5

continued overleaf   *continued overleaf*

(a) $yx - 2y = 1$
contd $yx = 1 + 2y$

$$x = \frac{1 + 2y}{y}$$

$$f^{-1}(x) = \frac{1 + 2x}{x}$$

(b) $yx - x = 5$
contd $x(y - 1) = 5$

$$x = \frac{5}{y - 1}$$

$$g^{-1}(x) = \frac{5}{x - 1}$$

Exercise 8.8

Find the inverses of the following functions

1. $f(x) = 3x + 1$
2. $f(x) = 2 - 5x$
3. $f(x) = \frac{1}{2}x + 1$
4. $f(x) = \frac{3}{4}(x + 2)$
5. $f(x) = \frac{5 - x}{2}$
6. $g(x) = \sqrt{x - 3},\ x \geq 3$
7. $g(x) = x^3 - 1$
8. $g(x) = \frac{1}{x + 1},\ x \neq -1$
9. $f(x) = \frac{x}{x + 1},\ x \neq -1$
10. $g(x) = 3\left(x - \frac{1}{2}\right)$

Example 16

$$f(x) = \frac{x + 1}{x - 4},\ x \neq 4$$

Find (a) $f(2)$ (b) $f(-3)$ (c) $f^{-1}(x)$ (d) $f^{-1}(3)$ (e) $f^{-1}(-1)$

Answer 16

(a) $f(2) = \frac{2 + 1}{2 - 4}$

$f(2) = \frac{3}{-2}$

$f(2) = \frac{-3}{2}$

(b) $f(-3) = \frac{-3 + 1}{-3 - 4}$

$f(-3) = \frac{-2}{-7}$

$f(-3) = \frac{2}{7}$

(c) $f(x) = \dfrac{x+1}{x-4}$ (d) $f^{-1}(3) = \dfrac{1+4\times 3}{3-1}$ (e) $f^{-1}(-1) = \dfrac{1+4\times {}^-1}{-1-1}$

$y = \dfrac{x+1}{x-4}$

$f^{-1}(3) = \dfrac{13}{2}$

$= \dfrac{-3}{-2}$

$yx - 4y = x + 1$

$= \dfrac{-3}{-2}$

$yx - x = 1 + 4y$

$f^{-1}(-1) = \dfrac{3}{2}$

$x(y - 1) = 1 + 4y$

$x = \dfrac{1+4y}{y-1}$

$f^{-1}(x) = \dfrac{1+4x}{x-1}$

An alternative method for (d) and (e) would be to solve for $f(x) = 3$ and $f(x) = -1$.

Composite Functions

Composite functions occur when one function is followed by another as in this diagram.

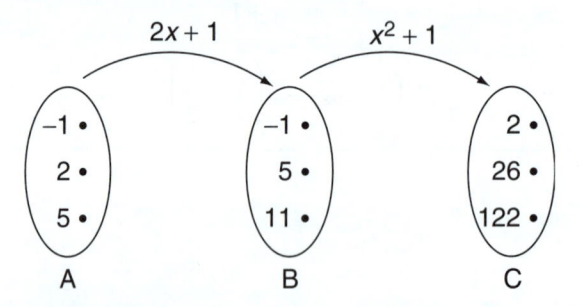

Mapping diagram: Set A {−1, 2, 5} maps via $2x+1$ to Set B {−1, 5, 11}, which maps via x^2+1 to Set C {2, 26, 122}.

If $f: x \longmapsto 2x + 1$, and $g: x \longmapsto x^2 + 1$,

Can we find a single function that will map set A directly onto set C?

$$f(x) = 2x + 1, \qquad g(x) = x^2 + 1$$

The notation gof or simply gf is used to indicate 'g after f', so gf(x) means the function f first and then the function g. This could also be written as g(f(x)) which is read as 'g of f of x'.

Using the example in the mapping diagram,

$f(x) = 2x + 1$ so $g(f(x)) = g(2x + 1)$

$g(x) = x^2 + 1$ so $g(2x + 1) = (2x + 1)^2 + 1$

$= 4x^2 + 4x + 1 + 1$

$= 4x^2 + 4x + 2$

$gf(x) = 4x^2 + 4x + 2$

To check this we find gf(–1), gf(2) and gf(5).
$$gf(-1) = 4 \times (-1)^2 + 4 \times (-1) + 2$$
$$= 2$$
$$gf(2) = 4 \times 2^2 + 4 \times 2 + 2$$
$$= 26$$
$$gf(5) = 4 \times 5^2 + 4 \times 5 + 2$$
$$= 122.$$
These agree with the mapping diagram so $gf(x) = 4x^2 + 4x + 2$ is correct.

Example 17
If $f(x) = 2x + 1$ and $g(x) = x^2 + 1$, find
(a) fg(x)　　(b) fg(–1)　　(c) fg(2)　　(d) fg(5)

Answer 17

(a) $fg(x) = f(g(x)) = f(x^2 + 1)$　　(b) $fg(-1) = 2 \times (^-1)^2 + 3$
　　$f(x^2 + 1) = 2(x^2 + 1) + 1$　　　　　　$= 5$
　　$fg(x) = 2x^2 + 3$

(c) $fg(2) = 2 \times 2^2 + 3$　　(d) $fg(5) = 2 \times 5^2 + 3$
　　$= 11$　　　　　　　　　　　$= 53$

The above example shows that the order does matter when calculating composite functions, and that in general, $fg(x) \neq gf(x)$.

The diagram might make the process clearer.

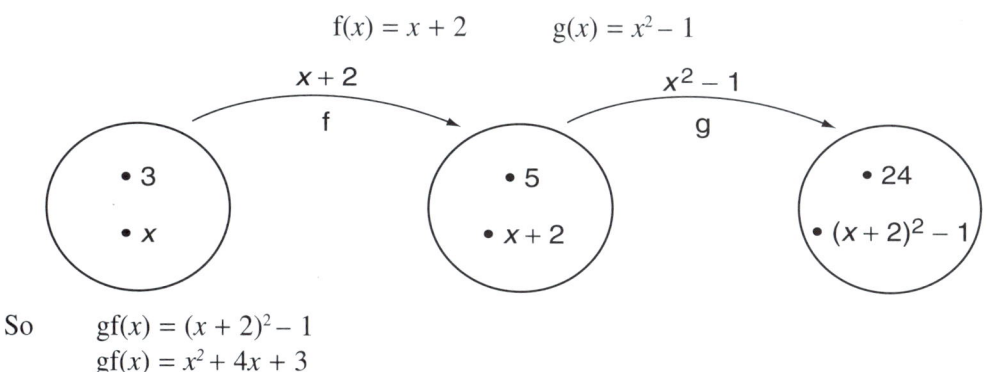

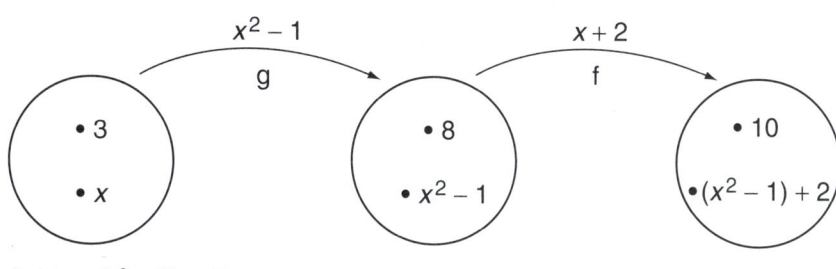

$$f(x) = x + 2 \qquad g(x) = x^2 - 1$$

So　$gf(x) = (x + 2)^2 - 1$
　　$gf(x) = x^2 + 4x + 3$

$fg(x) = (x^2 - 1) + 2$
$fg(x) = x^2 + 1$

Exercise 8.9

1. $f(x) = 3x - 1$
   $g(x) = 5x + 2$, find
   (a) $fg(x)$
   (b) $gf(x)$
   (c) $fg(2)$
   (d) $gf(-1)$

2. $g(x) = 2x^2 + 3$
   $h(x) = x^2 + 1$, find
   (a) $gh(3)$ NOTE: For part (a) find h(3) first and then find g of the result.
   (b) $hg(3)$ NOTE: Find g(3) first.

3. $f(x) = x + 4$
   $g(x) = x^2$
   (i) Find
       (a) $fg(x)$
       (b) $gf(x)$
       (c) $fg(10)$
   (ii) Solve $fg(x) = gf(x)$

4. $g(x) = x^2 + 2$
   $h(x) = \dfrac{1}{x^2 + 2}$ find
   (a) $gh(1)$
   (b) $hg(1)$

5. $f(x) = 3x$
   $g(x) = x + 3$, find
   (a) $fg(x)$
   (b) $gf(x)$
   (c) $ff(x)$
   (d) $gg(x)$
   (e) $g^{-1}(x)$
   (f) $gg^{-1}(x)$
   (g) $f^{-1}(x)$
   (h) $f^{-1}f(x)$

6. $f(x) = x^2 + 1$, find
   (a) $ff(x)$
   (b) $fff(x)$

Exercise 8.10

1. If $f(x) = 2x^2 + 3x - 1$, find
   (a) $f(0)$
   (b) $f(-1)$
   (c) x when $f(x) = -2$

2. $f(x) = x^2 + 1$
   $g(x) = \dfrac{1}{x+1}$, $x \neq -1$. Find, simplifying where necessary
   (a) $gf(x)$
   (b) $gg(x)$
   (c) $fg(-3)$
   (d) $fg\left(\dfrac{1}{2}\right)$
   (e) $g^{-1}(x)$

3. $h(x) = \dfrac{x+2}{x+3}$, $x \neq -3$, find
   (a) $h\left(\dfrac{1}{2}\right)$
   (b) $h\left(\dfrac{3}{5}\right)$
   (c) $h^{-1}(x)$
   (d) x when $h(x) = 0$
   (e) $h^{-1}(-2)$
   (f) $h^{-1}(0)$
   (g) x when $h(x) = -2$

Graphs of Inequalities

We have already found solutions to linear inequalities in one variable, and shown the results on a number line or one dimensional graph. We have also illustrated linear equations in two variables as straight lines on graphs.

We now need to illustrate linear inequalities in two variables on graphs. You will see that they have to be shown as areas on graphs.

Taking a simple example, $y \geqslant x + 1$, we can find pairs of values for x and y from the set of real numbers which satisfy this inequality. We will pick a few at random.

	$4 > 1 + 1$	so $x = 1, y = 4$	satisfies the inequality
and	$2.5 > 0 + 1$	$x = 0, y = 2.5$	
	$-1 > -3 + 1$	$x = -3, y = -1$	
	$-1.5 > -3 + 1$	$x = -3, y = -1.5$	
	$1 > -2 + 1$	$x = -2, y = 1$	
	$3.5 > -1 + 1$	$x = -1, y = 3.5$	all also satisfy the inequality.

If we draw the line $y = x + 1$ on a graph and then plot these points we will see in which region of the graph the points lie.

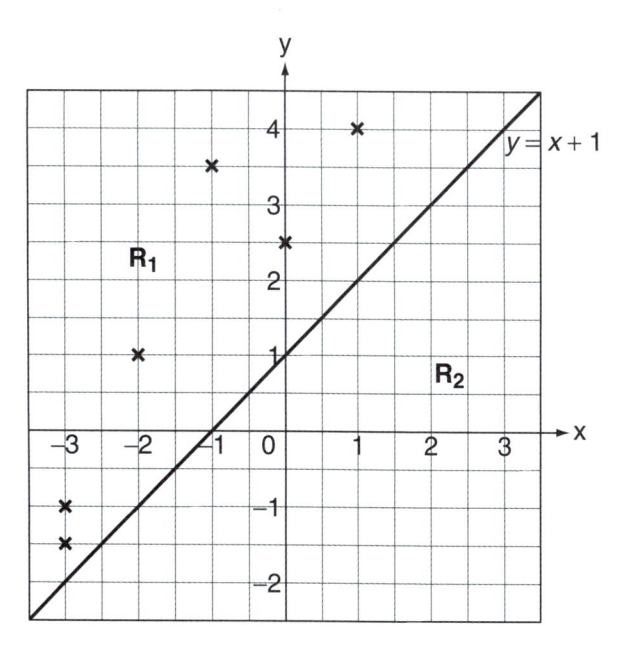

The line $y = x + 1$ joins all the points with x and y coordinates which satisfy the equation. So the y-coordinate of every point is equal to its x-coordinate plus one. The line also divides the graph into two regions, marked R_1 and R_2. In the region R_1, all the y-coordinates are greater than $x + 1$, and in the region R_2 all the y-coordinates are less than $x + 1$.
The graph is redrawn to make this clear.

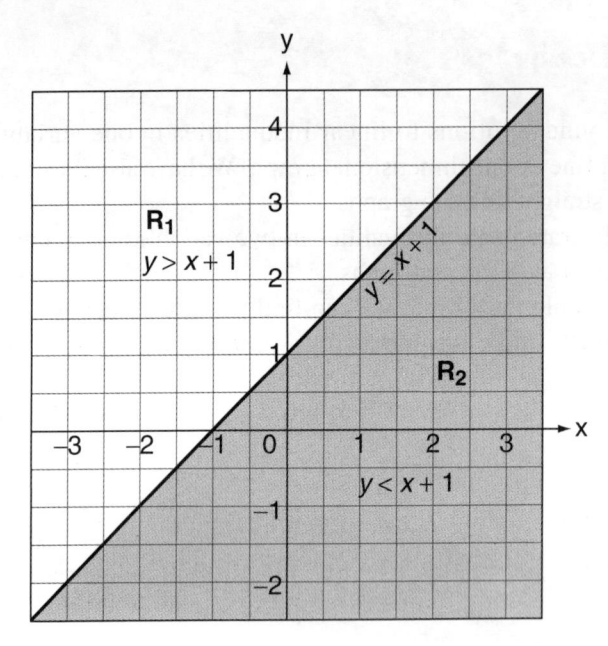

The inequality we are illustrating is $y > x + 1$, so the line $y = x + 1$ is included. The region R_2 is not included, and is shaded to show that it is unwanted.

For a strict inequality, for example, $y < -x + 1$, we first draw the line $y = -x + 1$, but because the line is not included (< rather than $\leqslant$) we draw a broken line as shown below.

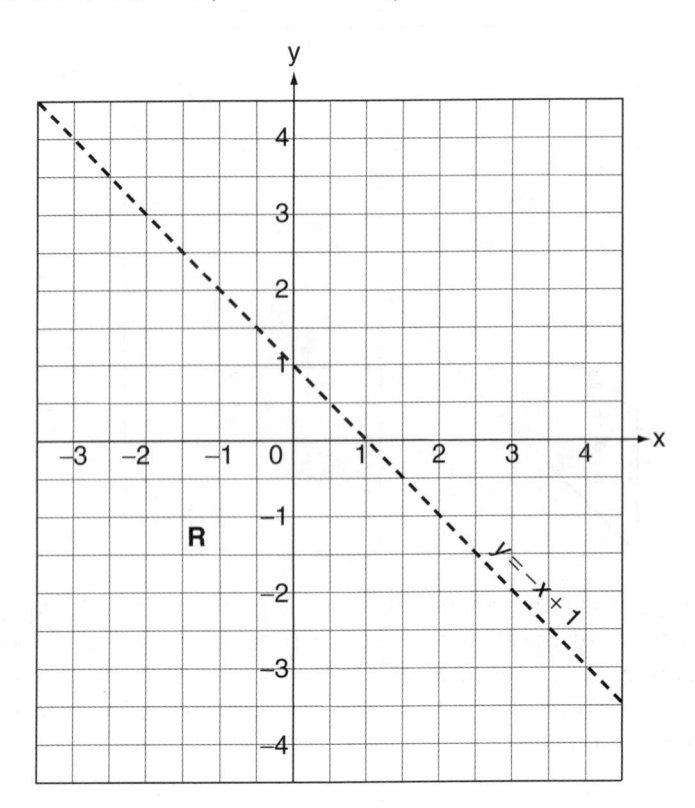

Further Algebra 195

We now have to decide which side of the line is the region we want, so we test a point which is not on the line. The origin (0, 0) is convenient.

$$y < x + 1$$
$$0 < 0 + 1$$

This is true: zero *is* less than zero plus one, so the origin lies in the region we want and we can mark **R** on that side of the line.

NOTE: Although it is easy to see which side of the line we require when the inequality is written in this form, it is always better to test with a point as above. Sometimes the inequality is written in a form which is deceptive.

The region R will usually be shown by shading the other side of the line, the unwanted region. You should read the question carefully to decide which side of the line to shade.

Example 18

By shading the **unwanted** side of the line, show the regions defined by these inequalities.

(a) $y \leqslant 2x + 1$ (b) $x < 2$ (c) $y \geqslant 0$

Answer 18

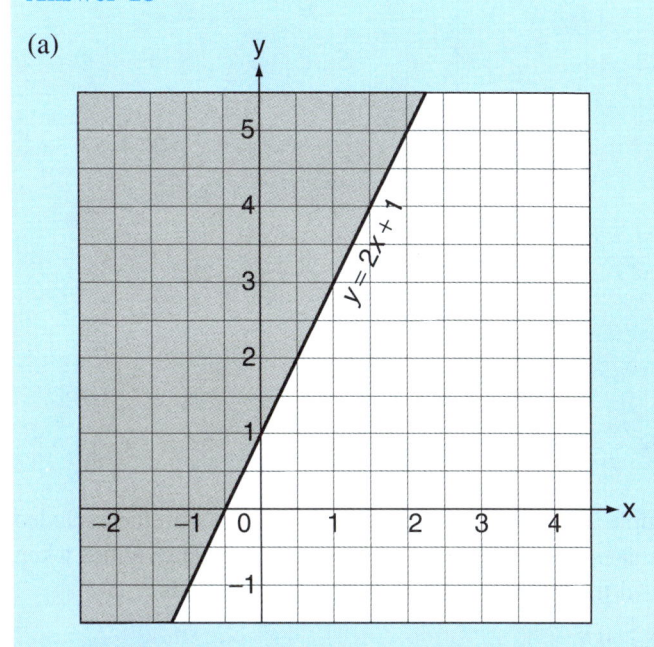

(a)

(b)

(c)

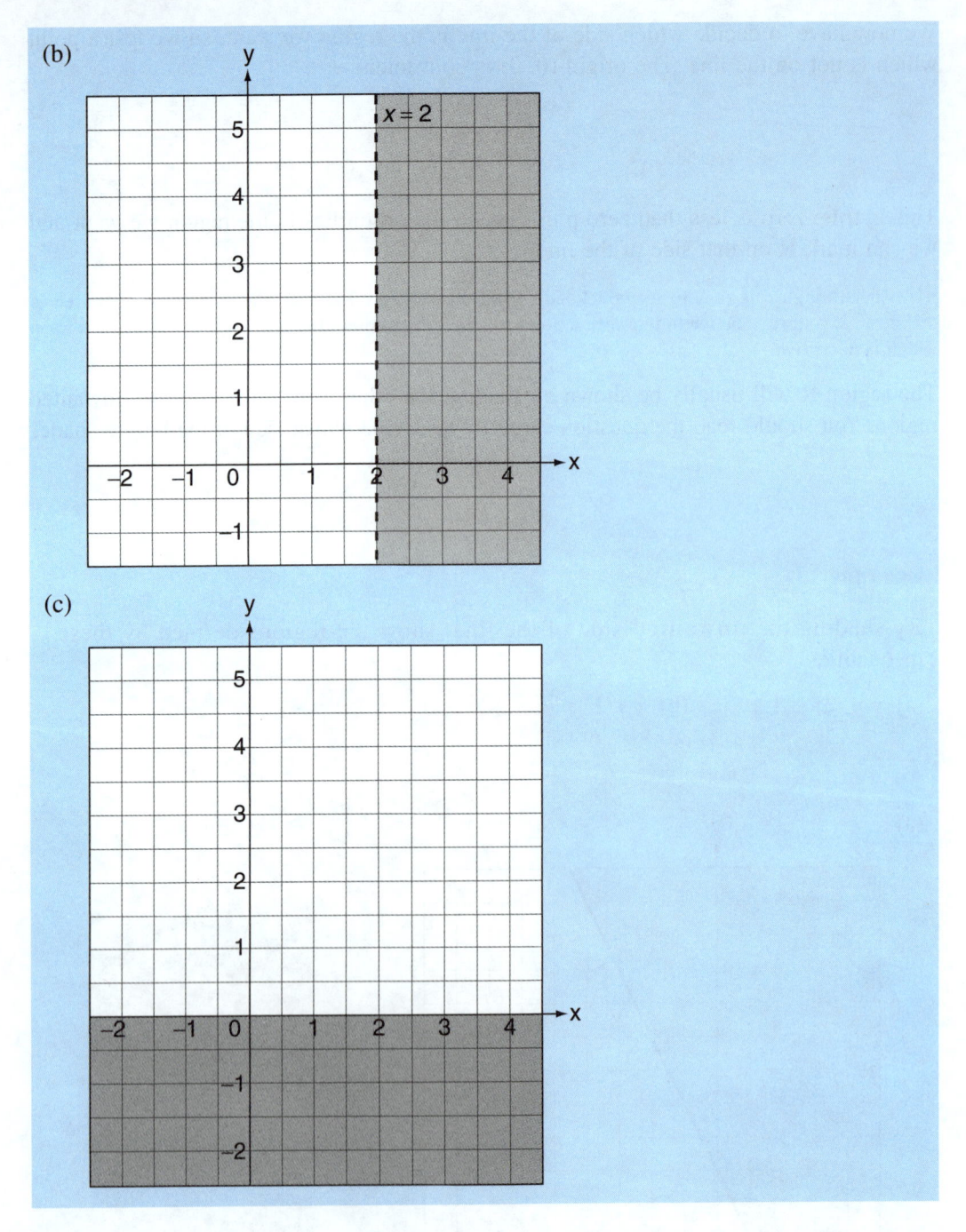

If all three inequalities in the example above are now drawn on one graph with the excluded regions shaded, the remaining unshaded region is the solution of the three inequalities taken simultaneously as shown below. All the points in the unshaded region satisfy

$y \leqslant 2x + 1$, $x < 2$ and $y \geqslant 0$

at the same time.

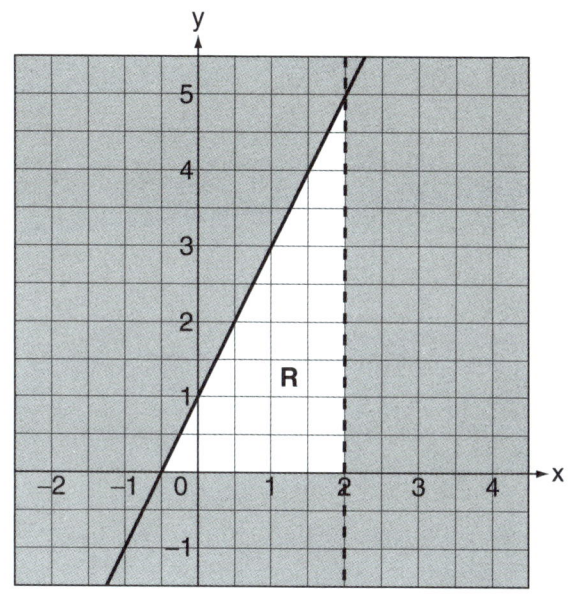

R represents the region defined by the inequalities.

Example 19
(a) Show the region defined by the inequalities
 $x + y < 5 \quad x \geqslant 0 \quad y \geqslant 1$ on a diagram. Shade the unwanted region.
(b) Mark with a cross on the diagram, all the points which have integer values of x and y, and are in the required region. That is $\{(x, y) : x, y \in \mathbf{Z}\}$
(c) List these points.

Answer 19
(a, b)

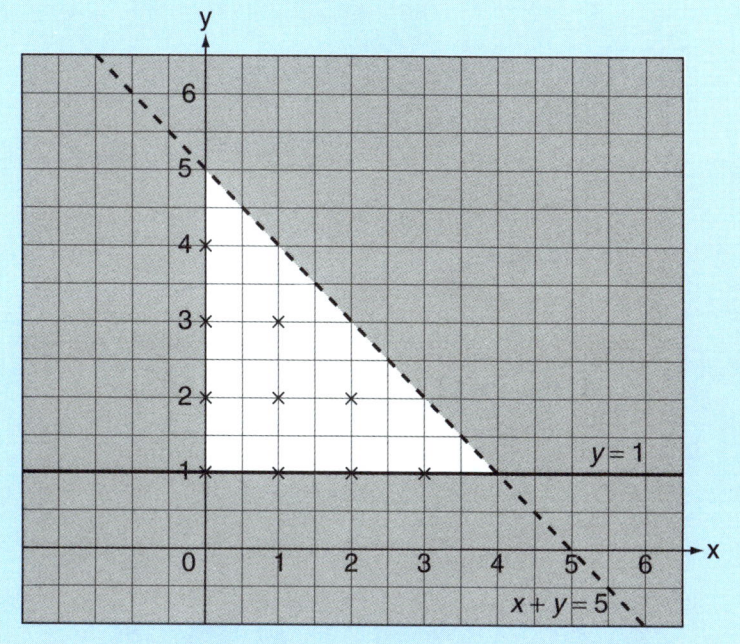

(c) (0, 1),   (0, 2),   (0, 3),   (0, 4),   (1, 1),   (1, 2),
     (1, 3),   (2, 2),   (2, 1),   (3, 1).

NOTE: Points such as (1, 4) are not included in the region because they are on the broken line representing the strict inequality $x + y < 5$.

Example 20

Show on a diagram the region which satisfies the inequalities

$-3 \leqslant x < 2,$ $\qquad$ $2 < y \leqslant 5.$

Shade the unwanted region.

Answer 20

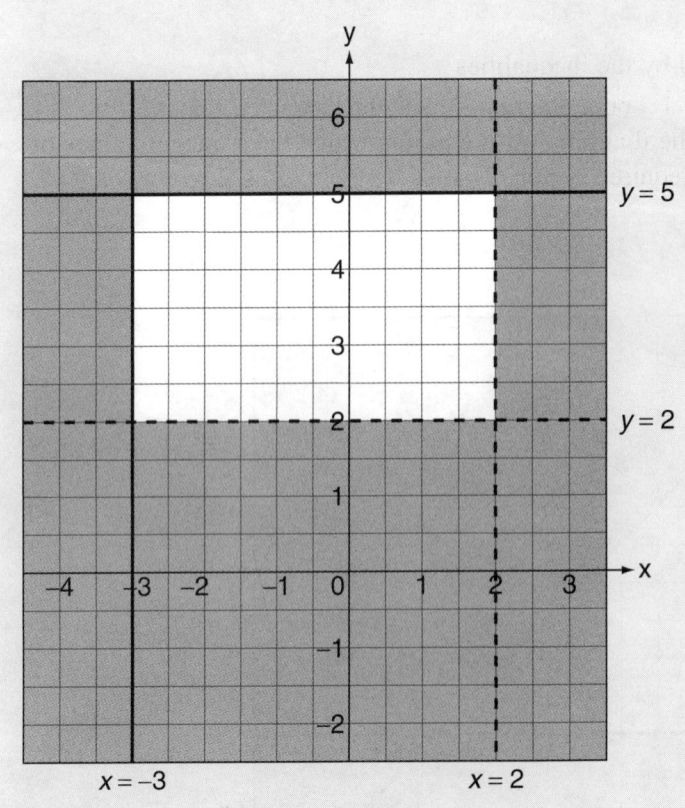

Example 21

Define the region **R** using simultaneous inequalities.

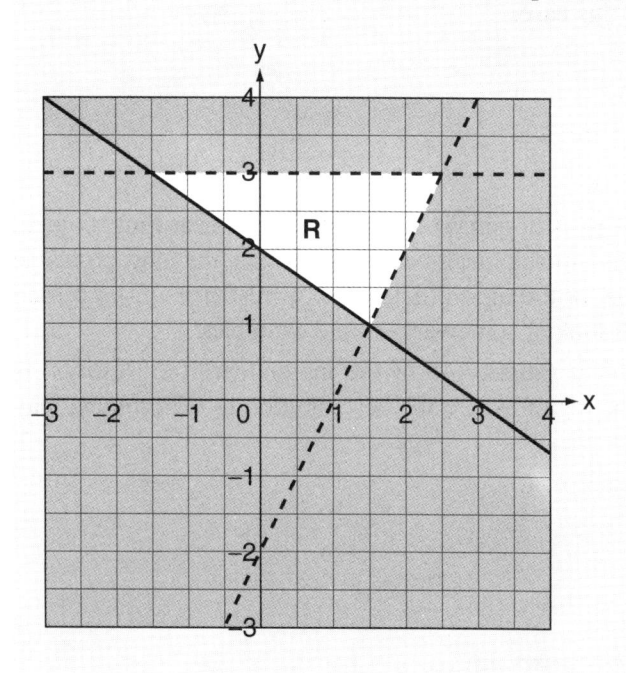

Answer 21

The lines surrounding **R** on the graph are $y = 3$, $y = \frac{-2}{3}x + 2$ and $y = 2x - 2$.

It is clear that the required side of the line $y = 3$ is $y < 3$.

Testing $(0, 0)$ for the line $y = \frac{-2}{3}x + 2$, $0 < 0 + 2$.

We want the side which does not contain the origin so the required inequality is $y \geqslant -\frac{2}{3}x + 2$.

For the line $y = 2x - 2$, testing the origin, $0 > 0 - 2$. For this line we want the side which *does* contain the origin so the required inequality is $y > 2x - 2$. Remember that the broken line represents a strict inequality.

The region **R** is defined by:

$$y < 3 \quad y \geqslant \frac{-2}{3}x + 2 \quad y > 2x - 2$$

Extended Mathematics for Cambridge IGCSE

Strictly speaking, the definition should also give the universal set of numbers. If $(x, y) \in Z$, we are only interested in integer values of x and y. If $(x, y) \in R$, we are interested in the entire area, as in this case.

Linear Programming

The intersection of these areas is the basis of **Linear Programming**. Linear Programming can be used by business and commerce to help determine, for example, the most profitable combination of items to manufacture, or the most efficient use of resources given certain restrictions, or **constraints**. We will look at some very simple examples.

In linear programming the region that satisfies all of the inequalities (constraints) is called the **feasible region**, and is the region where the best solution to the problem may be found.

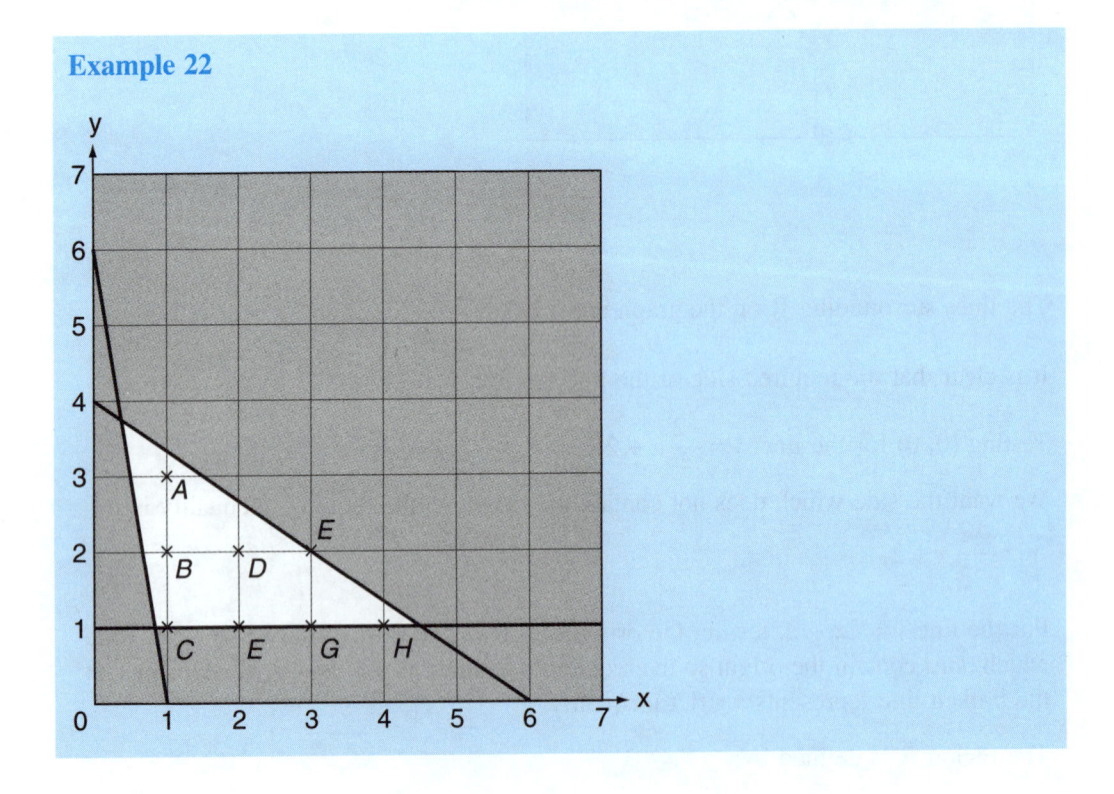

Example 22

The graph shows the feasible region for the manufacture of certain items. We will call them item x and item y. Crosses on the graph mark the integer values of x and y within the feasible region.

(a) List the coordinates of the marked points.

(b) Find the most profitable combination of items to manufacture if the profits on x and y are

 (i) profit on x = $2, profit on y = $3,

 (ii) profit on x = $1, profit on y = $4,

 (iii) profit on x = $5, profit on y = $3.

Answer 22

(a) $A(1, 3)$ $B(1, 2)$ $C(1, 1)$ $D(2, 2)$ $E(2, 1)$

  $F(3, 2)$ $G(3, 1)$ $H(4, 1)$

(b) (i) A: total profit = $1 \times 2 + 3 \times 3 = \$11$

  B: total profit = $1 \times 2 + 2 \times 3 = \$8$

  Using the same method:

  $C = \$5$ $D = \$10$ $E = \$7$ $F = \$12$

  $G = \$9$ $H = \$11$

F shows the most profitable combination of items. That is 3 of item x and 2 of item y, giving a total profit of $12.

  (ii) $A = \$13$ $B = \$9$ $C = \$5$ $D = \$10$

  $E = \$6$ $F = \$11$ $G = \$7$ $H = \$8$

A shows the most profitable combination. That is 1 of item x and 3 of item y, with a total profit of $13.

  (iii) $A = \$14$ $B = \$11$ $C = \$8$ $D = \$16$

  $E = \$13$ $F = \$21$ $G = \$18$ $H = \$23$

H is the most profitable combination. That is 4 of item x and 1 of item y, total profit $23.

This example shows how the best combination of items and thus the total profits can vary according to the profits on individual items.

Exercise 8.11

1.

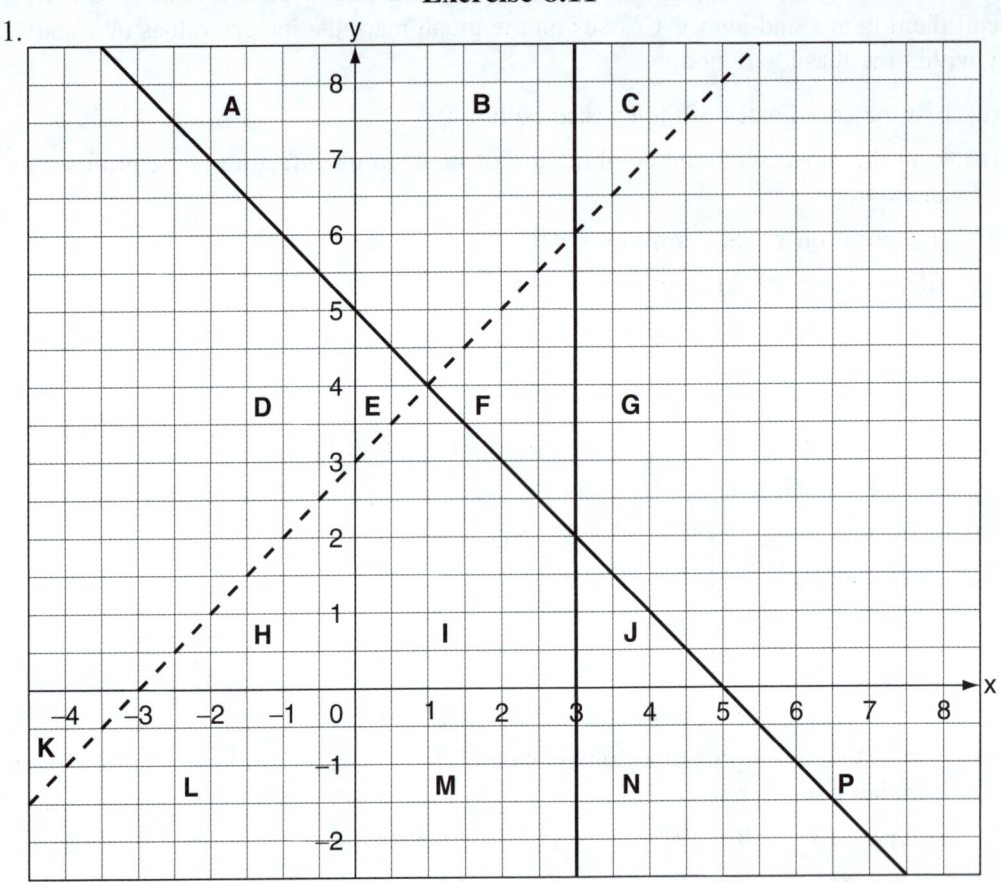

The lines $x = 0$, $x = 3$, $y = 0$, $y = x + 3$ and $x + y = 5$ divide the graph into 15 separate regions. Use inequalities to define the following regions

(a) **E** (b) **J** (c) **I** (d) **N** (e) **G** and **P** together
(f) **A**, **B** and **C** together (g) **D** and **K** together.

NOTE: The regions may not all be 'closed'. For example, region A is bounded only on two sides.

2. Show the solutions of these sets of inequalities on a graph, shading the unwanted regions and labelling the required region with an **R**.

   (a) $y \geq 0$, $x + y \geq 4$, $x \geq 0$
   (b) $4x + 3y < 12$, $y \leq 2x$, $y \geq 0$

3. (a) Show the solution set of these inequalities on a graph by shading the unwanted regions.

   $y < 3x + 6$, $0 < y < 4$, $x \leq 0$.

   (b) Mark with a cross all the integer solutions.
   (c) List the points which are integer solutions.

4. A firm has to manufacture two types of tractors: the Mini and the Maxi.
   They already have orders for 4 Maxi tractors and 1 Mini tractor. They can only manufacture 10 tractors in a month, and they want to make more Maxis than Minis.

Let the number of Mini tractors be x and the number of Maxi tractors be y.
(a) Write down four inequalities to show the above constraints.
(b) Draw a graph and shade the unwanted regions.
(c) Mark with crosses all the points with integer values of x and y which lie in the feasible region.
(d) The profit on the Mini tractors is $1500 and on the Maxi tractors is $2000. Determine the most profitable combination of tractors the firm should manufacture in the month and write down the total profit.

5. A car dealer sells Reliable and Gofaster cars. There is room at the showrooms for no more than 10 cars. The dealer knows that more Reliables will be sold than Gofasters, so he needs to have more Reliables than Gofasters in stock. However, the dealer wants to have at least 3 Gofasters to show potential customers.
Let the number of Reliables be x and the number of Gofasters be y.

(a) Write down three inequalities in x and y which define the information above.
(b) Copy the diagram and shade the unwanted regions.
(c) Mark with a cross the integer solutions in the required region.

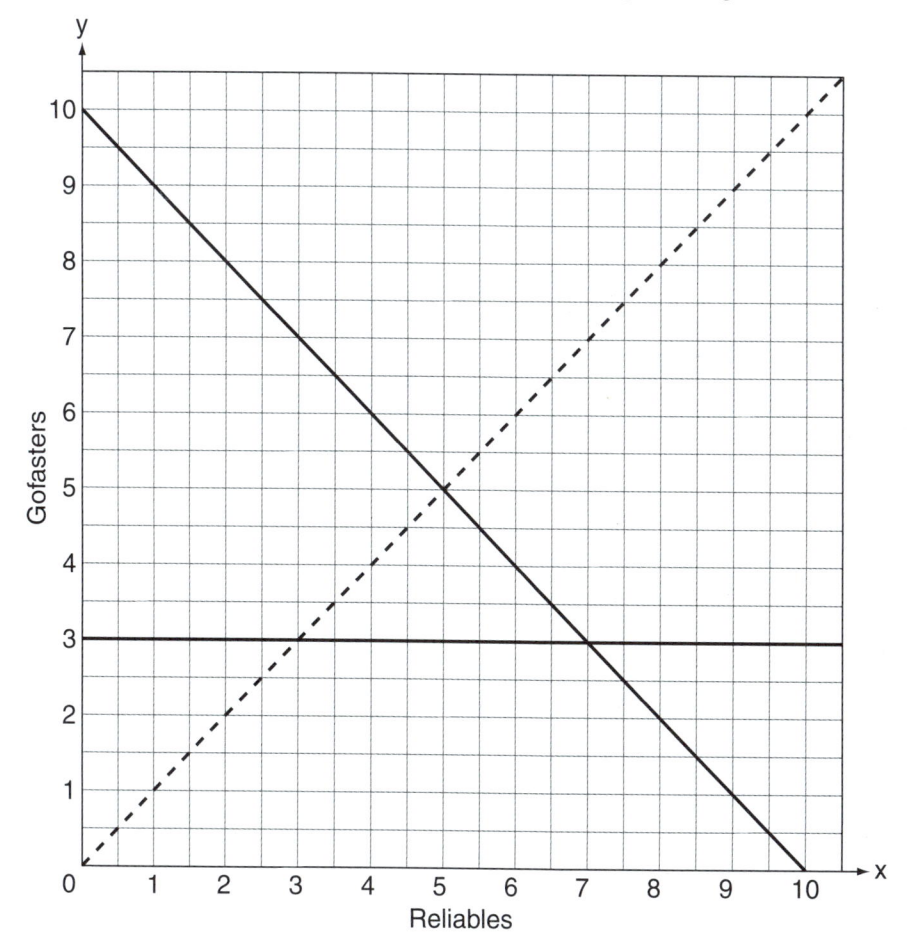

(d) The profit on Reliables is $1500 and the profit on Gofasters is $1800.

   (i) Copy and complete the two matrices below, showing the number of cars in the required region, and the profit on each model. The first matrix will need as many rows as the crosses in your graph.

$$\begin{pmatrix} x & y \\ & \\ & \end{pmatrix} \begin{pmatrix} 1500 \\ 1800 \end{pmatrix}$$

$$\ldots \times 2 \qquad 2 \times 1$$

   (ii) By carrying out the matrix multiplication in the order shown, find out which combination of Reliables and Gofasters would make the largest profit.

Exercise 8.12

Mixed Exercise

1. $\mathbf{A} = \begin{pmatrix} 1 & 0 \\ 0 & 1 \end{pmatrix}$ $\mathbf{B} = \begin{pmatrix} 2 & 1 \\ 3 & -1 \end{pmatrix}$ $\mathbf{C} = \begin{pmatrix} 1 & 3 & -5 \\ 2 & 0 & 4 \end{pmatrix}$

    $\mathbf{D} = (3 \ -4)$ $\mathbf{E} = \begin{pmatrix} 1 & 2 \\ 3 & 4 \\ 5 & 6 \end{pmatrix}$ $\mathbf{F} = \begin{pmatrix} 1 \\ 2 \end{pmatrix}$

(a) Write down the order of each matrix.
(b) Write down the orders of the products.
   (i) **CE**   (ii) **EC**   (iii) **DF**   (iv) **FD**
(c) Write down C transpose (**C′**).
(d) Find
   (i) Det **A**   (ii) **AB**   (iii) **BA**   (iv) $\mathbf{A}^{-1}$ (v) $\mathbf{B}^{-1}$

2. $f(x) = \dfrac{1}{x^2},$ $g(x) = \dfrac{1}{x},$ $h(x) = x^2 + 1$

  (a) Find
    (i) $f(-1)$ (ii) $h(0)$ (iii) x when $g(x) = 0.25$
    (iv) $g^{-1}(x)$ (v) $g^{-1}\left(\dfrac{1}{4}\right)$

  (b) Find
    (i) $fg(x)$ (ii) $gf(x)$ (iii) $fh(3)$ (iv) $fgh(x)$

3. (a) $f(x) = \dfrac{1}{2x+3}$. Find $f^{-1}(x)$.

   (b) $g(x) = \dfrac{x}{2x+3}$. Find $g^{-1}(x)$.

4. (a) Show, by shading the unwanted regions, the solution set of these inequalities on a graph.
   $-1 \leqslant x < 4$
   $x + y < 5$
   $0 \leqslant y < 3$

   (b) List all the points which are integer solutions to the set of inequalities.

Examination Questions

5. A shop buys x pencils and y pens.
   Pencils cost 15 cents each and pens cost 25 cents each.
   (a) There is a maximum of $20 to spend.
       Show that $3x + 5y \leqslant 400$.
   (b) The number of pens must not be greater than the number of pencils.
       Write down an inequality, in terms of x and y, to show this information.
   (c) There must be at least 35 pens.
       Write down an inequality to show this information.
   (d) (i) Using a scale of 1 cm to represent 10 units on each axis, draw an x-axis for $0 \leqslant x \leqslant 150$ and a y-axis for $0 \leqslant y \leqslant 100$.
       (ii) Draw three lines on your graph to show the inequalities in **parts (a), (b), and (c)**. Shade the **unwanted** regions.
   (e) When 70 pencils are bought, what is the largest possible number of pens?
   (f) The profit on each pencil is 5 cents and the profit on each pen is 7 cents.
       Find the largest possible profit.

   (0580/04 May/June 2004 q 9)

6. (a) The determinant of the matrix $\begin{pmatrix} k & 5 \\ -1 & 2 \end{pmatrix}$ is 14. Find k.

   (b) Find the inverse of the matrix $\begin{pmatrix} 3 & -1 \\ -4 & 2 \end{pmatrix}$.

   (4024/01 Oct/Nov 2004 q 15)

7. (a) Multiply $\begin{pmatrix} 5 & 4 \\ -3 & -2 \end{pmatrix} \begin{pmatrix} 2 & 1 & -4 \\ 0 & 3 & 6 \end{pmatrix}$.

   (b) Find the inverse of $\begin{pmatrix} 5 & 4 \\ -3 & -2 \end{pmatrix}$.

   (0580/02 May/June 2003 q 14)

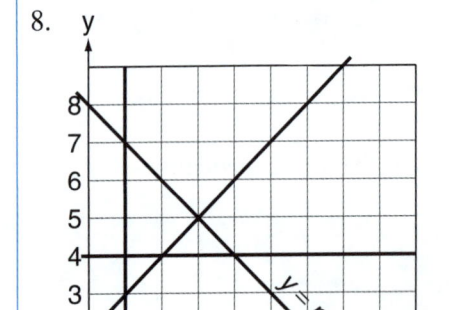

8.

   (a) One of the lines in the diagram is labelled $y = mx + c$.
   Find the values of m and c.
   (b) Show, by shading all the **unwanted** regions on the graph, the region defined by the inequalities
   $x \geqslant 1$, $y \leqslant mx + c$, $y \geqslant x + 2$ and $y \geqslant 4$.
   Write the letter **R** in the region required.

   (0580/02 May/June 2006 q 20)

9. (a) Make k the subject of the formula $\sqrt{\dfrac{h}{k}} = 3$.

   (b) The matrix **Y** satisfies the equation
   $$4\mathbf{Y} - 2\begin{pmatrix} 12 & 6 \\ -9 & 0 \end{pmatrix} = \mathbf{Y}.$$
   Find **Y**, expressing it in the form $\begin{pmatrix} a & b \\ c & d \end{pmatrix}$.

   (4024/02 May/June 2004 q 2 (part))

10. $f(x) = \dfrac{x+3}{x}$, $x \neq 0$.

    (a) Calculate $f\left(\dfrac{1}{4}\right)$.

    (b) Solve $f(x) = \dfrac{1}{4}$.

    (0580/02 May/June 2006 q 11)

11. A taxi company has 'SUPER' taxis and 'MINI' taxis. One morning a group of 45 people needs taxis. For this group the taxis company uses x 'SUPER' taxis and y 'MINI' taxis
   A 'SUPER' taxi can carry 5 passengers, and a 'MINI' taxi can carry 3 passengers.

   So $5x + 3y \geqslant 45$.

   (a) The taxi company has 12 taxis.
   Write down **another** inequality in x and y to show this information.
   (b) The taxi company always uses at least 4 'MINI' taxis.
   Write down an inequality in y to show this information.
   (c) Draw x- and y-axes from 0 to 15 using 1 cm to represent 1 unit on each axis.
   (d) Draw three lines on your graph to show the inequality $5x + 3y \geqslant 45$ **and** the inequalities from **parts (a) and (b)**.
   Shade the **unwanted** regions.
   (e) The cost to the taxi company of using a 'SUPER' taxi is $20 and the cost of using a 'MINI' taxi is $10.

   The taxi company wants to find the cheapest way of providing 'SUPER' and 'MINI' taxis for this group of people.
   Find the **two** ways in which this can be done.
   (f) The taxi company decides to use 11 taxis for this group.
      (i) The taxi company charges $30 for the use of each 'SUPER' taxi and $16 for the use of each 'MINI' taxi.
      Find the two possible **total** charges.
      (ii) Find the largest possible **profit** the company can make, using 11 taxis.

   (0580/04 May/June 2005 q 9)

12. $f(x) = x^2 - 4x + 3$ and $g(x) = 2x - 1$.

    (a) Solve $f(x) = 0$.
    (b) Find $g^{-1}(x)$.
    (c) Solve $f(x) = g(x)$, giving your answers correct to 2 decimal places.
    (d) Find the value of $gf(-2)$.
    (e) Find $fg(x)$. Simplify your answer.

    (0580/04 May/June 2005 q 8)

13. (a) In the diagram, the unshaded region, **R**, is defined by three inequalities.
    Two of these are
    $y \leqslant 2x + 2$ and $y \leqslant 5 - x$.
    Write down the third inequality.

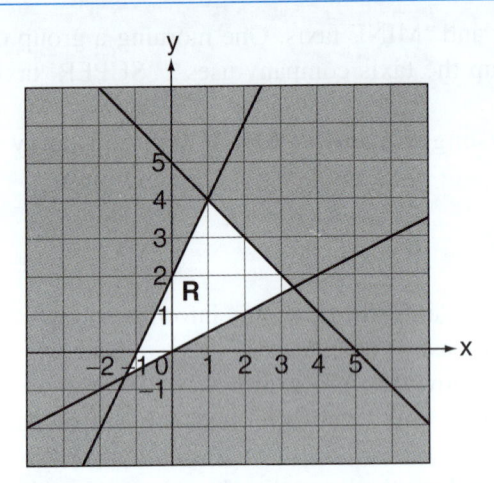

(b) Find the integer values of x which satisfy the following.
$4 \leqslant 2x + 13 < 9$

(4024/01 Oct/Nov 2005 q 14)

14.

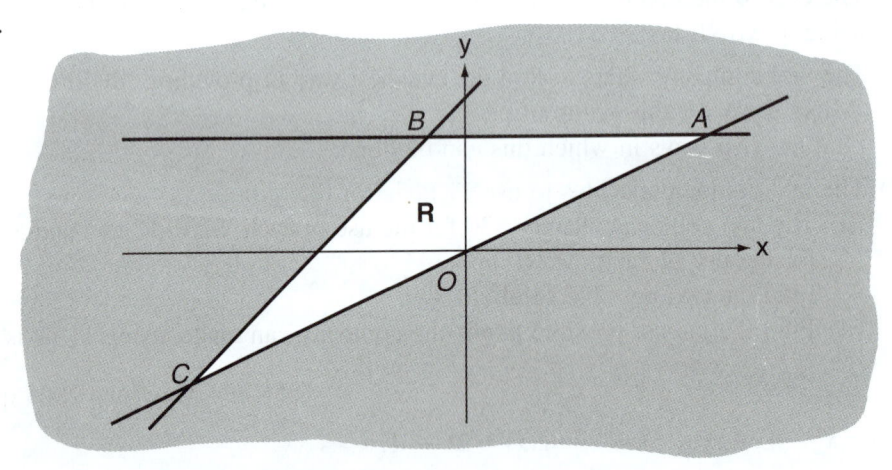

In the diagram, A is the point $(6, 3)$ and C is the point $(-8, -4)$.

The equation of AB is $y = 3$, and the equation of CB is $y = x + 4$.

(a) Find the coordinates of B.

(b) The unshaded region **R** inside triangle ABC is defined by three inequalities.

One of these is $y < x + 4$.

Write down the other two inequalities.

(4024/01 May/June 2007 q 10)

15. Given that $f(x) = \dfrac{5x - 4}{3}$, find

(a) $f\left(1\dfrac{1}{5}\right)$,

(b) $f^{-1}(x)$.

(4024/01 May/June 2007 q 16)

16.

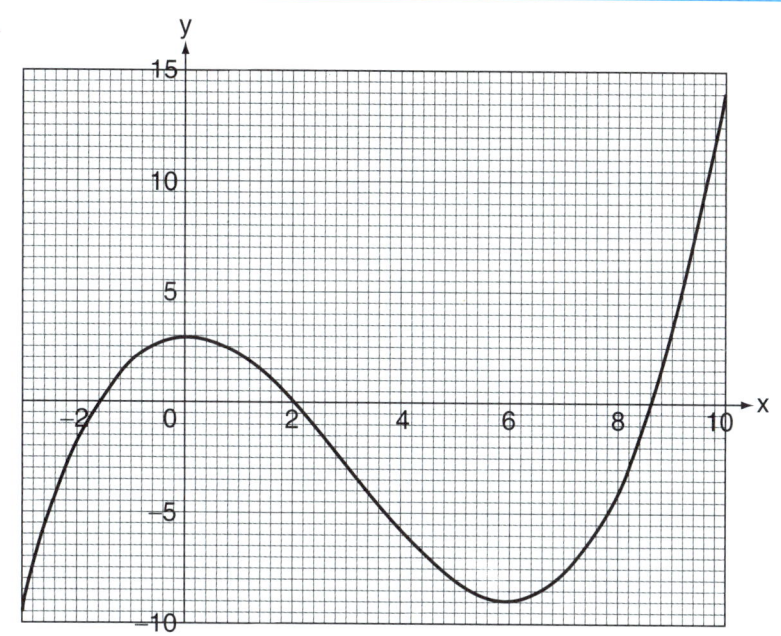

The diagram shows the accurate graph of $y = f(x)$.
(a) Use the graph to find
   (i) $f(0)$,   (ii) $f(8)$.
(b) Use the graph to solve
   (i) $f(x) = 0$,   (ii) $f(x) = 5$.
(c) k is an integer for which the equation $f(x) = k$ has exactly two solutions.
   Use the graph to find the values of k.
(d) Write down the range of values of x for which the graph of $y = f(x)$ has a negative gradient.
(e) The equation $f(x) + x - 1 = 0$ can be solved by drawing a line on the grid.
   (i) Write down the equation of this line.
   (ii) How many solutions are there for $f(x) + x - 1 = 0$?

(0580/04 May/June 2007 q 4)

17.

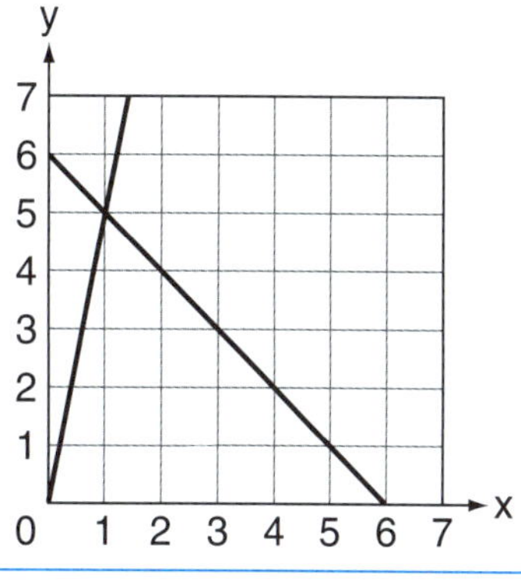

By shading the **unwanted** parts of the grid above, find and label the region R which satisfies the following three inequalities

$y \geqslant 3$ $\qquad$ $y \geqslant 5x$ $\qquad$ and $\qquad$ $x + y \leqslant 6$

(0580/02 May/June 2007 q 12)

18. The function f(x) is given by
    f(x) = 3x − 1.
    Find, in its simplest form,
    (a) f⁻¹ f(x),
    (b) ff(x).

(0580/02 May/June 2007 q 16)

19. $f(x) = 10^x$.
    (a) Calculate f(0.5).
    (b) Write down the value of f⁻¹(1).

(0580/02 Oct/Nov 2005 q 7)

20. $A = (5 \ -8) \qquad B = \begin{pmatrix} 2 & 6 \\ 5 & -4 \end{pmatrix} \qquad C = \begin{pmatrix} 4 & 6 \\ 5 & -2 \end{pmatrix} \qquad D = \begin{pmatrix} 4 \\ -2 \end{pmatrix}$

(a) Which one of the following matrix calculations is **not** possible?
(i) **AB** $\qquad$ (ii) **AD** $\qquad$ (iii) **BA** $\qquad$ (iv) **DA**
(b) Calculate **BC**.
(c) Use your answer to **part (b)** to write down **B⁻¹**, the inverse of **B**.

(0580/02 May/June 2004 q 22)

21. $A = (1 \ 2 \ 3) \qquad B = \begin{pmatrix} 2 & 0 \\ 1 & 4 \\ -1 & -3 \end{pmatrix} \qquad C = \begin{pmatrix} 2 & -1 \\ 2 & 2 \\ -1 & 0 \end{pmatrix}$

Find
(a) B − C, $\qquad$ (b) AB.

(4024/01 Oct/Nov 2005 q 15)

22. (a) Evaluate $\begin{pmatrix} 4 & 2 \\ 1 & 1 \end{pmatrix} \begin{pmatrix} 1 & -2 \\ -1 & 4 \end{pmatrix}$.

(b) **Write down** the inverse of $\begin{pmatrix} 1 & -2 \\ -1 & 4 \end{pmatrix}$.

(4024/01 May/June 2005 q 3)

23. Work out $\begin{pmatrix} 2 & 1 & 2 \\ 1 & 5 & 0 \\ 3 & -2 & 4 \end{pmatrix} \begin{pmatrix} 4 \\ -3 \\ -8 \end{pmatrix}$.

(0580/21 May/June 2008 q 15)

24. $\mathbf{A} = \begin{pmatrix} 1 & 2 \\ 1 & 1 \end{pmatrix}$ $\mathbf{I} = \begin{pmatrix} 1 & 0 \\ 0 & 1 \end{pmatrix}$

   (a) The matrix $\mathbf{B} = \mathbf{A}^2 - 2\mathbf{A} - \mathbf{I}$.
       Calculate $\mathbf{B}$. Show all your working.
   (b) Simplify $\mathbf{A}\mathbf{A}^{-1}$.

   (0580/02 Oct/Nov 2007 q 22)

25. $\mathbf{A} = \begin{pmatrix} x & 8 \\ 2 & x \end{pmatrix}$

   (a) Find $|\mathbf{A}|$, the determinant of $\mathbf{A}$, in terms of x.
   (b) Find the values of x when $|\mathbf{A}| = 9$.

   (0580/02 May/June 2007 q 11)

26. $f(x) = x^3 - 3x^2 + 6x - 4$ and $g(x) = 2x - 1$.
    Find
    (a) $f(-1)$ (b) $gf(x)$ (c) $g^{-1}(x)$

   (0580/21 May/June 2008 q 18)

27. $\begin{pmatrix} 1 & -2 \\ 0 & 1 \\ 5 & 6 \end{pmatrix} \begin{pmatrix} 3 & 4 & 8 & 7 \\ 1 & 1 & 3 & 3 \end{pmatrix}$

   The answer to this matrix multiplication is of order $a \times b$.
   Find the values of a and b.

   (0580/21 Oct/Nov 2008 q 2)

28. $\mathbf{A} = \begin{pmatrix} -2 & 3 \\ -4 & 5 \end{pmatrix}$

   Find $\mathbf{A}^{-1}$, the inverse of $\mathbf{A}$.

   (0580/21 May/June 2009 q 5)

29. $\mathbf{A} = \begin{pmatrix} x & 6 \\ 4 & 3 \end{pmatrix}$ $\mathbf{B} = \begin{pmatrix} 2 & 3 \\ 2 & 1 \end{pmatrix}$

   (a) Find $\mathbf{AB}$.   (b) When $\mathbf{AB} = \mathbf{BA}$, find the value of x.

   (0580/21 May/June 2009 q 21)

30. $f(x) = 2x - 1$ $g(x) = x^2 + 1$ $h(x) = 2^x$
    (a) Find the value of
        (i) $f\left(-\dfrac{1}{2}\right)$,   (ii) $g(-5)$ (iii) $h(-3)$
    (b) Find the inverse function $f^{-1}(x)$.
    (c) $g(x) = z$.
        Find x in terms of z.
    (d) Find $gf(x)$ in its simplest form.

(e) h(x) = 512. Find the value of x.
(f) Solve the equation 2f(x) + g(x) = 0, giving your answers correct to 2 decimal places.
(g) Sketch the graph of
  (i) y = f(x),   (ii) y = g(x).

(0580/04 May/June 2009 q 10)

31. Tiago does some work during the school holidays.
In one week he spends x hours cleaning cars and y hours repairing cycles.
The time he spends repairing cycles is at least equal to the time he spends cleaning cars.
This can be written as y ⩾ x.
He spends no more than 12 hours working.
He spends at least 4 hours cleaning cars.
(a) Write down two more inequalities in x and/or y to show this information.
(b) Draw x and y axes from 0 to 12, using a scale of 1 cm to represent 1 unit on each axis.
(c) Draw three lines to show the three inequalities. Shade the **unwanted** regions.
(d) Tiago receives $3 each hour for cleaning cars and $1.50 each hour for repairing cycles.
  (i) What is the least amount he could receive?
  (ii) What is the largest amount he could receive?

(0580/04 Oct/Nov 2006 q 9)

32. $\mathbf{M} = \begin{pmatrix} 1 & 1 \\ 1 & 2 \end{pmatrix}$ $\mathbf{M}^2 = \begin{pmatrix} 2 & 3 \\ 3 & 5 \end{pmatrix}$ $\mathbf{M}^3 = \begin{pmatrix} 5 & 8 \\ 8 & 13 \end{pmatrix}$

Find $\mathbf{M}^4$.

(0580/02 May/June 2007 q 7)

33. f : x ↦ 5 − 3x
(a) Find f(−1).   (b) Find f^{-1}(x).   (c) Find ff^{-1}(8).

(0580/02 Oct/Nov 2006 q 15)

34. A new school has x day students and y boarding students.
The fees for a day student are $600 a term.
The fees for a boarding student are $1200 a term.
The school needs at least $720 000 a term.
(a) Show that this information can be written as x + 2y ⩾ 1200.
(b) The school has a maximum of 900 students.
Write down an inequality in x and y to show this information.

(c) Draw two lines on a copy of the grid below and write the letter **R** in the region which represents these two inequalities.

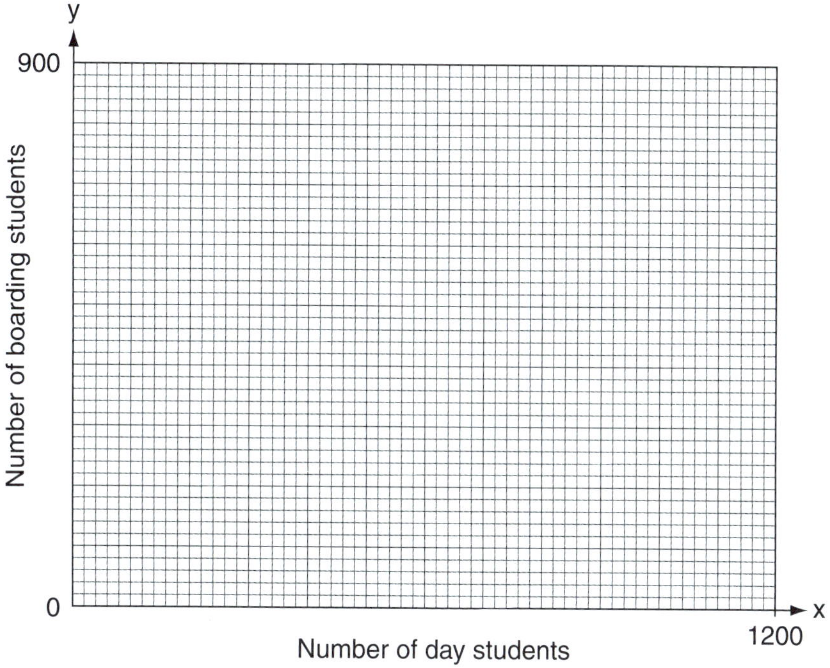

(d) What is the least number of **boarding** students at the school?

(0580/21 Oct/Nov 2008 q 20)

Chapter 9

Trigonometry

We now extend the use of trigonometry to cover triangles which do not have right angles. We look at angles greater than 90°, and use trigonometry in three dimensions.

Core Skills

1. Use the sine, cosine and tangent ratios and Pythagoras' Theorem to find the sides and angles marked with letters in the following diagrams. All lengths are in centimetres.

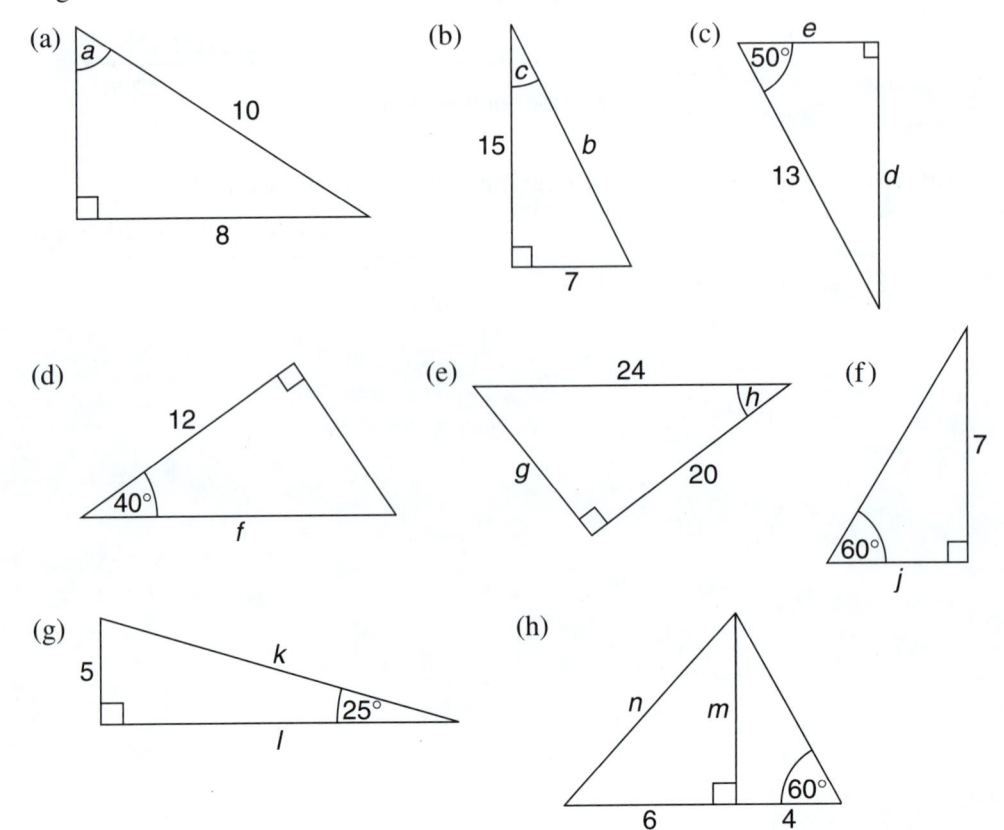

(a) Right triangle with angle a, hypotenuse 10, and legs 8 (bottom).

(b) Right triangle with angle c at top, side 15, side b, and base 7.

(c) Right triangle with 50° angle at top, side e along top, side 13, and side d.

(d) Triangle with 40° angle, side 12, and side f (with a right angle marked).

(e) Triangle with top side 24, side g, side 20, angle h, and a right angle at the bottom.

(f) Right triangle with 60° angle, hypotenuse 7, and side j.

(g) Right triangle with side 5, side k, 25° angle, and base l.

(h) Triangle with sides n and m, split by a vertical into base segments 6 and 4, with a 60° angle.

2. The bearing of A from B is 065°. Find the bearing of B from A.
3. The bearing of C from D is 200°. Find the bearing of D from C.
4. Write the following as three figure bearings.
   (a) SE   (b) NNW

Angles of Elevation and Depression

If you are sitting on the ground and looking at the base of a tall vertical pole you will need to lift your eyes to see the top of the pole. The angle through which you raise your eyes is called the *angle of elevation*, and is illustrated in the diagram below. For simplicity the diagram ignores the fact that your eyes will not be at ground level.

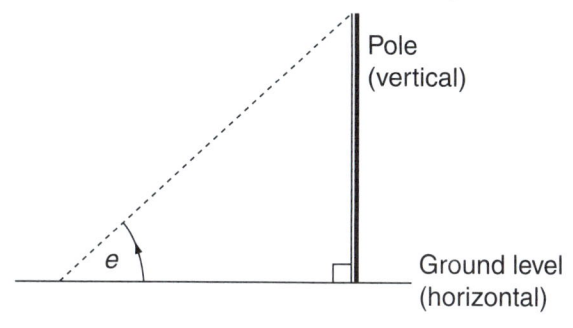

The angle of elevation is labelled e.

If the ground is horizontal and the pole is vertical, the angle that the pole makes with the ground is a right angle, so we are able to do calculations using right-angled triangles.

Similarly, if you are on the top of a cliff looking out to sea, the angle through which you have to lower your eyes from the horizontal to look at a ship is called the *angle of depression*, and is illustrated below.

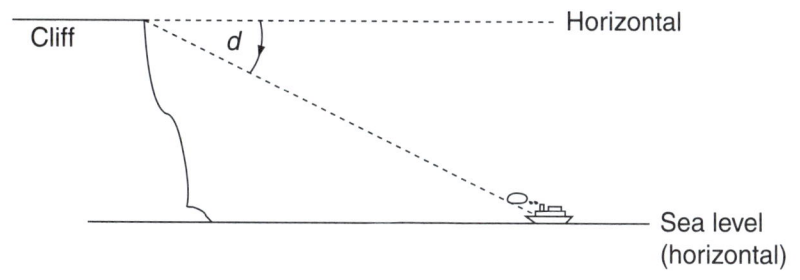

The angle of depression is labelled d.

In the diagram above you can see that, because the two horizontal lines must be parallel, the angle of elevation from the ship to the top of the cliff is also d, (alternate angles).

Example 1

The angle of elevation of the top of a building, seen by an observer from a distance of 10 metres away on horizontal ground is 60°. How tall is the building? (Ignore the height of the observer.)

Answer 1

The diagram shows a sketch of the building and the angle of elevation.

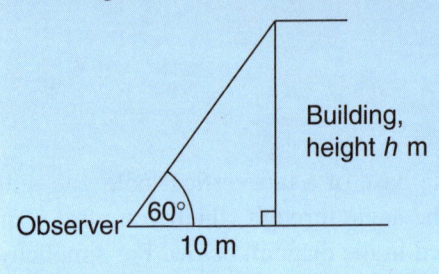

Observer — 60° — 10 m — Building, height h m

Let h be the height of the building.

$\tan 60° = \dfrac{h}{10}$

$10 \times \tan 60° = h.$

$h = 17.320\ldots$

The height of the building is 17.3 metres, to 3 significant figures.

Exercise 9.1

1. Ramiro is surveying a building. He is using a theodolite, which is an instrument for measuring angles. The theodolite is on a pole 170 centimetres above ground level. The building is 25 metres away on horizontal ground. Ramiro measures the angle of elevation of the top of the building. It is 50°.
   How tall is the building?

2. The given diagram shows the position of a theodolite on the top of a hill. It is being used to measure the height of a tower 70 metres away in a horizontal direction. The angle of depression of the bottom of the tower is 25°, and the angle of elevation of the top of the tower is 30°.

   Calculate
   (a) the height of the hill,   (b) the height of the tower.

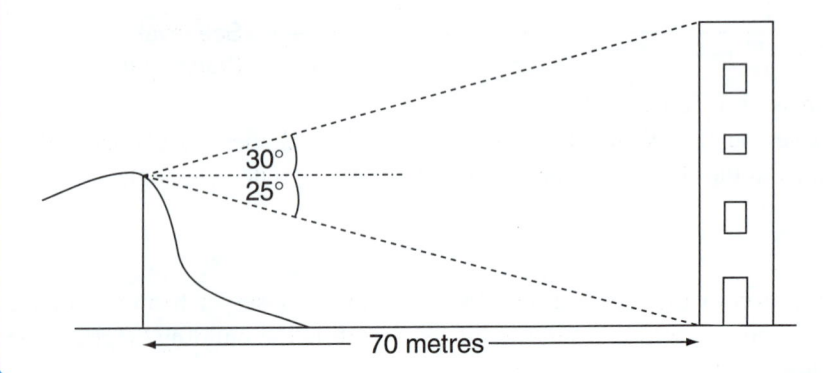

If the horizontal distance to the point vertically under the top of an object cannot be measured, as would happen, for example, if the height of a mountain was required, two measurements of angles of elevation can be taken.

The diagram below shows one example, with the two observations, *A* and *B*, being on a direct line with the peak of the mountain.

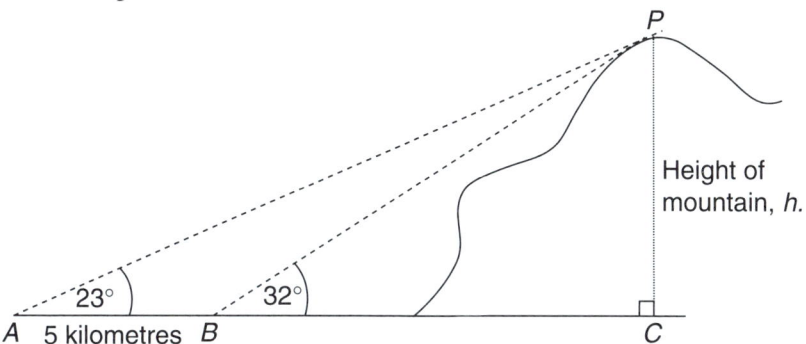

It is possible to work out the height of the mountain using the two right-angled triangles *APC* and *BPC*, but it requires quite a lot of algebra. We will now look at methods which can be used in triangles without right angles, such as triangle *APB* above.

Trigonometry with Angles between 0° and 180°

The diagram below shows an angle, α. This is the Greek letter alpha, which is often used in Mathematics. The angle is made by the positive *x*-axis and a line, *OP*, which can rotate *anticlockwise* through 180° about the origin. The line is 1 unit long (measured by the scale on the axes). A semicircle with the same radius is drawn with its centre at the origin.

The diagram shows two *quadrants*, (quarter circles) of the complete circle. You will go on to study the other two quadrants if you take Mathematics beyond IGCSE.

The quadrant in which both the *x*- and the *y*-coordinates are positive is always called the *first quadrant*, and the quadrant with negative *x*-coordinates and positive *y*-coordinates is always called the *second quadrant*.

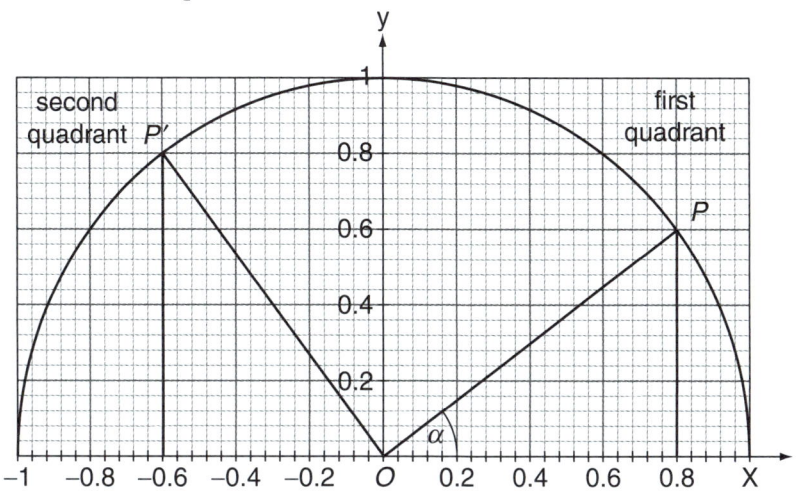

In the Core Book we defined the sine ratio in a right-angled triangle as $\dfrac{\text{opposite side}}{\text{hypotenuse}}$
In the diagram above, using the same definition, we see that

$$\sin \angle XOP = \sin \alpha = \dfrac{y}{OP},$$

where y is the y-coordinate of the point P. But we have made OP to be of unit length, so $\sin \alpha = \dfrac{y}{1}$, which is the same as $\sin \alpha = y$.

Now we have a new definition: $\sin \alpha = y$ when α is the angle the line OP makes with the positive direction of the x-axis and P is the point (x, y).

Measure the angle XOP in the diagram. It should be about 37°. The y-coordinate of P is 0.6. The calculator value for $\sin 37°$ is 0.60 to two decimal places, so allowing for experimental error and the limitations of the printing process, this is a good agreement.

All this is no different from the earlier work you did on the trigonometric ratios, except that it gives us a method for extending the angles to include obtuse angles.

P' is in the second quadrant in the diagram above, so $\angle XOP'$ is obtuse. According to the new definition $\sin \angle XOP' = y$, where y is the y-coordinate of the point P'. Measure the angle XOP' in the diagram. It should be about 127°. The y-coordinate is 0.8. The calculator value for $\sin 127°$ is 0.80 to two decimal places, so again there is a good agreement.

By the same argument, $\cos \alpha = \dfrac{\text{adjacent}}{\text{hypotenuse}} = \dfrac{x}{OP} = \dfrac{x}{1} = x.$

In the diagram $\angle XOP$ is about 37°, and the x-coordinate of P is 0.8. The calculator value for $\cos 37°$ is 0.80 to two decimal places, which is a good agreement as we would expect. Now look at the x-coordinate of P'. It is -0.6, and angle XOP' is about 127°.

The calculator value of $\cos 127°$ is -0.60 to two decimal places, so our new definition of cosine works for obtuse angles too. Remember that the cosines of angles between 90° and 180° are negative because the x-coordinates of points on the second quadrant are negative.

Practical investigation

Copy the diagram above, and draw a few more angles of your own. Complete this table with your extra examples. Add the calculator values of the sines and cosines of your angles to your table. You should find quite a good agreement between your experimental values and the calculator values.

angle	35°	118°					
sine (y-coordinate)	0.58	0.88					
cosine (x-coordinate)	0.81	−0.48					
calculator value for sine	0.57	0.88					
calculator value for cosine	0.82	−0.47					

Investigation of sine and cosine curves

Copy and complete the table, using your calculator and giving the sines and cosines to 2 decimal places.

$\alpha°$	0	20	40	60	80	90	100	120	140	160	180
sin α	0		0.64						0.64		
cos α	1		0.77						−0.77		

Using the values in the table copy and complete the graphs below, drawing a smooth continuous curve through all the points.

Sine curve

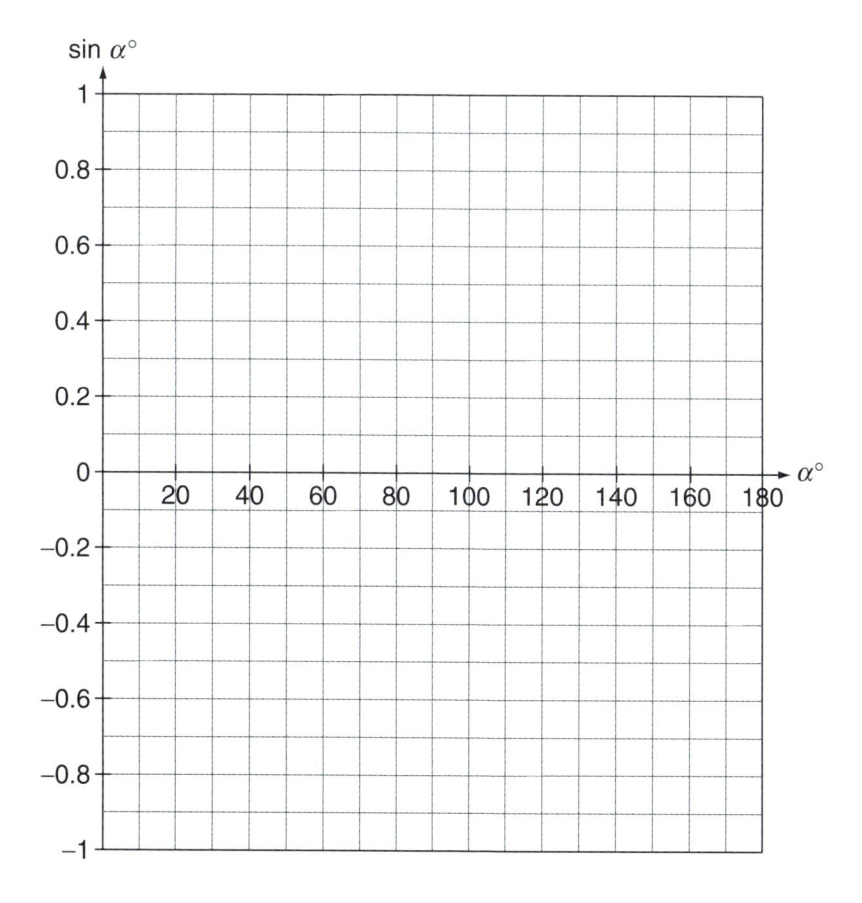

220 *Extended Mathematics for Cambridge IGCSE*

Cosine curve

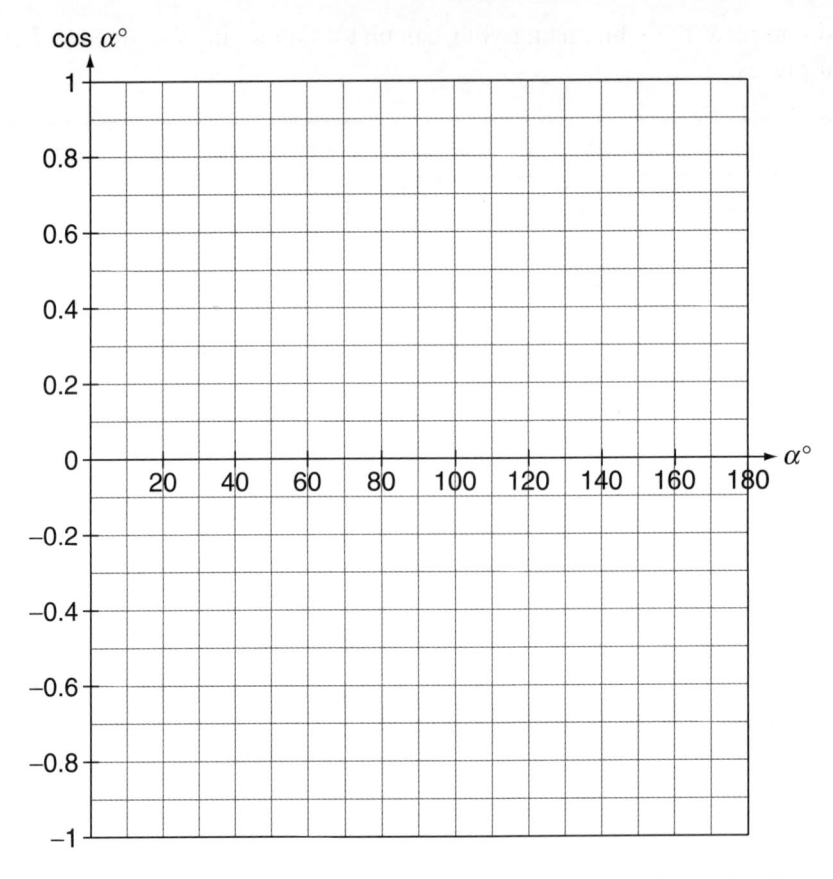

If you have drawn these graphs correctly they should look something like the following diagrams.

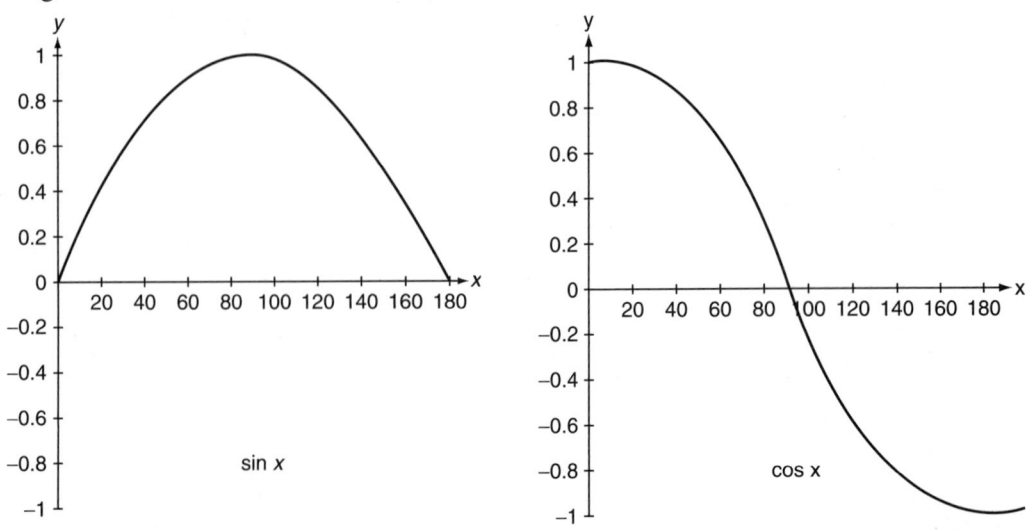

sin x

cos x

You can now use your graphs to find angles when given their sines or cosines.

To find the angle whose sine is 0.5 draw a horizontal line through 0.5 on the y-axis. Mark the point where the line meets the curve, and read off the x-coordinates of those points. You should find that the two points are (30, 0.5) and (150, 0.5).

Use your calculator to find $\sin^{-1}(0.5)$. It will give you 30°, which, as you see, is only one of the answers.

Your calculator will always give you the acute angle when you enter $\sin^{-1}$. In a right-angled triangle this is no problem, because if one angle is 90° both the other must be acute. However, in other triangles it is possible to have an obtuse angle, and you should always be aware that there might be two possible solutions when you enter $\sin^{-1} x$ into your calculator. The second solution (the obtuse angle) is found by subtracting the acute angle from 180°.

If you look at the cosine curve you will see that there is no problem. $\cos^{-1} 0.5 = 60°$ and $\cos^{-1}(-0.5) = 120°$.

Now that you have drawn these graphs of the sine and cosine curves you will see that we should think of sines and cosines of x as *functions* of x rather than ratios.

If you continue your study of Mathematics beyond this course you will find that we can work out the sine, cosine and tangent of any angle, no matter how large. You might like to continue your sine and cosine curves, using your calculator, to angles up to 360°. If you are also studying Physics or Science you might recognise these curves as waves.

Finding acute and obtuse angles from their sines and cosines

The signs of sines and cosines in the first two quadrants are summarised in the diagram.

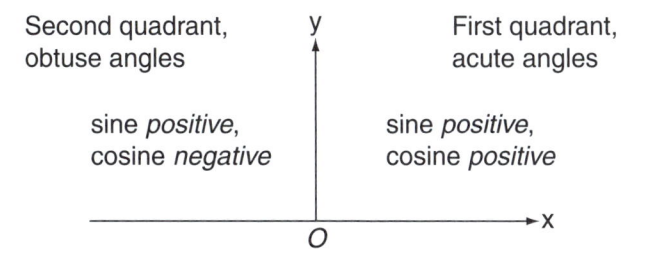

```
Second quadrant,         y      First quadrant,
obtuse angles            |      acute angles

sine positive,                  sine positive,
cosine negative                 cosine positive
                         |
─────────────────────────O─────────────────→ x
```

Using your calculator to find angles.

- Your calculator will automatically give you an acute angle if you enter, for example, $\cos^{-1}(0.4)$, and an obtuse angle if you include a negative sign, for example, $\cos^{-1}(-0.4)$.
- Your calculator will give you an acute angle if, for example, you enter $\sin^{-1}(0.4)$. To find the obtuse angle you need to subtract the acute angle from 180°.

You might try entering $\sin^{-1}(-0.4)$ into your calculator. Your calculator will probably give you a negative angle. Check this with your own calculator. This value is outside the range we are studying in this course.

Exercise 9.2

1. Find the following
   (a) sin 40° (b) sin 140° (c) sin 150°
   (d) sin 75° (e) cos 40°   (f) cos 140°
   (g) cos 150° (h) cos 75°

2. Find x in the following given that x lies between 0° and 180°.
   (a) cos x = 0.27   (b) cos x = 0.59
   (c) cos x = −0.27 (d) cos x = −0.59

3. Find x given that x is obtuse
   (a) sin x = 0.28 (b) sin x = 0.83
   (c) sin x = 0.57 (d) sin x = 0.77

4. For the following find two possible values of x, given that x lies between 0° and 180°
   (a) sin x = 0.5643 (b) sin x = 0.1254
   (c) sin x = 0.8432 (d) sin x = 0.5333

The Sine Rule

In the triangle ABC shown below, AD is drawn perpendicular to BC, and the length of AD is h cm.

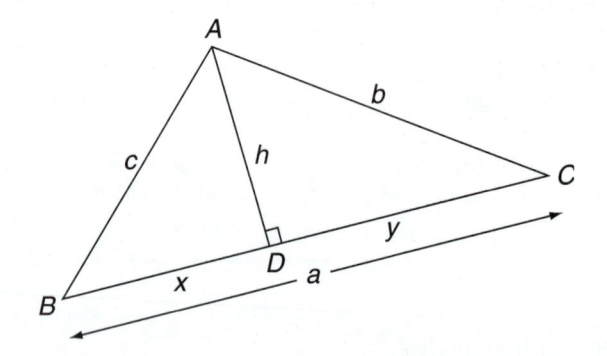

In triangle ADC, $\sin \angle ACD = \dfrac{h}{b}$

$$h = b \times \sin \angle ACD \quad \ldots \text{(i)}$$

In triangle ABD, $\sin \angle ABD = \dfrac{h}{c}$

$$h = c \times \sin \angle ABD \quad \ldots \text{(ii)}$$

Since the right hand sides of both equations (i) and (ii) are equal to h, they must be equal to each other.

$$b \times \sin \angle ACD = c \times \sin \angle ABD$$

$$b = \frac{c \times \sin \angle ABD}{\sin \angle ACD}$$

$$\frac{b}{\sin \angle ABD} = \frac{c}{\sin \angle ACD}$$

Rewriting sin ∠ACD as sin C, and sin ∠ABD as sin B, the equation becomes:

$$\frac{b}{\sin B} = \frac{c}{\sin C}$$

In the same way it can be shown that:

$$\frac{a}{\sin A} = \frac{b}{\sin B}$$

This is known as the **Sine Rule**, and can be written in two ways:

$$\frac{a}{\sin A} = \frac{b}{\sin B} = \frac{c}{\sin C}$$

or

$$\frac{\sin A}{a} = \frac{\sin B}{b} = \frac{\sin C}{c}$$

For convenience, if you need to calculate the length of a side you will use the first arrangement, but to calculate an angle use the second.

You can pick any pair of ratios for the calculation. For example, $\frac{a}{\sin A} = \frac{c}{\sin C}$.

Example 2

Using the measurements shown in the diagram of triangle *ABC* calculate the length of side *AC*.

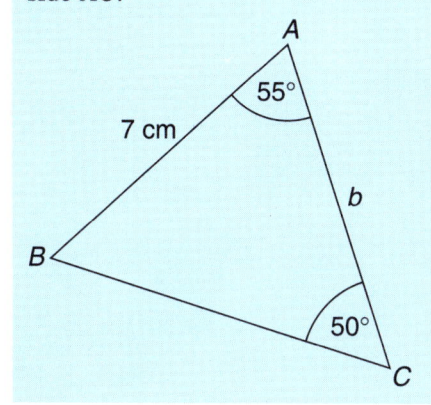

Answer 2

To use the Sine Rule we need pairs of opposite sides and angles, so to find b we need angle ABC.

Angle $ABC = 180 - 50 - 55 = 75°$

Using $\dfrac{b}{\sin B} = \dfrac{c}{\sin C}$

$\dfrac{b}{\sin 75} = \dfrac{7}{\sin 50}$

$b = \dfrac{7 \times \sin 75}{\sin 50}$

$b = 8.82648...$

The side $AC = 8.83$ cm

The ambiguous case of the Sine Rule

Example 3

In triangle DEF, angle $EDF = 20°$, $ED = 10$ cm, and $EF = 4$ cm.

Calculate angle EFD.

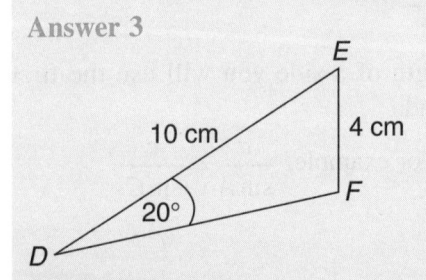

Answer 3

Using the Sine Rule,

$\dfrac{\sin F}{f} = \dfrac{\sin D}{d}$

$\dfrac{\sin F}{10} = \dfrac{\sin 20}{4}$

$\sin F = \dfrac{10 \times \sin 20}{4}$

$\sin F = 0.85505...$

$F = \sin^{-1} 0.85505...$

$F = 58.7652...$

Angle $EFD = 58.8°$

The answer to the above example appears to be 58.5°. But is this right?
In the diagram above angle *EFD* looks obtuse, although this could be because it is not drawn to scale.
Look at the diagram below.

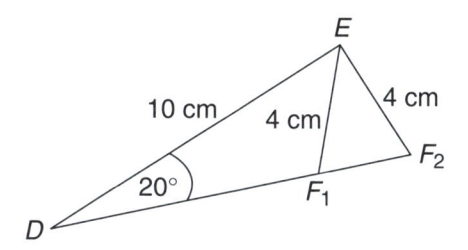

There are two possible triangles that can be drawn with the given measurements. They are shown as triangles EDF_1 and EDF_2 in the diagram.
Angle $EF_2D = 58.8°$ and angle $EF_1D = 180° - 58.8° = 121.2°$.
This is called the ambiguous case and unless further measurements are available it is not possible to say which is the required answer, so both should be given.
Notice that 20° + 58.8° = 78.8°, and 20° + 121.2° = 141.2°, both pairs of angles add up to less than 180° so both are possible as two of the angles in a triangle.
In each of the two triangles there is one obtuse angle.

Example 4

In triangle *PQR*, *PQ* = 8.4 cm, *QR* = 6.7 cm and angle *PRQ* = 71°.
Calculate angle *QPR*.

Answer 4

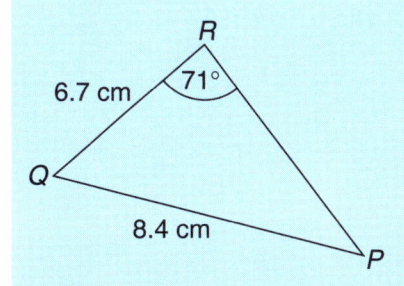

Using the Sine Rule,

$$\frac{\sin P}{p} = \frac{\sin R}{r}$$

$$\frac{\sin P}{6.7} = \frac{\sin 71}{8.4}$$

$$\sin P = \frac{6.7 \times \sin 71}{8.4}$$

$\sin P = 0.75416...$

$\sin^{-1} 0.75416... = 48.9523...$

Angle $QPR = 49.0°$

There is no ambiguity over the value of angle QPR because $49.0° + 71° = 120°$, leaving the third angle as $60°$.

If angle $QPR = 180° - 49° = 131°$, then $131° + 71° = 202°$, which is not possible in a triangle!

Using the Sine Rule in a practical situation

On page 217 we looked at the problem of finding the height of a mountain when you cannot measure the horizontal distance to the point vertically below the peak.
For convenience, this diagram is reproduced here, and we will now use the Sine Rule to help solve the problem.

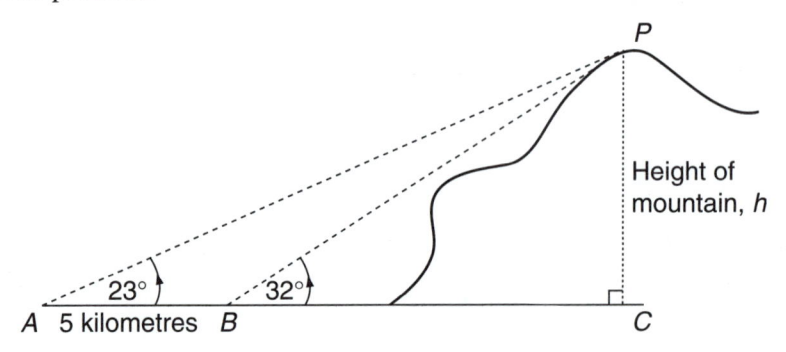

In triangle ABP, $\angle ABP = 180° - 32° = 148°$,

$$\angle APB = 180° - 23° - 148° = 9°$$

Using the Sine Rule, $\dfrac{AP}{\sin 148°} = \dfrac{5}{\sin 9°}$

$$AP = \frac{5 \times \sin 148°}{\sin 9°}$$

$$AP = 16.9374...$$

In triangle ACP, $\sin 23° = \dfrac{PC}{AP}$

$PC = AP \times \sin 23°$

$PC = 16.9374... \times \sin 23°$

$PC = 6.61797...$

The height of the mountain is 6.62 kilometres to 3 significant figures.

Exercise 9.3

Find the value represented by the letter in each diagram. The diagrams are not to scale.

1. 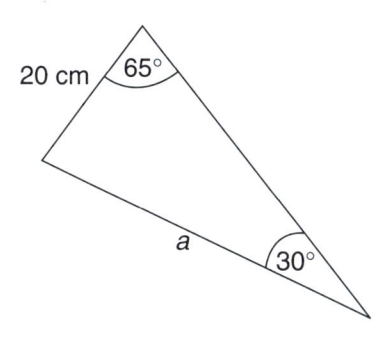 [Triangle with sides 20 cm, angle 65°, side a, angle 30°]

2. 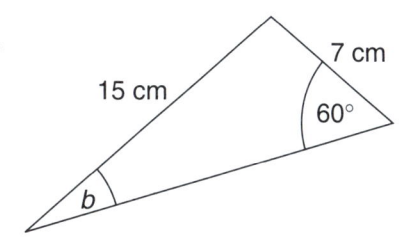 [Triangle with side 15 cm, side 7 cm, angle 60°, angle b]

3. 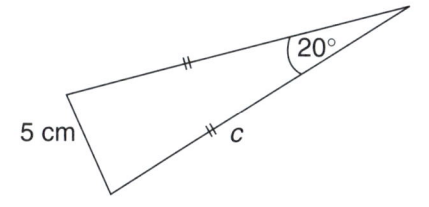 [Triangle with 5 cm, angle 20°, side c, two sides marked equal]

4. 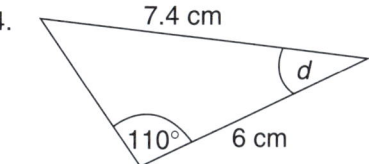 [Triangle with 7.4 cm, 6 cm, angle 110°, angle d]

   NOTE: Find the remaining angle first.

5. 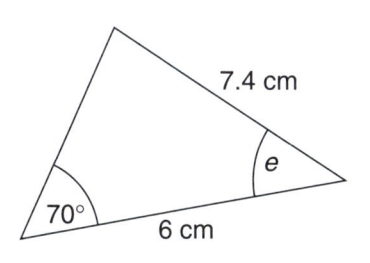 [Triangle with 7.4 cm, 6 cm, angle 70°, angle e]

6. 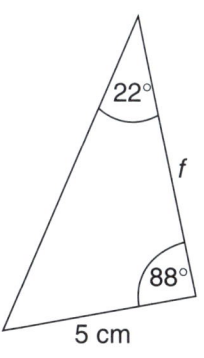 [Triangle with angle 22°, angle 88°, side 5 cm, side f]

7. [Triangle with 10 cm, 10 cm, angle 20°, angle g]

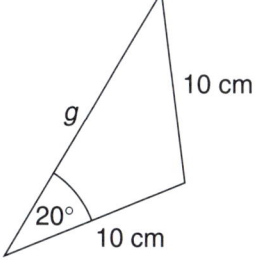

8. In triangle *ABC*, *AB* = 4.2 cm, *BC* = 5 cm and angle *ACB* = 50°. Calculate the two possible values of angle *BAC*.

228 *Extended Mathematics for Cambridge IGCSE*

The Cosine Rule

The Sine Rule involves pairs of opposite sides and angles. For example, look at the diagrams below.

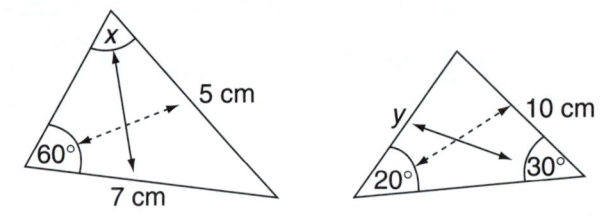

If, however, we are given any of the *following* triangles we cannot use the Sine Rule. (Try if you are not sure. In each case you will find you need another angle.)

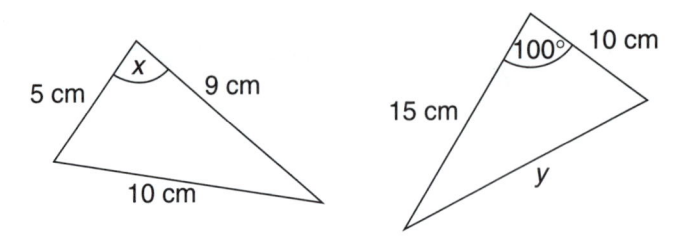

The alternative method uses the Cosine Rule.
This is slightly more complicated to derive, so it will just be stated here. It involves three sides and one angle.

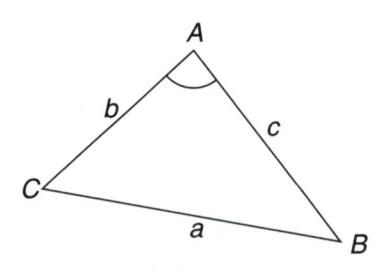

- To find an angle use: $\cos A = \dfrac{b^2 + c^2 - a^2}{2bc}$

- To find a side use: $a^2 = b^2 + c^2 - 2bc \cos A$

These are both the same formula, but arranged differently.
It is worth studying these two formulae carefully. You will see some symmetry in the way they are used. The two sides (lengths b and c) which include the angle (angle A) are always squared and added. Each letter (either in capital or lower case) appears twice. The second arrangement reminds us of Pythagoras' Theorem, but includes a 'correction term' ($2bc \cos A$) to allow for the fact that angle A is not a right angle.

Using the Cosine Rule and avoiding the very common errors

Practice the Cosine Rule using your calculator.
To find the angle, use brackets as shown here:

$$\cos A = \frac{(b^2 + c^2 - a^2)}{(2bc)}$$

To find the side, enter the terms into your calculator exactly as shown in the formula. You will have to enter the multiplication symbol in between each number in the final term, but your calculator will deal with the plus and minus signs correctly. One of the most common mistakes is to calculate the $2bc \cos A$ first and then get confused with the signs.

There is no ambiguity in the answers using the Cosine Rule because the calculator will distinguish between acute and obtuse angles according to whether the cosine is positive or negative.

Before going any further try these two questions to make sure that you understand the correct use of the calculator, and to check that you understand the logic your calculator uses.

1. Using your calculator find angle A, given that

$$\cos A = \frac{6^2 + 5^2 - 3^2}{2 \times 6 \times 5}$$

The correct answer is $A = 29.9°$
- If you were correct try the next question
- If you got 0.866666… you have found the cosine of the angle. Press $\boxed{\text{SHIFT}}$ $\boxed{\text{COS}}$ $\boxed{\text{ANS}}$ to get the angle.
- If you got 0.52 your calculator is set in radians. Change to degrees and recalculate.
- If you got 780, or 60.85 or any other impossible result you have probably not followed the advice above to put brackets round the numerator and denominator.

2. Using your calculator find side a, given that

$$a^2 = 7^2 + 5^2 - 2 \times 5 \times 7 \times \cos 150°$$

The correct answer is $a = 11.6$

- If you were correct you are ready to proceed with the rest of the chapter!
- If you got 134.62… You have found a^2 not a. Find the square root to finish the question.
- If you got 5.01 or 25.05 your calculator is in radians. Change to degrees.
- If you got −3.46, or any other impossible answer you have not followed the advice to enter the values into your calculator exactly as they appear in the question. Do not work out part of the calculation first or put brackets into the calculator.

Example 5

With reference to the triangle *PQR* shown

(a) If $p = 7$ cm, $q = 8$ cm and $r = 11$ cm, calculate $\angle PRQ$.

(b) If $p = 6$ cm, $r = 4$ cm and $\angle Q = 100°$, calculate *PR*.

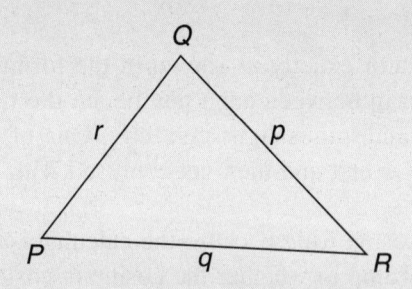

Answer 5

(a) $\cos R = \dfrac{p^2 + q^2 - r^2}{2pq}$

$\cos R = \dfrac{(7^2 + 8^2 - 11^2)}{(2 \times 7 \times 8)}$

$\cos R = -0.071428...$

$R = \cos^{-1} -0.071428...$

$R = 94.0960...$

Angle $PRQ = 94.1°$ to 1 decimal place.

(b) $q^2 = p^2 + r^2 - 2pr\cos Q$.

$q^2 = 6^2 + 4^2 - 2 \times 6 \times 4 \times \cos 100°$

$q^2 = 60.33511...$

$q = 7.76756...$

$PR = 7.77$ cm to 3 significant figures

NOTE: In both of the above parts of the example the calculator has dealt with the obtuse angle without any further input.

NOTE: Remember to take square root to get the length of the side.

Exercise 9.4

Calculate the values represented by letters in the following triangles. All lengths are in centimetres.

1. 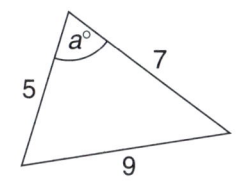
2. 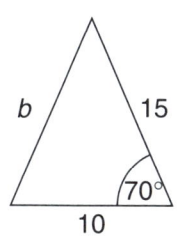
3. 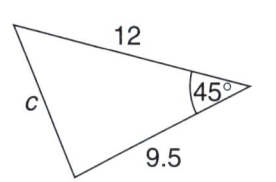
4. 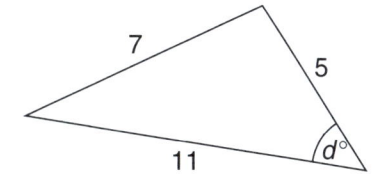
5. 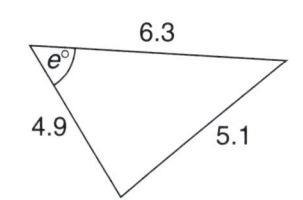
6.

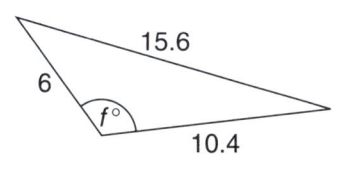

7. In triangle *ABC*, *AC* = 15 cm, *BC* = 12 cm and angle *ACB* = 30°. Calculate the length of *AB*.
8. In triangle *DEF*, *DF* = 5 cm, *EF* = 11 cm and angle *DFE* = 112°. Calculate the length of *DE*.
9. In triangle *GHJ*, *GH* = 3.1 cm, *GJ* = 6.7 cm and *HJ* = 4.9 cm. Calculate angle *GHJ*.
10. In triangle *KLM*, *KM* = 35 cm, *KL* = 15.1 cm and angle *LKM* = 130°. Calculate the length of *LM*.

The next exercise mixes Sine Rule and Cosine Rule so that you get practise in deciding which to use, and also provides some examples in which you will have to use both.

Exercise 9.5

1. Calculate the values represented by letters in these questions. You may not have to use all the information in the questions. All the lengths are in centimetres.

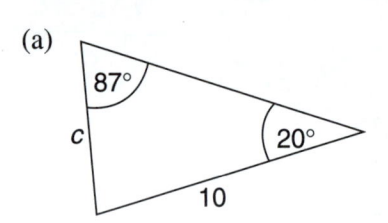

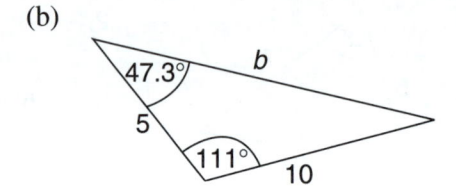

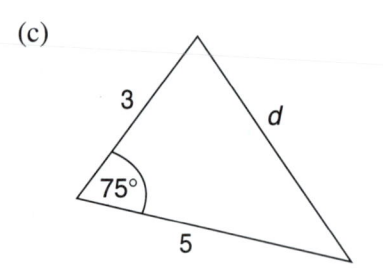

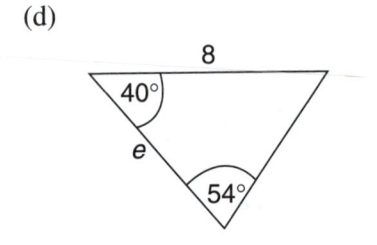

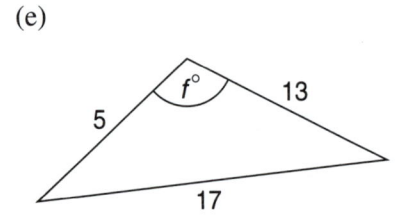

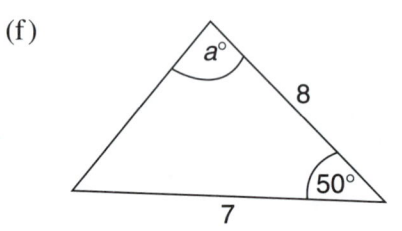

(a) Triangle with angles 87° and 20°, side 10 opposite to 87°, and side c opposite to 20°.

(b) Triangle with angles 47.3° and 111°, side 5 between them, side b opposite 111°, side 10 opposite 47.3°.

(c) Triangle with angle 75°, sides 3 and 5 adjacent to it, and side d opposite.

(d) Triangle with angles 40° and 54°, side 8 opposite 54°, side e opposite 40°.

(e) Triangle with sides 5, 17, 13 and angle f° opposite side 17.

(f) Triangle with angle a°, sides 8 and 7, and angle 50°.

NOTE: Work out the other side first.

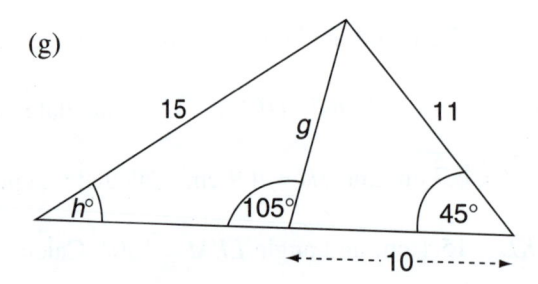

(g) Triangle with sides 15, 11, base 10, angle g, angles 105°, 45°, h°.

2. A ship leaves a port P and sails for 21 kilometres on a bearing of 073°. It then alters course to a bearing 125° and sails 13 kilometres.

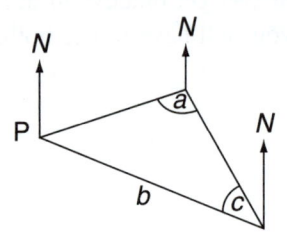

Copy the diagram and fill in the information given.

(a) Find the angle a.
(b) Calculate the distance b.
(c) Find the angle c.
(d) Find the bearing on which the ship must sail to return to port P.

Area of a Triangle

You already know how to calculate the area of a triangle if you are given, or can find, the base and the height.
However, there is another useful formula which can be used if you know two sides and the *included* angle. The included angle is the angle between the two known sides.
Look at the diagram.

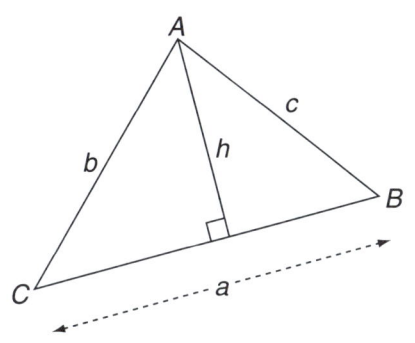

Suppose you are given the lengths of AC (b) and CB (a), and the size of angle C.
Drop a perpendicular line from A to BC. If BC is the base of the triangle then this perpendicular is the height.

$$\sin C = \frac{opp}{hyp} = \frac{h}{b}$$

$$h = b \times \sin C$$

The area of triangle $ABC = \frac{1}{2} \times BC \times h$

$$= \frac{1}{2} \times a \times b \times \sin C$$

The area of triangle $ABC = \frac{1}{2} ab \sin C$

This formula is often used and does not have to be derived each time.
Remember the area is 'half the product of the two sides times the sine of the angle between them'.
It does not matter if the angle is obtuse, the area will still be correct.

234   Extended Mathematics for Cambridge IGCSE

Example 6
Calculate the area of triangle *PQR*, where *PQ* = 15.1 cm, *QR* = 17.2 cm and angle *PQR* = 50°.

Answer 6

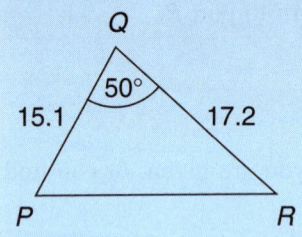

Area of triangle $PQR = \frac{1}{2}pr \sin Q = \frac{1}{2} \times 17.2 \times 15.1 \times \sin 50°$

$= 99.4785...$

Area of triangle $PQR = 99.5$ cm^2 to 3 significant figures.

Exercise 9.6
1. Calculate the areas of the following triangles.

    (a) 5 cm, 75°, 11 cm

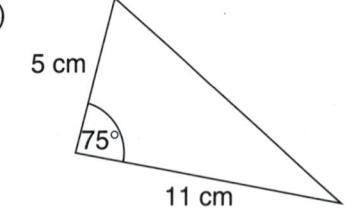

    (b) 7.5 cm, 100°, 12.2 cm

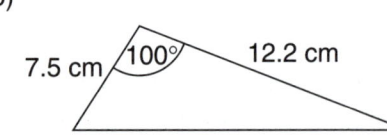

    (c) 40 km, 30°, 100 km

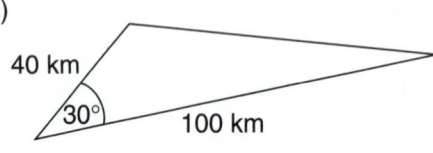

    (d) 8.5 m, 50°, 55°, 7.9 m

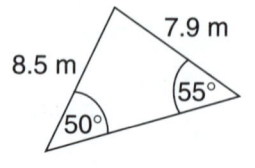

    (e) In triangle *ABC*, *AB* = *AC* = 10 cm. ∠*ABC* = 60°.
        Calculate the area of triangle *ABC*.

2. Using the following triangles, calculate
    (i) the angle marked *x*,
    (ii) the angle marked *y*,
    (iii) the area of the triangle.

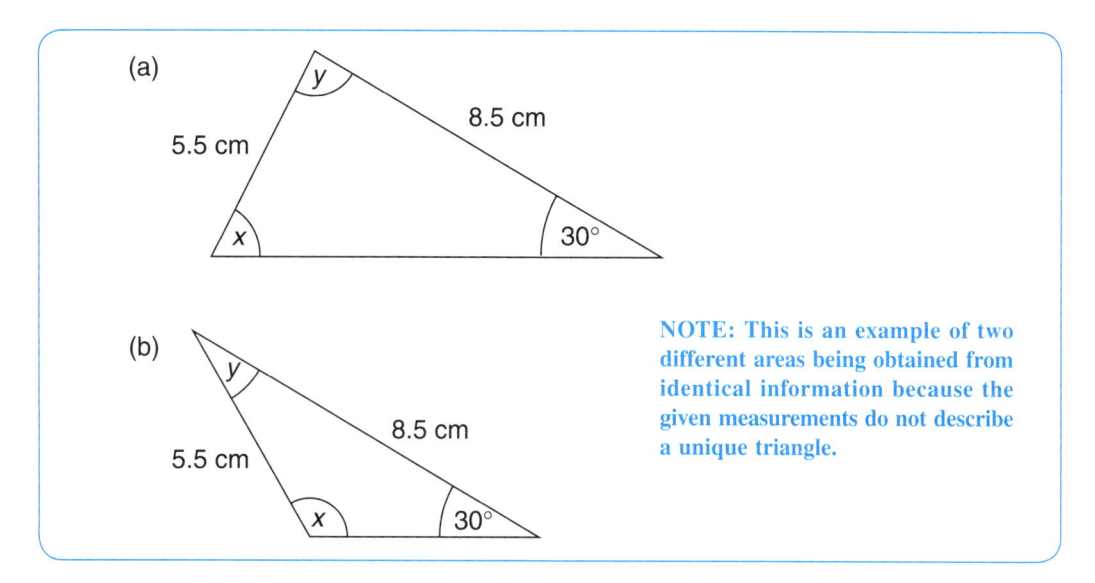

(a)

5.5 cm, 8.5 cm, angle y, angle x, 30°

NOTE: This is an example of two different areas being obtained from identical information because the given measurements do not describe a unique triangle.

(b)

5.5 cm, 8.5 cm, angle y, angle x, 30°

3-D Trigonometry

You need to be able to apply Pythagoras' Theorem, sine, cosine and tangent ratios and the Sine and Cosine Rules to three-dimensional objects such as pyramids.

One new concept is the angle between a line and a plane.

A plane is a flat surface such as a piece of paper.

The diagram shows a piece of paper lying horizontally, with a pencil point resting on it so that the pencil is at an angle. Immediately above the pencil is a light shining directly down on the paper. The pencil casts a shadow on the paper, and the angle between the pencil and its shadow is the angle between the pencil and the paper. The important point is that the light must be directly above the pencil. The line of the shadow is called the *projection* of the pencil onto the paper.

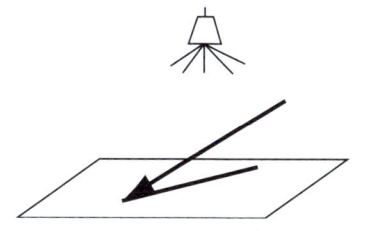

The pencil is perpendicular to the paper if it is at right angles to every single line that could be drawn on the paper. This is shown in the next diagram.

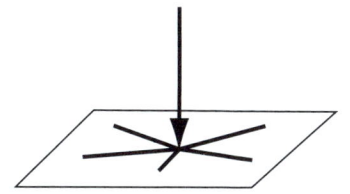

To calculate the angle between a line and a plane drop a perpendicular from any point on the line to the plane to make a right-angled triangle as shown in the next diagram.

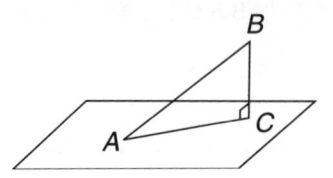

AB is the line, *BC* is the perpendicular from *B* to the plane, and *AC* is the projection of the line in the plane. *BC* is perpendicular to the plane, so it is perpendicular to every line drawn in the plane, and in particular it is perpendicular to *AC*, making triangle *ABC* a right-angled triangle. The angle between the line and the plane is angle *BAC*.

Example 7

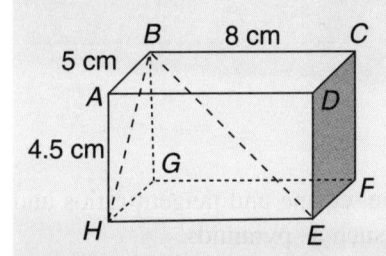

NOTE: This example shows the differences in the accuracies obtained when you carry forward rounded answers rather than calculator values to subsequent parts of the question. The *best* practice is to carry forward unrounded, calculator values, but *either method is acceptable in your examination.*

ABCDEFGH is a cuboid.

$AB = 5$ cm, $BC = 8$ cm and $AH = 4.5$ cm.

(a) Calculate
 (i) *BH* (ii) *GE* (iii) *BE* (iv) angle *GBE* (v) angle *HBE*
(b) Calculate the size of the angle between the diagonal *BE* and the plane *EFGH*.
(c) Write down the size of the angle between *BE* and the plane *ABGH*.

Answer 7

(a) (i) *BH* is the diagonal of the rectangle *ABGH*

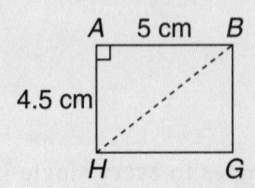

Using Pythagoras' Theorem,

$$BH^2 = 5^2 + 4.5^2$$

$$BH^2 = 45.25$$

$$BH = 6.72681\ldots$$

$$BH = 6.73 \text{ cm}$$

(ii) *GE* is the diagonal of rectangle *EFGH*.

Using Pythagoras' theorem,

$GE^2 = 8^2 + 5^2$

$GE^2 = 89$

$GE = \sqrt{89} = 9.43398\ldots$

$GE = 9.43$ cm

(iii) *BE* is the hypotenuse of triangle *BEG*.

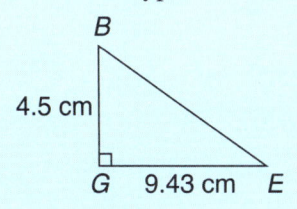

NOTE: In this part of the question it is best to use your calculator value for *GE*, (or you could use $GE^2 = 89$). If you still have the answer to the previous part displayed on your calculator use the ANS key.

Best practice

$BE^2 = 4.5^2 + 9.43398^2 = 109.25$

$BE = 10.45227\ldots$

$BE = 10.5$ cm

Alternative method

$BE^2 = 4.5^2 + 9.43^2 = 109.1749$

$BE = 10.44867\ldots$

$BE = 10.4$ cm NOTE: You must not round progressively to 10.45 then 10.5.

(iv) $\tan \angle GBE = \dfrac{GE}{BG} = \dfrac{9.43398\ldots}{4.5}$

$\tan \angle GBE = 2.09644\ldots$

$\angle GBE = 64.4989\ldots$

$\angle GBE = 64.5°$

NOTE: In the rest of the question either method will produce the same answer when rounded to 3 significant figures or 1 decimal place. Try both methods yourself, using $\sqrt{89}$ (or 9.43398…) or 9.43

(v) Angle *HBE* is in triangle *BEH*.

EH is perpendicular to plane *ABGH* so it is perpendicular to *BH*.
We now know all the sides of triangle *BEH* so any of the ratios can be used to find $\angle HBE$.

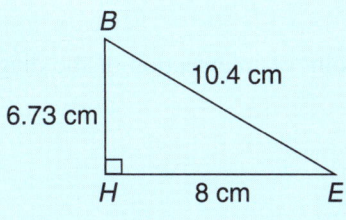

$\tan \angle HBE = \dfrac{HE}{BH} = \dfrac{8}{\sqrt{45.25}} = 1.18927\ldots$ NOTE: Using $\sqrt{45.25}$ or 6.72681… from part (a) (i).

> $\angle HBE = 49.94114...$
>
> $\angle HBE = 49.9°$
>
> (b) The angle between the diagonal *BE* and the plane *EFGH* is $\angle BEG$.
>
> The triangle is sketched in the answer to (a) (iii).
>
> $\tan \angle BEG = \dfrac{BG}{GE} = \dfrac{4.5}{9.43398} = 0.47699...$
>
> $\angle BEG = 25.50109...$
>
> $\angle BEG = 25.5°$
>
> The angle between the diagonal and the plane is 25.5°
>
> (c) The angle between *BE* and the plane is the angle *HBE*.
>
> $\angle HBE = 49.9°$

Make sure that you have worked through this example carefully and understood it. You should see that it can be a help to draw diagrams as you go along.

The questions you will be asked will not be as long as this, but this example should have helped you understand what is required.

Exercise 9.7

1.

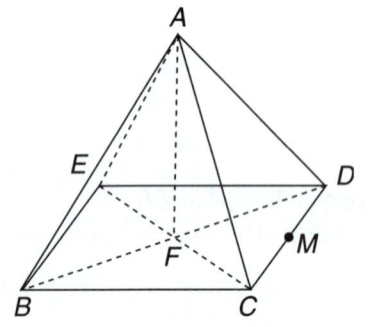

ABCDE is a square-based pyramid. *F* is the point of intersection of the diagonals of the base. *M* is the midpoint of the side *CD*.

BC = 5 centimetres, *AF* = 6 centimetres.

(a) Calculate *EC*.
(b) Calculate *AC*.
(c) Calculate the angle between *AC* and the base.
(d) Calculate angle *AMF*.

2.

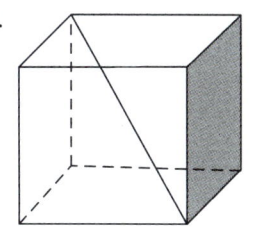

The diagram shows a cube with each side of length 8 centimetres. Calculate the length of the diagonal shown.

3.

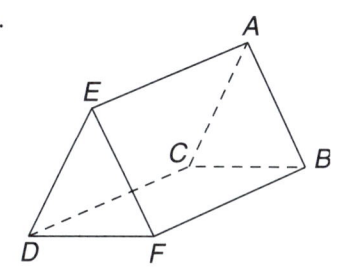

ABCDEF is an isosceles triangular prism with *DE* = *EF* = *AC* = *AB* = 7 centimetres. The length of the prism is 15 centimetres. Angle *DEF* is 40°.

(a) Calculate the length of *DF*.
(b) Calculate the length of *DB*.
(c) Calculate the length of *EB*.
(d) Use the cosine rule in triangle *BDE* to calculate the angle *EBD*.

Exercise 9.8

Mixed Exercise

1.

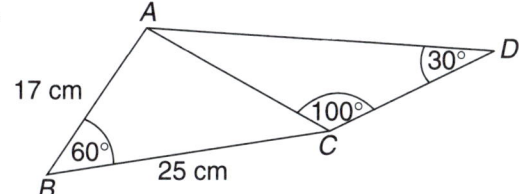

Using the diagram calculate
(a) *AC* (b) *AD* (c) the area of *ABCD* (d) angle *BAD*.

2.

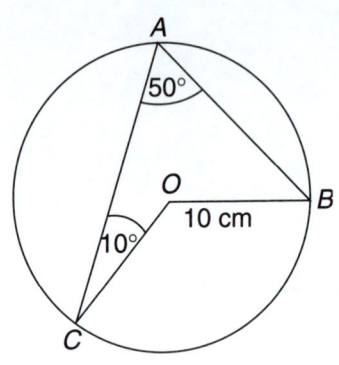

The diagram shows a circle, centre O, and radius 10 centimetres. Angle $CAB = 50°$.
(a) Find angle BOC.
(b) Calculate the length of the chord BC.
(c) Calculate the area of the quadrilateral $ABCO$.

NOTE: Draw in the radius OA.

3.

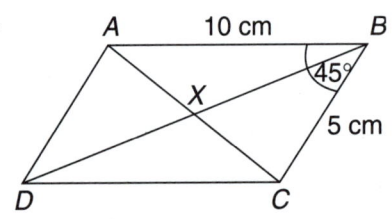

$ABCD$ is a parallelogram, and X is the point of intersection of the diagonals. Angle ABC is $45°$.

Calculate

(a) AC (b) DB (c) angle AXB (d) the area of triangle ACD.

4.

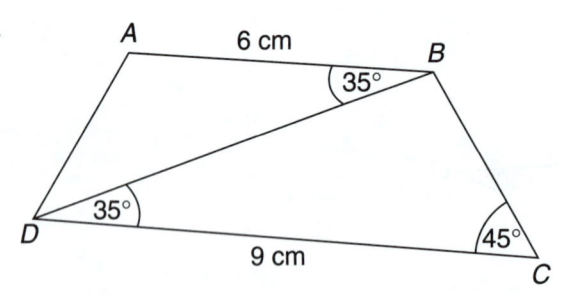

Using the diagram above calculate
(a) BD (b) AD (c) angle DAB (d) the area of the quadrilateral $ABCD$.

5.

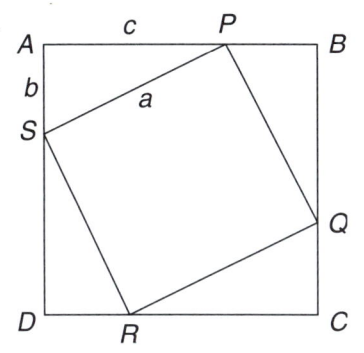

ABCD is a square, with sides of length $(b + c)$ centimetres.
PQRS is another square with sides of length a centimetres.

(a) Write down the area of the square ABCD in terms of b and c.
(b) Write down the area of the square PQRS.
(c) Write down the area of triangle APS in terms of b and c.
(d) The area of triangle $ASP = \dfrac{ABCD - PQRS}{4}$.

   Write $\dfrac{ABCD - PQRS}{4}$ in terms of a, b, and c.

(e) Put these two expressions for the area of the triangle ASP equal to each other. By multiplying out the brackets and simplifying show that $a^2 = b^2 + c^2$.
(f) What have you just proved by algebraic means?

Examination Questions

6. The diagram represents a framework.
   $BC = 1.3$ m, $BD = 1.9$ m and $BE = 1.5$ m.
   $B\hat{C}D = 76°$, $B\hat{A}E = 68°$ and $B\hat{E}D = 90°$.
   Calcualte

   (a) $D\hat{B}E$,
   (b) AE,
   (c) $B\hat{D}C$.

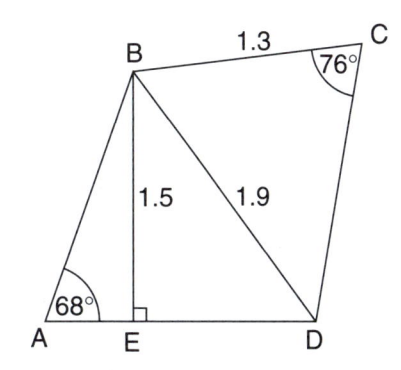

(4024/02 May/June 2007 q 2)

7.

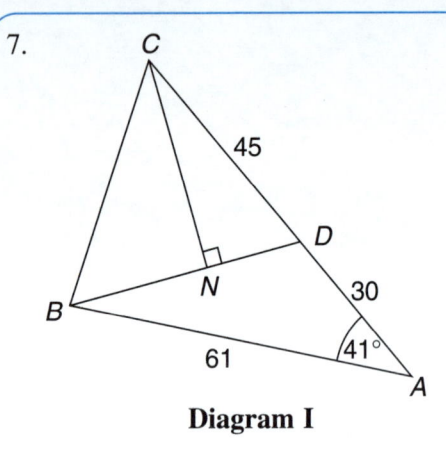

Diagram I

In Diagram I, the point D lies on AC and N is the foot of the perpendicular from C to BD.
AB = 61 m, AD = 30 m and DC = 45 m.
Angle BAC = 41°.

(a) Calculate BD.

(b) Show that, correct to the nearest square metre, the area of triangle BDA is 600 m².

(c) Explain why $\dfrac{\text{area of }\triangle BCD}{\text{area of }\triangle BDA} = \dfrac{3}{2}$.

(d) Calculate the area of triangle BCD.

(e) Hence calculate CN.

(f)

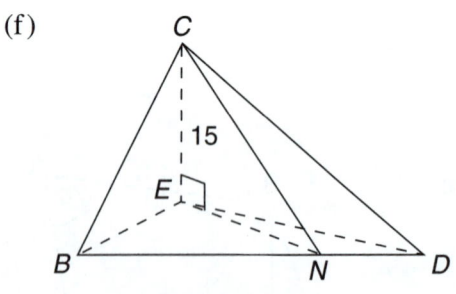

Diagram II

The same points B, C, D and N lie on a sloping plane.
The point E is 15 m vertically below C.
The points B, E, D and N lie on a horizontal plane.
Diagram II represents this information.
Calculate the angle of elevation of C from N.

(4024/02 May/June 2007 q 9)

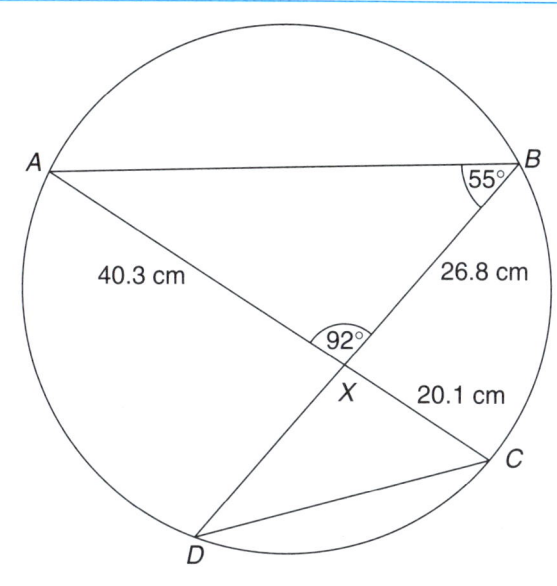

8.

A, *B*, *C* and *D* lie on a circle.
AC and *BD* intersect at *X*.
Angle *ABX* = 55° and angle *AXB* = 92°.
BX = 26.8 cm, *AX* = 40.3 cm and *XC* = 20.1 cm.

(a) Calculate the area of triangle *AXB*. **You must show your working**.
(b) Calculate the length of *AB*. **You must show your working**.
(c) Write down the size of angle *ACD*. Give a reason for your answer.
(d) Find the size of angle *BDC*.
(e) Write down the geometrical word which completes the statement
    "Triangle *AXB* is _____ to triangle *DXC*."
(f) Calculate the length of *XD*. **You must show your working**.

(0580/04 May/June 2007 q 3 a)

9. (a) Use your calculator to work out

$$\frac{1 - (\tan 40°)^2}{2(\tan 40°)}.$$

(b) Write your answer to **part (a)** in standard form.

(0580/02 May/June 2007 q 2)

10. Write the following in order of size, **smallest** first.

    cos 100°   sin 100°   tan 100°

(0580/02 May/June 2007 q 4)

11.

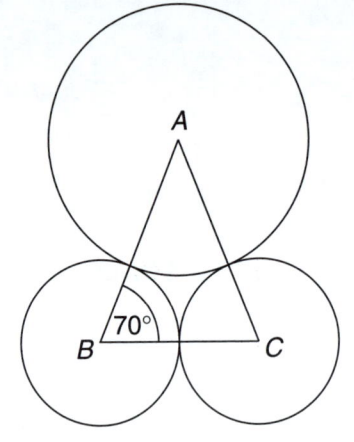

The diagram shows three touching circles.
A is the centre of a circle of radius x centimetres.
B and C are the centres of circles of radius 3.8 centimetres. Angle $ABC = 70°$. Find the value of x.

(0580/02 May/June 2007 q 14)

12. Calculate the value of $(\cos 40°)^2 + (\sin 40°)^2$.

(0580/02 May/June 2005 q 4)

13.

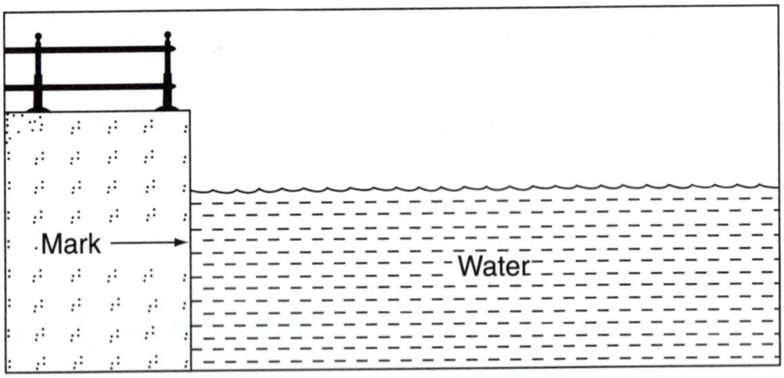

The height, h metres, of the water, above a mark on a harbour wall, changes with the tide.
It is given by the equation
$$h = 3\sin(30t)°$$
where t is the time in hours after midday.
(a) Calculate the value of h at midday.
(b) Calculate the value of h at 19 00.
(c) Explain the meaning of the negative sign in your answer.

(0580/02 May/June 2005 q 17)

14. A plane flies from Auckland *(A)* to Gisborne *(G)* on a bearing of 115°.
    The plane then flies onto Wellington *(W)*. Angle *AGW* = 63°.

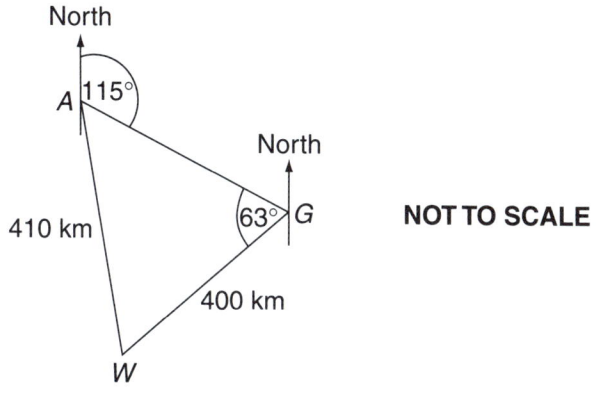

    NOT TO SCALE

    (a) Calculate the bearing of Wellington from Gisborne.
    (b) The distance from Wellington to Gisborne is 400 kilometres.
        The distance from Auckland to Wellington is 410 kilometres.
        Calculate the bearing of Wellington from Auckland.

    (0580/02 May/June 2005 q 20)

15.

    NOT TO SCALE

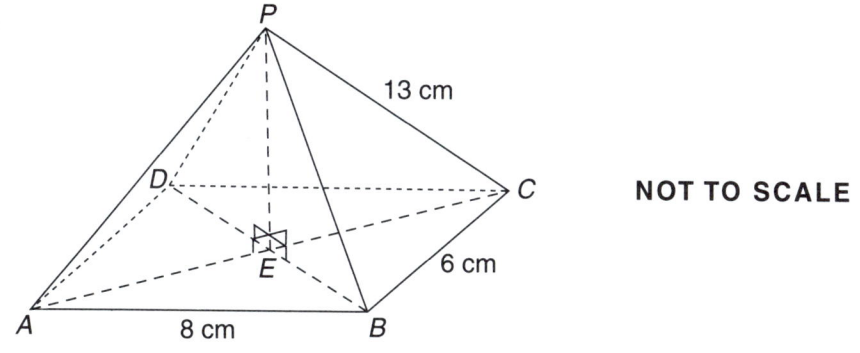

    The diagram shows a pyramid on a horizontal rectangular base *ABCD*.
    The diagonals of *ABCD* meet at *E*. P is vertically above *E*. *AB* = 8 cm, *BC* = 6 cm
    and *PC* = 13 cm.
    (a) Calculate *PE*, the height of the pyramid.
    (b) Calculate the volume of the pyramid.

    [The volume of a pyramid is given by $\frac{1}{3}$ × area of base × height.]

    (c) Calculate angle *PCA*.
    (d) *M* is the midpoint of *AD* and *N* is the midpoint of *BC*.
        Calculate angle *MPN*.
    (e) (i) Calculate angle *PBC*.
        (ii) *K* lies on *PB* so that *BK* = 4 cm.
        Calculate the length of *KC*.

    (0580/04 May/June 2006 q 6)

16.

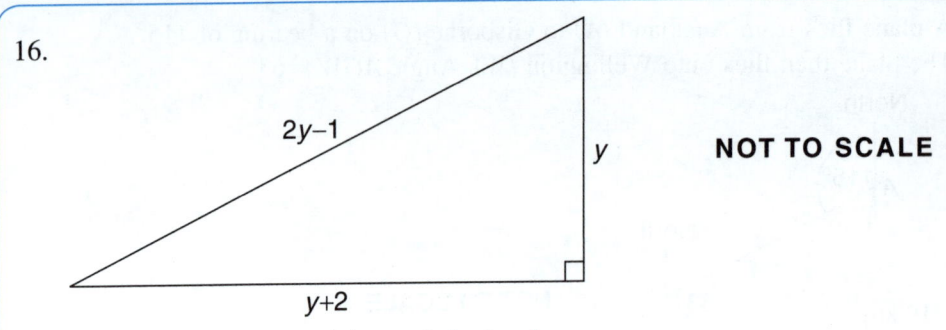

The diagram shows a right-angled triangle.
The lengths of the sides are given in terms of y.
  (i) Show that $2y^2 - 8y - 3 = 0$.
  (ii) Solve the equation $2y^2 - 8y - 3 = 0$, giving your answers to 2 decimal places.
  (iii) Calculate the area of the triangle.

(0580/04 May/June 2006 q 8 b)

17.

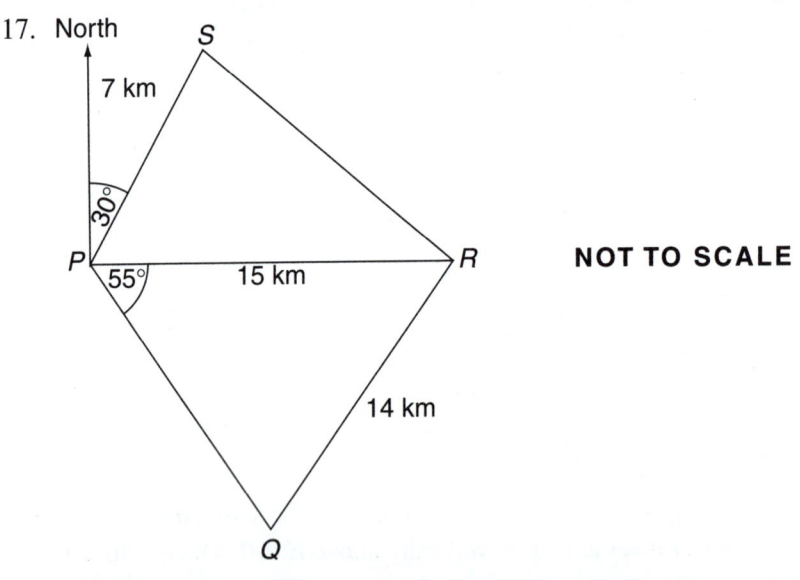

The quadrilateral *PQRS* shows the boundary of a forest.
A straight 15 kilometre road goes due east from *P* to *R*.
  (a) The bearing of *S* from *P* is 030° and *PS* = 7 km.
      (i) Write down the size of angle *SPR*.
      (ii) Calculate the length of *RS*.
  (b) Angle *RPQ* = 55° and *QR* = 14 km.
      (i) Write down the bearing of *Q* from *P*.
      (ii) Calculate the acute angle *PQR*.
      (iii) Calculate the length of *PQ*.
  (c) Calculate the area of the forest, correct to the nearest square kilometre.

(0580/04 Oct/Nov 2005 q 3)

18.

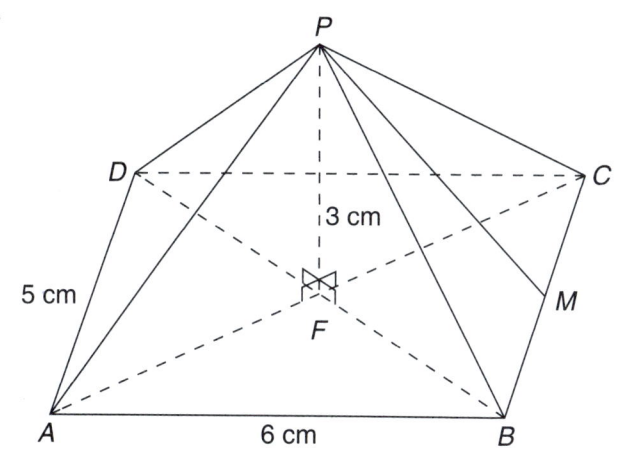

The diagram shows a pyramid on a rectangular base *ABCD*, with *AB* = 6 cm and *AD* = 5 cm. The diagonals *AC* and *BD* intersect at *F*.
The vertical height *FP* = 3 cm.

(a) How many planes of symmetry does the pyramid have?
(b) Calculate the volume of the pyramid.

   [The volume of a pyramid is $\frac{1}{3}$ × area of base × height.]

(c) The midpoint of *BC* is *M*. Calculate the angle between *PM* and the base.
(d) Calculate the angle between *PB* and the base.
(e) Calculate the length of *PB*.

(0580/04 Oct/Nov 2005 q 6)

19.

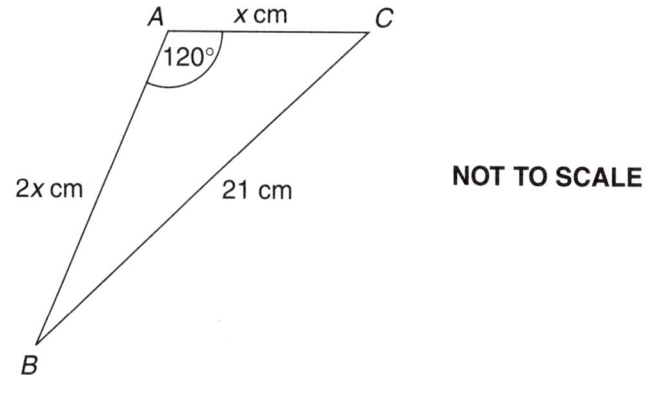

NOT TO SCALE

In triangle *ABC*, *AB* = 2*x* cm, *AC* = *x* cm, *BC* = 21 cm and angle *BAC* = 120°.
Calculate the value of *x*.

(0580/21 May/June 2008 q 11)

20. sin $x°$ = 0.86603 and $0 \leqslant x \leqslant 180$.
    Find the two values of x.
    (0580/21 Oct/Nov 2008 q 6)

21. $f(x) = \cos x°$, $g(x) = 2x + 4$
    Find   (a) $f(60)$,   (b) $fg(88)$,   (c) $g^{-1}(f(x))$.
    (0580/21 Oct/Nov 2008 q 15)

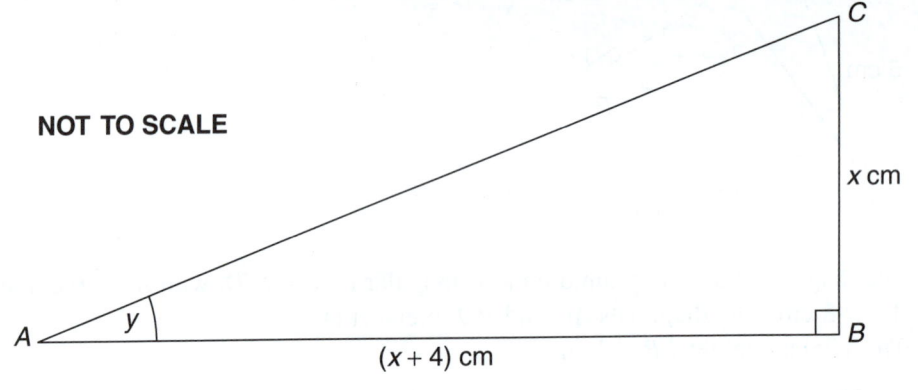

22. **NOT TO SCALE**

    (a) When the area of triangle ABC is 48 cm^2,
        (i) show that $x^2 + 4x - 96 = 0$,
        (ii) solve the equation $x^2 + 4x - 96 = 0$,
        (iii) write down the length of AB.
    (b) When $\tan y = \dfrac{1}{6}$, find the value of x.
    (c) When the length of AC is 9 cm,
        (i) show that $2x^2 + 8x - 65 = 0$,
        (ii) solve the equation $2x^2 + 8x - 65 = 0$,
            (**Show your working** and give your answers correct to 2 decimal places.)
        (iii) Calculate the perimeter of triangle ABC.
    (0580/04 Oct/Nov 2008 q 2)

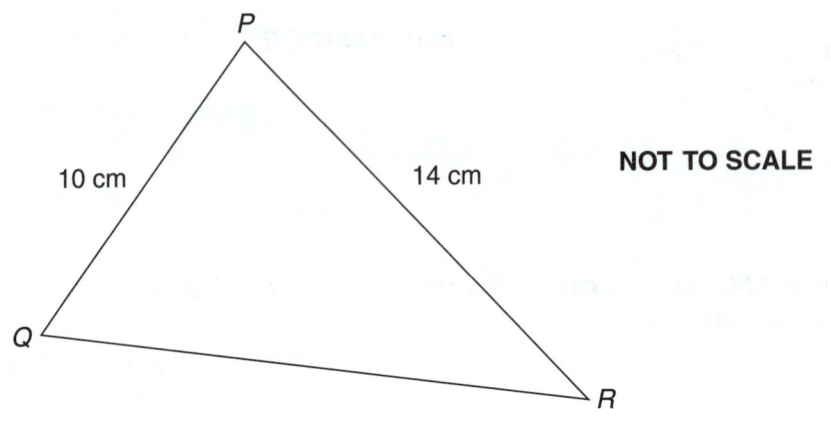

23. **NOT TO SCALE**

    In triangle PQR, angle QPR is acute, PQ = 10 cm and PR = 14 cm.

(a) The area of triangle *PQR* is 48 cm².
   Calculate angle *QPR* and show that it rounds to 43.3°, correct to 1 decimal place. You must show all your working.
(b) Calculate the length of the side *QR*.

(0580/04 May/June 2009 q 3)

24.

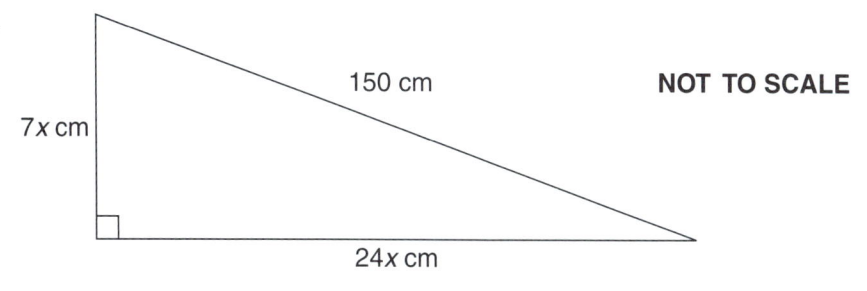

150 cm

7*x* cm

24*x* cm

NOT TO SCALE

The right-angled triangle shown in the diagram has sides of length 7*x* cm, 24*x* cm and 150 cm.
(a) Show that $x^2 = 36$.
(b) Calculate the perimeter of the triangle.

(0580/02 Oct/Nov 2006 q 10)

25.

T

h

B   25° A

80 m

18°

C

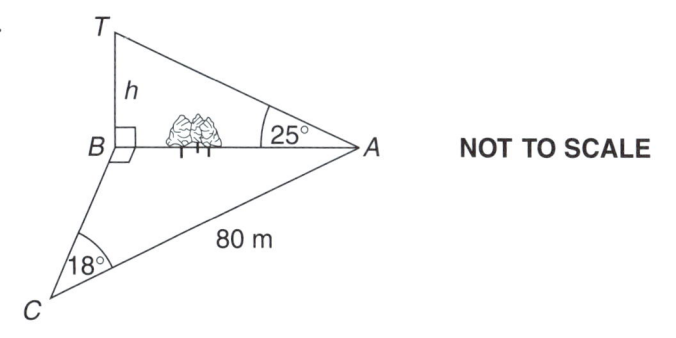

NOT TO SCALE

Mahmoud is working out the height, *h* metres, of a tower *BT* which stands on level ground.
He measures the angle *TAB* as 25°.
He cannot measure the distance *AB* and so he walks 80 m from *A* to *C*, where angle *ACB* = 18° and angle *ABC* = 90°.
Calculate
(a) the distance *AB*,
(b) the height of the tower, *BT*.

(0580/21 May/June 2009 q 15)

26.

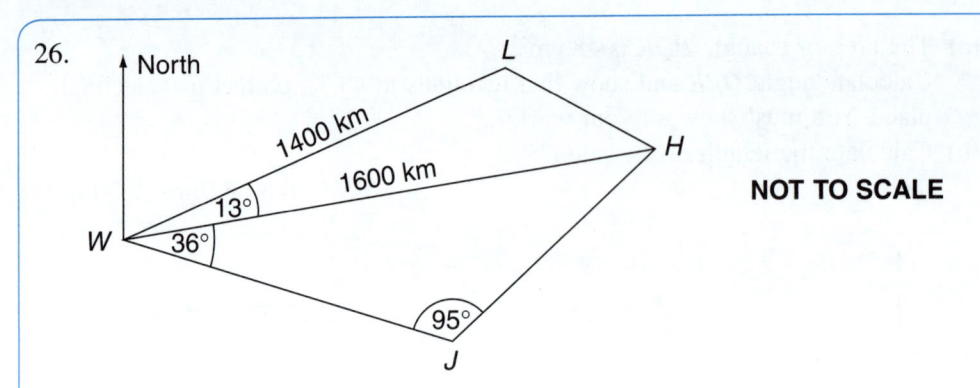

The diagram shows the positions of four cities in Africa, Windhoek (W), Johannesburg (J), Harari (H) and Lusaka (L).
WL = 1400 km and WH = 1600 km.
Angle LWH = 13°, angle HWJ = 36° and angle WJH = 95°.
(a) Calculate the distance LH.
(b) Calculate the distance WJ.
(c) Calculate the area of quadrilateral WJHL.
(d) The bearing of Lusaka from Windhoek is 060°.
Calculate the bearing of
(i) Harari from Windhoek,
(ii) Windhoek from Johannesburg.
(e) On a map the distance between Windhoek and Harari is 8 cm.
Calculate the scale of the map in the form 1:n.

(0580/04 Oct/Nov 2006 q 2)

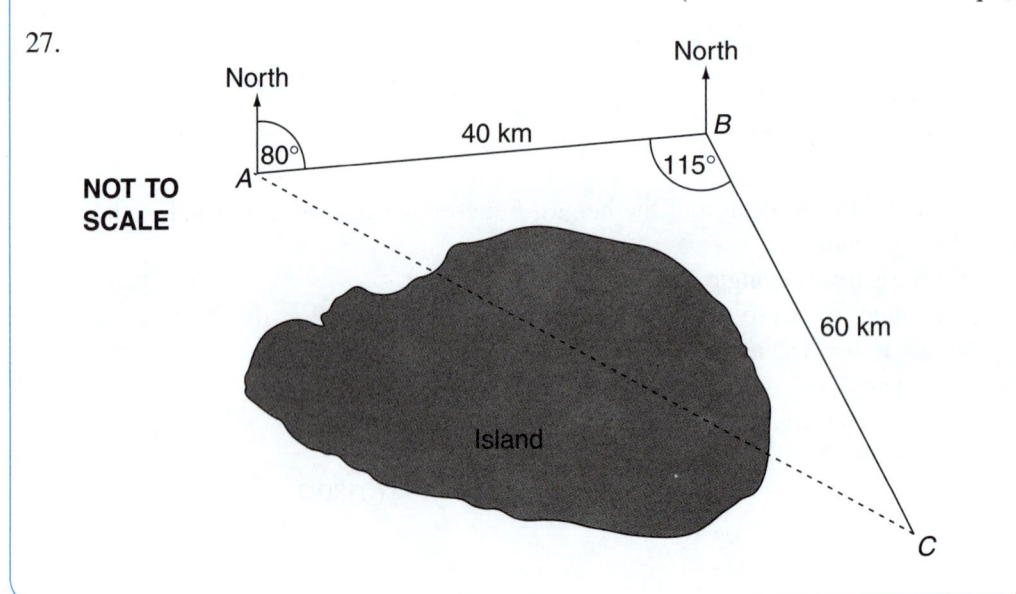

27.

To avoid an island, a ship travels 40 kilometres from A to B and then 60 kilometres from B to C.
The bearing of B from A is 080° and angle ABC is 115°.
(a) The ship leaves A at 11 55.
It travels at an average speed of 35 km/h.
Calculate, to the nearest minute, the time it arrives at C.
(b) Find the bearing of
   (i) A from B,
   (ii) C from B.
(c) Calculate the straight line distance AC.
(d) Calculate angle BAC.
(e) Calculate how far C is **east** of A.

(0580/04 Oct/Nov 2008 q 5)

28.

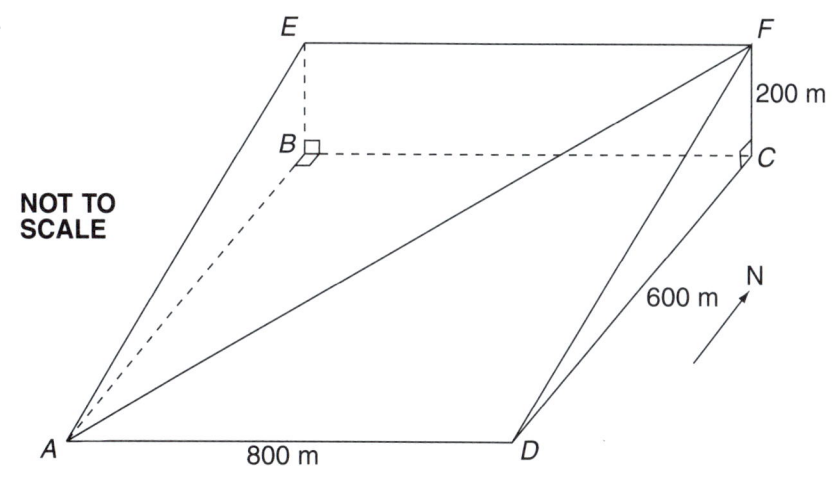

NOT TO SCALE

ABCD, BEFC and AEFD are all rectangles.
ABCD is horizontal, BEFC is vertical and AEFD represents a hillside.
AF is a path on the hillside.
AD = 800 m, DC = 600 m and CF = 200 m
(a) Calculate the angle that the path AF makes with ABCD.
(b) In the diagram D is due south of C.
Jasmine walks down the path from F to A in bad weather. She cannot see the path ahead.
The compass bearing she must use is the bearing of A from C.
Calculate this bearing.

(0580/21 May/June 2008 q 21)

29.

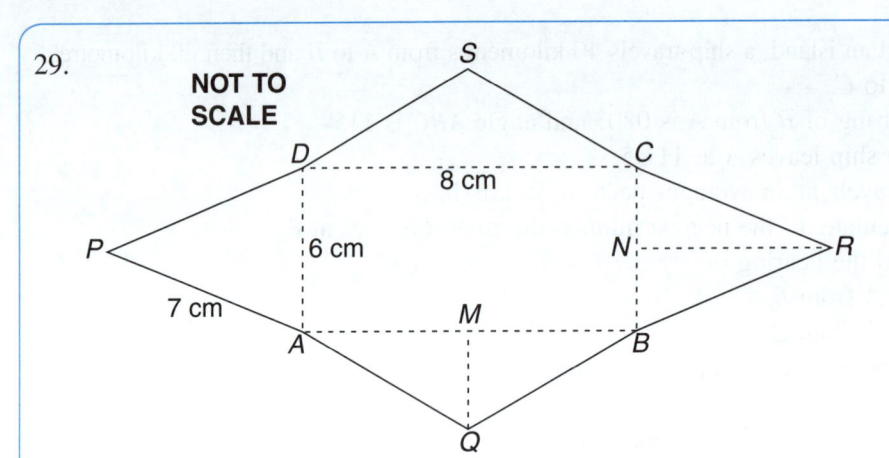

The diagram above shows the net of a pyramid.
The base *ABCD* is a rectangle 8 cm by 6 cm.
All the sloping edges of the pyramid are of length 7 cm.
M is the midpoint of *AB* and *N* is the midpoint of *BC*.
(a) Calculate the length of
   (i) *QM*,   (ii) *RN*.
(b) Calculate the surface area of the pyramid.
(c)

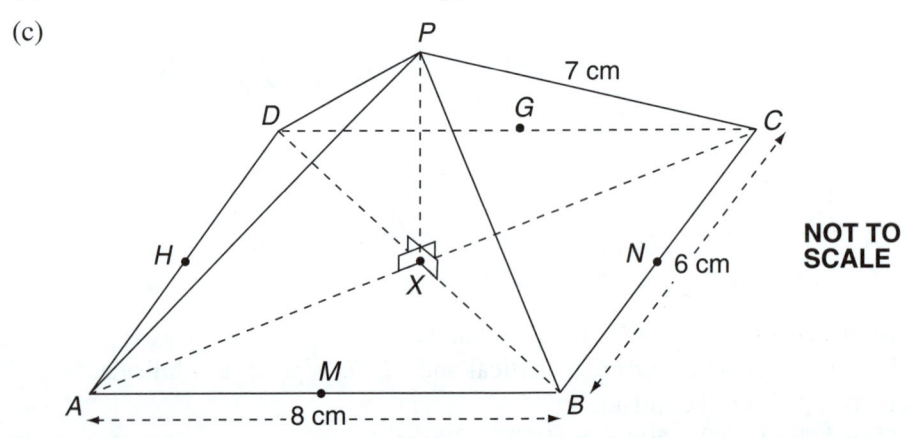

The net is made into a pyramid, with *P*, *Q*, *R* and *S* meeting at *P*.
The midpoint of *CD* is *G* and the midpoint *DA* is *H*.
The diagonals of the rectangle *ABCD* meet at *X*.
 (i) Show that the height, *PX*, of the pyramid is 4.90 cm, correct to 2 decimal places.
 (ii) Calculate angle *PNX*.
 (iii) Calculate angle *HPN*.
 (iv) Calculate the angle between the edge *PA* and the base *ABCD*.
 (v) Write down the vertices of a triangle which is a plane of symmetry of the pyramid.

(0580/04 Oct/Nov 2007 q 5)

Chapter 10

Transformations, Vectors and Matrices

We now look further at vectors and transformations. We will study two new transformations, stretch and shear, and how matrices can be used to describe transformations.

Core Skills

1. Describe fully each of the following single transformations. In each case *A* is mapped to *B*.

   (a)

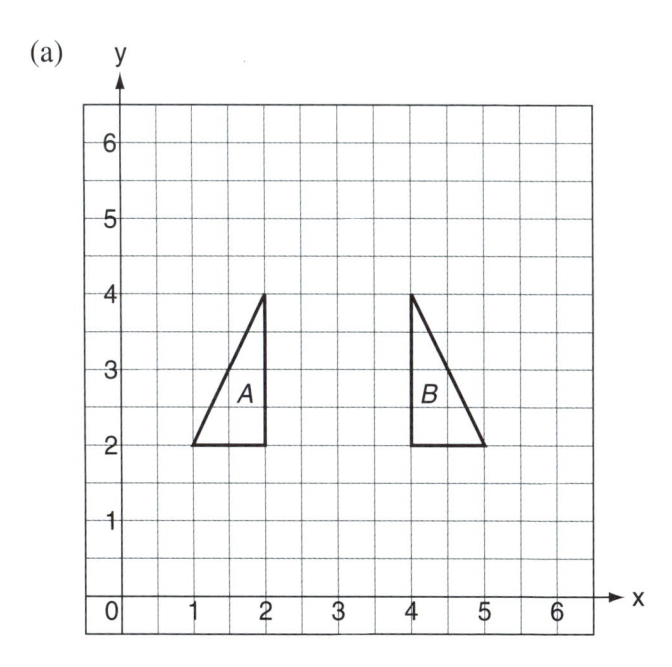

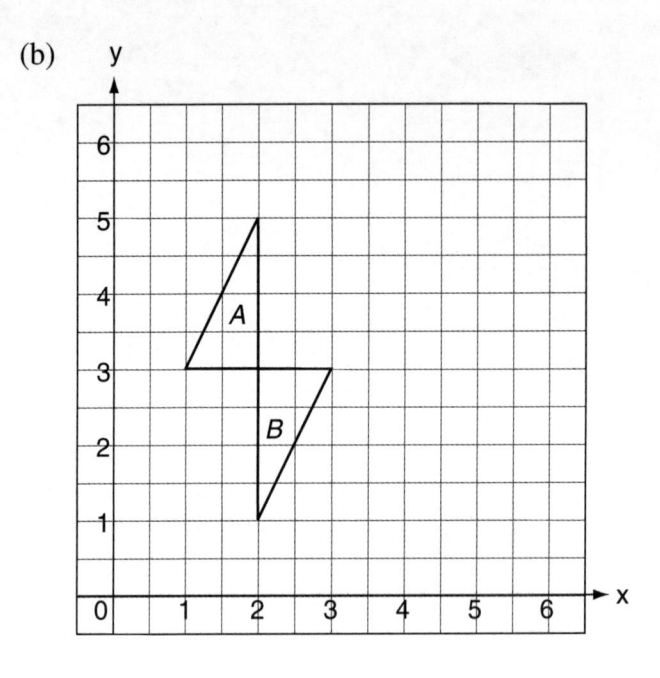

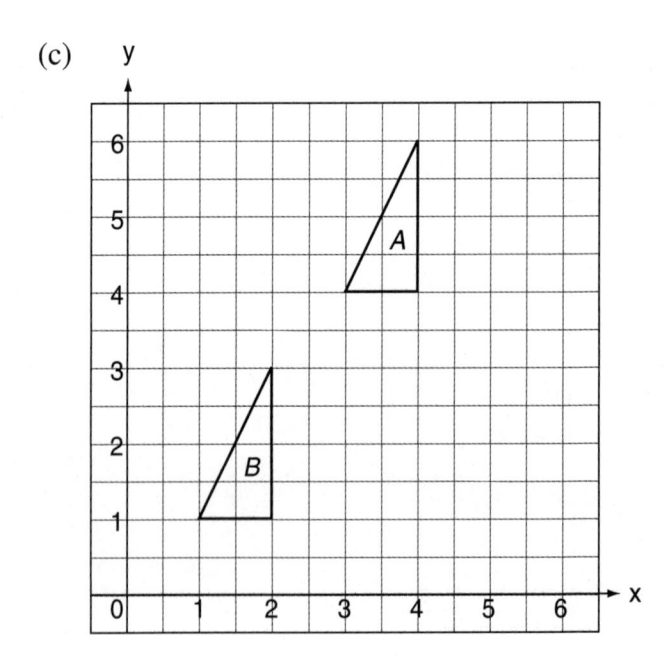

(b)

(c)

(d)

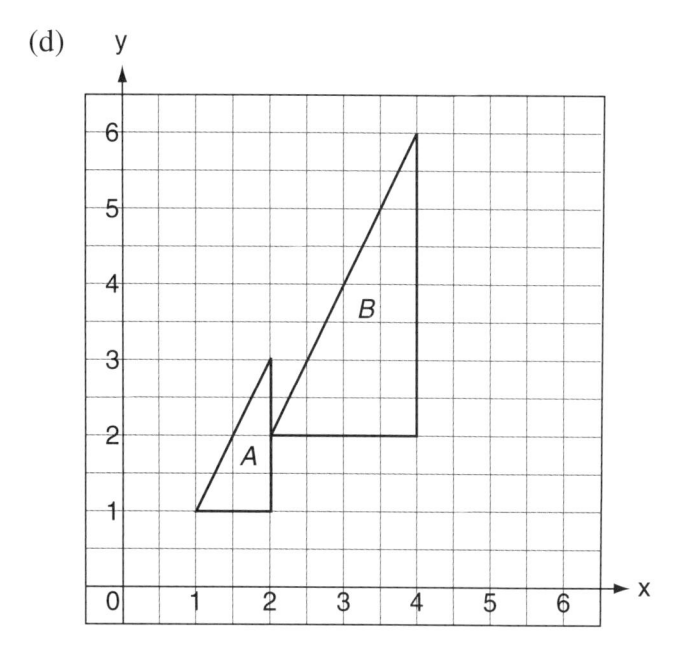

2. Simplify the following vectors.

   (a) $4\begin{pmatrix}-1\\0\end{pmatrix}+\begin{pmatrix}3\\5\end{pmatrix}$

   (b) $3\begin{pmatrix}1\\-4\end{pmatrix}-2\begin{pmatrix}1\\-6\end{pmatrix}$

3. Find x and y.

   (a) $\begin{pmatrix}x\\3y\end{pmatrix}+\begin{pmatrix}5\\7\end{pmatrix}=\begin{pmatrix}-1\\-2\end{pmatrix}$

   (b) $\begin{pmatrix}2x\\-5y\end{pmatrix}+\begin{pmatrix}2y\\x\end{pmatrix}=\begin{pmatrix}10\\-7\end{pmatrix}$

4. Which of these vectors are parallel?

   $a=\begin{pmatrix}-3\\4\end{pmatrix}$ $b=\begin{pmatrix}-3\\-4\end{pmatrix}$ $c=\begin{pmatrix}3\\4\end{pmatrix}$

   $d=\begin{pmatrix}4\\3\end{pmatrix}$ $e=\begin{pmatrix}-6\\-8\end{pmatrix}$ $f=\begin{pmatrix}-4\\3\end{pmatrix}$

5. What three things can you say about the relationship between these two vectors?

   $v=\begin{pmatrix}6\\-15\end{pmatrix}$ $w=\begin{pmatrix}-2\\5\end{pmatrix}$

More about Vectors

Lengths of vectors

Vectors have a direction and a length. The diagram shows two vectors, *a* and *b* which have the same length, but different directions, and two vectors *c* and *d* which have the same direction but different lengths.

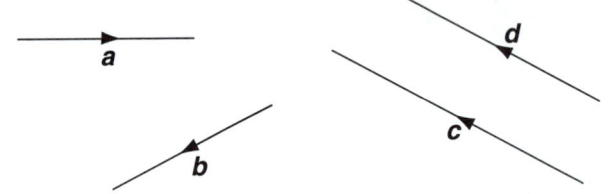

Each of these vectors would produce an image in a different place when applied to an object.

The length of a vector is easy to calculate using Pythagoras' Theorem.

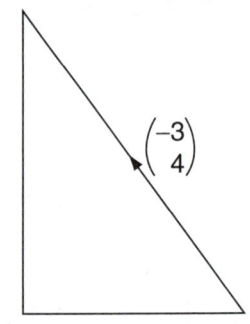

The diagram shows how the vector can be part of a right-angled triangle. The two parts of the vector make up two sides of the triangle, and the vector itself is the hypotenuse.

The length of the vector $\begin{pmatrix} -3 \\ 4 \end{pmatrix}$ as shown in the diagram is $\sqrt{(-3)^2 + 4^2} = \sqrt{9+16} = 5$.

Notice how the fact that the *x*-component is negative 3 makes no difference when we are calculating length and not direction, since -3 becomes $+9$ when squared.

The following vectors all have different directions but the same lengths (5 units).

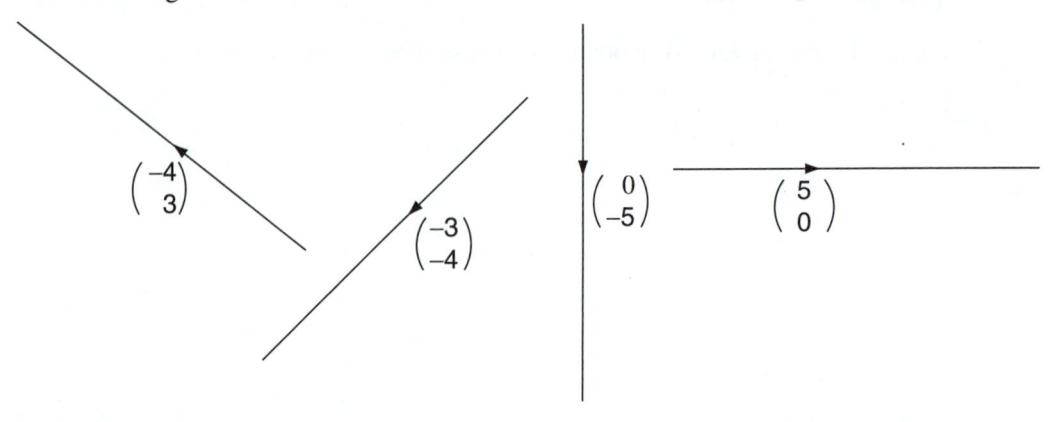

Transformations, Vectors and Matrices

The shorthand notation for the length of the vector a is $|a|$, which is read 'the **modulus** of a', or simply '**mod** a'. The length of a vector may also be referred to as the *magnitude* of the vector.

So if $a = \begin{pmatrix} x \\ y \end{pmatrix}$, then $|a| = \sqrt{x^2 + y^2}$.

The particular vector which joins the point A to the point B is written $\overrightarrow{AB}$, and is called a **directed line segment**. *'Directed'* means it has a *direction*, that is from A to B, (shown by the arrow) and *'line segment'* means that it is only part of a line, that is, the part between A and B. The line itself can go on indefinitely.

The length (or magnitude) of $\overrightarrow{AB}$, or the modulus of $\overrightarrow{AB}$, is written $|\overrightarrow{AB}|$, and $|\overrightarrow{AB}| = |\overrightarrow{BA}|$.

Combining vectors

A vector is a movement of the whole *plane*. (A plane in the mathematical sense is a flat surface.)

This may sound confusing, but one way to imagine it is to think of a piece of paper (the plane) on a flat, shiny surface such as glass. You could move that piece of paper across the glass without turning it, through a certain distance and in a certain direction, by putting your finger on *any* point on the paper.

This means that a vector may be drawn anywhere on your diagram and will always have the same effect as long as it is pointing in the same direction and is the same length.

The first diagram shows two vectors a and b drawn on two edges of a rectangle $PQRS$. They are the same length, but have directions at right angles to each other. The second diagram shows that these two vectors may also be represented on the other sides of the rectangle, because the opposite sides are parallel and of equal length. Notice that the corresponding arrows must point in the same direction in the two diagrams.

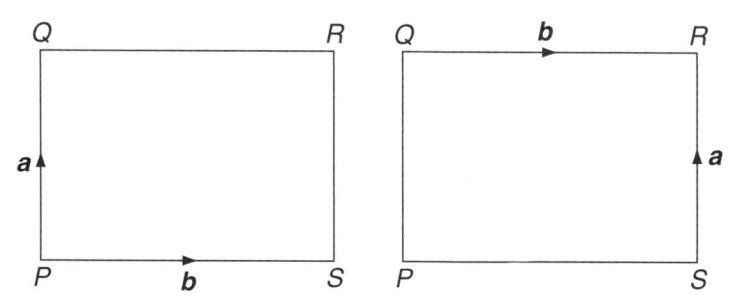

Suppose we wanted to get from P to R on the rectangle. We could go from:

- P to Q and then from Q to R,
- from P to S and then from S to R, or
- directly from P to R.

In terms of directed line segments:

$$\overrightarrow{PQ} + \overrightarrow{QR} = \overrightarrow{PS} + \overrightarrow{SR} = \overrightarrow{PR}$$

and in terms of vectors:

$$a + b = b + a = \overrightarrow{PR}$$

This shows that vectors may be combined to give a single vector if they are joined 'head to tail', that is, if the arrows follow round, without a break.

It is convenient to think of combining vectors as going on a journey. There are often several routes, but the start and the end of the journey are the same.

The next three diagrams show that the vectors may be joined in any order as long as the 'head to tail' rule is followed. The double arrow shows in each case the single vector which is equivalent to the addition. This is often called the **resultant vector**.

You will see that the resultant is the same regardless of the order in which the vectors are added.

In each case S is the Start and F the Finish.

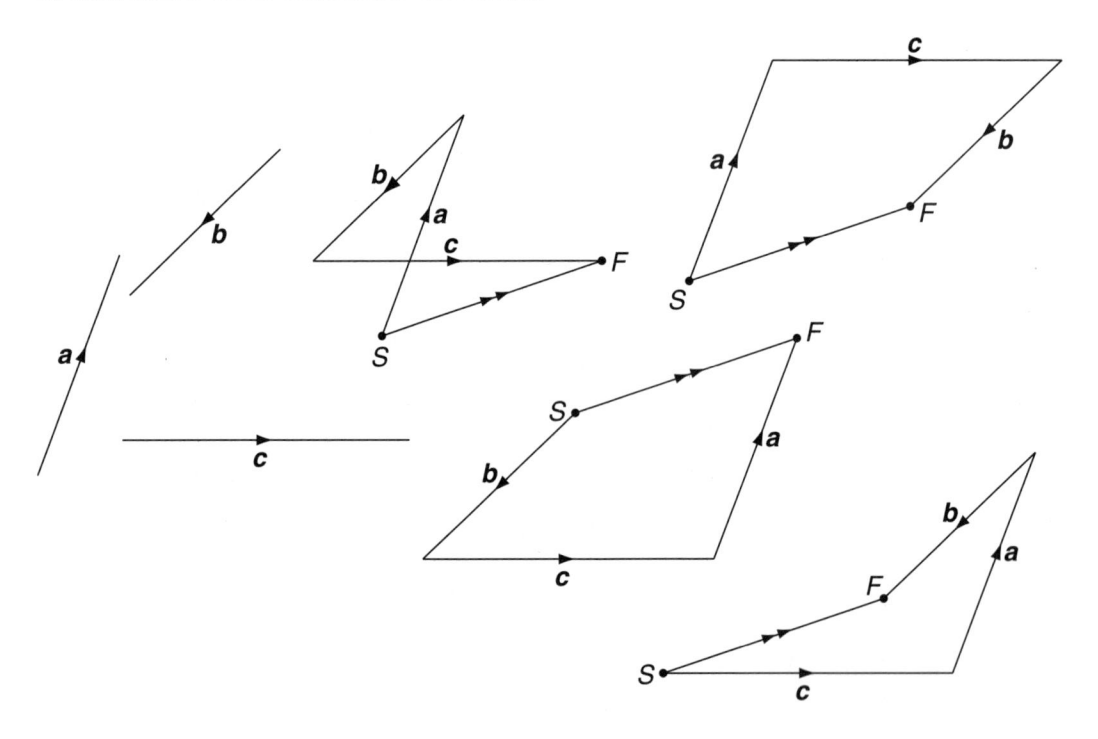

In terms of column vectors, if $a = \begin{pmatrix} 1 \\ 3 \end{pmatrix}$, $b = \begin{pmatrix} -2 \\ -2 \end{pmatrix}$ and $c = \begin{pmatrix} 4 \\ 0 \end{pmatrix}$, then

$$a + b + c = \begin{pmatrix} 1-2+4 \\ 3-2+0 \end{pmatrix} = \begin{pmatrix} 3 \\ 1 \end{pmatrix}, \quad b + c + a = \begin{pmatrix} -2+4+1 \\ -2+0+3 \end{pmatrix} = \begin{pmatrix} 3 \\ 1 \end{pmatrix}$$

and so on.

Transformations, Vectors and Matrices 259

Example 1

ABCDEF is a regular hexagon.

Copy the diagram.

(a) Find *x* in terms of *p* and/or *q*.
(b) Find *y* in terms of *p* and/or *q* and/or *r*.
(c) Find *z* in terms of *p* and/or *q* and/or *r*.

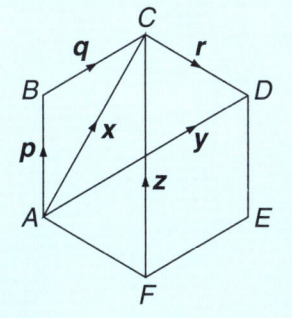

NOT TO SCALE

Answer 1

(a) $\vec{AC} = \vec{AB} + \vec{BC}$
$x = p + q$ NOTE: *x* goes from *A* to *C*. In terms of a journey you could also go from *A* to *B* and then on to *C*. So $\vec{AC} = \vec{AB} + \vec{BC}$.

(b) $\vec{AD} = \vec{AB} + \vec{BC} + \vec{CD}$
$y = p + q + r$ NOTE: Notice that when you join head to tail, $\vec{AB} + \vec{BC} + \vec{CD} = \vec{AD}$.

(c) $z = \vec{FC} = \vec{FA} + \vec{AB} + \vec{BC}$
$= -r + p + q$ NOTE: $\vec{FA} = \vec{DC} = -r$

You should be aware that, while we can directly add vectors when they are pointing in different directions, and joined head to tail, the same does not apply to lengths of lines.

For example, looking at the triangle below, where *a* is the vector $\vec{PQ}$ and *b* is the vector $\vec{QR}$.

$\vec{PR} = a + b$ $PR \neq 3 + 4$

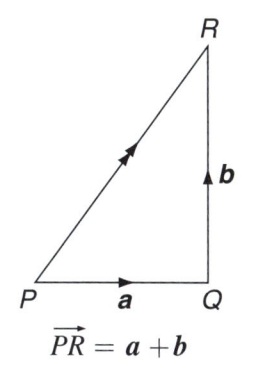

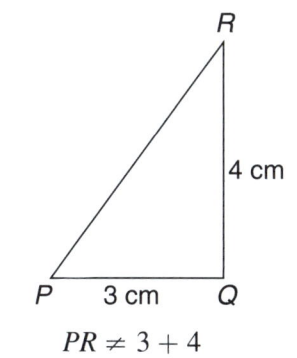

Subtraction of one vector from another is dealt with by adding the negative of the vector to be subtracted.
For example, look at the two diagrams below.
In the first, $x = a + b$, and in the second, $y = a - b$.

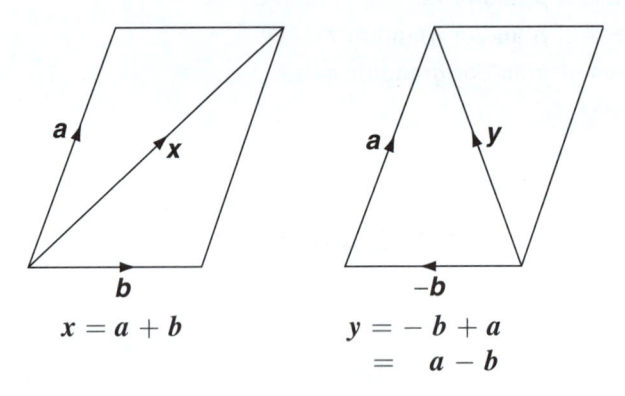

$$x = a + b$$

$$y = -b + a = a - b$$

Multiplying a vector by a number

Multiplying a vector by a positive number alters its length but not its direction. The first line drawn below shows the directed line segment $\overrightarrow{AB}$, which is a representative of the vector v. The second line drawn shows M, the midpoint of the line AB. It can be seen that the directed line segment $\overrightarrow{AM}$ is equal to the directed line segment $\overrightarrow{MB}$, which is equal to $\frac{1}{2}v$.

The third line shows the line extended to C, where $AB = BC$. It can be seen that $\overrightarrow{AC} = 2v$.

$$\overrightarrow{AB} = v \qquad \overrightarrow{AM} = \frac{1}{2}v \qquad \overrightarrow{AC} = 2v$$

Example 2

Draw a parallelogram $ABCD$.
Mark M, the midpoint of BD.
$\overrightarrow{AD} = p$, $\overrightarrow{AB} = q$.
Find the following directed line segments in terms of p and q.

(a) $\overrightarrow{BD}$

(b) $\overrightarrow{AC}$

(c) $\overrightarrow{AB} + \overrightarrow{BM}$

(d) $\overrightarrow{AM}$

(e) What can you say about $\overrightarrow{AM}$ and $\overrightarrow{AC}$ and hence about the diagonals of a parallelogram?

Answer 2

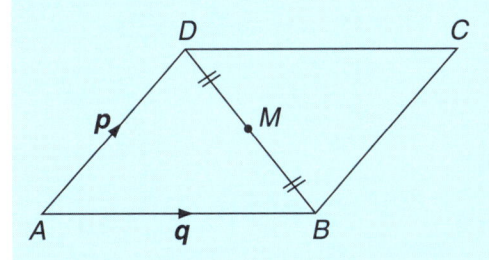

(a) $\overrightarrow{BD} = \overrightarrow{BA} + \overrightarrow{AD} = -q + p = p - q$
(b) $\overrightarrow{AC} = \overrightarrow{AB} + \overrightarrow{BC} = q + p$ NOTE: $\overrightarrow{BC} = \overrightarrow{AD} = p$
(c) $\overrightarrow{AB} + \overrightarrow{BM} = \overrightarrow{AB} + \frac{1}{2}\overrightarrow{BD} = q + \frac{1}{2}(p-q) = \frac{1}{2}(p+q)$
(d) $\overrightarrow{AM} = \overrightarrow{AB} + \overrightarrow{BM} = \frac{1}{2}(p+q)$
(e) $\overrightarrow{AM} = \frac{1}{2}\overrightarrow{AC} = \frac{1}{2}(p+q)$

Therefore M is also the midpoint of AC, and hence the diagonals of a parallelogram bisect each other.

Dividing a line in a given ratio

The line AB is to be divided in the ratio $2 : 3$ by the point C.
Just as in ordinary ratio questions we think of dividing the line into 5 equal parts, (because $2 + 3 = 5$).
Then $AC = 2$ parts and $CB = 3$ parts,

$AC = \frac{2}{5}AB$ and also $\overrightarrow{AC} = \frac{2}{5}\overrightarrow{AB}$

It is important to notice that dividing the line BA in the same ratio would move C nearer to B. The order of the letters specifying the line segment matters.

262 Extended Mathematics for Cambridge IGCSE

Example 3

(a) Draw a triangle *ABC*. Mark the point *D* which divides *BC* in the ratio 3 : 2. $\overrightarrow{AB} = p$ and $\overrightarrow{AC} = q$.

(b) Find $\overrightarrow{AD}$ in terms of *p* and *q*.

Answer 3

(a)

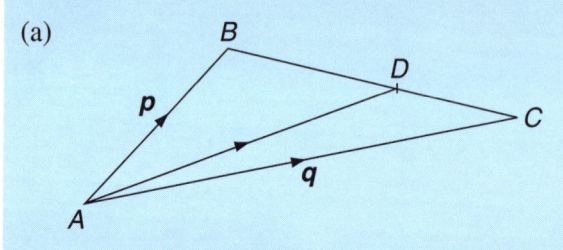

(b) $\overrightarrow{AD} = \overrightarrow{AB} + \overrightarrow{BD}$

To find $\overrightarrow{BD}$:

$$\overrightarrow{BD} = \frac{3}{5}\overrightarrow{BC}$$

$$\overrightarrow{BD} = \frac{3}{5}(-p + q)$$

so $\overrightarrow{AD} = p + \frac{3}{5}(-p + q)$

$$\overrightarrow{AD} = \frac{2}{5}p + \frac{3}{5}q$$

Parallel vectors

As seen in Chapter 10 of the Core Book, if one vector is a multiple of another, either positive or negative, then the two vectors are parallel to each other.
However, they could also be in the same straight line. For example,

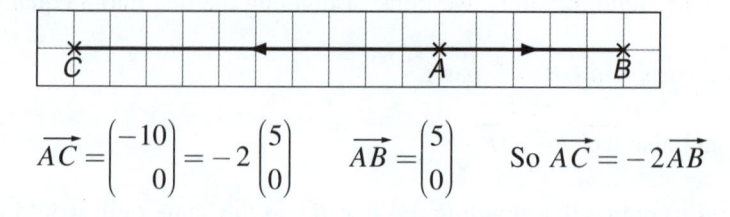

$$\overrightarrow{AC} = \begin{pmatrix} -10 \\ 0 \end{pmatrix} = -2 \begin{pmatrix} 5 \\ 0 \end{pmatrix} \qquad \overrightarrow{AB} = \begin{pmatrix} 5 \\ 0 \end{pmatrix} \qquad \text{So } \overrightarrow{AC} = -2\overrightarrow{AB}$$

Therefore $\overrightarrow{AB}$ and $\overrightarrow{AC}$ are parallel (have the same direction, but in opposite senses), but they have the point *A* in common, so this can only be true if they are in the same straight line. We say that "A is common to both", meaning that A belongs to both. The factor −2 shows that they are pointing in opposite senses, and that $\overrightarrow{AC}$ is twice as long as $\overrightarrow{AB}$.

Example 4

(a) Show that the vectors $u = \begin{pmatrix} 2 \\ 6 \end{pmatrix}$, $v = \begin{pmatrix} -6 \\ -18 \end{pmatrix}$ and $w = \begin{pmatrix} 1 \\ 3 \end{pmatrix}$ are all parallel to each other.

(b) What else can you say about these three vectors?

Answer 4

(a) $u = \begin{pmatrix} 2 \\ 6 \end{pmatrix} = 2\begin{pmatrix} 1 \\ 3 \end{pmatrix} = 2w \quad v = \begin{pmatrix} -6 \\ -18 \end{pmatrix} = -6\begin{pmatrix} 1 \\ 3 \end{pmatrix} = -6w$

$\frac{1}{2}u = -\frac{1}{6}v = w$

Therefore the vectors are all parallel.

(b) u is twice as long as w.

v is six times as long as w.

v is in the opposite direction (sense) to u and w.

v is three times as long as u.

$u = 2w, \quad v = -6w, \quad v = -3u.$

Exercise 10.1

1. Find the lengths of these vectors. Leave your answers in surd (square root) form, simplified where possible.

   (a) $\begin{pmatrix} 3 \\ 9 \end{pmatrix}$ (b) $\begin{pmatrix} -5 \\ -5 \end{pmatrix}$ (c) $\begin{pmatrix} 6 \\ -4 \end{pmatrix}$ (d) $\begin{pmatrix} -2 \\ 5 \end{pmatrix}$

2. $u = \begin{pmatrix} 7 \\ 1 \end{pmatrix}, v = \begin{pmatrix} 3 \\ 6 \end{pmatrix}, w = \begin{pmatrix} -2 \\ -5 \end{pmatrix}$

   (a) Find
   (i) $u + v + w$ (ii) $2u - 3w$ (iii) $u - v - w$

   (b) Find
   (i) $|u + v|$ (ii) $|u - v|$ (iii) $|u - v + w|$

3. $ABCD$ is a parallelogram.

   $\overrightarrow{AB} = u$ and $\overrightarrow{AD} = v$.

   (a) Find $\overrightarrow{AC}$ in terms of u and v.   (b) Find $\overrightarrow{BD}$ in terms of u and v.

4. PQRS is a kite.
   M is the midpoint of QS.

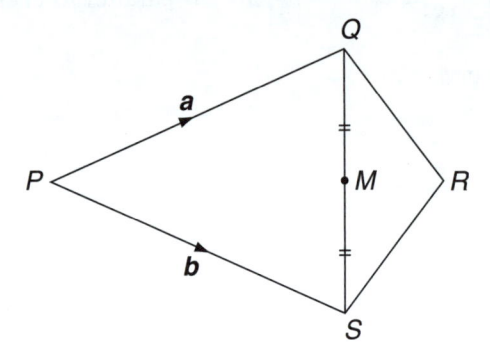

   (a) Why is $\overrightarrow{PR}$ not equal to $a + b$?
   (b) Write $\overrightarrow{QS}$ in terms of a and b.
   (c) Write $\overrightarrow{PM}$ in terms of a and b.

5. ABCDEF is a regular hexagon.
   M is the centre of the hexagon, and p, q and r are vectors as shown.

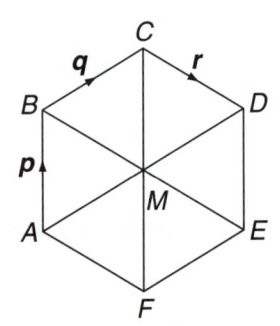

   NOT TO SCALE

   (a) Copy the figure and mark all the other representatives of p, q and r. Remember that the figure is a regular hexagon. For example, $\overrightarrow{MC} = p$.

   Find, in terms of p and/or q and/or r the following directed line segments, simplifying where necessary.

   (b) (i) $\overrightarrow{BA}$ (ii) $\overrightarrow{BD}$

   (iii) $\overrightarrow{AD}$ (iv) $\overrightarrow{FM}$

   (v) $\overrightarrow{ME} + \overrightarrow{CB}$ (vi) $\overrightarrow{CF} - \overrightarrow{AB}$

   (vii) $\overrightarrow{ED} + \overrightarrow{MA} + \overrightarrow{FM}$

Position vectors

The position vectors of points on a plane are the vectors drawn from a common origin to the points. It is convenient to label the position vectors with the same letters as the corresponding points, but with the vectors in lower case, and in **bold** (underline if hand written).
The diagram shows the points A, B and C and their position vectors **a**, **b** and **c**.

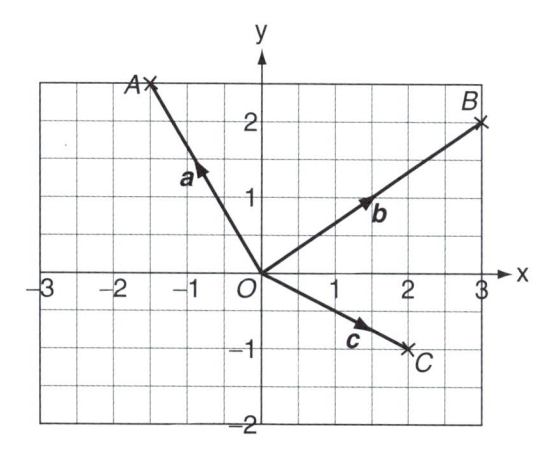

These position vectors are on a graph, so the vector for each point is the same as the coordinates of the point, expressed as a column vector in each case.
However, the position vectors do not have to be on a graph, and any point on the diagram can be taken as the origin. It should be labelled O to avoid any confusion.
For example, the diagram shows a parallelogram, OABC.

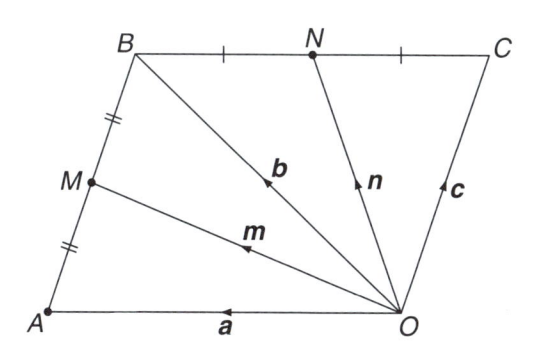

$\overrightarrow{OA} = \boldsymbol{a}$, $\overrightarrow{OB} = \boldsymbol{b}$ and $\overrightarrow{OC} = \boldsymbol{c}$.

M is the midpoint of AB, and N is the midpoint of BC.
m is the position vector of M and **n** is the position vector of N.
Vectors provide a powerful tool for proving geometrical facts. For example, using the above diagram we can prove that MN is parallel to AC, and that $MN = \frac{1}{2}AC$ as you will see in the worked example.

Example 5
Using the diagram of parallelogram *OABC* shown above, find the relationship between the lines *AC* and *MN*.

Answer 5
$\overrightarrow{AC} = -a + c = c - a$

$\overrightarrow{MN} = -m + n = n - m$

$m = a + \frac{1}{2}c, \qquad n = c + \frac{1}{2}a$ 　　　NOTE: Remember that $\overrightarrow{AB} = \overrightarrow{OC}$ and $\overrightarrow{CB} = \overrightarrow{OA}$.

$\overrightarrow{MN} = c + \frac{1}{2}a - (a + \frac{1}{2}c) = c + \frac{1}{2}a - a - \frac{1}{2}c = \frac{1}{2}c - \frac{1}{2}a = \frac{1}{2}(c - a)$

So $\overrightarrow{MN} = \frac{1}{2}\overrightarrow{AC}$

Hence *MN* is parallel to *AC* and $MN = \frac{1}{2}AC$

Exercise 10.2

1. *ABC* is a triangle. The position vectors relative to an origin *O*, and the vectors *p*, *q* and *r* are shown on the diagram.

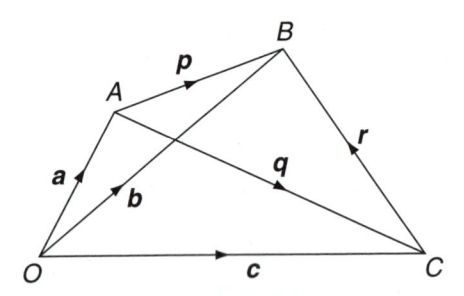

   Express in terms of *a*, *b* and/or *c*.
   (a) *p*　　(b) *p* + *q*　　(c) *r* − *q*　　(d) *p* + *r*

2. *a* and *b* are the position vectors of *A* and *B* as shown on the diagram.
   M is the midpoint of the line *AB*.

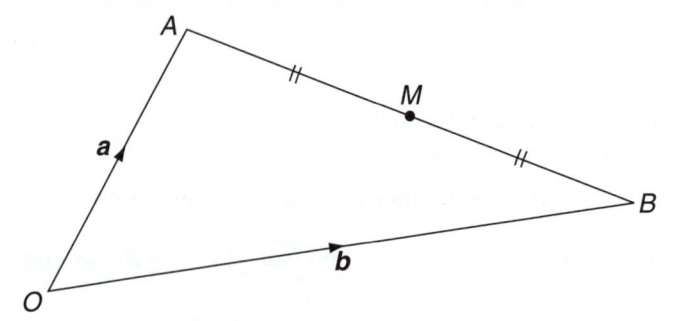

Express, in terms of *a* and *b*,
(a) $\overrightarrow{AB}$ 	(b) $\overrightarrow{AM}$ 	(c) $\overrightarrow{OM}$

3. *PQ* is a straight line.
   N divides *PQ* in the ratio 2 : 1.
   Draw a diagram, showing *P*, *Q* and *N*, and their position vectors, *p*, *q* and *n*, relative to an origin, *O*.
   Express, in terms of *p* and *q* the following directed line segments and vectors.
   (a) $\overrightarrow{PQ}$ 	(b) $\overrightarrow{PN}$ 	(c) *n* 	(d) $\overrightarrow{NQ}$

4. *ABC* is a triangle.
   The side *BC* is extended to *D*, where *BC* = *CD*, as shown in the diagram.

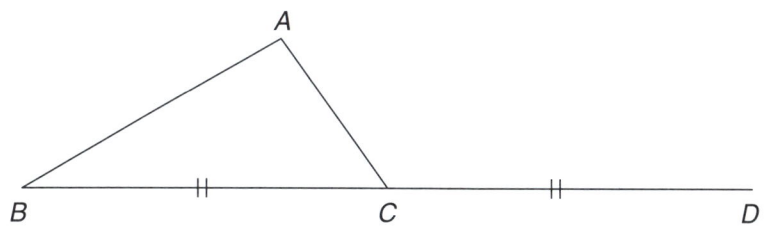

   Copy the diagram, draw an origin, *O*, and show the position vectors of the points *A*, *B*, *C* and *D*.
   Express the following in terms of *a*, *b* and *c*
   (a) $\overrightarrow{BA}$ 	(b) $\overrightarrow{BD}$ 	(c) $\overrightarrow{CD}$ 	(d) $\overrightarrow{DA}$

5. *ABC* is a triangle.
   $\overrightarrow{AB} = \begin{pmatrix} 3 \\ 3 \end{pmatrix}$, $\overrightarrow{BC} = \begin{pmatrix} 4 \\ -4 \end{pmatrix}$

   (a) Find $\overrightarrow{AC}$.
   (b) Calculate $|\overrightarrow{AB}|$, $|\overrightarrow{BC}|$ and $|\overrightarrow{AC}|$ as simplified surds.
   (c) Using $|\overrightarrow{AB}|^2$, $|\overrightarrow{BC}|^2$ and $|\overrightarrow{AC}|^2$ show that angle *ABC* = 90°.

6. Draw *x*- and *y*-axes from −5 to +5 with 1 centimetre per unit.
   Plot the following points
   A(−1, −4), *B*(0, 5) and *C*(−4, 5).
   (a) Find as column vectors
   (i) $\overrightarrow{CB}$ 	(ii) $\overrightarrow{BA}$ 	(iii) $\overrightarrow{CA}$
   (iv) $\overrightarrow{BC} - \overrightarrow{CA}$ 	(v) $\overrightarrow{BC} - \overrightarrow{AC} - \overrightarrow{BA}$
   (b) Calculate
   (i) $|\overrightarrow{AC}|$ 	(ii) $|\overrightarrow{BA}|$ 	(iii) $|\overrightarrow{BC}|$

268　*Extended Mathematics for Cambridge IGCSE*

More about Transformations

Translation

As we know, translations are defined by column vectors. The abbreviation **T** is used to represent a translation.

Example 6

T represents the translation $\begin{pmatrix} 3 \\ -5 \end{pmatrix}$, and A is the point $(1, 4)$, as in the diagram.

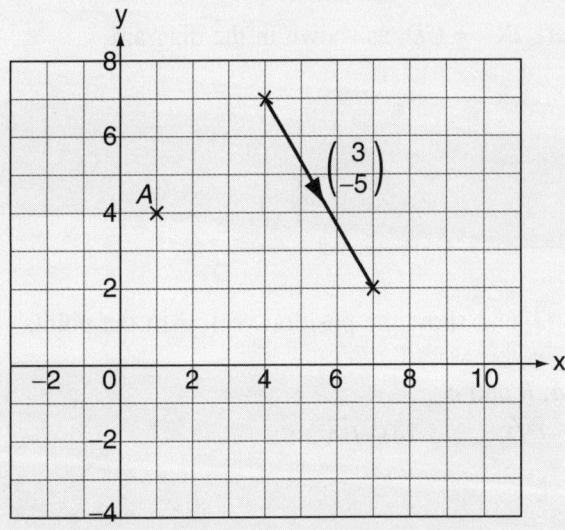

Find the image A' of A under the translation **T**.

NOTE: A', or A_1 are often used to represent the image of A after a transformation.

Answer 6

You can use the diagram as shown below.

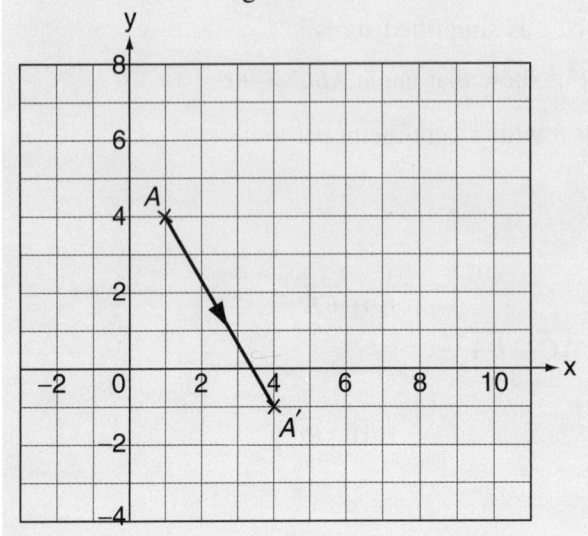

Alternatively A' can be calculated as shown.

$$A' = \mathbf{T}(A) = \begin{pmatrix} 3 \\ -5 \end{pmatrix}(1,4) = (3+1, -5+4) = (4, -1)$$

Either way **T** maps $A(1, 4)$ to $A'(4, -1)$ as in the diagram.

NOTE: This hardly needs a diagram, as you can see that the components of T are added to the coordinates of A, but a diagram does makes this clearer. Notice that this use of the vector notation is *not* the same as the multiplication of matrices.

It is important that you should understand that the vecctor notation $\mathbf{T}(1, 4)$, or $\begin{pmatrix} 3 \\ -5 \end{pmatrix}(1, 4)$, does **not** represent matrix multiplication. The (1, 4) represents a pair of coordinates, not a matrix. The translation is calculated by addition, not multiplication. If you are in any doubt draw a diagram.

Exercise 10.3

1. If $\mathbf{T}_1 = \begin{pmatrix} 3 \\ -1 \end{pmatrix}$ and $\mathbf{T}_2 = \begin{pmatrix} -5 \\ -1 \end{pmatrix}$ find the images of the following points after the translations indicated.

   (a) $\mathbf{T}_1(-1, 0)$ (b) $\mathbf{T}_2(-1, 0)$

   (c) $\mathbf{T}_2(16, 7)$ (d) $\mathbf{T}_1(5, 1)$

2. Find the column vectors associated with the following translations.

   (a) $A(4, 12)$ maps to $B(3, -8)$

   (b) $P(-6, 7)$ maps to $Q(-15, 0)$

   (c) $Y(-5, -4)$ maps to $Z(-7, 4)$

   NOTE: Remember that you can write down each vector by asking yourself, for example, "How do I get from A to B?" The answer is "back(−) 1 and down (−) 20".

Enlargement

What happens if the scale factor for an enlargement is negative?
The following diagrams show the enlargement of triangle ABC, scale factor -2 and centre $(1, -1)$. The image of triangle ABC under the enlargement is triangle $A'B'C'$.

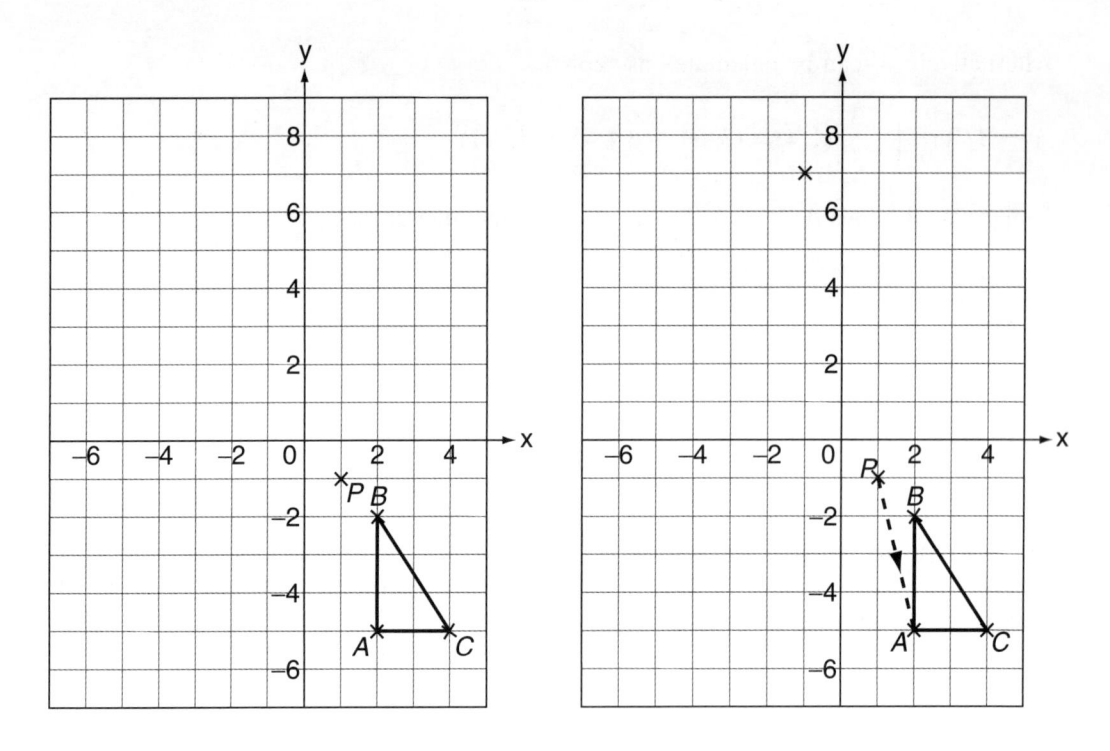

Diagrams showing the process of enlarging triangle ABC, centre P, $(1, -1)$, scale factor -2.

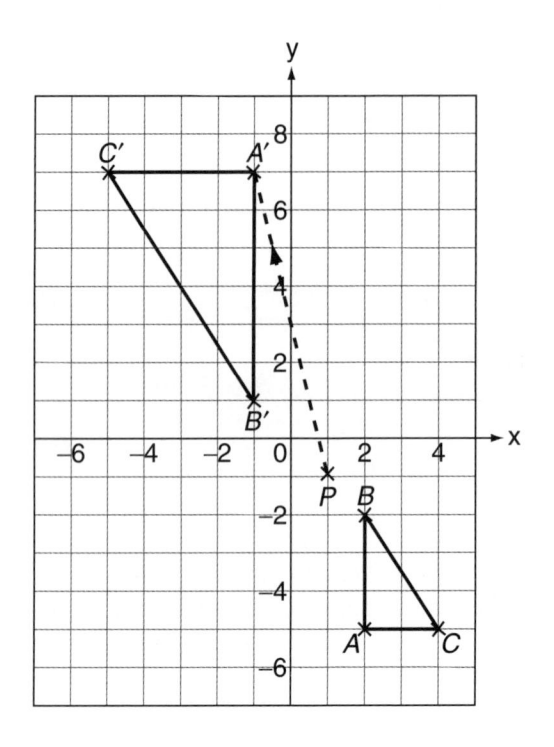

It is convenient to plot the points for the enlargement by using vectors.
In the diagram above the point $P(1, -1)$ is the centre of the enlargement. The vector $\overrightarrow{PA}$ is $\begin{pmatrix} 1 \\ -4 \end{pmatrix}$ so multiplying by the scale factor -2 gives $-2\begin{pmatrix} 1 \\ -4 \end{pmatrix} = \begin{pmatrix} -2 \\ 8 \end{pmatrix}$.

Hence vector $\overrightarrow{PA'} = \begin{pmatrix} -2 \\ 8 \end{pmatrix}$ and A' is the point $(-1, 7)$, and so on.

Shear

This is a transformation you have not yet met. It is shown in the following diagrams.

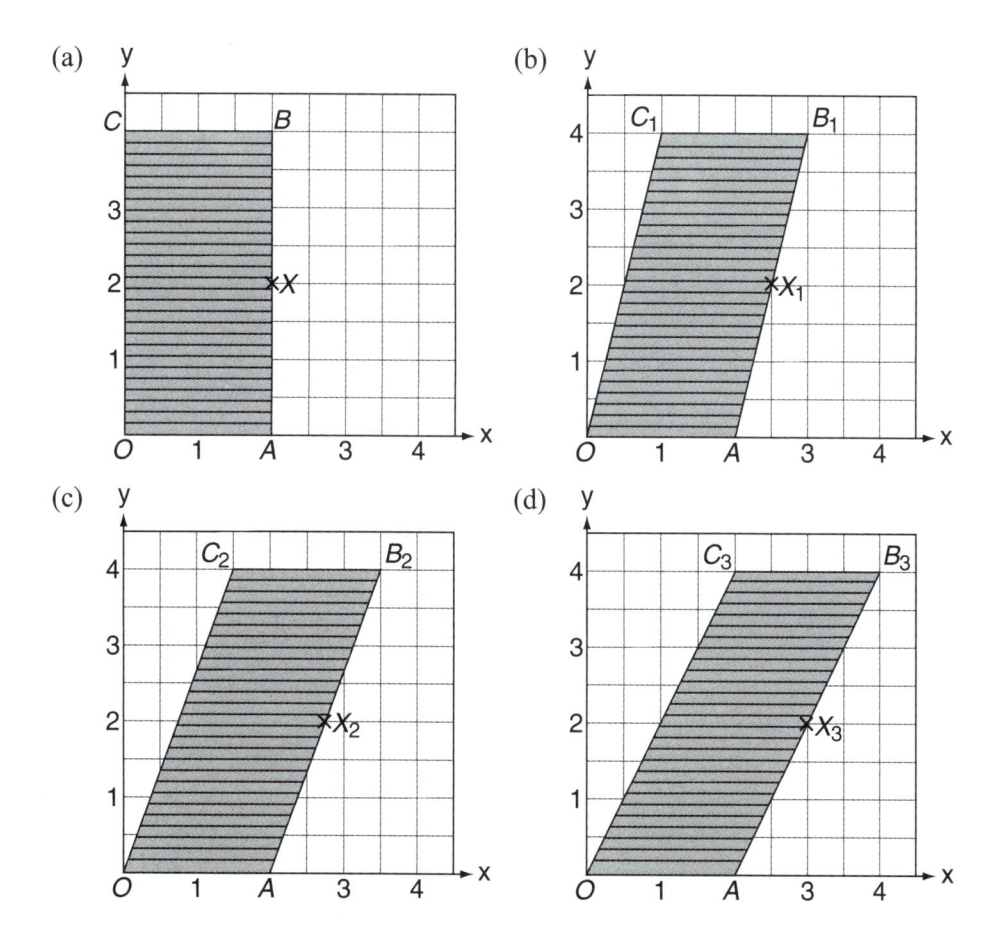

Imagine that the rectangle $OABC$ is the side view of a layered stack of smooth cards. The layers are free to slide over each other as shown in the diagrams. The height of the stack will not change, only the angle of the sides.
This transformation is called a **shear**, and is defined by the direction, a **shear factor**, and a fixed or **invariant** line.

Diagram (a) above shows the stack before any transformation, and the following diagrams show the stack under shears with the same invariant line (the x-axis) but increasing shear factors. The x-axis is the invariant line because OA, on the axis, does not move under the transformation.

The shear factor is calculated by dividing the distance moved by any point on the diagram by the perpendicular distance from that point to the invariant line. This sounds complicated, and it is easier to look at an example.

The shear which maps diagram (a) to diagram (b) for example, maps the point $B(2, 4)$ to the point $B_1(3, 4)$.

The distance moved by B is $3 - 2 = 1$ in the x-direction, and the perpendicular distance to the invariant line is 4.

The shear factor is given by $(3 - 2) \div 4 = \dfrac{1}{4}$.

This is true for any point on the two diagrams, so check for yourself that X to X_1 is also $\dfrac{1}{4}$.

Also, for practice, find the shear factors for the transformations mapping diagram (a) to (c), and diagram (a) to (d). They should be $\dfrac{3}{8}$ and $\dfrac{1}{2}$ respectively.

- Shear factor is

$$\frac{\text{distance any point moves in the direction parallel to invariant line}}{\text{perpendicular distance of that point to the invariant line}}$$

Stretch

A stretch is easier to visualise, as shown in the following diagrams.

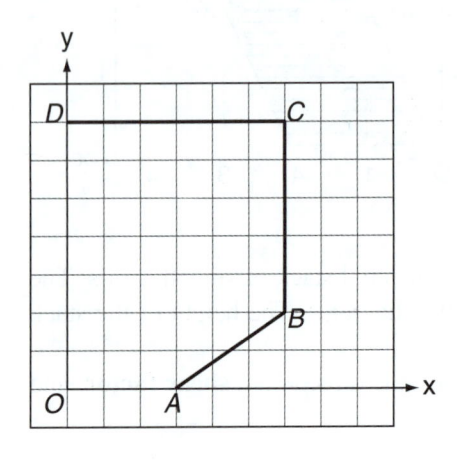

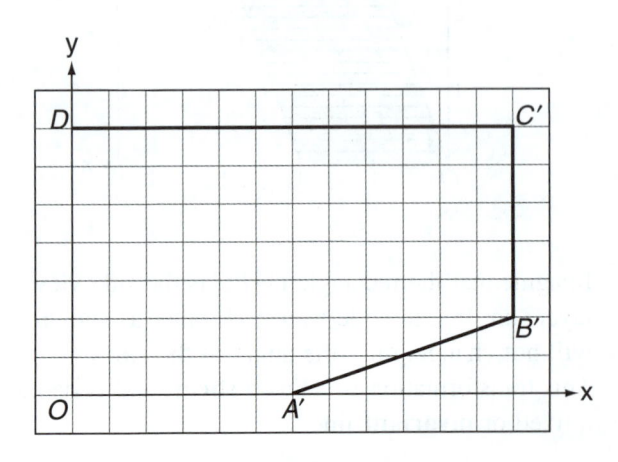

The stretch is defined by the direction, a scale factor and an invariant line.
- The scale factor is

$$\frac{\text{the distance of a point on the image from the invariant line}}{\text{distance of the corresponding point on the object from the invariant line}}$$

For example, in the diagram above the y-axis is the invariant line.

Scale factor is $\dfrac{DC'}{DC} = \dfrac{12}{6} = 2$

Direction is the positive x-direction.

Example 7

(a) Describe fully these two transformations.

  (i) Triangle A maps to triangle A' on the diagram.

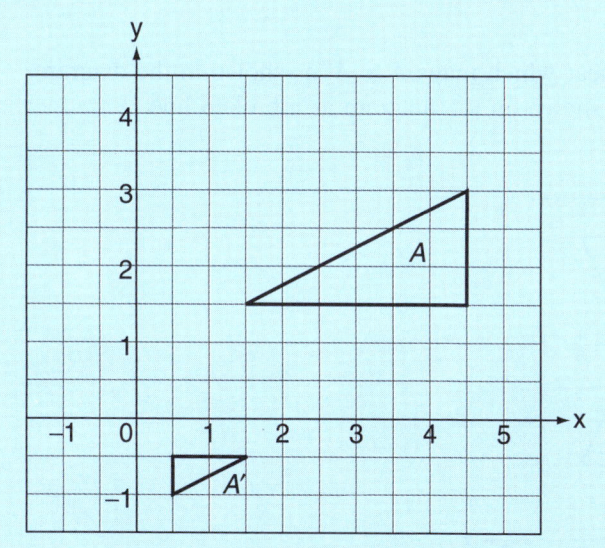

  (ii) A transformation which maps $A(1, 2)$ to $A'(1, 3)$, and $B(4, 5)$ to $B'(4, 7.5)$.

(b) Draw a diagram to show the image of $A(1, 4)$, $B(4, 4)$ and $C(2, 6)$ under a shear, the line $y = 2$ invariant, and shear factor 1 in the x-direction.

(c) Find the invariant line and shear factor for the shear shown below. Triangle PQR maps to triangle $P'Q'R'$.

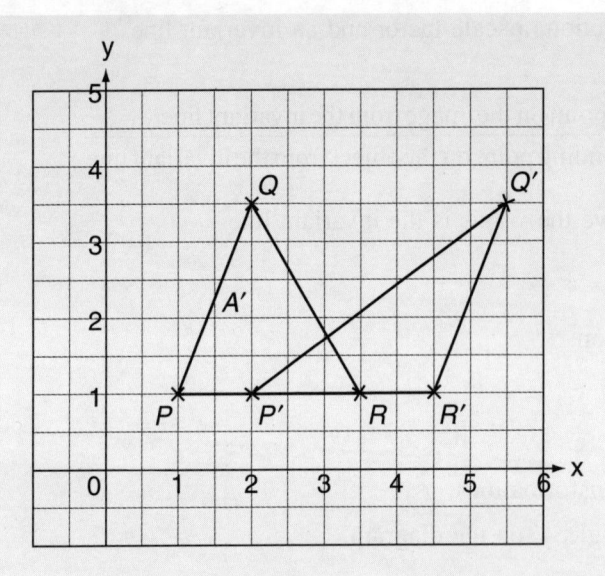

[Diagram: coordinate grid showing triangle PQR with P(1,1), Q(2,4), R(4,1) and image triangle P'Q'R' with P'(2,1), Q'(5,4), R'(5,1); A' marked near (2,2)]

NOTE: Find the invariant line by producing QP and $Q'P'$ until they meet. Repeat with QR and $Q'R'$. Both pairs of lines will meet on the invariant line. To find the shear factor use

$$\frac{PP'}{\text{distance from invariant line}}.$$

(d) (i) Describe fully the shear which maps A to A' as shown in the diagram. The invariant line is shown on the diagram as a broken line.

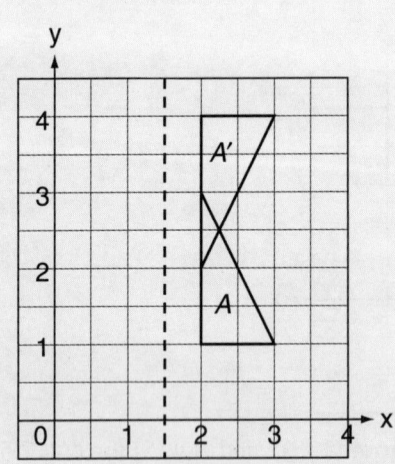

[Diagram: coordinate grid with shape A and image A', with broken invariant line at $x = 1.5$]

NOTE: This is deceptive. The point $(2, 1)$ maps to $(2, 2)$ and $(3, 1)$ maps to $(3, 4)$.

(ii) Describe fully another single transformation which would map A to A'.

Answer 7

(a) (i) Enlargement, scale factor $-\dfrac{1}{3}$, centre $(1.5, 0)$

(ii) Stretch, x-axis invariant, scale factor 1.5, in the y-direction.

NOTE: Plot the points on squared paper to check this.

(b)

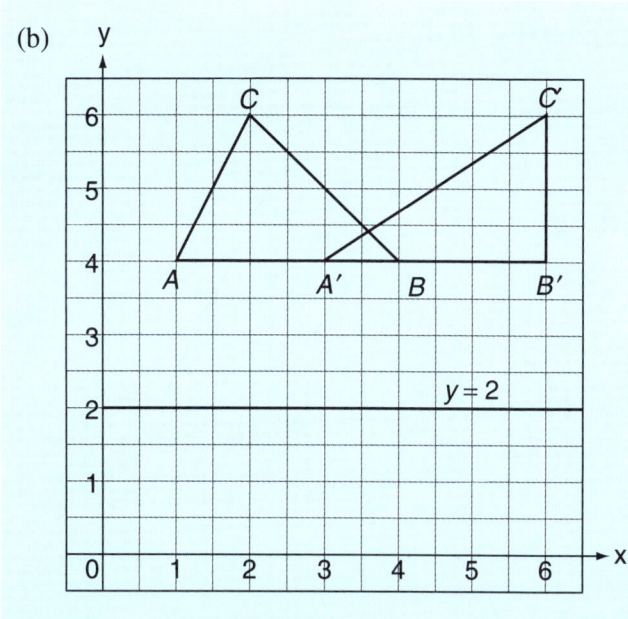

[Graph showing triangle ABC with A at (1,4), B at (3,4), C at (2,6), sheared to A' at (3,4), B' at (5,4), C' at (6,6), with line y = 2 marked as invariant]

NOTE: Take the distance of *A* from the invariant line (2 units), multiply it by the shear factor (1) and then move *A* 2×1 units to the right to give *A'*. Repeat with *B* and *C*.

(c) Shear factor 1 in the *x*-direction, with *x*-axis invariant

(d) (i) Shear factor 2, in the *y*-direction, with the line $x = 1.5$ invariant

   (ii) Reflection in the line $y = 2.5$

Rotation

At Core level rotations are usually about the Origin or a point on the object. At Extended level the centre of rotation may be any point on the diagram.

The sense, or direction, of the rotation may be described as *clockwise* or *anticlockwise*, or the notation you learned in the chapter on Trigonometry may be used. You will remember that anticlockwise rotations are taken as positive and clockwise rotations as negative. So for example, a rotation of $-90°$ about the origin is the same as $90°$ anticlockwise.

The centre of the rotation can be found with a tracing as described in the Core Book, or by a geometric method which you can practise in question 1 in the next exercise.

The next exercise is designed to improve your skills with and understanding of transformations. You should work through it very carefully and make sure you try every question. Work as accurately as you can.

Exercise 10.4

1.

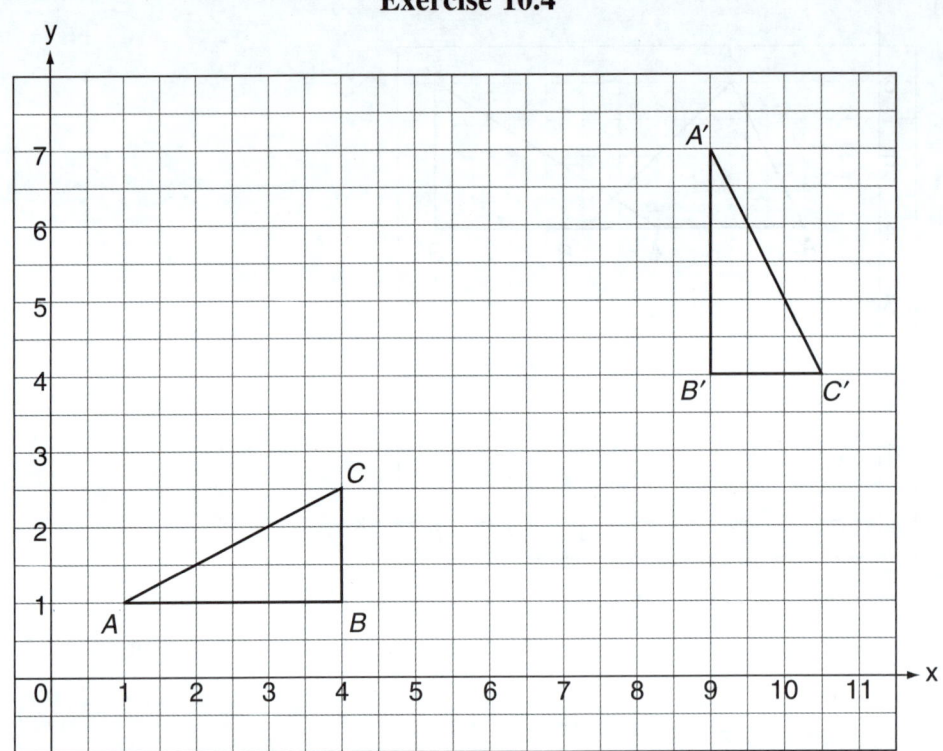

Copy the diagram showing the object, triangle *ABC*, and its image, triangle *A'B'C'* after a single transformation.

(a) Accurately join a pair of corresponding vertices on the object and image, such as *A* and *A'*.
(b) Using the method learned in Chapter 6 of the Core Book, with straight edge and compasses only construct the perpendicular bisector of the line *AA'*.
(c) Repeat (a) and (b) for another pair of vertices such as *C* and *C'*.
    Mark and label the point where the two bisectors cross *P*.
(d) With centre *P* and radius *PA* draw an arc from *A* to *A'*.
(e) With centre *P* and radius *PB* draw an arc from *B* to *B'*.
(f) What is special about the point *P*, and why?   **NOTE: Think back to your work on circles in Chapter 6 of the Core Book.**
(g) Describe fully the single transformation that maps triangle *ABC* to triangle *A'B'C'*.

2. On 5 millimetre squared paper, or graph paper, draw a rectangle 15 centimetres by 6 centimetres as shown in the diagram. The diagram here is reduced to fit on the page. Mark a point *X* in the centre of the rectangle. Copy triangle *A*.

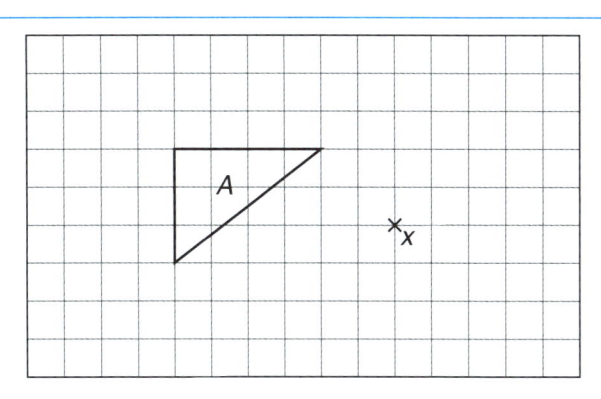

(a) With centre X and the scale factors below, enlarge triangle A.
   (i) Scale factor 2. Label the image A_1.
   (ii) Scale factor $\frac{1}{2}$. Label the image A_2.
   (iii) Scale factor -2. Label the image A_3.
   (iv) Scale factor $-\frac{1}{2}$. Label the image A_4.

(b) Describe the single transformation which would map triangle A_3 onto triangle A_2.

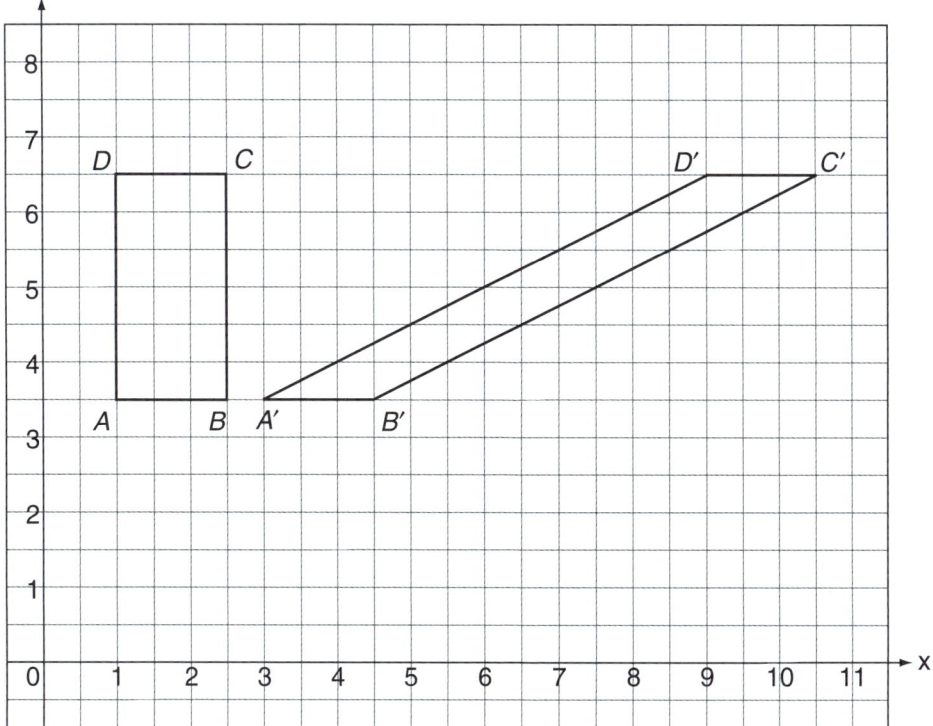

3.

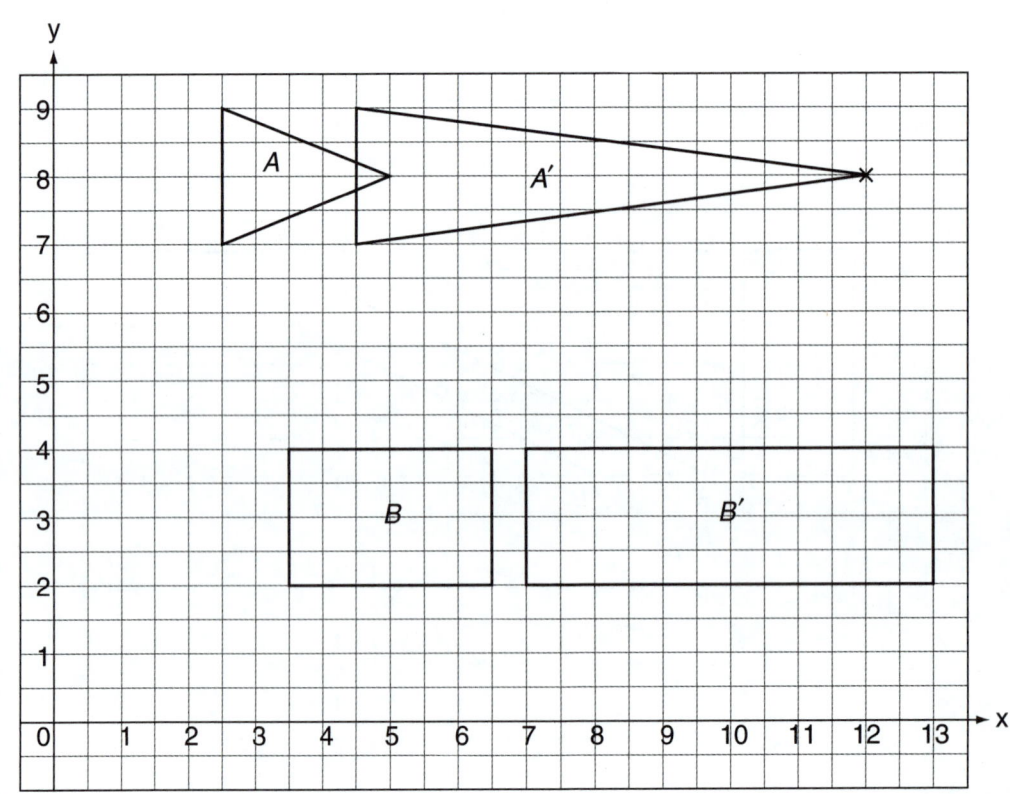

Copy the diagram.
(a) Produce *DA* and *D'A'*. (Produce *DA* means make *DA* longer, in the direction *D* to *A*.)
(b) Produce *CB* and *C'B'*.
(c) These two pairs of extended lines meet on the invariant line.
Describe fully the single transformation that maps *ABCD* onto *A'B'C'D'*.

4. (a) $f(x) = \dfrac{6x}{x+2}$.

   (i) Find $f^{-1}(x)$.
   (ii) Draw a set of axes with $0 \leqslant x \leqslant 6$ and also $0 \leqslant y \leqslant 6$.
   (iii) Draw accurately the line $y = x$.
   (iv) Plot the curves $y = f(x)$ (domain $0 \leqslant x \leqslant 6$) and $y = f^{-1}(x)$ (domain $0 \leqslant x \leqslant 4.5$) on the same graph.
   (v) What single transformation would map $y = f(x)$ onto $y = f^{-1}(x)$?

   (b) Repeat questions (a) (i) to (a) (v) with $f(x) = \dfrac{2}{3}x + 1$, domain $0 \leqslant x \leqslant 6$, and $f^{-1}(x)$, domain $1 \leqslant x \leqslant 5$.

5.

Copy the diagram.
Using the method described in question 3 to find the invariant line, describe fully the following single transformations.
(a) Triangle A to triangle A'.
(b) Rectangle B to rectangle B'.

NOTE: Draw in and produce the diagonals of both the rectangles to find the invariant line.

Exercise 10.5

1. Describe fully the single transformation which maps $A(2, 4)$ to $A'(-1, -2)$ and $B(-4, 2)$ to $B'(2, -1)$.

   NOTE: Plot the points and join AB and $A'B'$.

2. Draw x- and y-axes with $-6 \leqslant x \leqslant 8$ and $-6 \leqslant y \leqslant 6$. Use 1 centimetre per unit. Plot the points $(1, -1)$, $(1, -4)$ and $(3, -4)$ to form triangle A.
   (a) Shear triangle A with the x-axis invariant, shear factor 1 in the positive x-direction. Label the image B.
   (b) Stretch triangle A with the line $y = -5$ invariant, scale factor 2, in the positive y-direction. Label the image C.
   (c) (i) Enlarge triangle A, centre the origin, scale factor -1.5. Label the image D.
      (ii) Describe fully the single transformation which would map D onto A.

3. Draw x- and y-axes with $0 \leqslant x \leqslant 18$ and $0 \leqslant y \leqslant 10$, with 2 units per centimetre. Plot the points $(1, 3)$, $(5, 3)$, $(5, 8)$ and $(1, 6)$ and join to form a trapezium. Label it A.
Plot the points $(3, 3)$, $(7, 3)$, $(17, 8)$ and $(9, 6)$ and join to form another trapezuim. Label it B.
Describe fully the transformation which maps A to B.

Matrices and Transformations

The diagram shows the reflection of $A(3, 2)$ and $B(2, 1)$ in the y-axis to $A'(-3, 2)$ and $B'(-2, 1)$.

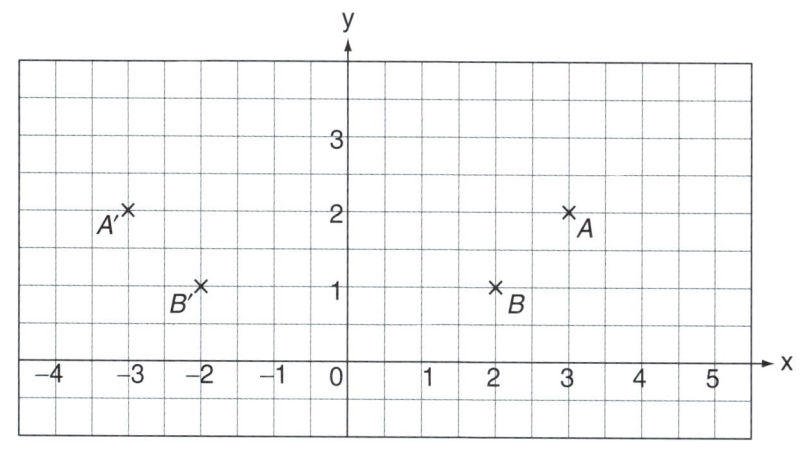

The points $A(3, 2)$ and $B(2, 1)$ can be represented by entering their *position vectors* $a = \begin{pmatrix} 3 \\ 2 \end{pmatrix}$ and $b = \begin{pmatrix} 2 \\ 1 \end{pmatrix}$ into a matrix: $\begin{pmatrix} 3 & 2 \\ 2 & 1 \end{pmatrix}$.

In the same way A' and B' can be written $\begin{pmatrix} -3 & -2 \\ 2 & 1 \end{pmatrix}$.

These matrices give us a powerful tool when working with transformation as we will see. Is there a matrix which could perform the reflection of A and B in the y-axis as shown above? We will look at three methods for finding a suitable matrix for this task.

First method for finding a matrix

Look at the following matrix equation:

$$\begin{pmatrix} a & b \\ c & d \end{pmatrix} \begin{pmatrix} 3 & 2 \\ 2 & 1 \end{pmatrix} = \begin{pmatrix} -3 & -2 \\ 2 & 1 \end{pmatrix}$$

and multiply out the left hand side according to the matrix rules (see Chapter 3).

$$\begin{pmatrix} 3a+2b & 2a+b \\ 3c+2d & 2c+d \end{pmatrix} = \begin{pmatrix} -3 & -2 \\ 2 & 1 \end{pmatrix}$$

NOTE: Compare the elements in the matrices.

$3a + 2b = -3$ ... (i)
$2a + b = -2$ ... (ii)
$3c + 2d = 2$ ... (iii)
$2c + d = 1$ ... (iv)

Solving (i) and (ii) simultaneously gives $a = -1$ and $b = 0$,
and solving (iii) and (iv) simultaneously gives $c = 0$ and $d = 1$.

So $\begin{pmatrix} a & b \\ c & d \end{pmatrix} = \begin{pmatrix} -1 & 0 \\ 0 & 1 \end{pmatrix}$, which is the matrix which maps A to A' and B to B' under reflection in the y-axis.

In general,

$$\begin{pmatrix} -1 & 0 \\ 0 & 1 \end{pmatrix} \begin{pmatrix} x \\ y \end{pmatrix} = \begin{pmatrix} -x \\ y \end{pmatrix}.$$

Check this yourself.

Second method for finding a matrix

Another method for finding the required matrix is to use an inverse matrix.

If $\quad A = \begin{pmatrix} 3 & 2 \\ 2 & 1 \end{pmatrix}$, $A' = \begin{pmatrix} -3 & -2 \\ 2 & 1 \end{pmatrix}$ and M is the required matrix,

then $\quad\quad\quad MA = A'$, multiplying both sides by A^{-1} gives

$\quad\quad\quad\quad MAA^{-1} = A'A^{-1}$.

$\quad\quad\quad\quad AA^{-1} = I$, the identity matrix, so

$\quad\quad\quad\quad M = A'A^{-1}$.

Since $A = \begin{pmatrix} 3 & 2 \\ 2 & 1 \end{pmatrix}$, $|A|$ or $\det A = -1$, and $A^{-1} = \begin{pmatrix} -1 & 2 \\ 2 & -3 \end{pmatrix}$.

$$\mathbf{A'} = \begin{pmatrix} -3 & -2 \\ 2 & 1 \end{pmatrix}, \qquad \text{so } \mathbf{A'A}^{-1} = \begin{pmatrix} -3 & -2 \\ 2 & 1 \end{pmatrix}\begin{pmatrix} -1 & 2 \\ 2 & -3 \end{pmatrix} = \begin{pmatrix} -1 & 0 \\ 0 & 1 \end{pmatrix},$$

which again gives the matrix representing reflection in the y-axis.
Again, you are advised to work through this to practise the method.

NOTE: Do not try this with more than two points because you only know how to find the inverse of a two by two matrix.

Third method for finding a matrix

The final, and quickest, method for basic transformations is shown in the following diagrams.

The first diagram shows two unit vectors, one along the x-axis and one along the y-axis, which form a unit square.

The vector in the x-direction is marked with a double arrow, because it is important to be able to distinguish the two vectors after the transformation. The second diagram shows what happens to these two vectors after reflection in the y-axis.

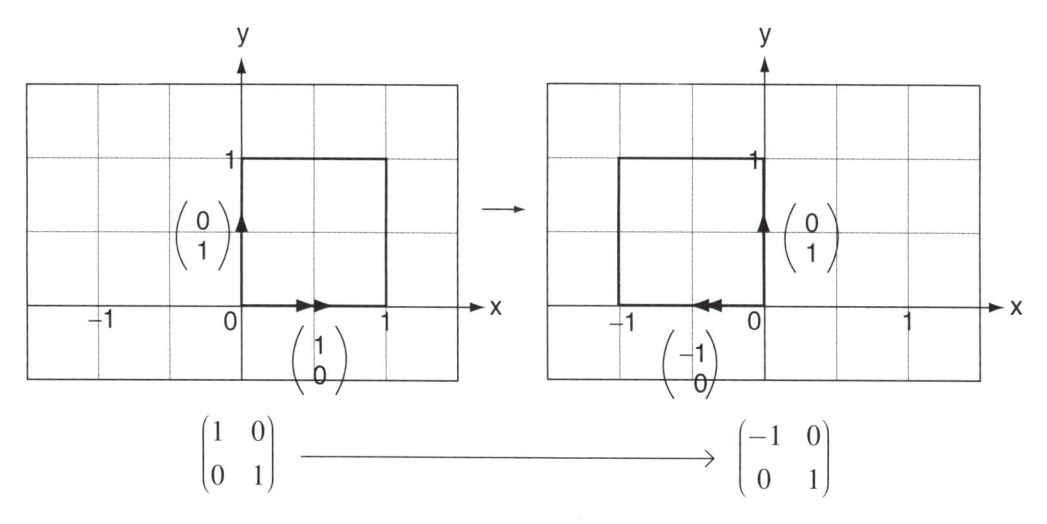

$$\begin{pmatrix} 1 & 0 \\ 0 & 1 \end{pmatrix} \longrightarrow \begin{pmatrix} -1 & 0 \\ 0 & 1 \end{pmatrix}$$

Underneath each diagram the two vectors are shown in matrix form, with the double arrowed vector first.

You will see that the first matrix is the unit matrix, and the second matrix represents reflection in the y-axis, as proved above.

Hence, for any point (x, y), $\begin{pmatrix} -1 & 0 \\ 0 & 1 \end{pmatrix}\begin{pmatrix} x \\ y \end{pmatrix} = \begin{pmatrix} -x \\ y \end{pmatrix}.$

Example 8

(a) Derive the matrices for
  (i) reflection in the line $y = x$,
  (ii) enlargement, scale factor 3, centre the origin.

282   *Extended Mathematics for Cambridge IGCSE*

(b) Describe fully the transformation represented by $\begin{pmatrix} -1 & 0 \\ 0 & -1 \end{pmatrix}$.

(c) **R** = rotation 90° clockwise about the origin.
   (i) Find the matrix representing **R**.   (ii) Find the matrix $\mathbf{R}^{-1}$.
   (iii) Describe fully the transformation represented by $\mathbf{R}^{-1}$.

Answer 8

(a) (i)

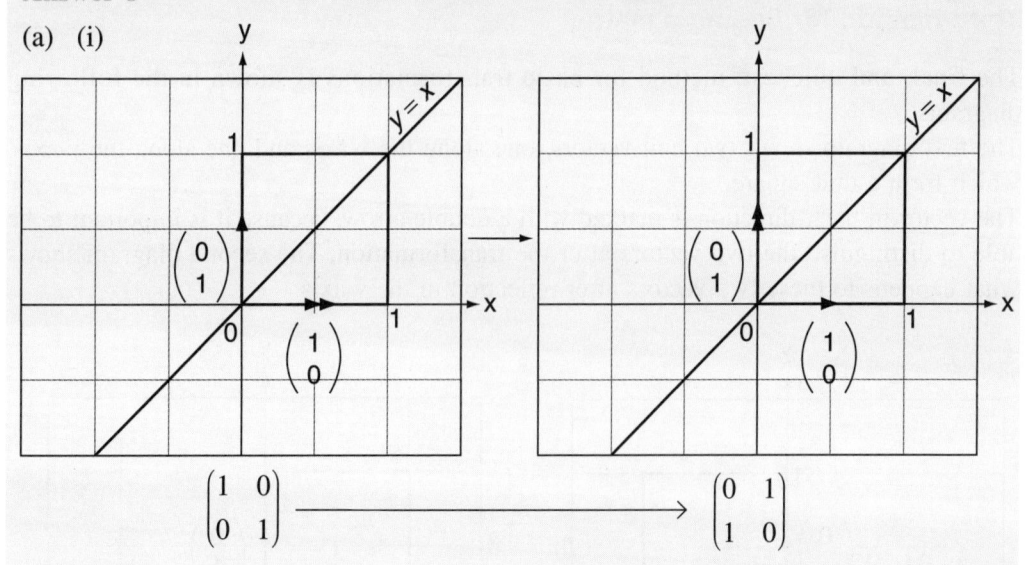

$\begin{pmatrix} 1 & 0 \\ 0 & 1 \end{pmatrix} \longrightarrow \begin{pmatrix} 0 & 1 \\ 1 & 0 \end{pmatrix}$

Reflection in the line $y = x$ is represented by the matrix $\begin{pmatrix} 0 & 1 \\ 1 & 0 \end{pmatrix}$.

(ii)

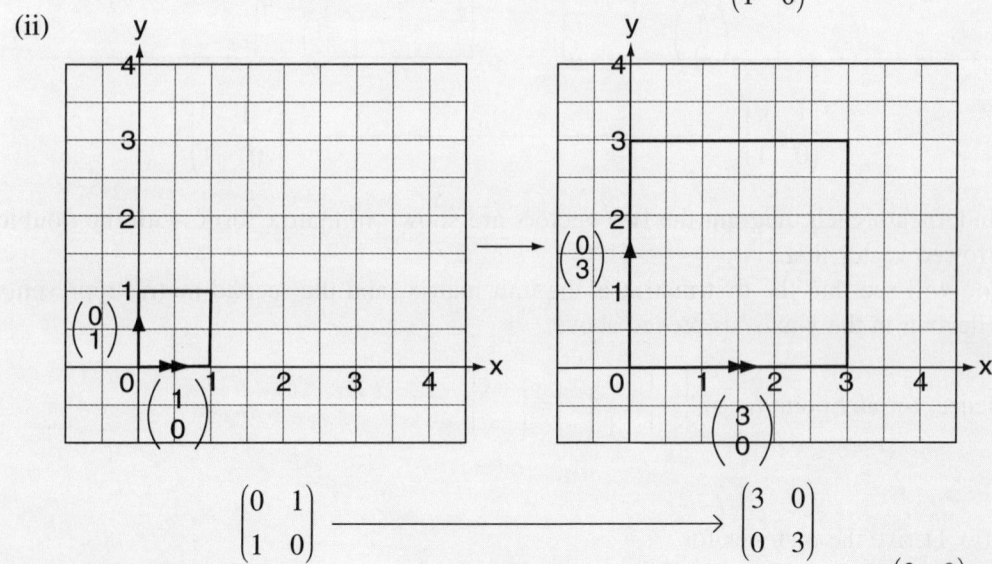

$\begin{pmatrix} 0 & 1 \\ 1 & 0 \end{pmatrix} \longrightarrow \begin{pmatrix} 3 & 0 \\ 0 & 3 \end{pmatrix}$

Enlargement, scale factor 3, centre the origin is represented by the matrix $\begin{pmatrix} 3 & 0 \\ 0 & 3 \end{pmatrix}$.

(b)

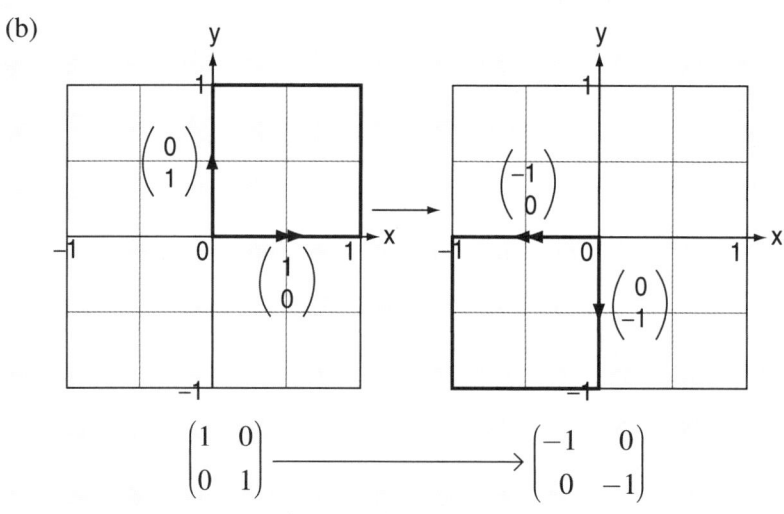

$$\begin{pmatrix} 1 & 0 \\ 0 & 1 \end{pmatrix} \longrightarrow \begin{pmatrix} -1 & 0 \\ 0 & -1 \end{pmatrix}$$

The transformation $\begin{pmatrix} -1 & 0 \\ 0 & -1 \end{pmatrix}$ is rotation about the origin 180°.

(c) (i)

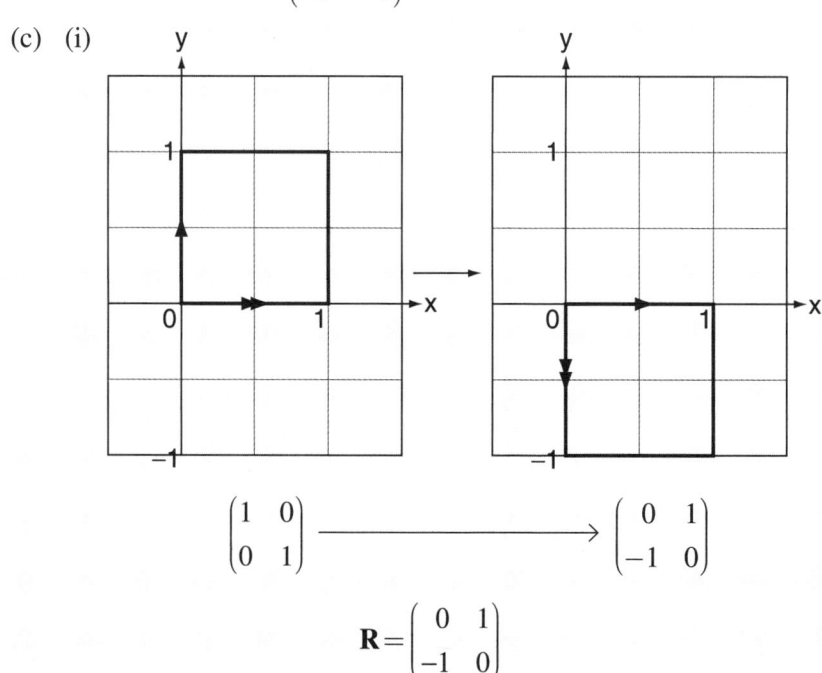

$$\begin{pmatrix} 1 & 0 \\ 0 & 1 \end{pmatrix} \longrightarrow \begin{pmatrix} 0 & 1 \\ -1 & 0 \end{pmatrix}$$

$$\mathbf{R} = \begin{pmatrix} 0 & 1 \\ -1 & 0 \end{pmatrix}$$

(ii) $\mathbf{R}^{-1} = \dfrac{1}{0 - ^-1} \begin{pmatrix} 0 & -1 \\ 1 & 0 \end{pmatrix} = \begin{pmatrix} 0 & -1 \\ 1 & 0 \end{pmatrix}$

$\mathbf{R}^{-1} = \begin{pmatrix} 0 & -1 \\ 1 & 0 \end{pmatrix}$

(iii) R^{-1} is rotation about the origin, 90° anticlockwise.

Example 9

(a) Find the coordinates of the image of the triangle ABC where A is the point $(-1, -2)$, B is $(3, 5)$ and C is $(4, -1)$ under the transformation represented by the matrix $\begin{pmatrix} 0 & 1 \\ -1 & 0 \end{pmatrix}$.

(b) Find matrices associated with
   (i) a shear, x-axis invariant, shear factor 2 in the x-direction,
   (ii) a stretch, x-axis invariant, scale factor 2 in the y-direction.

Answer 9

(a) $\begin{pmatrix} 0 & 1 \\ -1 & 0 \end{pmatrix} \begin{pmatrix} -1 & 3 & 4 \\ -2 & 5 & -1 \end{pmatrix} = \begin{pmatrix} -2 & 5 & -1 \\ 1 & -3 & -4 \end{pmatrix}$

(b) (i)

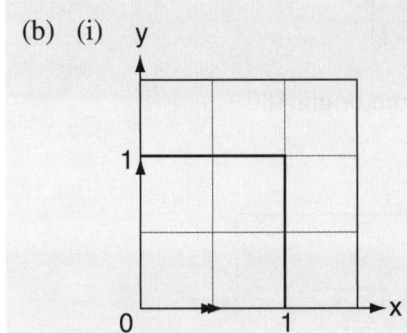

$\begin{pmatrix} 1 & 0 \\ 0 & 1 \end{pmatrix}$ 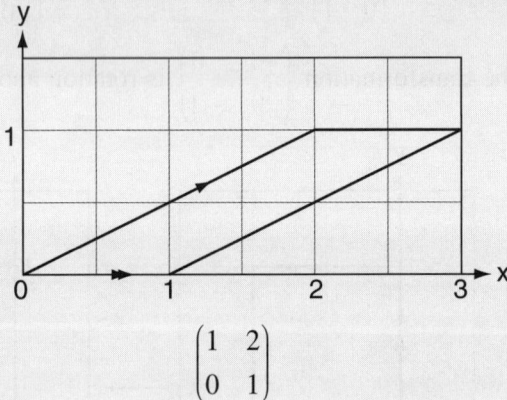 $\qquad$ $\begin{pmatrix} 1 & 2 \\ 0 & 1 \end{pmatrix}$

The required matrix is $\begin{pmatrix} 1 & 2 \\ 0 & 1 \end{pmatrix}$

(ii)

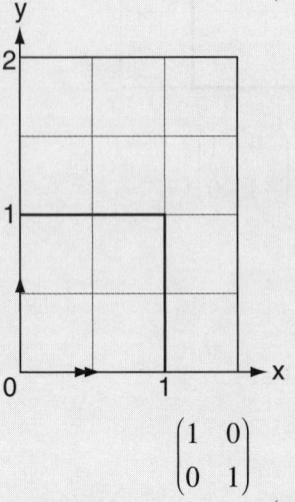

$\begin{pmatrix} 1 & 0 \\ 0 & 1 \end{pmatrix}$ 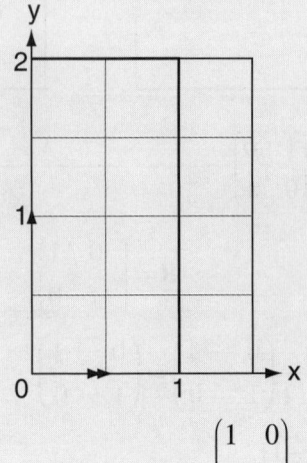 $\qquad$ $\begin{pmatrix} 1 & 0 \\ 0 & 2 \end{pmatrix}$

The required matrix is $\begin{pmatrix} 1 & 0 \\ 0 & 2 \end{pmatrix}$

Exercise 10.6

1. Derive matrices for the following transformations
   (a) reflection in the line $y = -x$,
   (b) shear, scale factor 2 in the positive y-direction with the y-axis invariant,
   (c) stretch, scale factor 3 in the positive x-direction, y-axis invariant,
   (d) enlargement, scale factor -2, centre the origin.

2. A transformation maps triangle A to triangle A', as shown in the diagram.
   (a) Find the matrix associated with this transformation.
   (b) Describe the transformation fully.

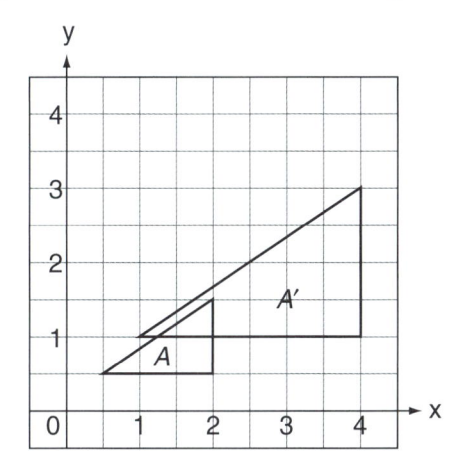

3. A transformation maps $A(2, 2)$ and $B(5, 3)$ to $A'(-2, -2)$ and $B'(-3, -5)$.
   (a) Find the matrix associated with this transformation.
   (b) Describe the transformation fully.
   (c) Use the matrix to find the image of $C(-1, 6)$ and $D(0, -4)$ under the transformation.

4. Find the vectors which produce the following translations
   (a) $(0, 2)$ to $(2, 0)$
   (b) $(1, 5)$ to $(-1, 5)$
   (c) $(-7, -3)$ to $(-4, -6)$

5. Find triangle $A'B'C'$ after reflection of triangle ABC in the y-axis, where $A = (-1, -3)$, $B = (5, 7)$ and $C = (0, 1)$.

Notation

The following abbreviations are commonly used in transformation work.

- **M**   Reflection   **Note: Think of Mirror**
- **R**   Rotation

286 *Extended Mathematics for Cambridge IGCSE*

- **E** Enlargement
- **T** Translation
- **H** Shear
- **S** Stretch

Any capital letter *may* be used as long as the transformation is defined in some way.

Successive transformations

You may be asked to do more than one transformation on an object. For example, reflection in the x-axis followed by a rotation of 90° clockwise about the origin. Under these transformations triangle A maps to triangle A_1, and this then maps to triangle A_2 as shown in the diagram.

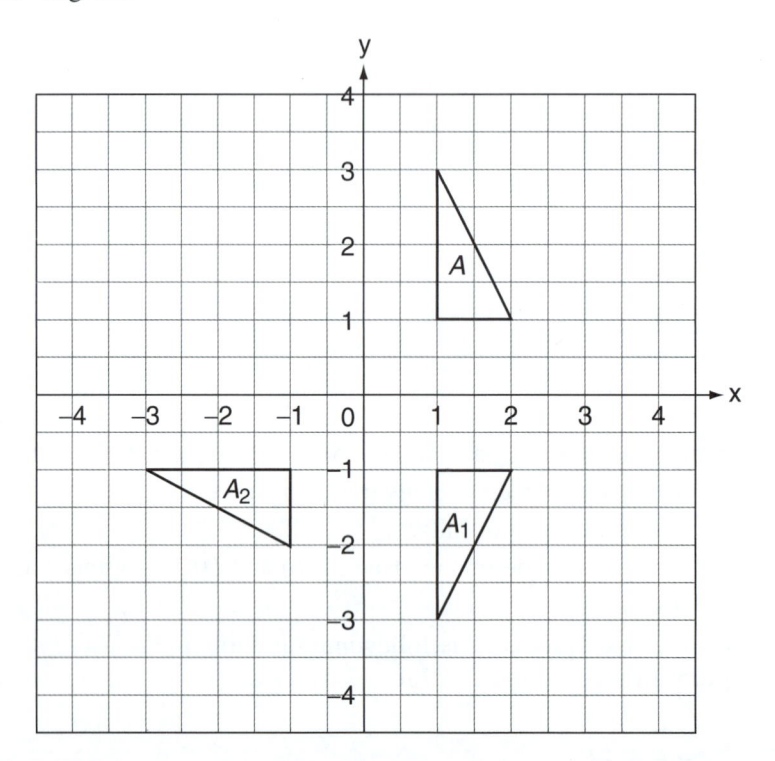

If $\mathbf{M}_x$ represents reflection in the x-axis, then $\mathbf{M}_x(A) = A_1$.
If $\mathbf{R}_{-90}$ represents rotation 90° clockwise about the origin, then $\mathbf{R}_{-90}(A_1) = A_2$.
In the same way as the functions you have already studied, where $fg(x)$ means do g first and then f, or $f(g(x))$, then we can write $\mathbf{R}_{-90}(\mathbf{M}_x(A)) = A_2$, or simply $\mathbf{R}_{-90}\mathbf{M}_x(A) = A_2$, both of which mean do the reflection first, then the rotation.

Looking at the diagram is there a single transformation that would map A directly to A_2?

You should see that reflection in the line $y = -x$ would map A to A_2.

So sometimes there is a single transformation which will perform the same operation as a succession of two or more transformations.

We can check this result with the matrices for the two transformations.

$$\mathbf{M}_x = \begin{pmatrix} 1 & 0 \\ 0 & -1 \end{pmatrix} \text{ and } \mathbf{R}_{-90} = \begin{pmatrix} 0 & 1 \\ -1 & 0 \end{pmatrix}.$$

M first and then **R** would be performed by the following matrix multiplication.

$$\mathbf{R}_{-90}\mathbf{M}_x = \begin{pmatrix} 0 & 1 \\ -1 & 0 \end{pmatrix}\begin{pmatrix} 1 & 0 \\ 0 & -1 \end{pmatrix} = \begin{pmatrix} 0 & -1 \\ -1 & 0 \end{pmatrix}$$

The matrix for reflection in the line $y = -x$ is $\begin{pmatrix} 0 & -1 \\ -1 & 0 \end{pmatrix}$, so we have proved the result.

Use the above methods to check the result for the transformations in the opposite order, that is, for $\mathbf{R}_{-90}$ followed by $\mathbf{M}_x$.
You should find that the combined transformation is the same as reflection in the line $y = x$, not $y = -x$ as before.
This is further proof that the order of transformations can make a difference to the outcome.

Example 10

(a) $\mathbf{M}_y$ = reflection in the y-axis, $\mathbf{R}_{180}$ = rotation of 180° about the origin and $\mathbf{M}_{x=1}$ = reflection in the line $x = 1$.
Find, either by drawing a sketch or by matrix multiplication where possible, single transformations to represent the following.
  (i) $\mathbf{R}_{180}\mathbf{M}_y$ (ii) $\mathbf{M}_y\mathbf{M}_{x=1}$

(b) (i) Find by matrix multiplication $\mathbf{M}_y^{\;2}$ where $\mathbf{M}_y$ = reflection in the y-axis as above.
  (ii) Explain this result.
  (iii) Hence write down $\mathbf{M}_y^{\;-1}$, the inverse of $\mathbf{M}_y$.

(c) $\mathbf{T}_1 = \begin{pmatrix} 1 \\ 3 \end{pmatrix}$, $\mathbf{T}_2 = \begin{pmatrix} -4 \\ -3 \end{pmatrix}$ and $\mathbf{T}_3 = \begin{pmatrix} 6 \\ 0 \end{pmatrix}$.

Find, by vector addition, single transformations to represent the following.
  (i) $\mathbf{T}_1\mathbf{T}_3$ (ii) $\mathbf{T}_3\mathbf{T}_1$ (iii) $\mathbf{T}_1\mathbf{T}_2\mathbf{T}_3$

(d) $\mathbf{M}_1$ = reflection in $x = y$, $\mathbf{M}_2$ = reflection in the x-axis, $\mathbf{R}$ = rotation 90° clockwise. A is the point (1, 4). Find
  (i) $\mathbf{M}_1\mathbf{M}_2(A)$ (ii) $\mathbf{M}_2\mathbf{M}_1(A)$ (iii) $\mathbf{M}_1\mathbf{R}(A)$ (iv) $\mathbf{R}\mathbf{M}_1(A)$

Answer 10

(a) (i) $\mathbf{R}_{180} = \begin{pmatrix} -1 & 0 \\ 0 & -1 \end{pmatrix}$ $\mathbf{M}_y = \begin{pmatrix} -1 & 0 \\ 0 & 1 \end{pmatrix}$

$$R_{180}\,M_y = \begin{pmatrix} -1 & 0 \\ 0 & -1 \end{pmatrix}\begin{pmatrix} -1 & 0 \\ 0 & 1 \end{pmatrix} = \begin{pmatrix} 1 & 0 \\ 0 & -1 \end{pmatrix}$$

$R_{180}\,M_y$ = reflection in the x-axis.

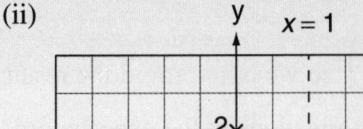

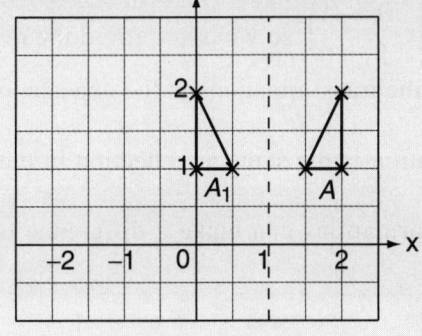

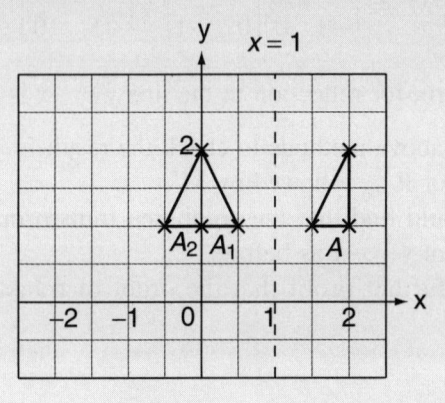

(ii)

$M_{x=1}$ maps A to A_1 $\qquad$ M_y maps A_1 to A_2

$$M_y M_{x=1} = \text{translation} \begin{pmatrix} -2 \\ 0 \end{pmatrix}$$

(b) (i) $M_y = \begin{pmatrix} -1 & 0 \\ 0 & 1 \end{pmatrix}$ $\qquad$ $M_y^2 = \begin{pmatrix} -1 & 0 \\ 0 & 1 \end{pmatrix}\begin{pmatrix} -1 & 0 \\ 0 & 1 \end{pmatrix} = \begin{pmatrix} 1 & 0 \\ 0 & 1 \end{pmatrix} = I$

(ii) M_y^2 maps the object to the image and then back to the object.

(iii) $M_y^{-1} = \begin{pmatrix} -1 & 0 \\ 0 & 1 \end{pmatrix}$

(c) (i) $T_1 T_3 = \begin{pmatrix} 7 \\ 3 \end{pmatrix}$ $\qquad$ (ii) $T_3 T_1 = \begin{pmatrix} 7 \\ 3 \end{pmatrix}$ $\qquad$ (iii) $T_1 T_2 T_3 = \begin{pmatrix} 3 \\ 0 \end{pmatrix}$

(d) $M_1 = \begin{pmatrix} 0 & 1 \\ 1 & 0 \end{pmatrix}$ $\qquad$ $M_2 = \begin{pmatrix} 1 & 0 \\ 0 & -1 \end{pmatrix}$ $\qquad$ $R = \begin{pmatrix} 0 & 1 \\ -1 & 0 \end{pmatrix}$

(i) $M_1 M_2(A) = \begin{pmatrix} 0 & 1 \\ 1 & 0 \end{pmatrix}\begin{pmatrix} 1 & 0 \\ 0 & -1 \end{pmatrix}\begin{pmatrix} 1 \\ 4 \end{pmatrix} = \begin{pmatrix} 0 & -1 \\ 1 & 0 \end{pmatrix}\begin{pmatrix} 1 \\ 4 \end{pmatrix} = \begin{pmatrix} -4 \\ 1 \end{pmatrix}$

$M_1 M_2(A) = (-4, 1)$

Alternative method:

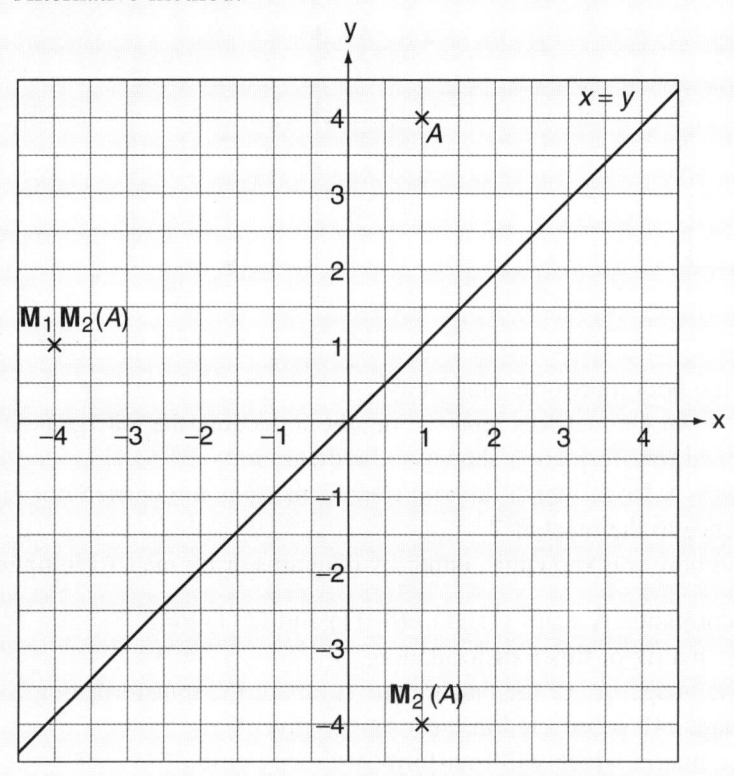

(ii) $\mathbf{M_2M_1}(A) = \begin{pmatrix} 1 & 0 \\ 0 & -1 \end{pmatrix}\begin{pmatrix} 0 & 1 \\ 1 & 0 \end{pmatrix}\begin{pmatrix} 1 \\ 4 \end{pmatrix} = \begin{pmatrix} 0 & 1 \\ -1 & 0 \end{pmatrix}\begin{pmatrix} 1 \\ 4 \end{pmatrix} = \begin{pmatrix} 4 \\ -1 \end{pmatrix}$

$\mathbf{M_2M_1}(A) = (4, -1)$

(iii) $\mathbf{M_1R}(A) = \begin{pmatrix} 0 & 1 \\ 1 & 0 \end{pmatrix}\begin{pmatrix} 0 & 1 \\ -1 & 0 \end{pmatrix}\begin{pmatrix} 1 \\ 4 \end{pmatrix} = \begin{pmatrix} -1 & 0 \\ 0 & 1 \end{pmatrix}\begin{pmatrix} 1 \\ 4 \end{pmatrix} = \begin{pmatrix} -1 \\ 4 \end{pmatrix}$

$\mathbf{M_1R}(A) = (-1, 4)$

(iv) $\mathbf{RM_1}(A) = \begin{pmatrix} 0 & 1 \\ -1 & 0 \end{pmatrix}\begin{pmatrix} 0 & 1 \\ 1 & 0 \end{pmatrix}\begin{pmatrix} 1 \\ 4 \end{pmatrix} = \begin{pmatrix} 1 & 0 \\ 0 & -1 \end{pmatrix}\begin{pmatrix} 1 \\ 4 \end{pmatrix} = \begin{pmatrix} 1 \\ -4 \end{pmatrix}$

$\mathbf{RM_1}(A) = (1, -4)$

Exercise 10.7

In this exercise the 'unit square' refers to the square with coordinates (0, 0), (1, 0), (1, 1) and (0, 1).

1. The transformation of the unit square shown in the diagrams consists of two separate transformations, a reflection and an enlargement.

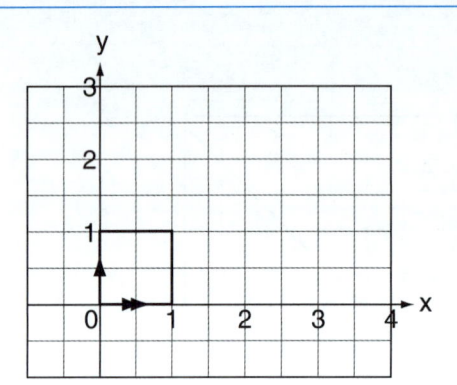

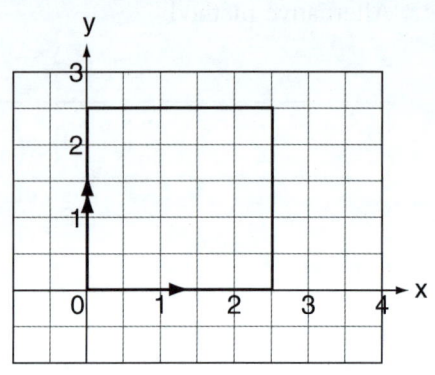

(i) Describe both transformations fully.
(ii) Write down the matrices representing the reflection, the enlargement and the combined transformation shown in the diagram.
(iii) Does the order in which the two transformations are carried out make any difference to the outcome?
Explain how you know this, either by diagrams or by matrix multiplication.

2. (a) A transformation **X** maps (1, 3) to (5, 11) and (2, 1) to (5, 7).
   Find the matrix of this transformation.
   (b) Draw the image, A', of the unit square, A, under the transformation **X**.
   (c) This image, A', is reflected in the x-axis to give A''.
   Find the matrix which will transform A directly onto A''.

3. (a) A is the point (1, 2), B is (1, 4) and C is (2, 2). Draw triangle ABC.
   (b) Draw triangle $A_2B_2C_2$, which is the image of triangle ABC after a reflection in the x-axis followed by a rotation of 90° clockwise about the origin.
   (c) What single transformation will map $A_2B_2C_2$ back to ABC?

4. List the matrices representing the following transformations. Use the unit square to find the matrices where necessary.
   (a) Identity, **I**
   (b) Rotation 90° clockwise about O, **R**$_1$
   (c) Rotation 180° about O, **R**$_2$
   (d) Reflection in x-axis, **M**$_1$
   (e) Reflection in y-axis, **M**$_2$
   (f) Reflection in $x = y$, **M**$_3$
   (g) Reflection in $y = -x$, **M**$_4$

   NOTE: This makes a useful reference. Check your answers and keep the list safely. If you can learn some of these matrices it will make answering questions quicker, but check that you have them right each time you use them.

5. Using **R**$_1$ = rotation 90° clockwise about O as in question 4, find **R**$_1^{-1}$, (the inverse of **R**$_1$).
   What does **R**$_1^{-1}$ represent?

6. Using the notation in questions 4 and 5 copy and complete this table showing the effects of two successive transformations.

Second transformation

	R_1	R_2	M_1	M_2	M_3	M_4
R_1	R_2	R_1^{-1}				
R_2		I				
M_1			I			
M_2						
M_3						
M_4						

First transformation

Exercise 10.8

Mixed Exercise

1. Solve this vector equation for x and y.

$$\frac{1}{2}\begin{pmatrix} x \\ y \end{pmatrix} + \begin{pmatrix} -1 \\ 3 \end{pmatrix} = \begin{pmatrix} -1 \\ 7 \end{pmatrix}$$

2. Draw a triangle OAB where A is (7, 2) and B is (2, 5).
   (a) Draw triangle $OA'B'$, under an enlargement, centre the origin and scale factor 1.5.
   (b) What can you say about triangles OAB and $OA'B'$?
   (c) If the area of triangle OAB is 32 square units calculate the area of the trapezium $ABB'A'$.

3. Copy the diagram.

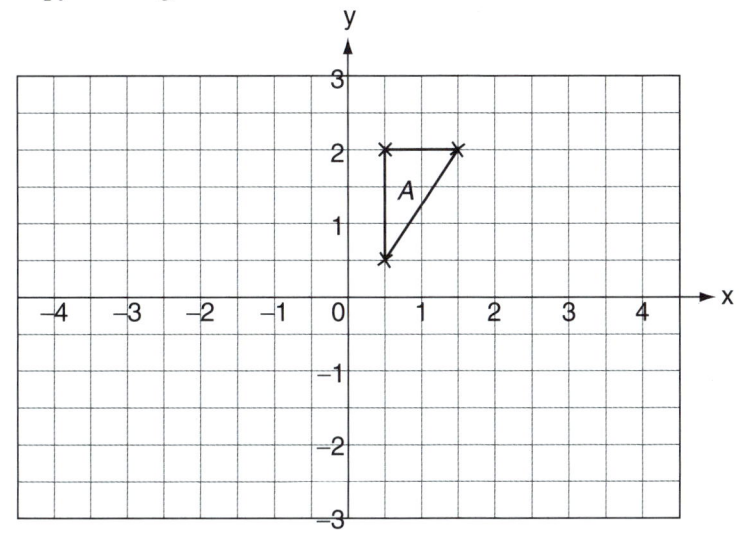

   (a) Draw the image of triangle A after reflection in the line $y = x - 1$. Label the image B.

(b) Draw the image of triangle A after reflection in the line $x = -1$. Label the image C.

(c) Describe fully the single transformation which will map B onto C.

4. A shear, x-axis invariant, maps the rectangle ABCD onto the parallelogram $A'B'C'D'$. A is the point $(1, -1)$, B is $(2, -1)$, C is $(2, 1)$, D is $(1, 1)$ and C' is $(3.5, 1)$.

    (a) Draw the rectangle and the parallelogram on a graph with 1 centimetre per unit on each axis.

    NOTE: The points below the x-axis (with negative y-coordinates) will map to the left, while those with positive y-coordinates will map to the right.

    (b) Write down the shear factor for the transformation.

5. Describe fully the single transformation which maps the rectangle to the parallelogram in the diagram below.

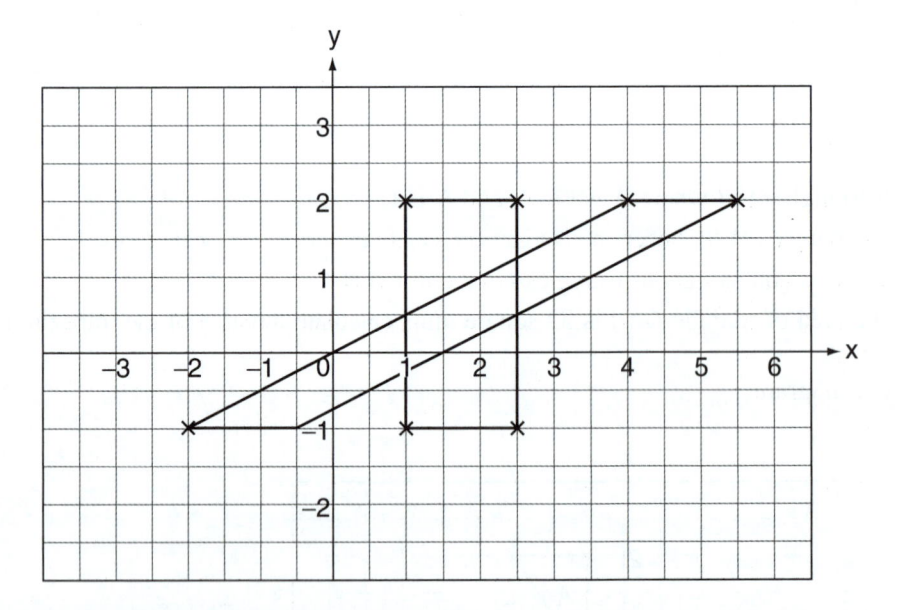

6. A stretch, with the line $x = 1$ invariant, maps the rectangle ABCD onto the rectangle $A'B'C'D'$.

    A is the point $(2, 3)$, B is $(3, 3)$, C is $(3, 1)$, A' is $(4, 3)$ and C' is $(7, 1)$

    (a) By drawing the rectangles on coordinate axes, find D, B' and D'.

    (b) Find the scale factor of the stretch.

7.

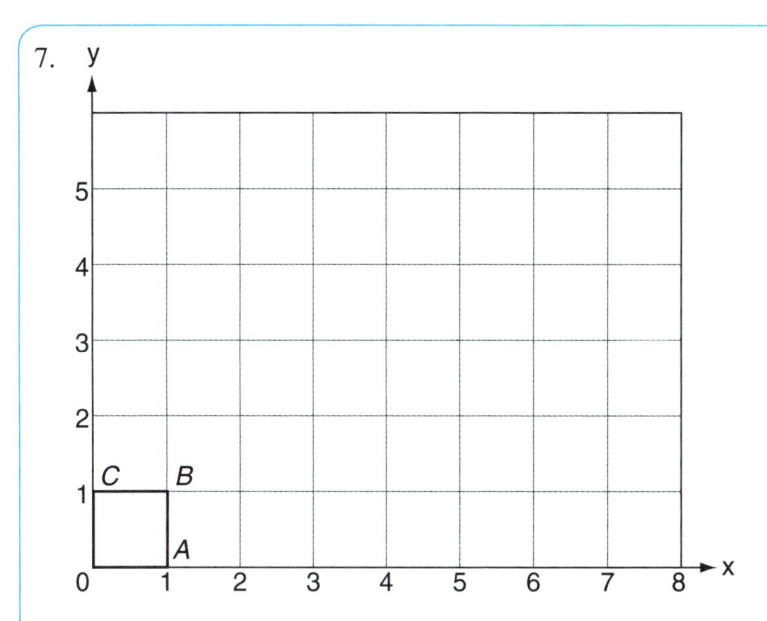

Copy the diagram.

(a) *OABC* is a unit square. By carrying out the matrix multiplication below find the image *OA'B'C'* of *OABC* under the transformation given by the matrix $\mathbf{M} = \begin{pmatrix} 5 & 2 \\ 1 & 4 \end{pmatrix}$.

$$\begin{array}{cccc} & O & A & B & C \end{array} \quad \begin{array}{cccc} O & A' & B' & C' \end{array}$$
$$\begin{pmatrix} 5 & 2 \\ 1 & 4 \end{pmatrix} \begin{pmatrix} 0 & 1 & 1 & 0 \\ 0 & 0 & 1 & 1 \end{pmatrix} = \begin{pmatrix} \ldots & \ldots & \ldots & \ldots \\ \ldots & \ldots & \ldots & \ldots \end{pmatrix}$$

(b) (i) Draw *OA'B'C'* on your copy of the diagram.
 (ii) What is the shape of *OA'B'C'*?

(c) By drawing a rectangle round *OA'B'C'*, and removing triangles, or otherwise, calculate the area of *OA'B'C'*.

(d) (i) Find det **M**.
 (ii) What do you notice about your answers to (c) and (d) (i)?

Examination Questions

8. (a) The diagrams show triangles *A*, *B*, *C* and *D*.
 (i) The single transformation **P** maps △*A* onto △*B*.
 Describe, fully, the transformation **P**.

294   *Extended Mathematics for Cambridge IGCSE*

[Graph showing triangles B and A in the first quadrant area]

(ii) The single transformation **Q** maps △A onto △C.
Describe, fully, the transformation **Q**.

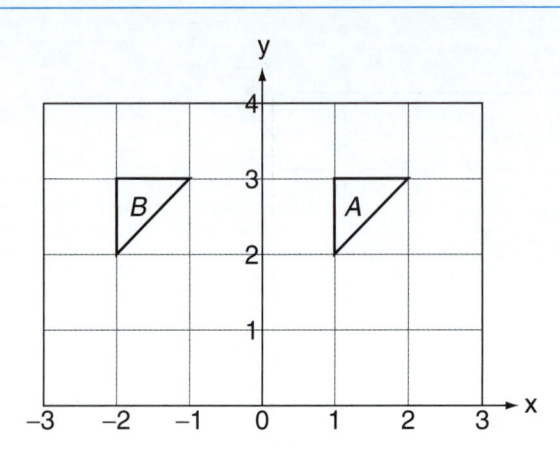

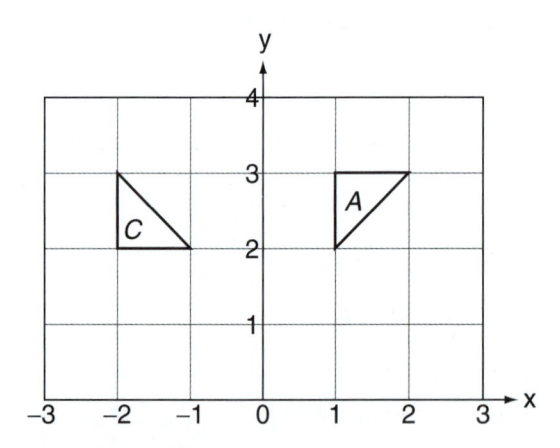

[Graph showing triangles C and A]

(iii) The reflection **R** maps △A onto △D.
Find the matrix that represents the reflection **R**.

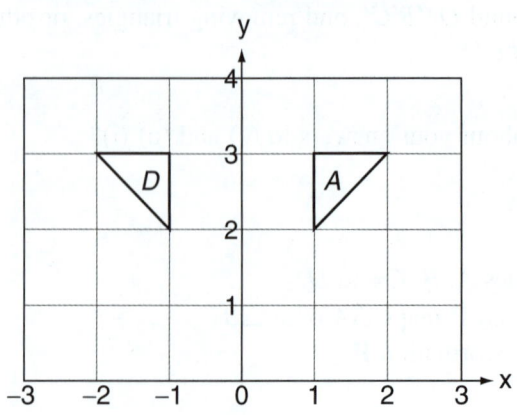

[Graph showing triangles D and A]

(b) The diagram shows the points $E(1, 3)$, $F(2, 3)$ and $G(-1, 3)$. An enlargement, centre E, maps F onto G.

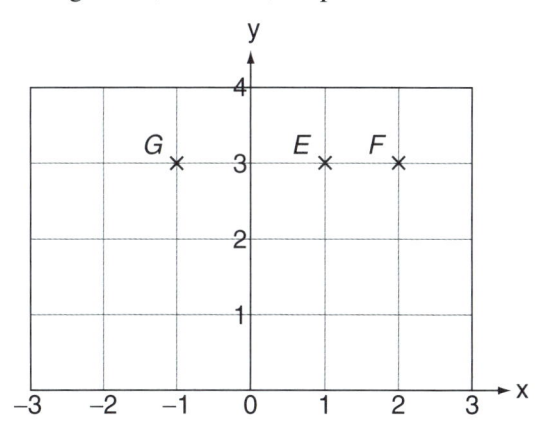

Write down
  (i) the scale factor,
  (ii) the coordinates of the image of $(0, 4)$.

(c) $\mathbf{M} = \begin{pmatrix} -1 & 3 \\ -2 & 4 \end{pmatrix}$

  (i) Find the determinant of $\mathbf{M}$.
  (ii) Write down the inverse of $\mathbf{M}$.
  (iii) Find the matrix $\mathbf{X}$, where $\mathbf{MX} = \begin{pmatrix} 4 \\ -2 \end{pmatrix}$.

(4024/02 May/June 2007 q 11)

9. (a) Draw and label x- and y-axes from -8 to $+8$, using a scale of 1 cm to 1 unit on each axis.
  (b) Draw and label triangle ABC with $A(2, 2)$, $B(5, 2)$ and $C(5, 4)$.
  (c) On your grid
    (i) translate **triangle ABC** by the vector $\begin{pmatrix} 3 \\ -9 \end{pmatrix}$ and label this image $A_1B_1C_1$,
    (ii) reflect **triangle ABC** in the line $x = -1$ and label this image $A_2B_2C_2$,
    (iii) rotate **triangle ABC** by $180°$ about $(0, 0)$ and label this image $A_3B_3C_3$.
  (d) A stretch is represented by the matix $\begin{pmatrix} 1.5 & 0 \\ 0 & 1 \end{pmatrix}$.
    (i) Draw the image of **triangle ABC** under this transformation. Label this image $A_4B_4C_4$.
    (ii) Work out the inverse of the matrix $\begin{pmatrix} 1.5 & 0 \\ 0 & 1 \end{pmatrix}$.
    (iii) Describe **fully** the single transformation represented by this inverse.

(0580/04 Oct/Nov 2005 q 2)

10. *OABC* is a parallelogram. $\overrightarrow{OA} = \mathbf{a}$ and $\overrightarrow{OC} = \mathbf{c}$.
    M is the midpoint of *OB*.
    Find $\overrightarrow{MA}$ in terms of $\mathbf{a}$ and $\mathbf{c}$.

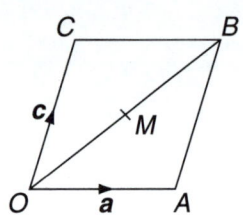

(0580/02 Oct/Nov 2005 q 8)

11. (a) Draw and label *x*-and *y*-axes from −6 to 6, using a scale of 1 cm to 1 unit.
    (b) Draw triangle *ABC* with *A*(2, 1), *B*(3, 3) and *C*(5, 1).
    (c) Draw the reflection of triangle *ABC* in the line $y = x$. Label this $A_1B_1C_1$.
    (d) Rotate **triangle $A_1B_1C_1$** about (0, 0) through 90° anti-clockwise. Label this $A_2B_2C_2$.
    (e) Describe fully the single transformation which maps triangle *ABC* onto triangle $A_2B_2C_2$.
    (f) A transformation is represented by the matrix $\begin{pmatrix} 1 & 0 \\ -1 & 1 \end{pmatrix}$.
       (i) Draw the image of triangle *ABC* under this transformation. Label this $A_3B_3C_3$.
       (ii) Describe fully the single transformation represented by the matrix $\begin{pmatrix} 1 & 0 \\ -1 & 1 \end{pmatrix}$.
       (iii) Find the matrix which represents the transformation that maps triangle $A_3B_3C_3$ onto triangle *ABC*.

(0580/04 May/June 2007 q 2)

12. Transformation **T** is translation by the vector $\begin{pmatrix} 3 \\ 2 \end{pmatrix}$.
    Transformation **M** is reflection in the line $y = x$.
    (a) The point *A* has co-ordinates (2, 1).
        Find the co-ordinates of
        (i) **T**(*A*),
        (ii) **MT**(*A*).
    (b) Find the 2 by 2 matrix **M**, which represents the transformation **M**.
    (c) Show that, for any value of *k*, the point *Q*(*k*−2, *k*−3) maps onto a point on the line $y = x$ following the transformation **TM**(*Q*).
    (d) Find **M**$^{-1}$, the inverse of the matrix **M**.

(e) **N** is the matrix such that $\mathbf{N} + \begin{pmatrix} 0 & 3 \\ 1 & 0 \end{pmatrix} = \begin{pmatrix} 0 & 4 \\ 0 & 0 \end{pmatrix}$.

   (i) Write down the matrix **N**.

   (ii) Describe completely the **single** transformation represented by **N**.

(0580/04 May/June 2006 q 7)

13. (a)

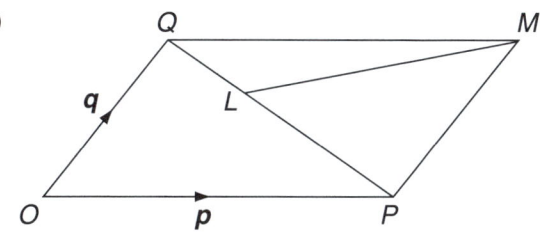

NOT TO SCALE

$OPMQ$ is a parallelogram and O is the origin.

$\overrightarrow{OP} = p$ and $\overrightarrow{OQ} = q$.

L is on PQ so that $PL : LQ = 2:1$.

Find the following vectors in terms of p and q. Write your answers in their simplest form.

   (i) $\overrightarrow{PQ}$,

   (ii) $\overrightarrow{PL}$,

   (iii) $\overrightarrow{ML}$,

   (iv) the position vector of L.

(b) R is the point $(1, 2)$. It is translated onto the point S by the vector $\begin{pmatrix} 3 \\ -4 \end{pmatrix}$.

   (i) Write down the co-ordinates of S.

   (ii) Write down the vector which translates S onto R.

(c) The matrix $\begin{pmatrix} 0 & 1 \\ -1 & 0 \end{pmatrix}$ represents a **single** transformation.

   (i) Describe fully this transformation.

   (ii) Find the co-ordinates of the image of the point $(5, 3)$ after this transformation.

(d) Find the matrix which represents a reflection in the line $y = x$.

(0580/04 Oct/Nov 2006 q 6)

14.

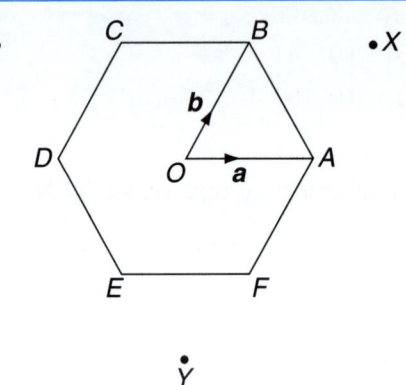

A regular hexagon, *ABCDEF*, has centre *O*.
$\overrightarrow{OA} = \mathbf{a}$ and $\overrightarrow{OB} = \mathbf{b}$.

(a) Express, as simply as possible, in terms of *a* and/or *b*.

  (i) $\overrightarrow{DO}$,

  (ii) $\overrightarrow{AB}$,

  (iii) $\overrightarrow{DB}$.

(b) Explain why $|\mathbf{a}| = |\mathbf{b}| = |\mathbf{b}-\mathbf{a}|$.

(c) The points *X*, *Y* and *Z* are such that

$\overrightarrow{OX} = \mathbf{a} + \mathbf{b}$, $\overrightarrow{OY} = \mathbf{a} - 2\mathbf{b}$ and $\overrightarrow{OZ} = \mathbf{b} - 2\mathbf{a}$.

  (i) Express, as simply as possible, in terms of *a* and/or *b*,

   (a) $\overrightarrow{AX}$,

   (b) $\overrightarrow{YX}$.

  (ii) What can be deduced about *Y*, *A* and *X*?

(d) Express, as simply as possible, in terms of *a* and/or *b*, the vector $\overrightarrow{XZ}$.

(e) Show that triangle *XYZ* is equilateral.

(f) Calculate $\dfrac{\text{Area of triangle } OAB}{\text{Area of triangle } XYZ}$.

(4024/02 Oct/Nov 2005 q 11)

15. On the grid below $\overrightarrow{OP} = \mathbf{p}$ and $\overrightarrow{OQ} = \mathbf{q}$.

(a) Given the $\overrightarrow{OR} = \mathbf{p} - \mathbf{q}$, mark the point *R* clearly on a copy of the grid.

(b) The point *S* is shown on the grid.

Given the $\overrightarrow{OS} = \mathbf{q} + h\mathbf{p}$, find h.

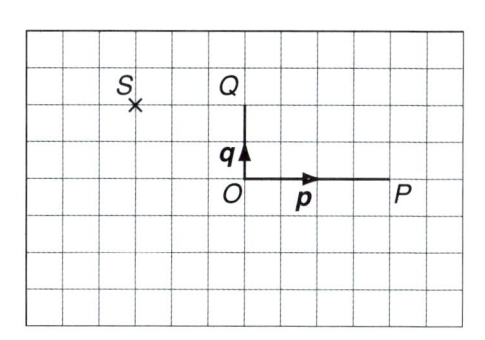

(4024/01 May/June 2007 q 8)

16. The diagram below shows the point P and triangles A, B, and C.

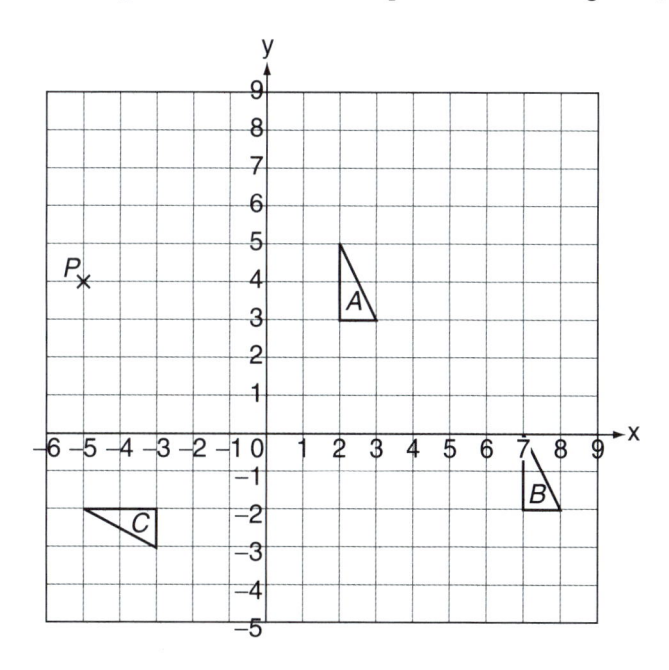

(a) The translation **T** maps $\triangle A$ onto $\triangle B$.

Given that $\mathbf{T}(P) = Q$, write down the coordinates of Q.

(b) Describe fully the single transformation which maps $\triangle A$ onto $\triangle C$.

(c) The matrix $\begin{pmatrix} 1 & 0 \\ -2 & 1 \end{pmatrix}$ represents the shear S.

Given that $S(A) = D$, draw and label ΔD on the diagram above.

(d) ΔA is mapped onto ΔE by a rotation of 90° clockwise about the point (4, 2). Draw and label ΔE on the diagram above.

(4024/01 Oct/Nov 2004 q 25)

17. (a) Describe fully the single transformation that maps ΔXYZ onto ΔXPQ.

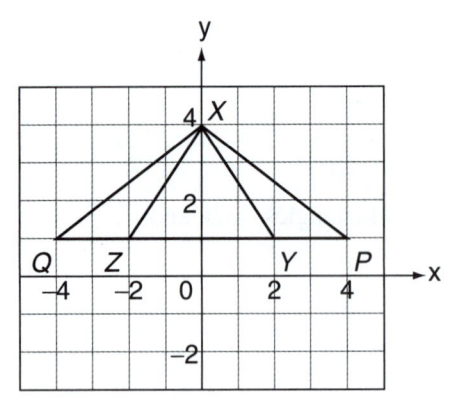

(b) The diagram shows ΔABC and the point B' (9, 2).
  (i) A translation maps B onto B'.
      Write down the column vector that represents this translation.
  (ii) A shear in which the x-axis is invariant maps ΔABC onto $\Delta A'B'C'$.
      (a) Draw $\Delta A'B'C'$ on a copy of the diagram.
      (b) State the shear factor.

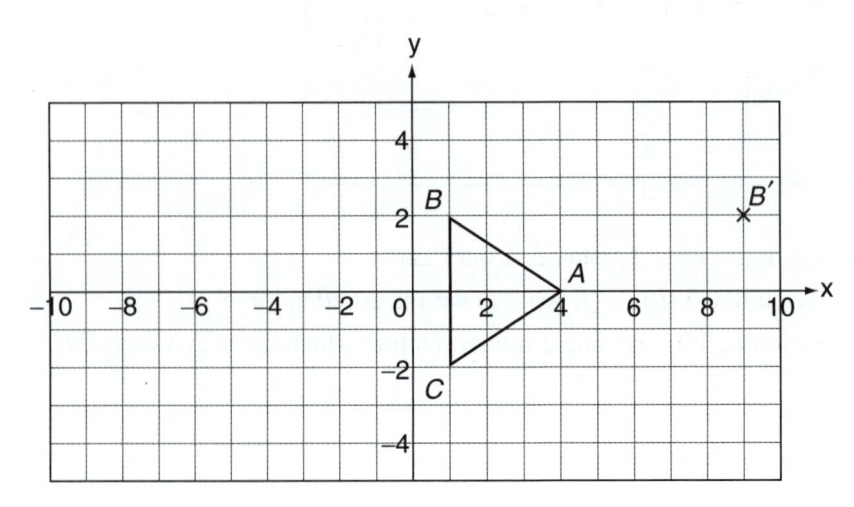

(4024/01 May/June 2004 q 26)

18.

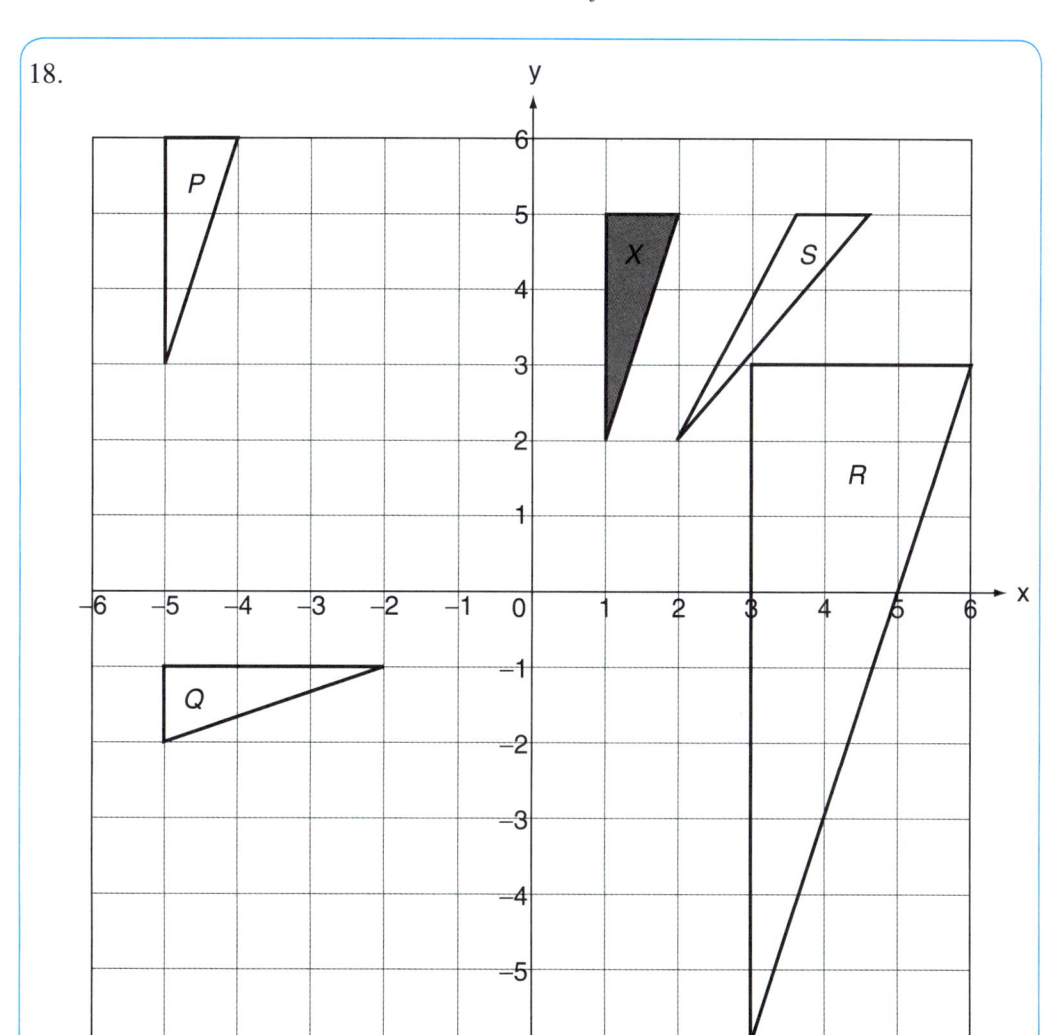

(a) Describe fully the single transformation which maps
   (i) triangle X onto triangle P,
   (ii) triangle X onto triangle Q,
   (iii) triangle X onto triangle R,
   (iv) triangle X onto triangle S.

(b) Find the 2 by 2 matrix which represents the transformation that maps
   (i) triangle X onto triangle Q,
   (ii) triangle X onto triangle S.

(0580/04 May/June 2005 q 3)

19.

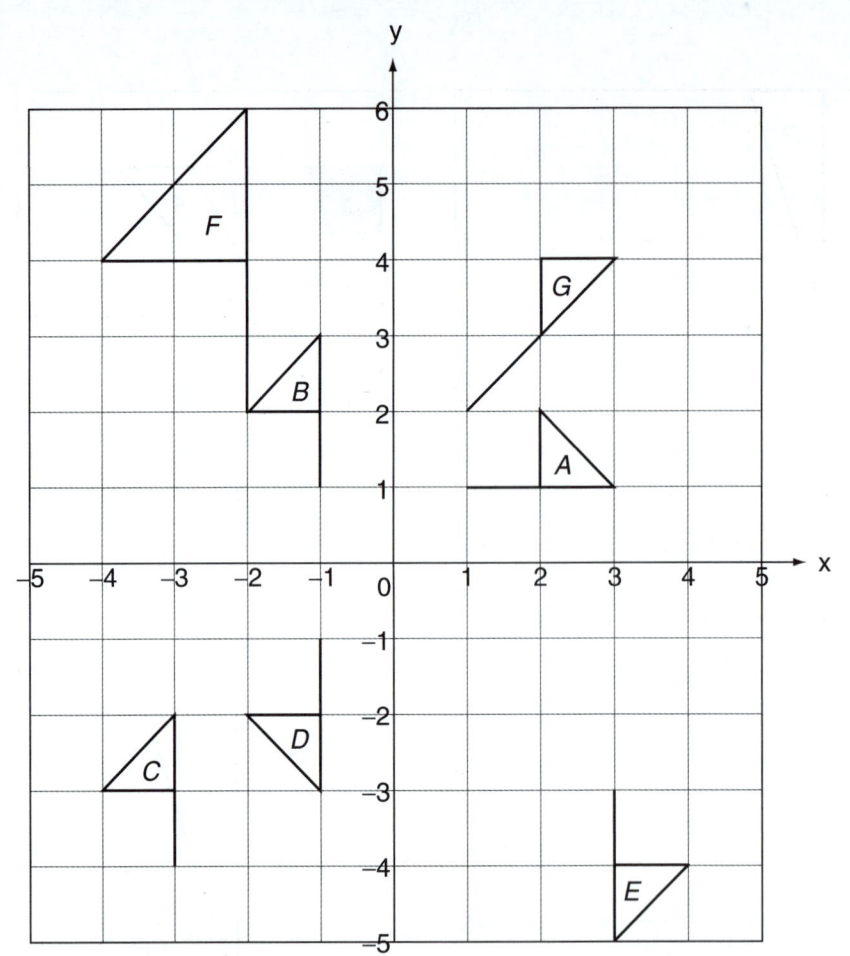

(a) Describe fully the **single** transformation which maps
  (i) shape A onto shape B,
  (ii) shape B onto shape C,
  (iii) shape A onto shape D,
  (iv) shape B onto shape E,
  (v) shape B onto shape F,
  (vi) shape A onto shape G.

(b) A transformation is represented by the matrix $\begin{pmatrix} 0 & -1 \\ 1 & 0 \end{pmatrix}$.

  Which shape above is the image of shape A after this transformation?

(c) Find the 2 by 2 matrix representing the transformation which maps
  (i) shape B onto shape D,
  (ii) shape A onto shape G.

(0580/04 May/June 2004 q 8)

20. (a) Draw x-and y-axes from 0 to 12 using a scale of 1 cm to 1 unit on each axis.
    (b) Draw and label triangle T with vertices (8, 6), (6, 10) and (10, 12).
    (c) Triangle T is reflected in the line $y = x$.
       (i) Draw the image of triangle T. Label this image P.
       (ii) Write down the matrix which represents this reflection.
    (d) A transformation is represented by the matrix $\begin{pmatrix} \frac{1}{2} & 0 \\ 0 & \frac{1}{2} \end{pmatrix}$
       (i) Draw the image of triangle T under this transformation. Label this image Q.
       (ii) Describe fully this single transformation.
    (e) Triangle T is stretched with the y-axis invariant and a stretch factor of $\frac{1}{2}$.
       Draw the image of triangle T. Label this image R.

    (0580/04 Oct/Nov 2007 q 7)

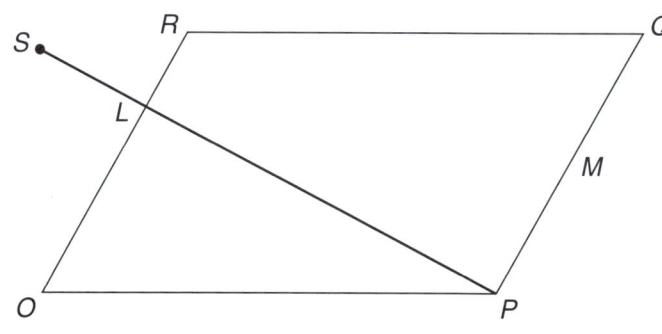

21.

NOT TO SCALE

OPQR is a parallelogram.
O is the origin.
$\overrightarrow{OP} = \mathbf{p}$ and $\overrightarrow{OR} = \mathbf{r}$.

M is the midpoint of PQ and L is on OR such that $LR = 2:1$.
The line PL is extended to the point S.

(a) Find, in terms of $\mathbf{p}$ and $\mathbf{r}$, in their simplest forms,
   (i) $\overrightarrow{OQ}$ (ii) $\overrightarrow{PR}$ (iii) $\overrightarrow{PL}$ (iv) the position vector of M.

(b) PLS is a straight line and $PS = \frac{3}{2}PL$.
   Find, in terms of $\mathbf{p}$ and/or $\mathbf{r}$, in their simplest forms,
   (i) $\overrightarrow{PS}$,   (ii) $\overrightarrow{QS}$.

(c) What can you say about the points Q, R and S?

(0580/04 May/June 2008 q 9)

22.

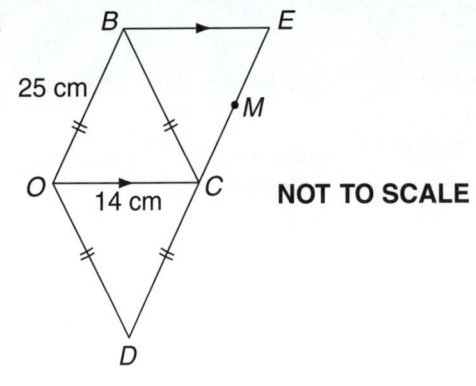

OBCD is a rhombus with sides of 25 cm. The length of the diagonal *OC* is 14 cm.

(a) Show, **by calculation**, that the length of the diagonal *BD* is 48 cm.
(b) Calculate, correct to the nearest degree,
   (i) angle *BCD*,
   (ii) angle *OBC*.
(c) $\overrightarrow{DB} = 2p$ and $\overrightarrow{OC} = 2q$.

   Find, in terms of *p* and *q*,
   (i) $\overrightarrow{OB}$,
   (ii) $\overrightarrow{OD}$.

(d) *BE* is parallel to *OC* and *DCE* is a straight line.
   Find, in its simplest form, $\overrightarrow{OE}$ in terms of *p* and *q*.

(e) *M* is the midpoint of *CE*.
   Find, in its simplest form, $\overrightarrow{OM}$ in terms of *p* and *q*.

(f) *O* is the origin of a co-ordinate grid, *OC* lies along the *x*-axis and $q = \begin{pmatrix} 7 \\ 0 \end{pmatrix}$.
   ($\overrightarrow{DB}$ is vertical and $|\overrightarrow{DB}| = 48$.)
   Write down as column vectors
   (i) *p*,
   (ii) $\overrightarrow{BC}$.

(g) Write down the value of $|\overrightarrow{DE}|$

(0580/04 May/June 2007 q 5)

23.

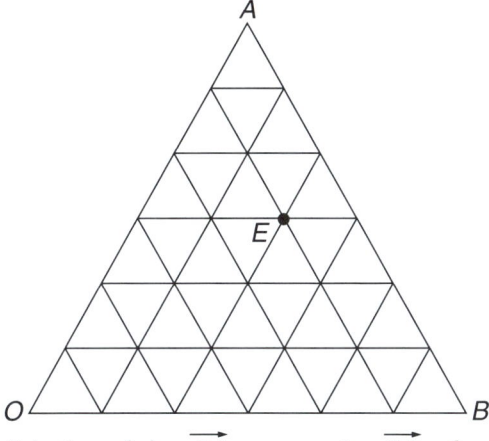

O is the origin, $\overrightarrow{OA} = \boldsymbol{a}$ and $\overrightarrow{OB} = \boldsymbol{b}$.

(a) C has position vector $\dfrac{1}{3}\boldsymbol{a} + \dfrac{2}{3}\boldsymbol{b}$.

Mark the point C on the diagram.

(b) Write down, in terms of $\boldsymbol{a}$ and $\boldsymbol{b}$, the position vector of the point E.

(c) Find, in terms of $\boldsymbol{a}$ and $\boldsymbol{b}$, the vector $\overrightarrow{EB}$.

(0580/02 Oct/Nov 2007 q 15)

24.

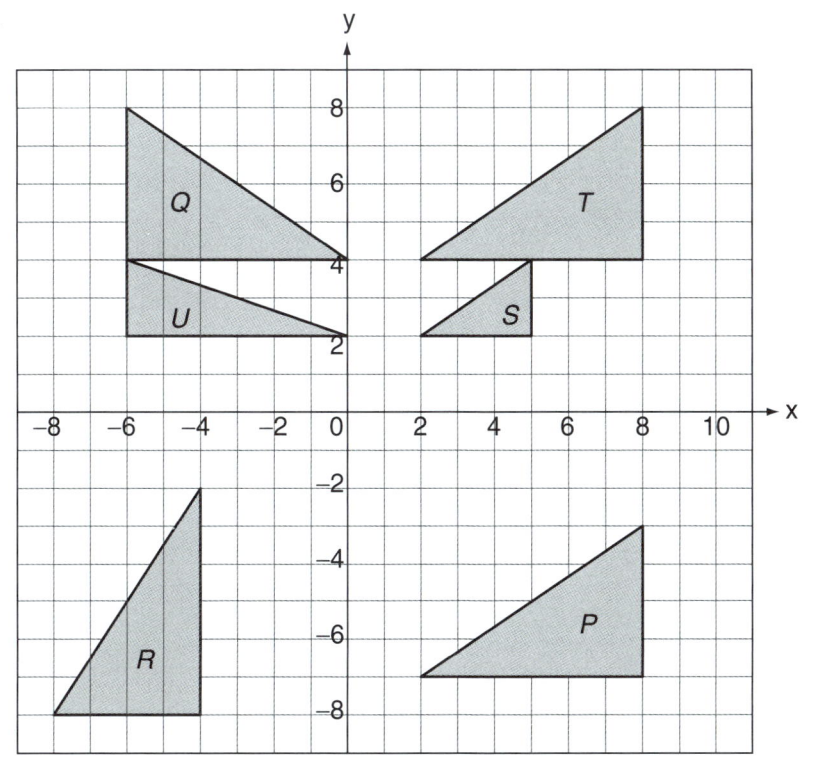

The diagram shows triangles, P, Q, R, S, T and U.

(a) Describe fully the **single** transformation which maps triangle

    (i) T onto P,
   (ii) Q onto T,
  (iii) T onto R,
  (iv) T onto S,
   (v) U onto Q.

(b) Find the 2 by 2 matrix representing the transformation which maps triangle

    (i) T onto R,
   (ii) U onto Q.

(0580/04 Oct/Nov 2008 q 7)

25.

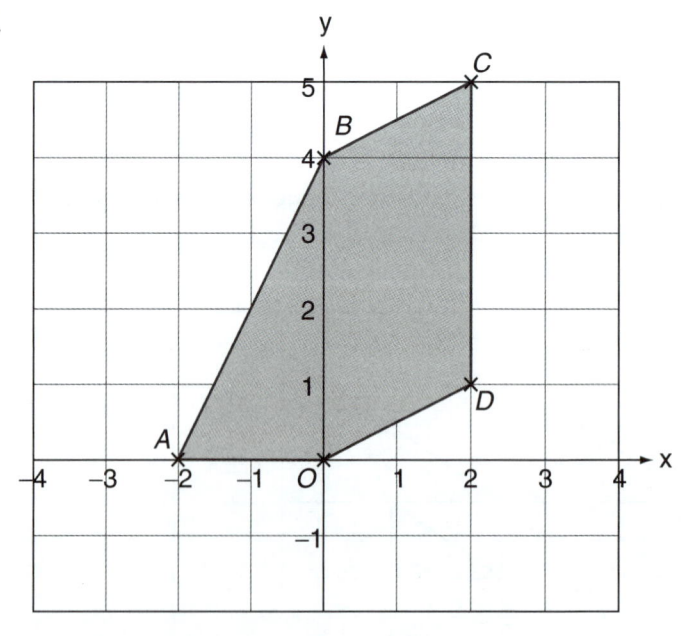

The pentagon OABCD is shown on the grid above.

(a) Write as column vectors

    (i) $\overrightarrow{OD}$,

   (ii) $\overrightarrow{BC}$.

(b) Describe fully the single transformation which maps the side BC onto the side OD.

(c) The shaded area inside the pentagon is defined by 5 inequalities.

One of these inequalities is $y \leqslant \dfrac{1}{2}x + 4$.

Find the other 4 inequalities.

(0580/04 May/June 2008 q 6)

26.

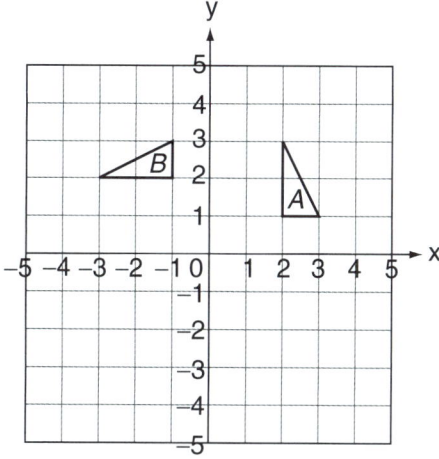

(a) A transformation is represented by the matrix $\begin{pmatrix} 0 & -1 \\ -1 & 0 \end{pmatrix}$

  (i) On a copy of the gird above, draw the image of triangle A after this transformation.
  (ii) Describe fully this transformation.

(b) Find the 2 by 2 matrix representing the transformation which maps triangle A onto triangle B.

(0580/21 May/June 2008 q 19)

Chapter 11

Statistics

This chapter introduces some more statistical diagrams, and more measures of spread.

Core Skills

1.
```
6  5  7  9  1  3  5  8  1  2  6
5  4  3  3  3  2  9  8  4  6  7
1  5  6  5  9  8  7  2  1  9  7
5  3  4  2  8  1
```
   (a) Construct a frequency table for the above data.
   (b) Find the mean, the median, the mode and the range for this data.

2.

Class	Frequency
$0 \leqslant n < 10$	10
$10 \leqslant n < 20$	15
$20 \leqslant n < 30$	17
$30 \leqslant n < 40$	11
$40 \leqslant n < 50$	6

Use the above grouped frequency table of some continuous data to draw a simple histogram.

3. The table below shows some data for a pie chart. Find the missing values (a), (b) and (c).

Colour	Frequency	Angle
red	10	(a)
blue	(b)	72°
green	5	(c)
yellow	9	108°

Statistics 309

4. Comment on the correlation shown in each of the following scatter diagrams.

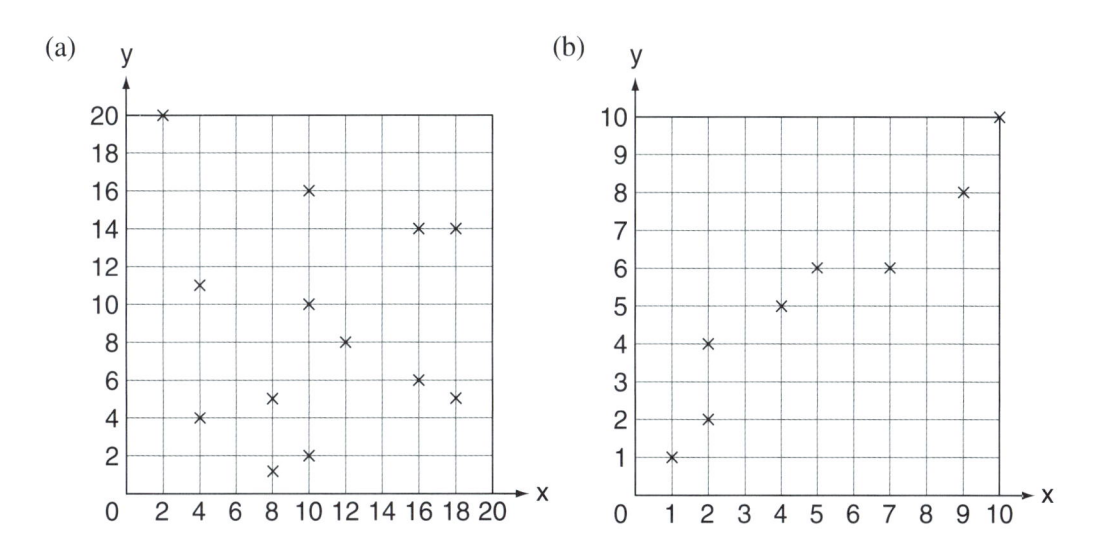

(a)

(b)

5. The table shows the heights of some students rounded to the nearest centimetre. Copy and complete the table by adding the class boundaries.

Heights of students (h cm)	Lower class boundary	Upper class boundary
141 to 150		
151 to 160		
161 to 170		

Histograms

You will have seen in the Core Book how to draw simple histograms. We will now look further at histograms.

Sometimes a set of data will be grouped into classes of different widths. In this case using the height of the bars as a measure of the frequency would be misleading. Instead we use the *area* of the bar to represent the frequency.

To work out the height to draw each bar we need to include the *class width* in the table. The height of each bar can now be calculated by dividing the frequency (area) by the class width (width). The height of the bar is called the **frequency density**.

- Frequency density = $\dfrac{\text{frequency}}{\text{class width}}$

310   *Extended Mathematics for Cambridge IGCSE*

- This is the same as:

$$\text{Height of bar} = \frac{\text{area of bar}}{\text{width of bar}}$$

The following examples will show the method. The frequency table includes a new column to show class width, and another to show the frequency density.

You should notice that both bar charts and simple histograms also have the area of each bar proportional to the frequency, but in those cases we do not have to calculate the frequency density because the bars are all of the same width.

You need to know that the **modal class** of a grouped frequency distribution is the *class* with the highest *frequency density*, which is not necessarily the one with the highest frequency.

Example 1

(a) Use the frequency table to draw a histogram.

Time (t minutes)	Frequency
$70 < t \leqslant 90$	8
$90 < t \leqslant 100$	7
$100 < t \leqslant 110$	28
$110 < t \leqslant 120$	11
$120 < t \leqslant 160$	16

(b) Write down the modal class.

Answer 1

Time (t minutes)	Class width	Frequency (Area)	Frequency density (2 dp) (Height = Area ÷ Class width)
$70 < t \leqslant 90$	20	8	0.4
$90 < t \leqslant 100$	10	7	0.7
$100 < t \leqslant 110$	10	28	2.8
$110 < t \leqslant 120$	10	11	1.1
$120 < t \leqslant 160$	40	16	0.4

(a)

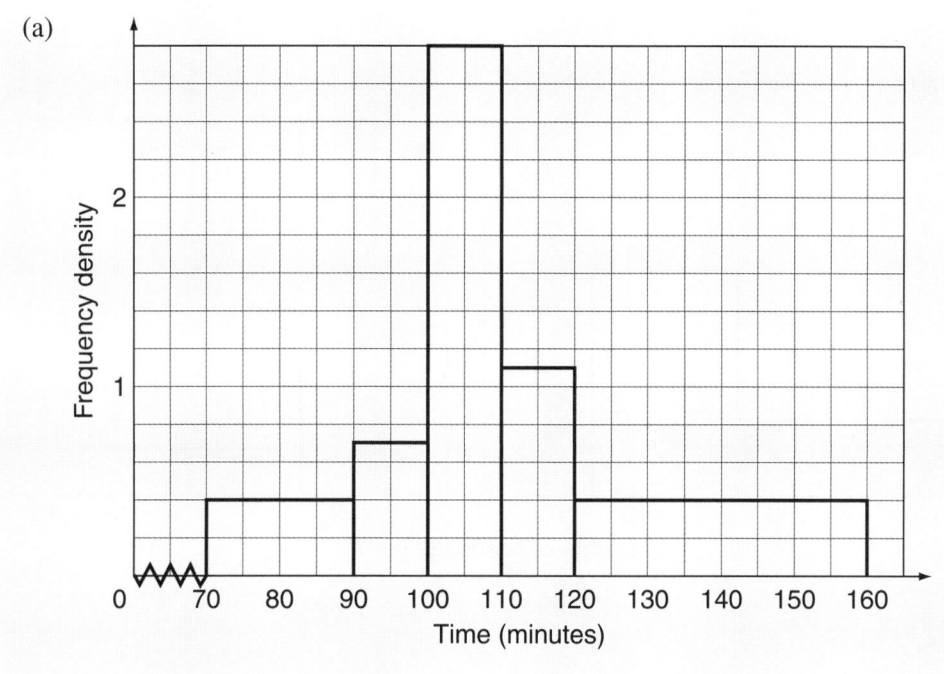

(b) The modal class is $100 < t \leqslant 110$.

The scale of the histogram may be shown by labelling the vertical axis with the frequency density, or by showing a key as in the next example.

Example 2

The table and the histogram show the heights of some seedlings measured to the nearest centimetre.

(a) Use the table to complete the histogram.
(b) Use the histogram to complete the table.

Height (h cm)	Frequency (f)	Class width (w)	Frequency density (f ÷ w)
1–3	3	3	1
6–7	5	2	2.5
8–9	4	2	2
10–12	2	3	0.67

312 *Extended Mathematics for Cambridge IGCSE*

Key ☐ = 1 seedling

Heights of seedlings (cm)

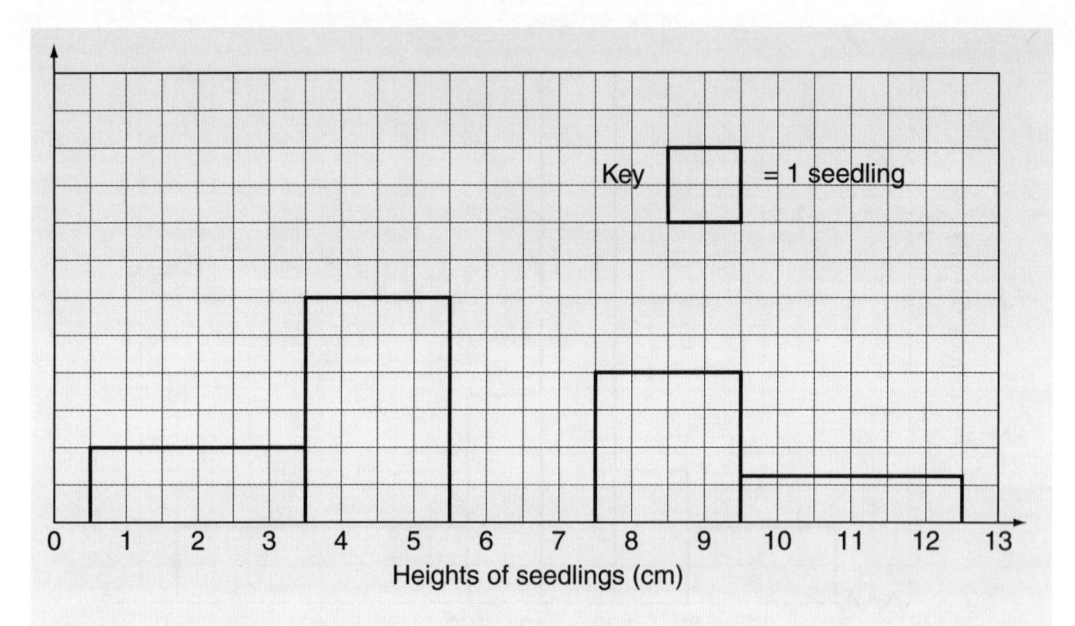

Answer 2

It may help to add the class boundaries to the table.

Height (h cm)	Frequency (f)	Class width (w)	Frequency density ($f \div w$)	Class boundaries	
1–3	3	3	1	0.5	3.5
6–7	5	2	2.5	5.5	7.5
8–9	4	2	2	7.5	9.5
10–12	2	3	0.67	9.5	12.5

NOTE: The class boundaries in, for example, the first class, are $0.5 \leqslant h < 3.5$ because the rounding of the heights of the seedlings to the nearest centimetre means that a seedling of height 3.5 centimetres rounds to 4 centimetres and goes into the next class.

(a) The missing bar is the class 6–7. From the table we see that the frequency is 5 and the class width is 2. The key shows that 5 cm² will represent 5 seedlings. The height of the bar will be 5 cm² ÷ 2 cm = 2.5 cm. We can now draw the bar and complete the scale on the vertical axis

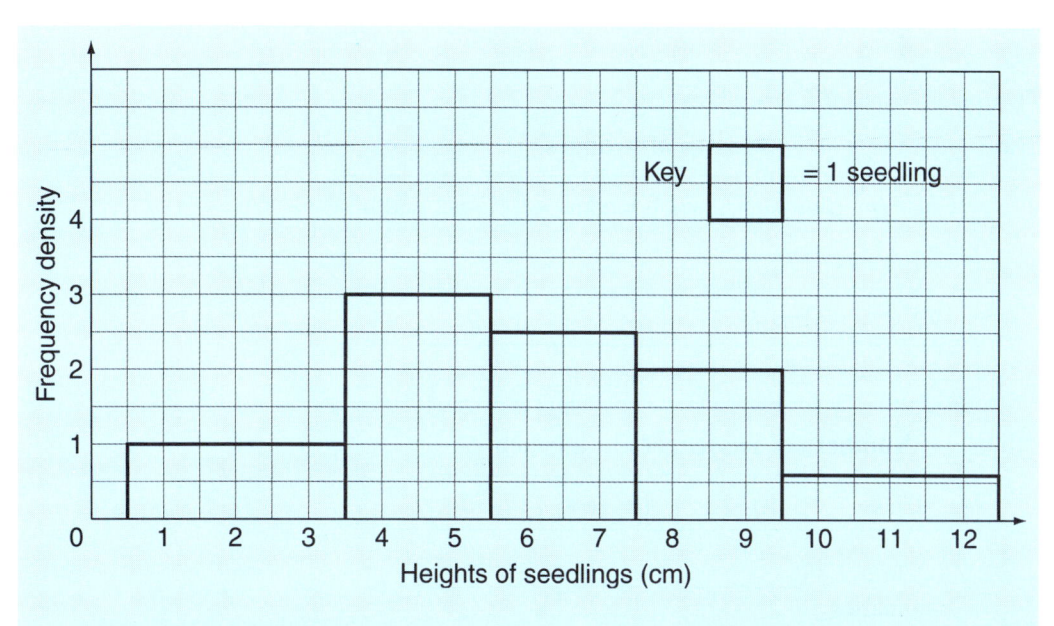

(b) The missing data in the table is for the class 4–5, which has a class width of 2 cm, and an area of 6 cm². The frequency density = 6 ÷ 2 = 3

Height (h cm)	Frequency (f)	Class width (w)	Frequency density (f ÷ w)
1–3	3	3	1
4–5	6	2	3
6–7	5	2	2.5
8–9	4	2	2
10–12	2	3	0.67

Some questions may give neither a key nor a scale on the frequency density axis. However, they will tell you the frequency that one of the bars represents, so that, knowing the corresponding class width, you can calculate the scale of the frequency density from that bar.

It is important to be able to work out the class width in every case. The class width is the distance between the class boundaries.

In the case of the heights of the seedlings in example 2 above, which were measured to the nearest centimetre, we know where each one belongs in the table but the rounding disguises the fact that the data is continuous, so we must remember the class boundaries when drawing the histogram.

Example 3

Complete each of these tables

(a) **Masses of oranges**

Class (m grams)	Class boundaries	Class width
$160 < m \leqslant 170$		
$170 < m \leqslant 190$		
$190 < m \leqslant 250$		

(b) **Numbers of passengers in a coach**

Class	10–12	13–16	17–30
Class boundaries			
Class width			

(c) Ages of people at a family celebration, given as a whole number of years. (Remember that ages are usually given to a whole number of years but not rounded up, so 9 years and 11 months would still be shown as 9, which is why the class is shown as $5 \leqslant age < 10$.)

Class age	Class boundaries	Class width
$5 \leqslant age < 10$		
$10 \leqslant age < 40$		
$40 \leqslant age < 60$		
$60 \leqslant age < 100$		

Answer 3

(a)

Class (m grams)	Class boundaries		Class width
$160 < m \leqslant 170$	160	170	10
$170 < m \leqslant 190$	170	190	20
$190 < m \leqslant 250$	190	250	60

(b) In this example, the *numbers* of passengers, the data is discrete. If we are to draw a histogram, which always has the bars touching, we have to use class boundaries *as if* the data was continuous.

Class	10–12	13–16	17–30
Class boundaries	9.5	12.5	16.5
	12.5	16.5	30.5
Class width	3	4	14

The class width may be found by the difference between the boundaries (12.5 − 9.5 = 3 in the first class above) or by counting the data values represented (10, 11, 12 = 3 in this case). However, the class boundaries are still needed when drawing the histogram, as we have seen.

(c)

Class age	Class boundaries		Class width
$5 \leqslant age < 10$	5	10	5
$10 \leqslant age < 40$	10	40	30
$40 \leqslant age < 60$	40	60	20
$60 \leqslant age < 100$	60	100	40

Exercise 11.1

1. Complete the following tables, and write down the modal classes.

   (a) **The lengths of some grass leaves**

Class (l cm)	Frequency (f)	Class boundaries	Class width (w)	Frequency density ($f \div w$)
$20 < l \leqslant 30$	1			
$30 < l \leqslant 50$	10			
$50 < l \leqslant 70$	15			
$70 < l \leqslant 100$	20			
$100 < l \leqslant 150$	3			

(b) **Numbers of letters posted each day in one letter box over 140 days**

Class	Frequency (f)	Class boundaries		Class width (w)	Frequency density ($f \div w$)
0–30	5	0	30.5		
31–50	25				
51–70	41				
71–100	50				
101–150	7				
151–250	2				

NOTE: The lower class boundary of the first class could be -0.5, which makes little sense. However, the class width is still 31 if you count the days with no envelopes as the first number in the class. Also, of course the numbers of letters are discrete.

2. Draw the histograms for the data sets in question 1. Label the vertical axes with the frequency density.

3. Draw up a frequency table for each of these histograms. Each table should include 5 columns showing Class, Class boundaries, Class width, Frequency density and Frequency.

(a)

Ages of people in a small village

(b)

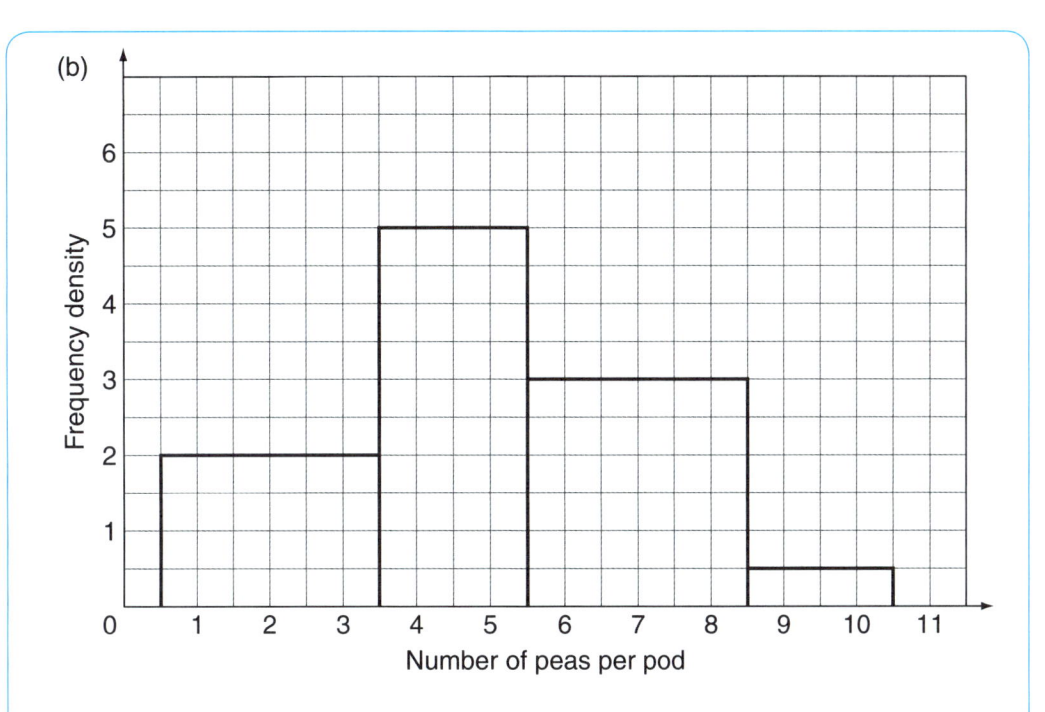

4. Using the table and histogram below
   (a) calculate and add the scale of the frequency density to the vertical axis,
   (b) complete the histogram,
   (c) complete the table,
   (d) identify and write down the modal class.

Number of steps (s) taken in one day by some students on a fitness course

Class	$0 < s \leqslant 1000$	$1000 < s \leqslant 5000$	$5000 < s \leqslant 10000$	$10000 < s \leqslant 20000$
Frequency	20	100	105	

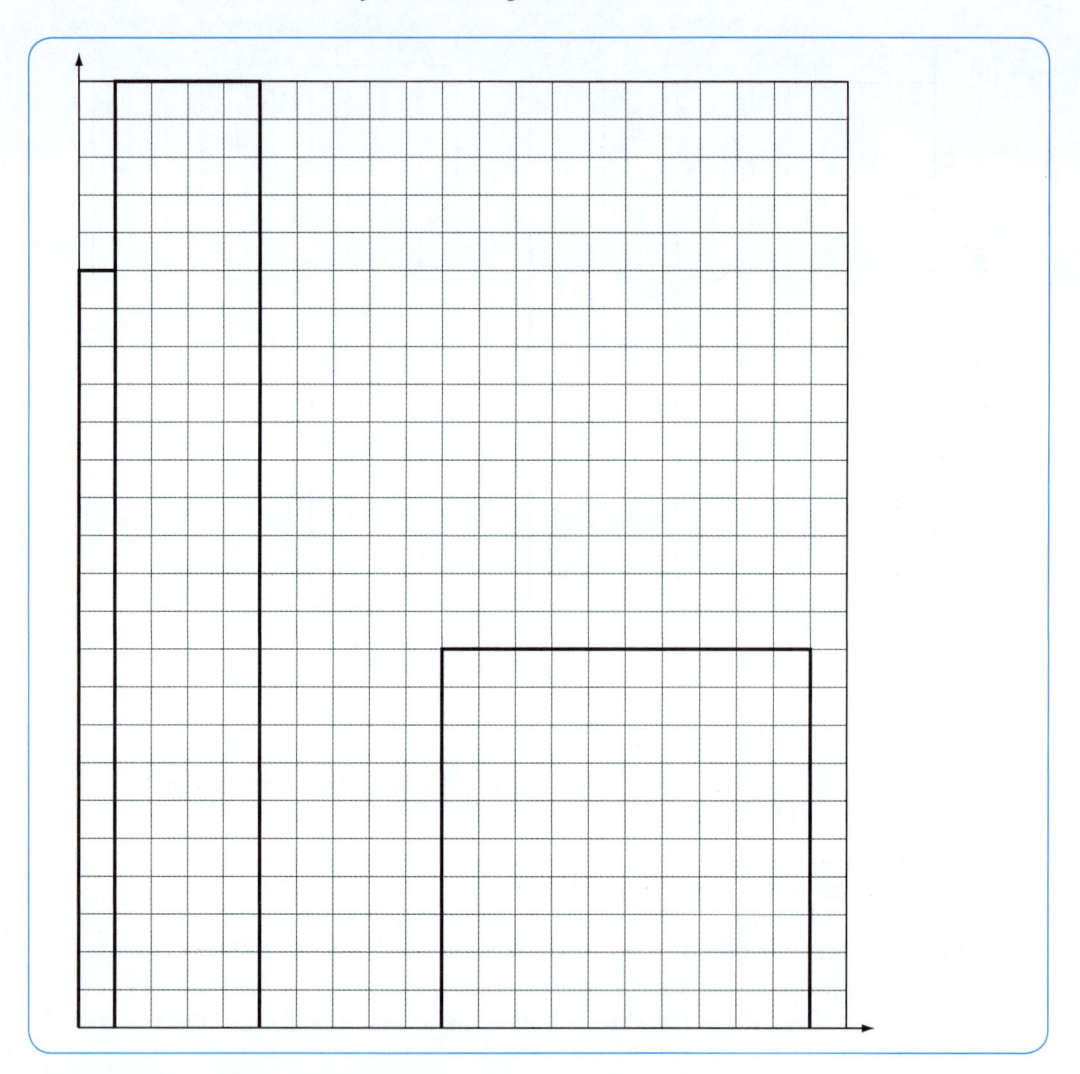

The Mean from a Grouped Frequency Table

In an ungrouped frequency table the mean can be calculated by multiplying each item of data by its frequency and then dividing by the total frequency, as we have seen in the Core Book. However, we do not know where each data value lies within the class intervals of a grouped frequency distribution, so we can only *estimate* the mean. This is done by using the class midpoint, on the assumption that the values in the class will be scattered approximately evenly throughout the class.

Statistics 319

The class midpoint is the number at the centre of each class, and is obtained by taking the mean of the class boundaries.

Example 4
Estimate the mean of the following distribution

Class	5–9	10–14	15–24	25–30
Frequency	3	5	7	3

Answer 4

Class	5–9	10–14	15–24	25–30	TOTAL
Frequency	3	5	7	3	18
Class midpoint	7	12	19.5	27.5	
Frequency × Midpoint	21	60	136.5	82.5	300

The mean = $\dfrac{\text{total(frequency} \times \text{midpoint)}}{\text{total frequency}} = \dfrac{300}{18} = 16.7$ to 3 significant figures.

Exercise 11.2

Estimate the mean of each of these data sets

1.

Class	Frequency
$10 < x \leqslant 15$	10
$15 < x \leqslant 25$	20
$25 < x \leqslant 40$	32
$40 < x \leqslant 60$	15

NOTE: Copy the table and add two more columns, one for Class midpoint and one for Frequency × Midpoint.

2.

Class	1–2	3–4	5–6	7–8	9–10
Frequency	5	6	3	8	2

3.

Class	0.5–0.9	0.9–1.3	1.3–2.5	2.5–6.5
Frequency	20	24	30	10

Cumulative Frequency

Data that is presented as a grouped frequency distribution loses its original individual data values. We have seen in the Core Book how to construct a grouped frequency table, and find the mean, median and mode from an ungrouped frequency table. Example 4 and Exercise 11.2 have shown how to estimate the mean from a grouped frequency table, and we can identify the modal class. We now will look at estimating the median from a grouped frequency distribution.

To do this we plot a **cumulative frequency curve**. Cumulative frequency is a *running total* of the frequencies (add them up as you go along), starting from the lowest values, as is shown in the following table.

Marks of 45 students in an examination

Class (%)	Frequency
$0 < mark \leqslant 15$	3
$15 < mark \leqslant 20$	2
$20 < mark \leqslant 25$	5
$25 < mark \leqslant 35$	10
$35 < mark \leqslant 45$	10
$45 < mark \leqslant 60$	13
$60 < mark \leqslant 80$	2

Mark	Cumulative frequency	
$\leqslant 15$	3	
$\leqslant 20$	5	(3 + 2)
$\leqslant 25$	10	(5 + 5)
$\leqslant 35$	20	(10 + 10)
$\leqslant 45$	30	(20 + 10)
$\leqslant 60$	43	(30 + 13)
$\leqslant 80$	45	(42 + 2)

The first thing to notice is that the cumulative frequencies in the cumulative frequency table show the number of data values in the corresponding class *and* all the previous classes. This means that the cumulative frequency is plotted against the *upper* class boundary as shown in the curve on the next page. (For example, there are 10 students who scored 25 *or less*).

When drawing a cumulative frequency curve:

- the cumulative frequency is always shown on the vertical axis,
- the points are plotted against the upper class boundaries,
- the curve is always increasing, or possibly horizontal, but never decreases,
- the top of the cumulative frequency curve represents the total frequency (100% of the data set).
- It is usual to draw a smooth curve connecting the data points unless you are asked to construct a **cumulative frequency polygon** in which case the points are joined by straight lines.

Cumulative frequency curve

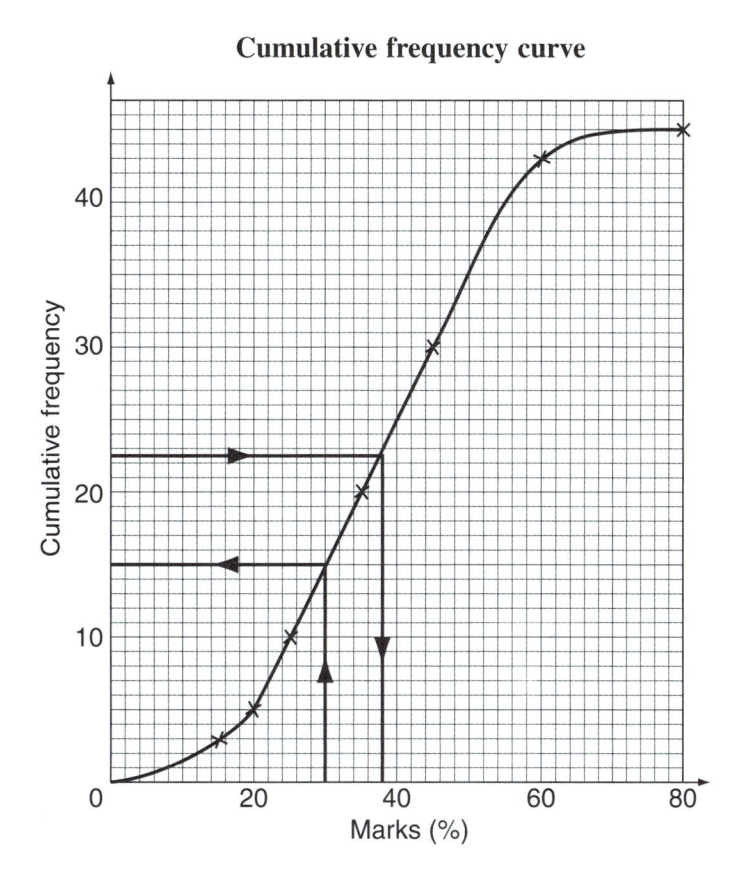

The curve can now be used to estimate the median of the data.

The median is the data value that divides the data set into two equal parts when the data values are arranged in order.

Since the table has already arranged the values into increasing order we read the data value that corresponds to 50% of the cumulative frequency, that is, half way to the highest plotted point up the cumulative frequency axis. From the axis rule a horizontal line to the curve, and then rule a vertical line down to the horizontal axis. Read the median from the scale on the horizontal axis.

Example 5

Use the curve above to estimate

(a) the median of the data set,

(b) how many students scored 30 or less.

Answer 5

(a) The median = 38

(b) 15 students scored 30 or less

Quartiles, Inter-Quartile Range and Percentiles

In the Core Book we looked at three sets of examination marks, and found that the mean was affected by the extreme, non-typical data values. These also affect the range, which is used as a rough measure of the spread of the data.

The **inter-quartile** range is another measure of spread which is not so affected by extreme values.

The median divides the data into two equal groups, so is a measure of the middle of the data. The **quartiles** divide each of these two equal groups again into two equal parts. So the median and quartiles divide the data into 4 equal parts.

Consider the following ordered data set:

2   4   4   5   5   5   6   7   8   8   8   9

There are 12 data values so the median is half way between the 6th and 7th values.

The median is $\dfrac{5+6}{2} = 5.5$

2   4   4   5   5   5   ...   6   7   8   8   8   9

The **lower quartile** divides the lower half into two equal parts. There are 6 data values so the lower quartile is between the 3rd and 4th.
The lower quartile is 4.5

Similarly the **upper quartile** is between the 9th and 10th value, so the upper quartile is 8. These are shown in the diagram below.

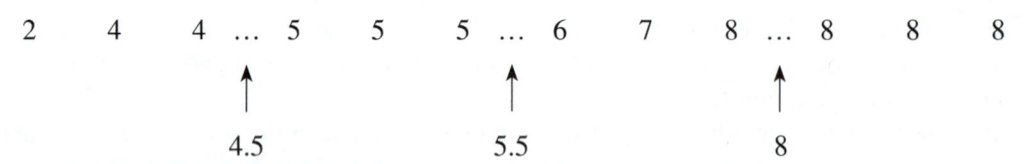

2   4   4   ...   5   5   5   ...   6   7   8   ...   8   8   8
          ↑   ↑   ↑
         4.5   5.5   8

The median and quartiles are often abbreviated as:
Lower quartile = q_1 or Q_1
Median = q_2 or Q_2
Upper quartile = q_3 or Q_3
So in this example $Q_1 = 4.5$ $Q_2 = 5.5$ $Q_3 = 8$

The inter-quartile range is the distance between the upper and lower quartiles.
In this case the inter-quartile range = $Q_3 - Q_1 = 8 - 4.5 = 3.5$.

NOTE: There are other methods of estimating the median and quartiles which may give slightly different results from those shown in this chapter, both here and in Chapter 11 of the Core Book, but this method is perfectly adequate and easy to visualise and understand.

When we need to estimate the inter-quartile range from a grouped frequency distribution we use the cumulative frequency curve. The following, typical curve will illustrate this.

The highest point on the cumulative frequency curve (which represents the total frequency) is 100%. The lower quartile is one quarter (25%) of the way up to 100% on the cumulative frequency axis, and the upper quartile is three quarters of the way up (75%).

Be aware that the cumulative frequency scale could go higher than the highest point on the curve, but that part should be ignored, as the diagram below shows.

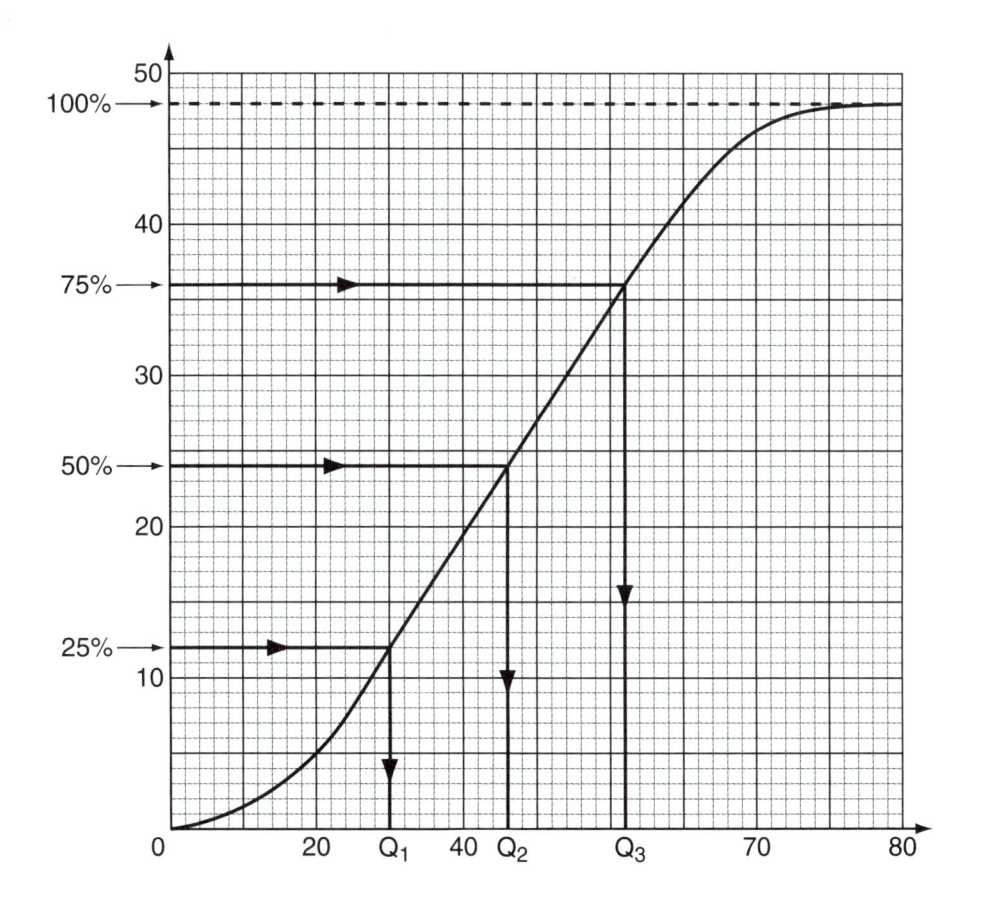

NOTE: **When you are using your cumulative frequency curve in this way ruled pencil lines are part of your working. Do not leave them out. It is also helpful to use arrows on these lines to show the direction of the working.**

Example 6

Use the curve in example 5 to estimate the inter-quartile range of the data.

Answer 6

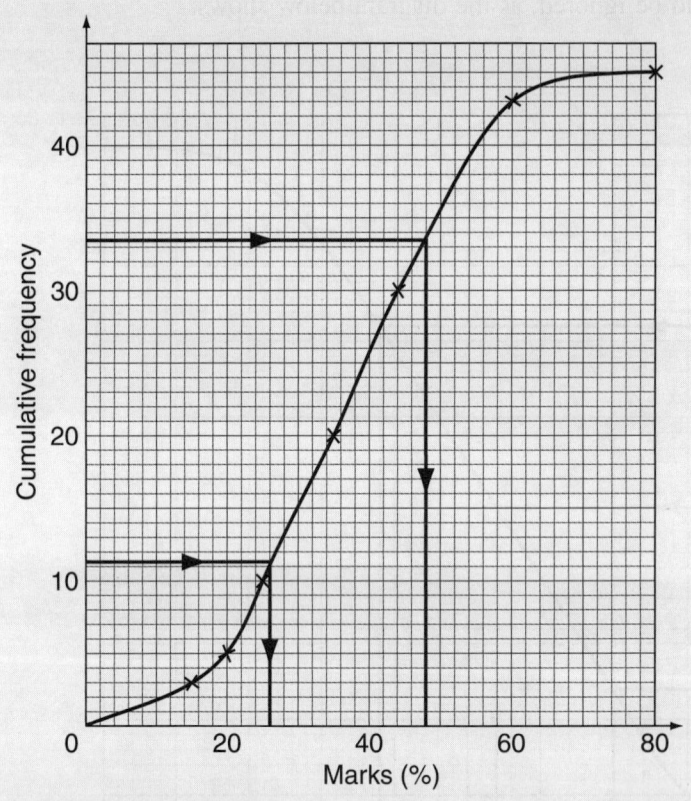

Lower quartile = 26
Upper quartile = 48
Inter-quartile range = 22

The data can also be divided into **percentiles**, which, as their name suggests, are each one hundredth of the way through the data set.

For example, to find the 30th percentile you go 30% up the vertical (cumulative frequency) axis and read across to the curve, and then down to the horizontal axis.

The last thing you have to know about the use of cumulative frequency curves is shown in the example below.

Example 7

Use the cumulative frequency curve of some students' percentage marks in a test (Example 5) to estimate

(a) how many students have marks of 60% or more,
(b) how many students have marks between 40% and 50%.
(c) The range of marks between the 40th and 60th percentiles.

NOTE: Think carefully where these data items lie! This question is often answered incorrectly in examinations.

Answer 7

We copy the curve and rule lines to show our working:

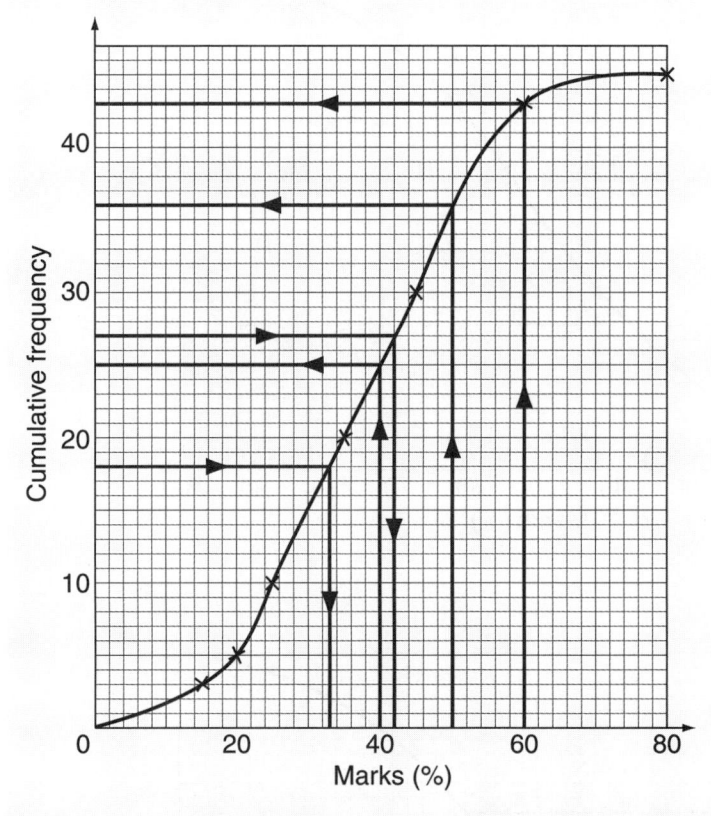

(a) The mark of 60% corresponds with 43 students so 45 − 43 = 2 students have marks of 60% or more
(b) 36 − 25 = 11.
   11 students have marks between 40% and 50%
(c) 40% of 45 = 18, 60% of 45 = 27.
   The 40th and 60th percentiles are at 18 and 27 on the cumulative frequency axis.
   The range of marks between 40th and 60th percentiles = 42 − 33 = 9 marks

NOTE: This is a very important Hint, and if ignored may well lead to a common error! Make sure that you know which axis you are reading from and which to. The quartiles and percentiles are not found by dividing the horizontal axis into equal parts! The horizontal axis represents the value of the data item, and the vertical axis represents its position in the ordered set of data. Think carefully every time you use a cumulative frequency curve.

Example 8

Use this cumulative frequency curve showing some more examination marks to find
(a) the 45th percentile mark,
(b) the number of candidates passing the examination if the pass mark is 60%.

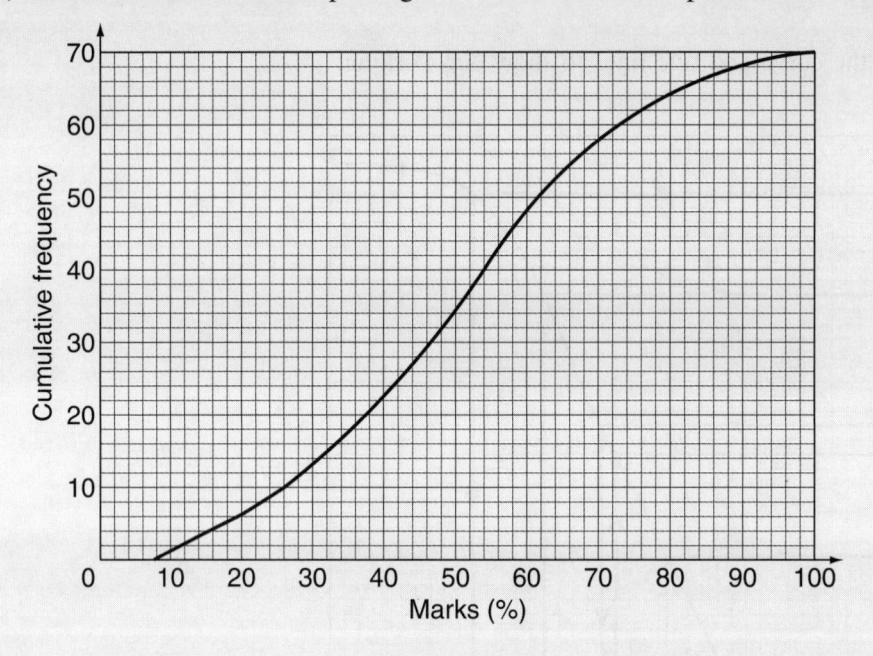

Answer 8

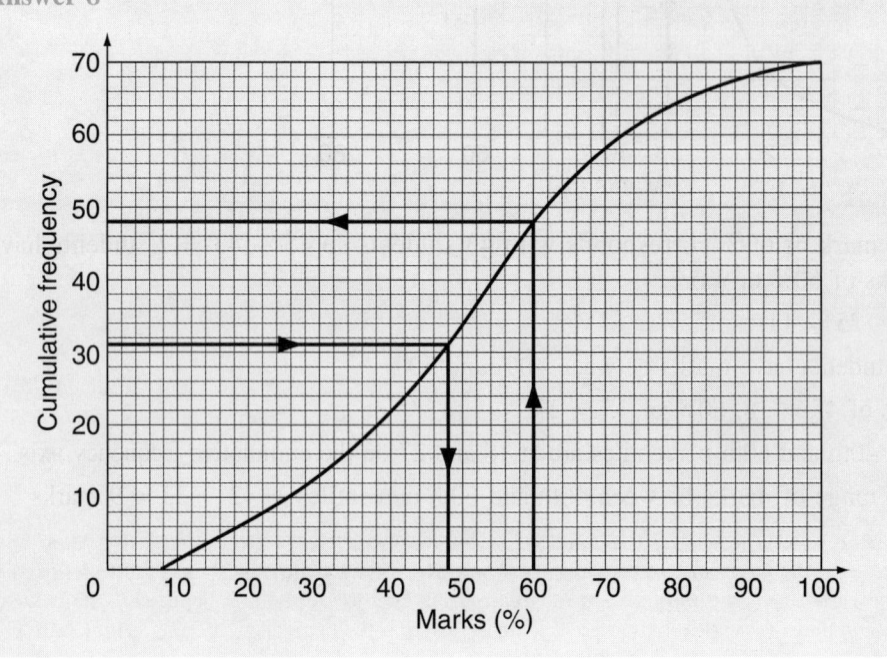

Statistics 327

(a) 45% of 70 = 31.5
So the 45th percentile lies between the 31st and 32nd items on the ordered list originally used to draw the curve. We find that 31.5 on the cumulative frequency axis corresponds with a mark of 48%.
The 45th percentile is the mark 48%

(b) 70 − 48 = 22.
22 students passed the examination

Example 9

(a) The tables below record the times, to the nearest minute, taken for students from two forms (classes) at a school to complete a memory test. Draw two cumulative frequency curves on the same grid to illustrate the two sets of data.

(b) Use the two curves to compare the performance of the two sets of students.

Form 3A

Time (t minutes)	Frequency
$10 < t \leqslant 15$	2
$15 < t \leqslant 20$	6
$20 < t \leqslant 25$	10
$25 < t \leqslant 35$	11
$35 < t \leqslant 45$	1

Form 3B

Time (t minutes)	Frequency
$10 < t \leqslant 20$	4
$20 < t \leqslant 25$	5
$25 < t \leqslant 30$	8
$30 < t \leqslant 40$	10
$40 < t \leqslant 60$	3

Answer 9

(a) **Form 3A**

Time (t minutes)	Frequency	Cumulative frequency
$10 < t \leqslant 15$	2	2
$15 < t \leqslant 20$	6	8
$20 < t \leqslant 25$	10	18
$25 < t \leqslant 35$	11	29
$35 < t \leqslant 45$	1	30

Form 3B

Time (t minutes)	Frequency	Cumulative frequency
$10 < t \leqslant 20$	4	4
$20 < t \leqslant 25$	5	9
$25 < t \leqslant 30$	8	17
$30 < t \leqslant 40$	10	27
$40 < t \leqslant 60$	3	30

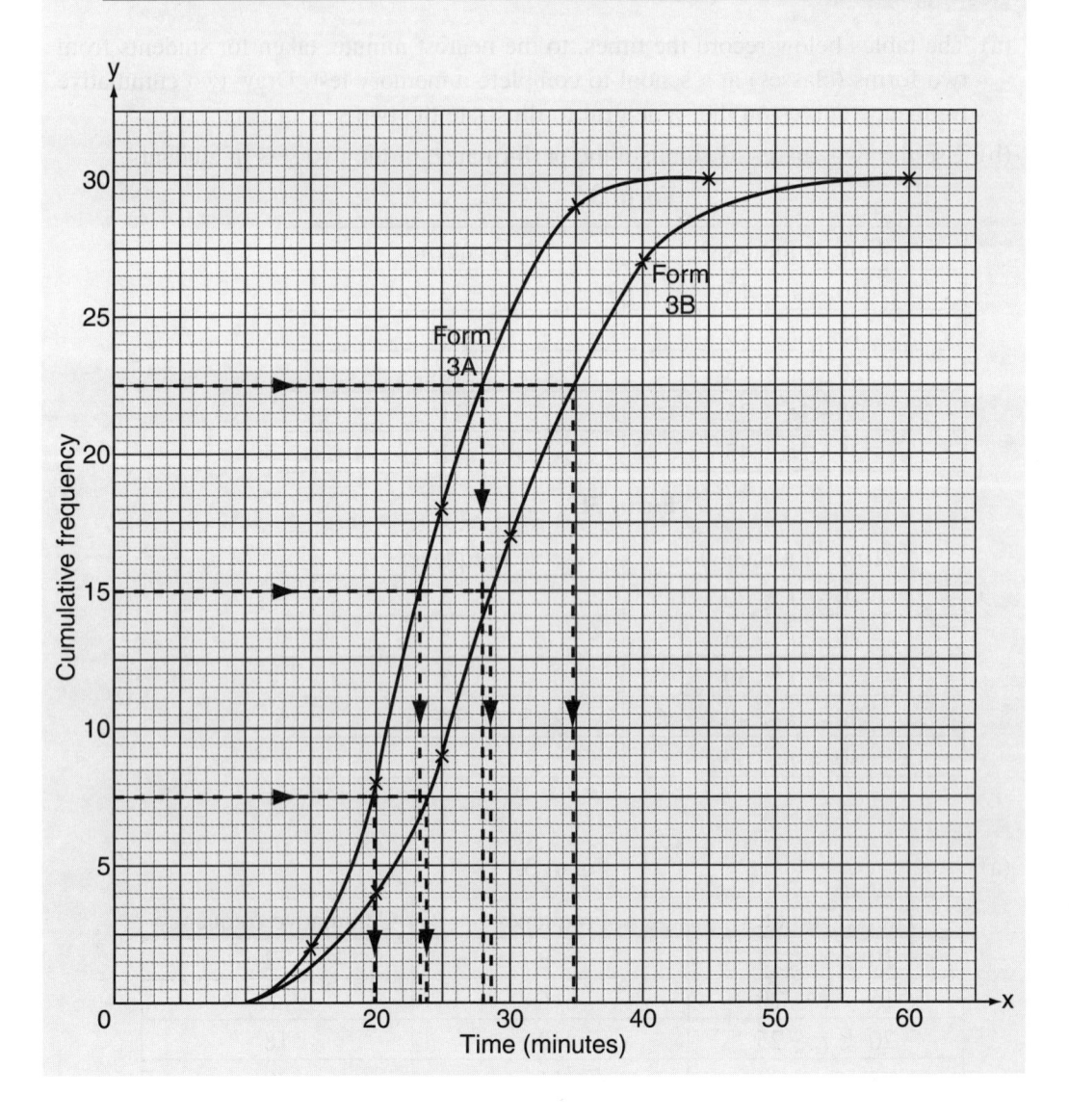

(b) From the cumulative frequency curves:

Form 3A $\quad Q_1 = 20$
$\quad\quad\quad\quad\quad Q_2 = 23.5$
$\quad\quad\quad\quad\quad Q_3 = 28$

Inter-quartile range $= 28 - 20 = 8$

Form 3B $\quad Q_1 = 24$
$\quad\quad\quad\quad\quad Q_2 = 28.5$
$\quad\quad\quad\quad\quad Q_3 = 35$

Inter-quartile range $= 35 - 24 = 11$

On average Form 3A completed the test quicker than Form 3B. The spread of Form 3A's marks was less than that of Form 3B.

Exercise 11.3

1. (a) Draw a cumulative frequency curve to illustrate the data given in the table.
   (b) Estimate the median, quartiles and inter-quartile range.
   (c) Estimate the 65th percentile.

Class	Frequency
$10 < x \leqslant 20$	9
$20 < x \leqslant 30$	16
$30 < x \leqslant 40$	27
$40 < x \leqslant 50$	13

2. Masses of 28 adults, (m kg).

74.8	74.9	90	83.3	94.5	68.7	70.1
84.5	88.0	78.5	69.1	70.5	72.9	69.4
75.7	79.9	81.4	92.5	88.5	76.8	83.4
82.8	79.8	90.5	88.8	83.1	82.5	78.5

(a) Use the data given above to complete the grouped frequency distribution below.

Class	$65 < m \leqslant 70$	$70 < m \leqslant 75$	$75 < m \leqslant 80$	$80 < m \leqslant 85$	$85 < m \leqslant 90$	$90 < m \leqslant 100$
Frequency						

(b) Draw a cumulative frequency curve.
(c) Estimate (i) the median mass,
(ii) the inter-quartile range,
(iii) the 40th percentile.

Exercise 11.4

Mixed Exercise

1. The table below shows the times some half marathon runners took to complete the course. The times are in minutes.

74.2	75.2	77.6	79.4	80.1	80.2	84.4	86.0
88.0	93.6	94.1	95.5	98.4	98.9	99.0	99.9
100.0	100.0	101.5	102.9	103.0	103.5	104.7	105.6
105.8	106.0	106.2	106.9	107.1	107.4	107.7	107.9
108.4	108.5	108.7	108.9	109.1	109.1	109.3	109.4
109.6	109.6	109.9	109.9	110.6	111.2	111.8	115.2
116.1	117.0	117.0	117.8	118.1	118.9	119.3	120.0
134.6	135.2	135.9	141.4	143.9	146.8	149.9	153.1

(a) Draw a grouped frequency table with the times grouped into classes $70 < t \leqslant 80$, $80 < t \leqslant 90$, $90 < t \leqslant 100$, $100 < t \leqslant 120$, and $120 < t \leqslant 160$. Include in your table, Class boundaries, Class width, Class midpoint, Frequency density and Cumulative frequency.
(b) Draw a histogram.
(c) Estimate the mean.
(d) Identify the modal class.
(e) Draw a cumulative frequency curve.
(f) Use the cumulative frequency curve to estimate the median, quartiles and 70th percentile.

2. The following Histogram shows the results of another Half marathon run.

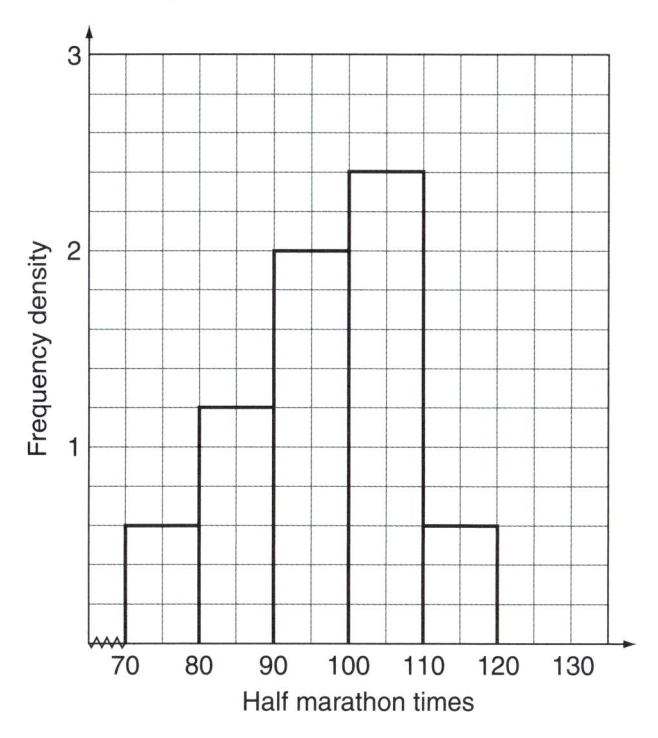

(a) Use the histogram to draw a grouped frequency table, including Class width, Frequency density, Class midpoint, Frequency and Cumulative frequency, with the times grouped as in question 1.
(b) Estimate the mean.
(c) Draw a cumulative frequency curve.
(d) Estimate the median and quartiles.
(e) How many people completed this marathon in less than 100 minutes?
(f) Estimate the number of runners who took more than 108 minutes to complete this course.

3. A girls' school and a boys' school are to merge to form one school.
The mean number of girls per class in the girls' school is 25.4 (to 3 significant figures) and there are 12 classes all together. In the boys' school the mean number of boys per class is 23.8, and there are 15 classes. The new school will have 22 classes.
Calculate the mean number of students per class in the new school, giving your answer to 3 significant figures. Show all your working.

4. A group of 20 students have a mean mark of 72% in their Maths exam. A new student joins the group. He had scored 68% in the same examination. What is the new mean mark of the enlarged group?

5. In another school the classes were being reorganised according to the marks the students had gained in their end of year examination. The top set of 25 students had a mean mark of 75% before the reorganisation. Three students with a mean mark of 61% were moved down to the second set, and two students with a mean mark of 80% were moved up into the top set.

   (a) How many students are there now in the top set?
   (b) What is the new mean mark of the top set?

Examination Questions

6. The heights of 40 children were measured.
   The results are summarized in the table below.

Height (h cm)	$105 < h \leqslant 115$	$115 < h \leqslant 125$	$125 < h \leqslant 135$	$135 < h \leqslant 145$
Frequency	5	10	20	5

   (a) (i) Identify the modal class.
       (ii) Calculate an estimate of the mean height.
   (b) The cumulative frequency curve representing this information is shown below.

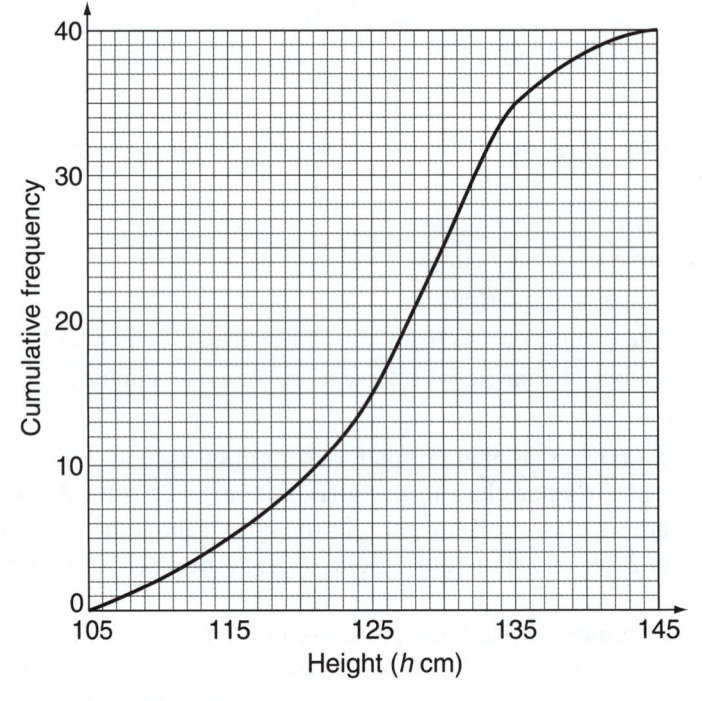

   Use the curve to find
   (i) the inter-quartile range,
   (ii) the number of children whose heights are in the range 120 cm to 130 cm.

   (4024/01 May/June 2007 q 25)

7. Kristina asked 200 people how much water they drink in one day.
   The table shows her results.

Amount of water (x litres)	Number of people
$0 < x \leqslant 0.5$	8
$0.5 < x \leqslant 1$	27
$1 < x \leqslant 1.5$	45
$1.5 < x \leqslant 2$	50
$2 < x \leqslant 2.5$	39
$2.5 < x \leqslant 3$	21
$3 < x \leqslant 3.5$	7
$3.5 < x \leqslant 4$	3

   (a) Write down the modal interval.
   (b) Calculate an estimate of the mean.
   (c) Make a cumulative frequency table for this data.
   (d) Using a scale of 4 cm to 1 litre of water on the horizontal axis and 1 cm to 10 people on the vertical axis, draw the cumulative frequency graph.
   (e) Use your cumulative frequency graph to find
       (i) the median,
       (ii) the 40th percentile,
       (iii) the number of people who drink at least 2.6 litres of water.
   (f) A doctor recommends that a person drinks at least 1.8 litres of water each day.
       What percentage of these 200 people do not drink enough water?

   (0580/04 May/June 2007 q 6)

8. (a) The numbers 0, 1, 1, 1, 2, k, m, 6, 9, 9 are in order ($k \neq m$).
       Their median is 2.5 and their mean is 3.6.
       (i) Write down the mode.
       (ii) Find the value of k.
       (iii) Find the value of m.
   (b) 100 students are given a question to answer.
       The time taken (t seconds) by each student is recorded and the results are shown in the table.

t	$0 < t \leqslant 20$	$20 < t \leqslant 30$	$30 < t \leqslant 35$	$35 < t \leqslant 40$	$40 < t \leqslant 50$	$50 < t \leqslant 60$	$60 < t \leqslant 80$
Frequency	10	10	15	28	22	7	8

   Calculate an estimate of the mean time taken.

(c) The data in **part (b)** is re-grouped to give the following table.

t	$0 < t \leqslant 30$	$30 < t \leqslant 60$	$60 < t \leqslant 80$
Frequency	p	q	8

(i) Write down the values of p and q.

(ii) Draw an accurate histogram to show these results.

Use a scale of 1 cm to represent 5 seconds on the horizontal time axis.

Use a scale of 1 cm to 0.2 units of frequency density (so that 1 cm² on your histogram represents 1 student).

(0580/04 May/June 2006 q 9 (part))

9. The heights (h cm) of 270 students in a school are measured and the results are shown in the table

h	Frequency
$120 < h \leqslant 130$	15
$130 < h \leqslant 140$	24
$140 < h \leqslant 150$	36
$150 < h \leqslant 160$	45
$160 < h \leqslant 170$	50
$170 < h \leqslant 180$	43
$180 < h \leqslant 190$	37
$190 < h \leqslant 200$	20

(a) Write down the modal group.
(b) (i) Calculate an estimate of the mean height.
   (ii) Explain why the answer to **part (b)(i)** is an estimate.
(c) The following table shows the cumulative frequencies for the heights of the students.

h	Cumulative frequency
$h \leqslant 120$	0
$h \leqslant 130$	p
$h \leqslant 140$	q
$h \leqslant 150$	r
$h \leqslant 160$	120
$h \leqslant 170$	170

table continued overleaf

h	Cumulative frequency
$h \leqslant 180$	213
$h \leqslant 190$	250
$h \leqslant 200$	270

Write down the values of p, q and r.

(d) Using a scale of 1 cm to 5 units, draw a horizontal h-axis, starting at $h = 120$. Using a scale of 1 cm to 20 units on the vertical axis, draw a cumulative frequency diagram.

(e) Use your diagram to find
  (i) the median height,
  (ii) the upper quartile,
  (iii) the inter-quartile range,
  (iv) the 60th percentile.

(f) All the players in the school's basketball team are chosen from the 30 tallest students.
Use your diagram to find the least possible height of any player in the basketball team.

(0580/04 Oct/Nov 2005 q 9)

10. The cumulative frequency curve shows the distribution of the masses of 100 people.

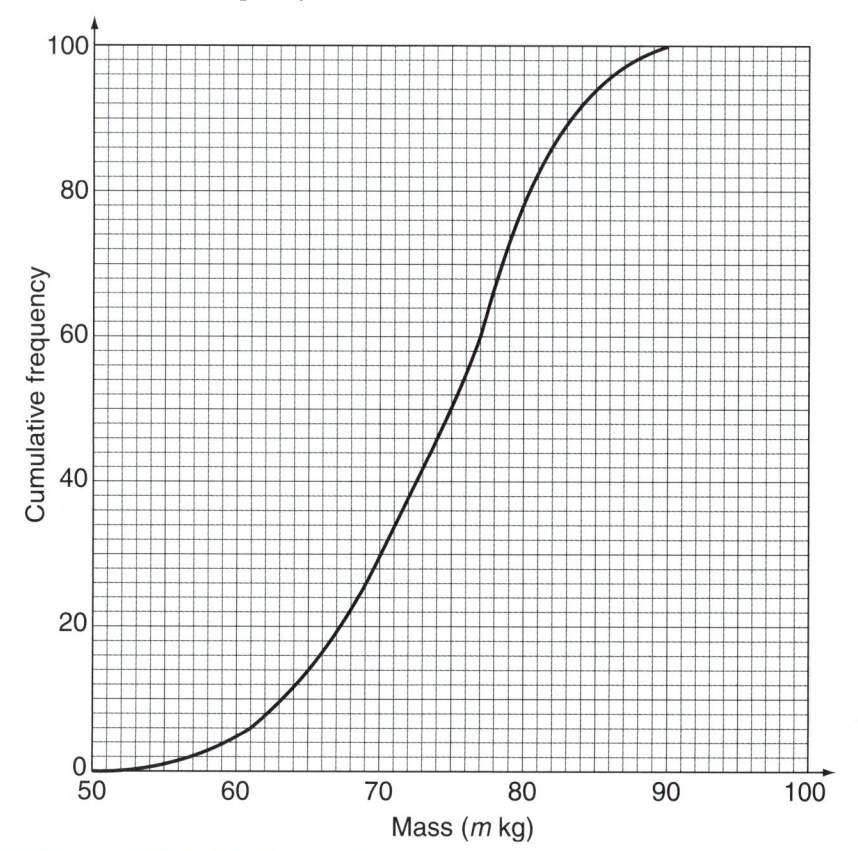

Find
(a) the median,
(b) the upper quartile,
(c) the number of people with masses in the range $65 < m \leqslant 72$.

(4024/01 May/June 2005 q 17)

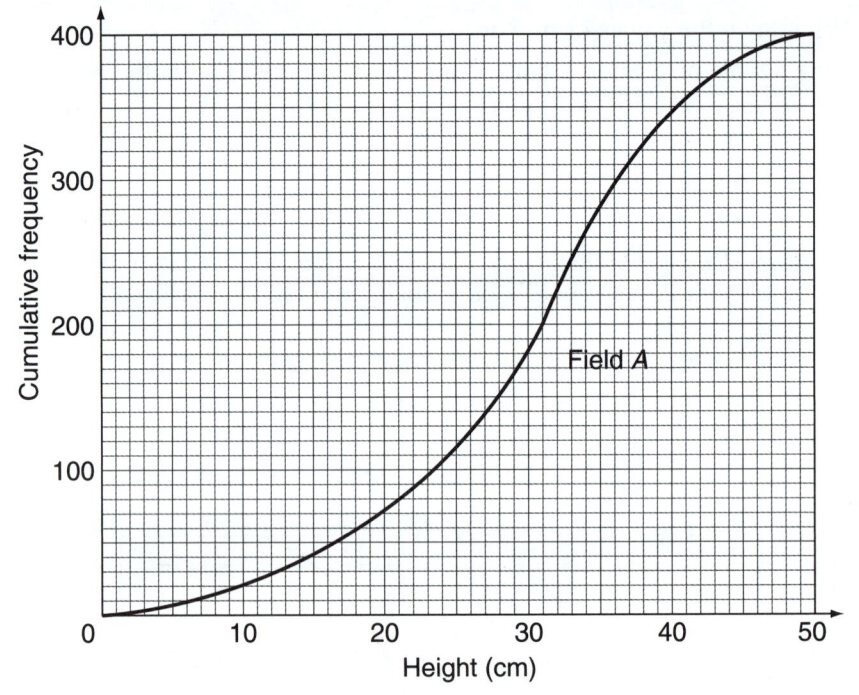

11.

The diagram above is the cumulative frequency curve for the heights of 400 plants which were grown in Field *A*.

Use the graph to find
(a) the number of plants that grew to a height of **more** than 30 cm,
(b) the inter-quartile range.
(c) Another 400 plants were grown in Field *B*.
  The cumulative frequency distribution of the heights of these plants is shown in the table.

Height (h cm)	$h \leqslant 10$	$h \leqslant 15$	$h \leqslant 20$	$h \leqslant 25$	$h \leqslant 30$	$h \leqslant 35$	$h \leqslant 40$	$h \leqslant 50$
Cumulative frequency	35	75	130	200	280	330	370	400

Copy the graph and **on the same axes as for Field *A***, draw the cumulative frequency curve for the plants grown in Field *B*.

(d) By comparing the two curves, state, with a reason, which Field produced the taller plants.

(4024/01 Oct/Nov 2005 q 19)

12. In a survey, 200 shoppers were asked how much they had just spent in a supermarket.
    The results are shown in the table.

Amount (x)	$0 < x \leqslant 20$	$20 < x \leqslant 40$	$40 < x \leqslant 60$	$60 < x \leqslant 80$	$80 < x \leqslant 100$	$100 < x \leqslant 140$
Number of shoppers	10	32	48	54	36	20

    (a) (i) Write down the modal class.
        (ii) Calculate an estimate of the mean amount, giving your answer correct to 2 decimal places.
    (b) (i) Make a cumulative frequency table for these 200 shoppers.
        (ii) Using a scale of 2 cm to represent $20 on the horizontal axis and 2 cm to represent 20 shoppers on the vertical axis, draw a cumulative frequency diagram for this data.
    (c) Use your cumulative frequency diagram to find
        (i) the median amount,
        (ii) the upper quartile,
        (iii) the inter-quartile range,
        (iv) how many shoppers spent at least $75.
    (0580/4 May/June 2003 q 8)

13. One hundred children were asked how far they could swim.
    The results are summarized in the table.

Distance (d metres)	$0 < d \leqslant 100$	$100 < d \leqslant 200$	$200 < d \leqslant 400$
Number of children	30	50	20

    (a) The histogram in the answer space represents part of this information.
    Complete the histogram.

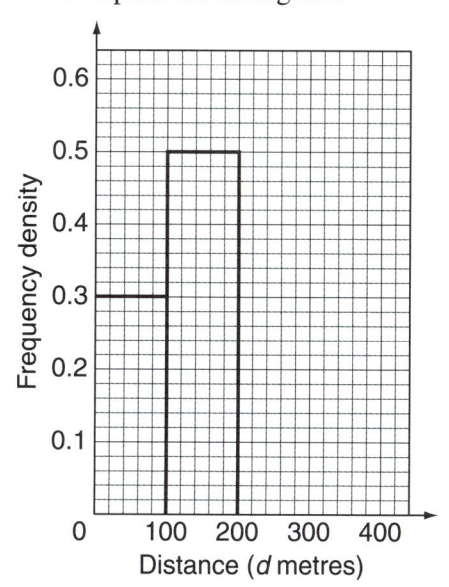

(b) A pie chart is drawn to represent the three groups of children.
Calculate the angle of the sector that represents the group of 20 children.
(4024/01 May/June 2005 q 11)

14. On a certain stretch of road, the speeds of some cars were recorded.
The results are summarized in the table.
Part of the corresponding histogram is shown alongside.

Speed (x km/h)	Frequency
$25 < x \leqslant 45$	q
$45 < x \leqslant 55$	30
$55 < x \leqslant 65$	p
$65 < x \leqslant 95$	12

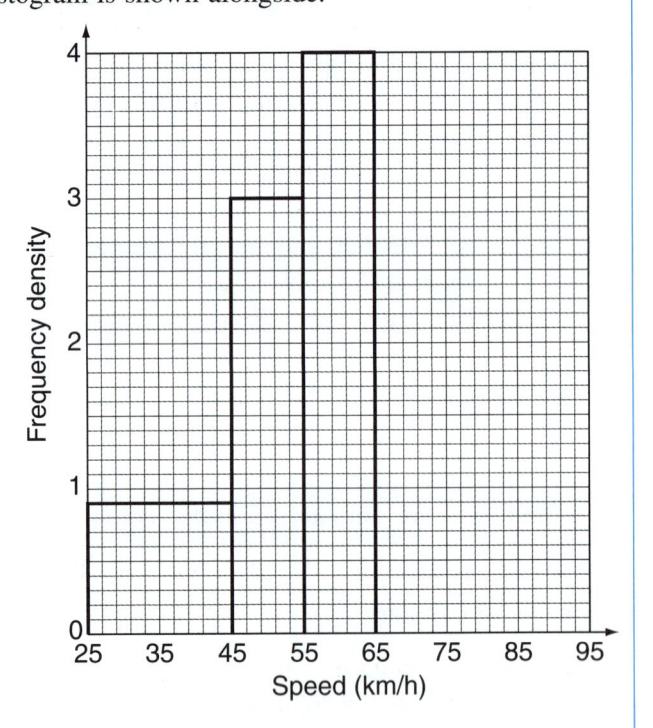

(a) Find the value of
   (i) p,
   (ii) q.
(b) Complete the histogram.

(4024/01 May/June 2007 q 11)

15. (a) Each student in a class is given a bag of sweets.
The students note the number of sweets in their bag.
The results are shown in the table, where $0 \leqslant x < 10$.

Number of sweets	30	31	32
Frequency (number of bags)	10	7	x

   (i) State the mode.
   (ii) Find the possible values of the median.
   (iii) The mean number of sweets is 30.65.
        Find the value of x.

(b) The mass, m grams, of each of 200 chocolates is noted and the results are shown in the table.

Mass(m grams)	$10 < m \leqslant 20$	$20 < m \leqslant 22$	$22 < m \leqslant 24$	$24 < m \leqslant 30$
Frequency	35	115	26	24

(i) Calculate an estimate of the mean mass of a chocolate.
(ii) On a histogram, the height of the column for the $20 < m \leqslant 22$ interval is 11.5 cm.
Calculate the heights of the other three columns.
Do not draw the histogram.

(0580/04 Oct/Nov 2008 q 6)

16. The mass of each of 200 tea bags was checked by an inspector in a factory.
The results are shown by the cumulative frequency curve.

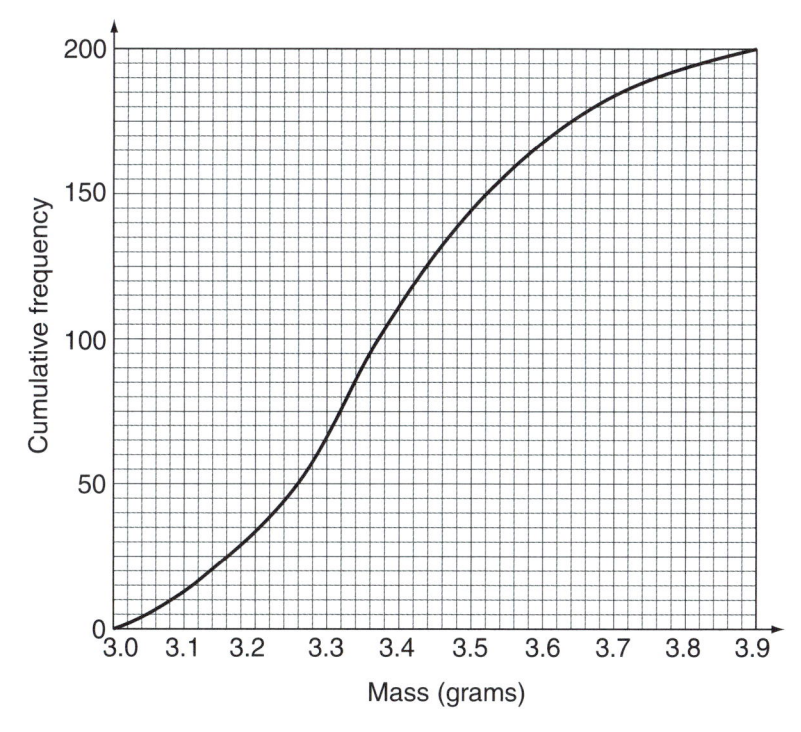

Use the cumulative frequency curve to find
(a) the median mass,
(b) the inter-quartile range,
(c) the number of tea bags with a mass greater than 3.5 grams.

(0580/02 Oct/Nov 2007 q 19)

17. (a) The quiz scores of a class of n students are shown in the table.

Quiz score	6	7	8	9
Frequency (number of students)	9	3	a	5

The mean score is 7.2 Find
  (i) a,
  (ii) n,
  (iii) the median score.

(b) 200 students take a mathematics test.
   The cumulative frequency diagram shows the results.

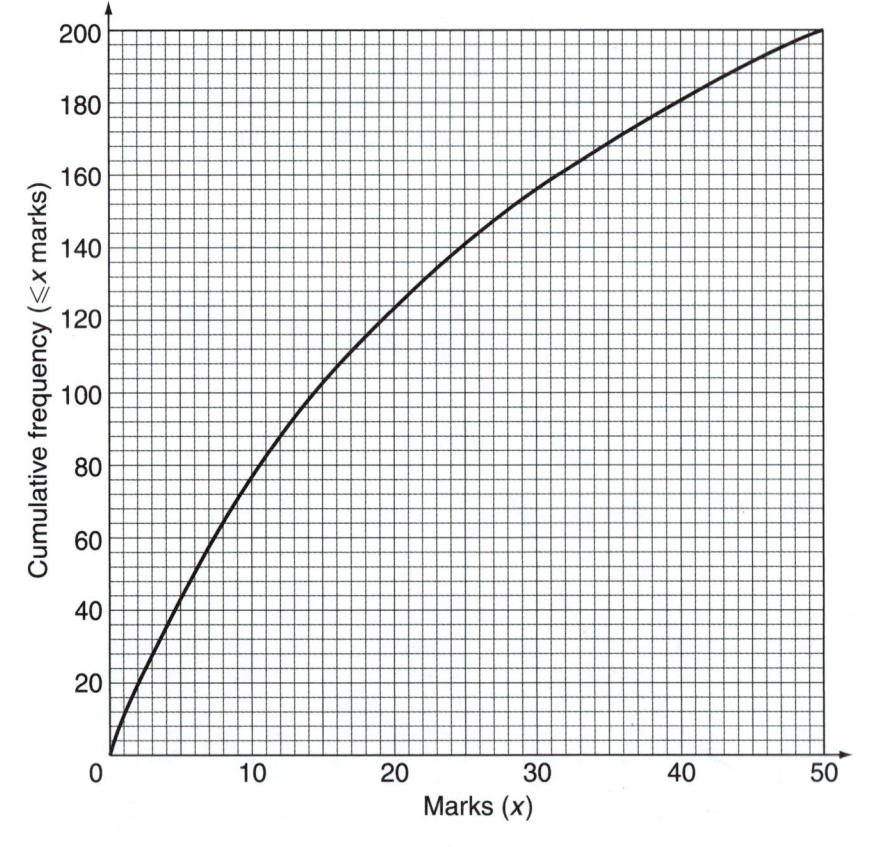

Write down
  (i) the median mark,
  (ii) the lower quartile,
  (iii) the upper quartile,
  (iv) the inter-quartile range,
  (v) the lowest possible mark scored by the top 40 students,
  (vi) the number of students scoring more than 25 marks.

(c) Another group of students takes an English test.
The results are shown in the histogram.

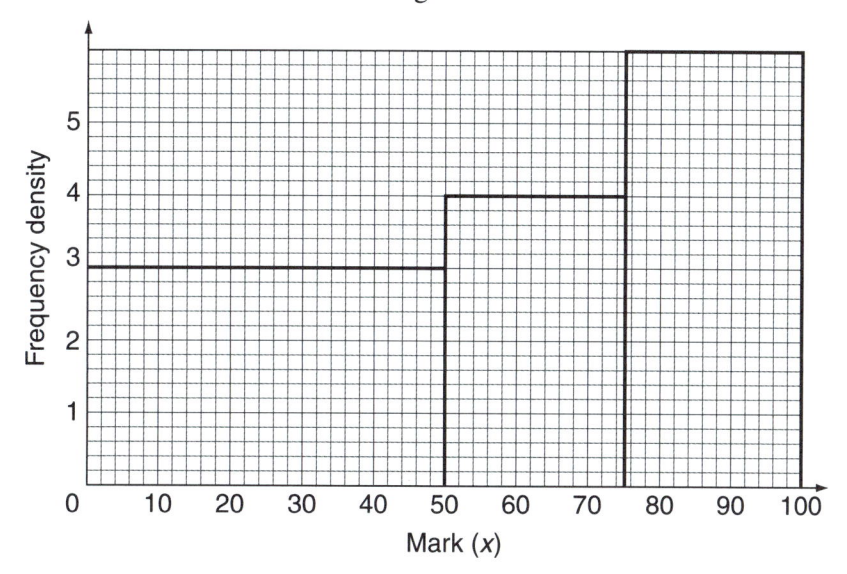

100 students score marks in the range $50 < x \leqslant 75$.
 (i) How many students score marks in the range $0 < x \leqslant 50$?
 (ii) How many students score marks in the range $75 < x \leqslant 100$?
 (iii) Calculate an estimate of the mean mark of this group of students.

(0580/04 Oct/Nov 2006 q 7)

18. A normal die, numbered 1 to 6, is rolled 50 times.

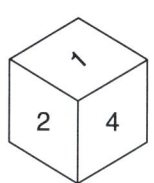

The results are shown in the frequency table.

Score	1	2	3	4	5	6
Frequency	15	10	7	5	6	7

(a) Write down the modal score.
(b) Find the median score.
(c) Calculate the mean score.
(d) The die is then rolled another 10 times.
    The mean score for the 60 rolls is 2.95.
    Calculate the mean score for the extra 10 rolls.

(0580/04 May/June 2009 q 2)

Chapter 12

Probability

This chapter completes your course for Extended Level IGCSE.
In the Core Book we looked at the probabilities of single events. In this chapter, we will look at the probabilities of combined events, using tree diagrams and Venn diagrams.

Core Skills

1. A bag contains 10 beads coloured red, blue and green.
   (a) If a bead is chosen at random it is known that the probability of it being red is $\frac{2}{5}$. How many red beads are in the bag?
   (b) There are three green beads in the bag. What is the probability of choosing a blue bead?
   (c) What is the probability of choosing a yellow bead?
2. A manufacturing company tests 50 components selected at random from a batch of 1500, and finds that 3 of them are faulty.
   (a) What is the relative frequency of faulty components? Give your answer as a percentage.
   (b) How many components could be expected to be faulty in the whole batch?

Combined Events

Amarus tosses an unbiased coin and rolls a fair five-sided spinner. The spinner is equally likely to stop on any of the numbers from 1 to 5.
To find P(heads *and* 4) he can list all the possible outcomes:

        H,1 H,2 H,3 H,4 H,5 T,1 T,2 T,3 T,4 T,5

There are ten possible outcomes and only one which is heads and 4, so P(heads *and* 4) $= \frac{1}{10}$.

This was a relatively easy list to write down, but some can be much longer.
A **possibility diagram** also known as a **probability space diagram** is an aid to drawing up the list so that no outcomes are forgotten.
For this example, where the events are independent of each other, a suitable diagram would be:

		Spinner				
		1	2	3	4	5
Coin	H	H,1	H,2	H,3	H,4	H,5
	T	T,1	T,2	T,3	T,4	T,5

It is easy to see that there are ten possible outcomes in the table. This is the number of outcomes from the coin times the number of outcomes from the spinner.

We are interested in one particular outcome, that is heads *and* 4, which occurs once only, so $P(H,4) = \frac{1}{10}$.

What is $P(H) \times P(4)$?

Can you see that this is $\frac{1}{2} \times \frac{1}{5} = \frac{1}{10}$?

Work out $P(T) \times P(\text{odd number})$. Does this agree with $P(T, \text{odd number})$ from the table?

We will find out more about this later in the chapter.

An alternative method is to mark the outcomes on a grid like a graph. Each outcome is represented by a grid point or coordinates.

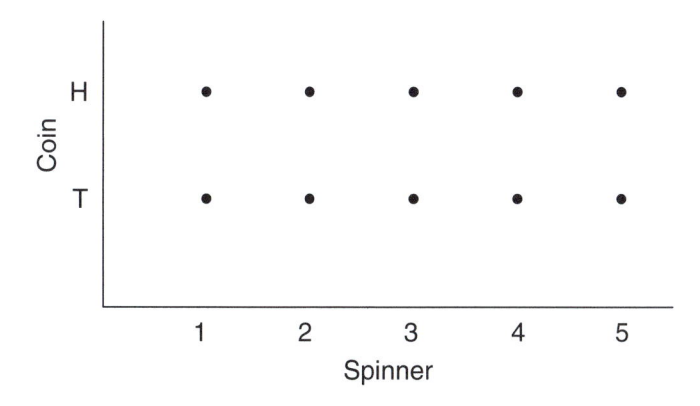

If the spinner had eight sides, with the numbers 1 to 8 on it, and the coin was tossed again, how many dots would there be on the grid?

Example 1

Suppose two spinners are used instead of a spinner and a coin.

One spinner (A) goes from 1 to 4, and the other (B) from 1 to 5.

(a) How many dots would be on the grid now?

(b) How many possible outcomes would there be?

(c) What would be the probability of scoring a double two (2 *and* 2) with these spinners?

(d) Check that P(2 on spinner A) × P(2 on spinner B) = P(2, 2).

Answer 1

(a) 20 dots (4 × 5). Draw the grid if you are not sure.

(b) 20

(c) $\dfrac{1}{20}$

(d) $\dfrac{1}{4} \times \dfrac{1}{5} = \dfrac{1}{20}$

Example 2

Two six sided dice are thrown, one is red and the other is blue.
The possible outcomes are shown below *with the scores on the two dice added together.*

Red die

+	1	2	3	4	5	6
1	2	3	4	5	6	7
2	3	4	5	6	7	8
3	4	5	6	7	8	9
4	5	6	7	8	9	10
5	6	7	8	9	10	11
6	7	8	9	10	11	12

Blue die

(a) How many possible outcomes are there?

(b) Find P(a total score of 6),

(c) find P(a double),

(d) find P(total score of *less* than 5),

(e) find P(both dice showing even numbers).

(f) What is the most likely score?

(g) Find P(total of 1).

(h) Find P(total less than 12).

Answer 2

NOTE: It is acceptable to leave your answers to probability questions as unsimplified fractions unless otherwise specified in the question.

(a) 36 possible outcomes

(b) There are 5 totals of six, and 36 possible outcomes, so P(six) = $\frac{5}{36}$

(c) There are six doubles (for example 1,1 and so on), so P(a double) = $\frac{6}{36}$

(d) There are six ways of getting a total score of less than 5. (They are 2,3,3,4,4,4).
So P(total less than 5) = $\frac{6}{36}$

(e) There are 9 ways in which both dice could show even numbers. (For example, 2,4).
So P(both showing even numbers) = $\frac{9}{36}$

(f) P(a total of seven) = $\frac{6}{36}$. All the others are less than this, so the most likely score is seven

(g) P(total of 1) = 0

(h) P(total less than 12) = 1 − P(total of 12) = 1 − $\frac{1}{36}$ = $\frac{35}{36}$

Example 3

Bakari tosses a coin twice and notes both outcomes.
He gets tails both times.
He says the probability of this happening is $\frac{1}{3}$, because there are three possible outcomes: two heads, a head and a tail, and two tails.
Seema says he is wrong, and that the probability should be $\frac{1}{4}$.
Who is right?

Answer 3

Seema is right.

There are 4 possible outcomes: H,H H,T T,H T,T

(If this does not seem right think of tossing two different coins, say a pound and a dollar.
Then the outcomes would be: £H,$H £H,$T £T,$H £T,$T)

Exercise 12.1

1. The possible outcomes of an experiment using two five sided spinners are shown on the grid below. Each outcome is represented by a point on the grid.

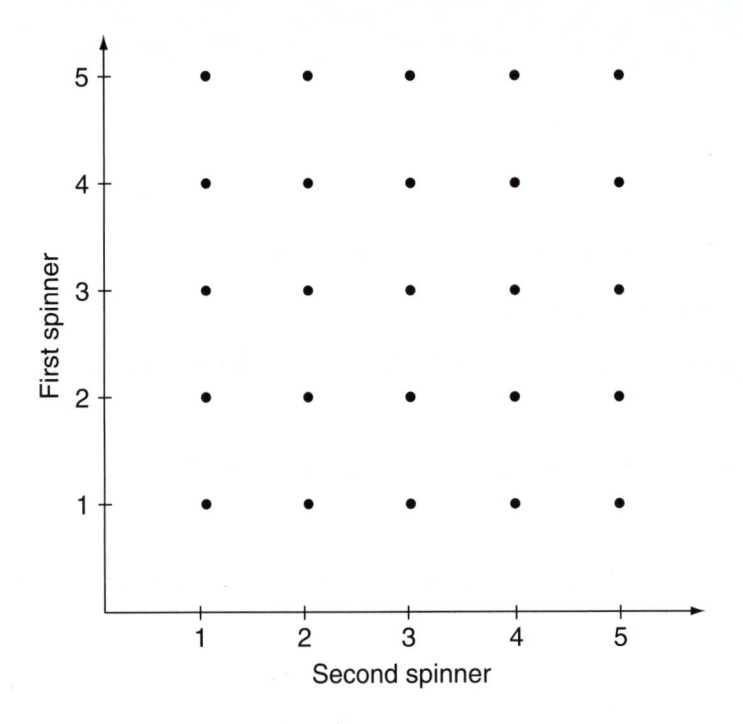

    (a) How many possible outcomes are there?
    (b) Find P(total score of 4).
    (c) Find P(total score greater than 6).
    (d) Find P(a double).
    (e) How many outcomes are there where the score on the first spinner is greater than the score on the second?
    (f) What is the most likely total score?

2. Patony is doing an experiment with two bags containing coloured counters.
    In one bag there are four counters, one each of red, blue, green and yellow.
    In the other bag there are five counters, one each of red, blue, green, purple and white.
    He picks one counter from each bag without looking in the bags.
    (a) Draw a possibility space diagram to show all the possible results of picking a counter from each bag.
    (b) What is the probability of drawing two counters of the same colour?

3. There are two boxes of counters, box A and box B.

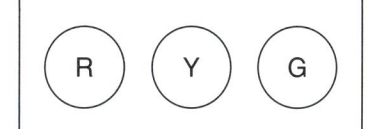

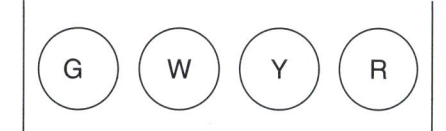

Box A contains 1 red, 1 yellow and 1 green counter.
Box B contains 1 green, 1 white, 1 yellow and 1 red counter.
Meer chooses one counter at random from each box.
Copy and complete the following possibility space diagram.

Box A

	R	Y	G
R	R,R		
Y			
G			
W			

Box B

What is the probability that he chooses

(a) two counters of the same colour,
(b) one green counter and one red counter,
(c) one yellow and one white counter,
(d) two white counters?

Tree Diagrams

Another useful diagram to use for combined events is a **tree diagram.**
A simple example of a tree diagram is a toy train set.
When this train comes to a junction (marked with a dot) there is a probability of $\frac{3}{4}$ that it will go straight on. This means that there is a probability of $\frac{1}{4}$ that it will turn.
The diagram shows the layout, with the probabilities marked on each branch.

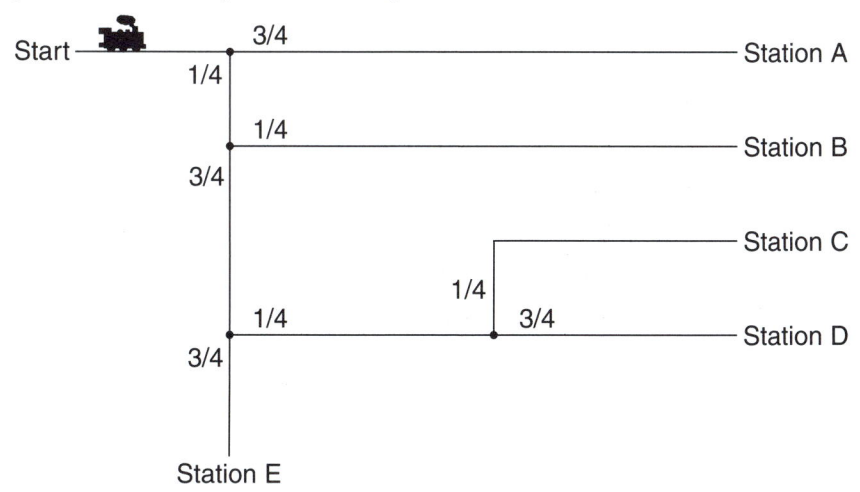

348 Extended Mathematics for Cambridge IGCSE

The probability of the train arriving at station A is $\frac{3}{4}$.

The probability of the train arriving at station B is found by *multiplying* the probabilities at each junction along the route from the start to station B: $\frac{1}{4} \times \frac{1}{4} = \frac{1}{16}$.

If you are not sure about this remember that this means that the train turns at the first junction *and* turns at the second junction. Look back to page 343 where we multiplied the probabilities when we wanted two outcomes together.

The probability of the train arriving at A *or* B (we do not mind which) is found by *adding* the probability of it arriving at station A and the probability of it arriving at station B. So:

$$P(A \text{ or } B) = P(A) + P(B) = \frac{3}{4} + \frac{1}{16}$$
$$= \frac{12}{16} + \frac{1}{16}$$
$$= \frac{13}{16}$$

Example 4

Use the diagram of the toy train layout to answer the following questions.
(a) What is the probability of the train arriving at station C?
(b) What is the probability of the train arriving at station D?
(c) What is the probability of the train arriving at stations C, D *or* E?
(d) What is the total of all the probabilities?
(e) What is the probability of the train *not* arriving at A?

Answer 4

(a) $P(C) = \frac{1}{4} \times \frac{3}{4} \times \frac{1}{4} \times \frac{1}{4} = \frac{3}{256}$

(b) $P(D) = \frac{1}{4} \times \frac{3}{4} \times \frac{1}{4} \times \frac{3}{4} = \frac{9}{256}$

(c) $P(C, D \text{ or } E) = P(C) + P(D) + P(E)$
$$= \frac{3}{256} + \frac{9}{256} + \left(\frac{1}{4} \times \frac{3}{4} \times \frac{3}{4}\right)$$
$$= \frac{3}{256} + \frac{9}{256} + \frac{9}{64}$$
$$= \frac{3}{256} + \frac{9}{256} + \frac{36}{256} = \frac{48}{256}$$

(d) $P(A) + P(B) + P(C) + P(D) + P(E) = 1$

(e) $P(\text{not } A) = 1 - P(A)$
$$= 1 - \frac{3}{4}$$
$$= \frac{1}{4}$$

Probability 349

A tree diagram will help you distinguish between **dependent and independent events**.
For example, suppose you are going to choose two coloured discs at random from a box containing five red and six blue discs, that is 11 discs altogether.
The first time you do this experiment you will *not* replace the first disc before choosing the second disc. This is called **selection without replacement**.
For the first choice P(R) = $\frac{5}{11}$, and P(B) = $\frac{6}{11}$. These probabilities are shown on the tree diagram below.
For the second choice the probabilities depend on what happened first.
If a red was picked the first time there are now only four reds, but still six blues in the box, making ten discs altogether. So now P(R) = $\frac{4}{10}$, and P(B) = $\frac{6}{10}$. The second choice is *dependent* on the first.
The diagram shows these probabilities.

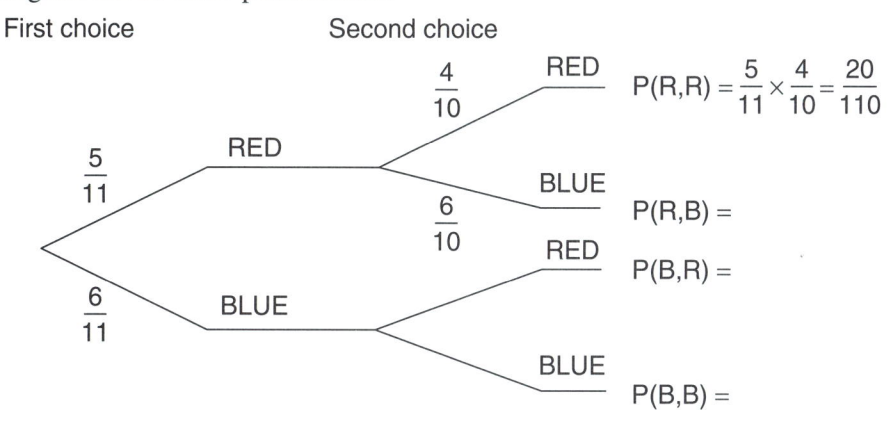

```
First choice           Second choice
                          4/10    RED      P(R,R) = 5/11 × 4/10 = 20/110
            5/11  RED
                          6/10    BLUE     P(R,B) =
                                  RED      P(B,R) =
            6/11  BLUE
                                  BLUE     P(B,B) =
```

Important points about tree diagrams:
- Read the diagram from left to right.
- Every time a choice is made the diagram splits into branches.
- The outcome of each choice is shown at the end of the branch.
- The probability of that outcome is shown on the branch.
- Every time there is a choice the probabilities on the branches at that point *add* up to 1.
- The probability of, for example two reds in the tree diagram above (Red *and* Red), is found by *multiplying* the probabilities along the branches that lead to P(R, R).
- The probability of getting one of each colour in the example above is found by *adding* the probabilities P(R, B) and P(B, R).
- The probabilities of all the possible outcomes are shown at the ends of the branches. They all add up to 1, because they cover all possibilities.
- It is useful to remember that if you travel along the branches from left to right you multiply the probabilities, and if you want to combine probabilities vertically you add them.

Example 5

(a) Copy the diagram above and fill in the rest of the probabilities.
(b) What is the probability of getting one of each colour?

Answer 5

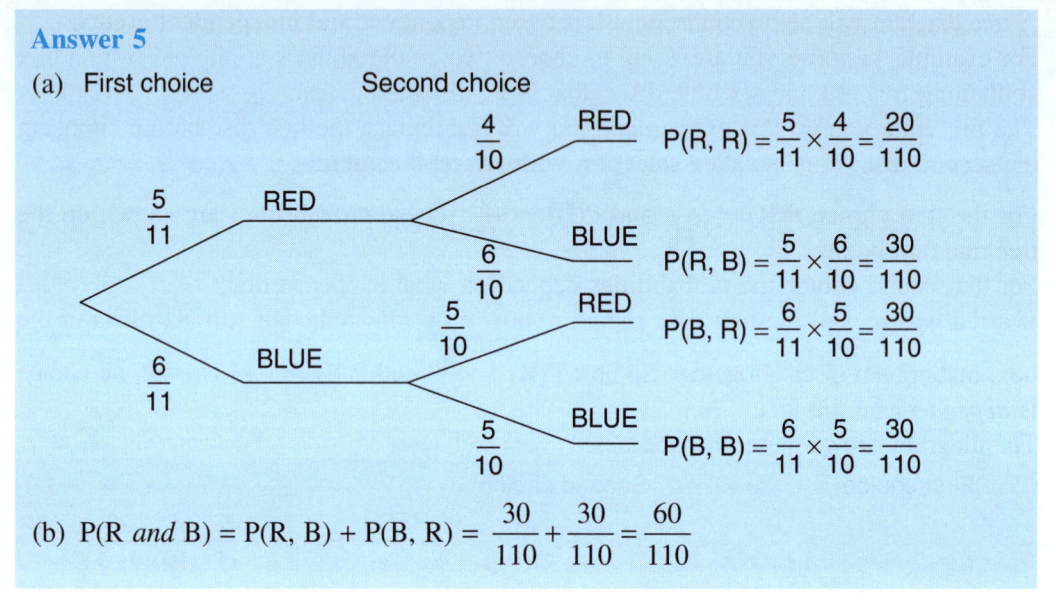

(a) First choice   Second choice

$$P(R, R) = \frac{5}{11} \times \frac{4}{10} = \frac{20}{110}$$

$$P(R, B) = \frac{5}{11} \times \frac{6}{10} = \frac{30}{110}$$

$$P(B, R) = \frac{6}{11} \times \frac{5}{10} = \frac{30}{110}$$

$$P(B, B) = \frac{6}{11} \times \frac{5}{10} = \frac{30}{110}$$

(b) $P(R \text{ and } B) = P(R, B) + P(B, R) = \frac{30}{110} + \frac{30}{110} = \frac{60}{110}$

Now, if this experiment were to be repeated *with* replacement the second choice will not be dependent on the first. The outcomes of the first and second choices are *independent* of each other.

The tree diagram below shows this situation.

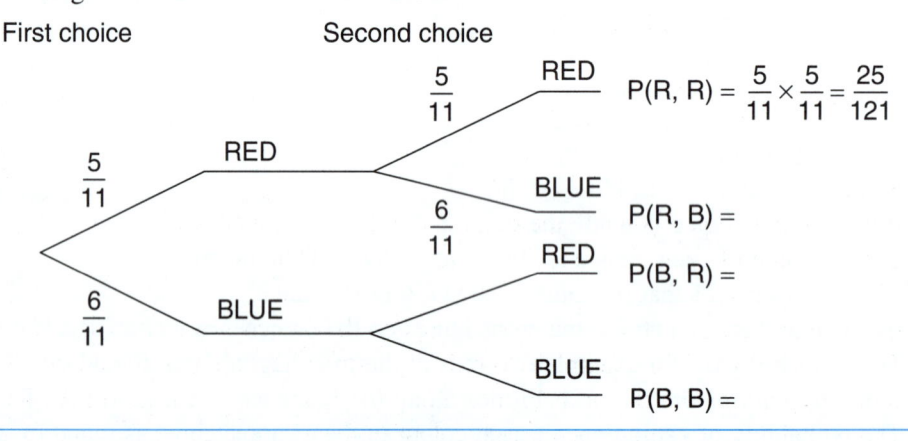

First choice   Second choice

$$P(R, R) = \frac{5}{11} \times \frac{5}{11} = \frac{25}{121}$$

P(R, B) =

P(B, R) =

P(B, B) =

Exercise 12.2

1. (a) Copy and complete the diagram above showing the probabilities of the remaining outcomes of selection *with* replacement.
   (b) What is the probability of getting one of each colour?
2. The average number of wet days in Mumbai in September is 13.
   The average number of wet days in New Delhi in September is 4.
   (a) What is the probability that any day chosen at random in Mumbai in September will be
       (i) wet,   (ii) dry.

(b) Draw a tree diagram to show the probabilities of wet and dry days in Mumbai and New Delhi in September.
(c) Find the probability that on September 5th next year it will be
   (i) wet in Mumbai *and* in New Delhi, (ii) wet in Mumbai and dry in New Delhi,
   (iii) both cities will be dry.
(d) What is the probability that it will be dry in New Delhi?

3. The probability of Sukatai passing her Maths exam at the first attempt is $\frac{3}{5}$.

   If she fails it and resits the probability of her passing the second time is $\frac{3}{4}$.

   Draw a tree diagram to show this. Remember that if she passes first time she does not have to take the exam again so there is no need for another branch in this part of the tree diagram.

Venn Diagrams

Venn diagrams and set notation are also useful tools in the study of probability.
You may want to have a quick look back at Chapter 1 to remind yourself about Venn diagrams before continuing.
The Venn diagram might show the numbers in each set or it can show the probabilities in each set.

Example 6
In a class of 40 students 20 play football, 24 play cricket and 9 play both cricket and football.
Draw a Venn diagram to show the numbers of students who play each game.
If a student is picked at random what is the probability that he or she plays
(a) neither football nor cricket, (b) cricket but not football?

Answer 6
The Venn diagram shows the numbers of students who play each game.

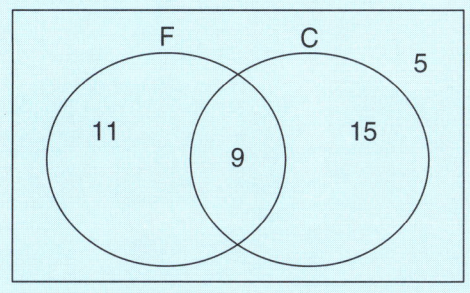

(a) There are 5 students who play neither football nor cricket, so the probability that the student plays neither is $\frac{5}{40}$

(b) There are 15 students who play cricket but not football, so the probability is $\frac{15}{40}$

Using Set Notation

Set notation provides a neat method for discussing probabilities.
Remember that:

$$P(A \cup B) = P(\text{A or B or both})$$
$$P(A \cap B) = P(\text{both A and B})$$
$$P(A') = P(\text{not A}) = 1 - P(A)$$
$$P(A \cup B)' = P(\text{neither A nor B})$$

If the events are **mutually exclusive** then $P(A \cap B) = 0$

The Venn diagram for the example above could have been drawn to show the probabilities in each set as in the Example below.

Remember that the total probability is 1.

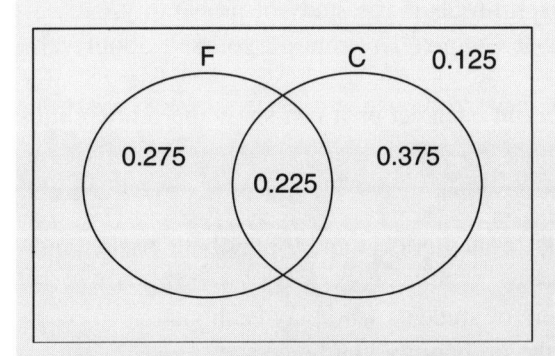

Example 7

```
    F           C       0.125

  0.275      0.375
        0.225
```

The Venn diagram from the example above has been redrawn to show the probabilities in each set.

(a) Are F and C mutually exclusive?

A student is picked at random from the class.
Find the following probabilities

(b) P(F)

(c) P(C)

(d) P(neither F nor C)

(e) P(both F and C)

(f) P(F or C or both)

A student is picked and found to play Football.

(g) Calculate the probability that he or she also plays Cricket.

Answer 7

First it is worth checking that the total Probability is 1:

$$0.275 + 0.225 + 0.375 + 0.125 = 1.$$

So there should be no probabilities missing, and each probability is correct.

(a) F and C are not mutually exclusive because a student may play both football and cricket

(b) Using the previous Venn diagram we can calculate P(F) by finding the total number who play football and dividing by the number in the class.

$$P(F) = (11 + 9) \div 40 = \frac{20}{40} = 0.5$$

It is perhaps easier to use the second Venn diagram and add the probabilities in F. So P(F) = 0.275 + 0.225 = 0.5

(c) P(C) = 0.225 + 0.375 = 0.6

(d) P(neither F nor C) = P(F ∪ C)′ = 0.125

(e) P(both F and C) = P(F ∩ C) = 0.225

(f) P(F or C or both) = P(F ∪ C) = 0.275 + 0.225 + 0.375 = 0.875

(g) For this part of the question we are only interested in the students who play football.

P(student plays cricket given that he or she plays football) = P(F ∩ C) ÷ P(F)
$$= 0.225 \div 0.5$$
$$= 0.45$$

Example 8

The Venn diagram shows the probabilities of two events represented by the sets A and B.

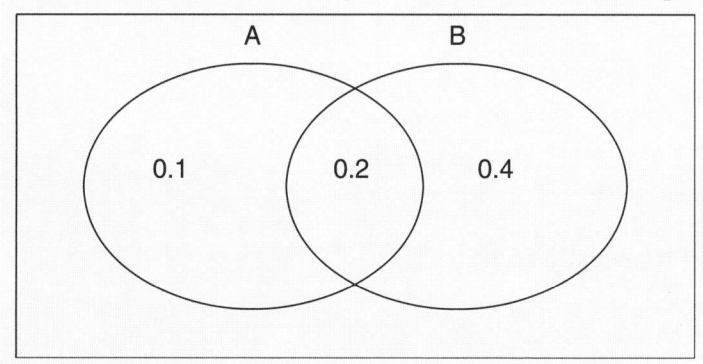

(a) Find P(A ∪ B)′.
(b) Find P(A).
(c) Find P(A *or* B *or* both).

(d) Find P(A *and* B).

(e) Find P(A or B but not both).

(f) Given that A has occurred, what is the probability of B also occurring?

Answer 8

(a) P(A ∪ B)′ = 1 − (0.1 + 0.2 + 0.4) = 1 − 0.7 = 0.3

(b) P(A) = 0.1 + 0.2 = 0.3

(c) P(A *or* B *or* both) = P(A ∪ B) = 0.1 + 0.2 + 0.4 = 0.7

(d) P(A *and* B) = P(A ∩ B) = 0.2

(e) P(A or B but not both) = 0.1 + 0.4 = 0.5

(f) P(B occurring given that A has already occurred) = P(A ∩ B) ÷ P(A)

$$= 0.2 \div (0.1 + 0.2)$$
$$= 0.2 \div 0.3$$
$$= \frac{2}{3}$$

Exercise 12.3

1. The Venn diagram shows two sets, D and E and some probabilities.

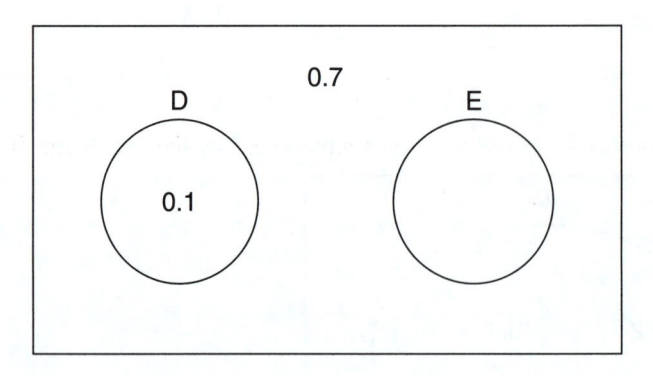

(a) Are the events D and E mutually exclusive?
(b) Find P(E).
(c) Write down P(D ∩ E).
(d) Find P(D ∪ E).
(e) Find P(D′).

Probability 355

2. In a class of 35 students, 20 wear glasses (G), 10 are tall for their age (T), and 2 are tall and wear glasses.
   (a) Draw two Venn diagrams, one showing the numbers in the sets G and T, and one showing the probabilities.
   (b) Find P(G ∪ T)'.
   (c) A student who wears glasses is picked at random. What is the probability that he or she is also tall?

Using set notation with tree diagrams

Example 9

Peter drives to work. His journey takes him through two sets of traffic lights which can be either red (stop) or green (go).
The probability that the first set are red when he reaches them is 0.4, and the probability that the second set are red is 0.7.

(a) Draw a tree diagram to show these probabilities.

(b) Do you think that the events that the first set are red and the second set are red are independent?

(c) Are the events that the first set are red and the second set are red mutually exclusive?

(d) Are the outcomes that the first set of lights are red and the first set of lights are green mutually exclusive?

(e) Find $P(R_1 \cap R_2)$, where R_1 is the event that the first set show red, and R_2 is the event that the second set show red.

(f) Find the probability that either both sets are red or both sets are green.

(g) Find the probability that the sets are different.

Answer 9

(a)   first set   second set
      of lights   of lights

```
            R₁            R₂      P(R₁ ∩ R₂) = 0.4 × 0.7 = 0.28
      0.4  ╱   0.7 ╱
          ╱       ╲ 0.3
         ╱          G₂    P(R₁ ∩ G₂) = 0.4 × 0.3 = 0.12
         ╲          0.7
      0.6 ╲       ╱ R₂    P(G₁ ∩ R₂) = 0.6 × 0.7 = 0.42
           G₁
              ╲ 0.3
                G₂        P(G₁ ∩ G₂) = 0.6 × 0.3 = 0.18
```

(b) They could be dependent if the signals are linked together to improve traffic flow. If they are not linked they are independent. These are independent because the probabilities of the second set being red or green are the same regardless of whether the first set were red or green. So the second set does not depend on what the first set were.

(c) These events are not mutually exclusive because they can both happen at once.

(d) These events are mutually exclusive because they cannot both happen at once.

(e) $P(R_1 \cap R_2) = 0.28$

(f) $P((R_1 \cap R_2) \cup (G_1 \cap G_2)) = 0.28 + 0.18 = 0.46$

(g) P(sets are different) = 1 − P(sets are the same) = 1 − 0.46 = 0.54

Exercise 12.4

Mixed Exercise

1. There are two classes in a school both with girls and boys.
   There are 10 girls in year 5, and 13 boys.
   There are 24 girls altogether, and a total of 49 pupils in the two classes.
   Copy and complete the diagram below to show the numbers of boys and girls in each class.

	Girls (G)	Boys (B)	Totals
Year Five (F)	10	13	
Year Six (S)			
Totals	24		49

   (a) If a pupil is picked at random from the two classes what is the probability that it will be a girl?
   (b) Find $P(G \cap S)$.
   (c) Find P(F).
   (d) A pupil is picked at random from year Six. What is the probability that it is a boy?

2. A bag contains 8 coloured discs, 3 red (R) and 5 yellow (Y). Two discs are selected without replacement.
   (a) Draw a tree diagram to show these events.
   (b) Is P(R) for the first and second selection the same or different?
   (c) Are the events of selecting the first and second discs independent?
   (d) What is the probability that the second disc is yellow given that the first disc was red?
   (e) Find $P(R_1 \cap Y_2)$ where R_1 is the event that the first disc was red, and Y_2 is the event that the second disc was yellow.
   (f) Calculate the probability that both discs are different colours.

3.

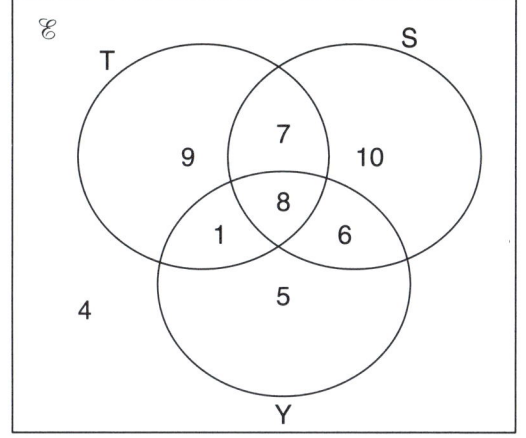

A packet of mixed flower seeds contains 50 seeds of various flowers. Some of the flowers will be tall (T), some will be yellow (Y) and some will be spring flowers (S). Some will have more than one of these characteristics.
The Venn diagram shows the distribution of these characteristics.

(a) If a seed is picked at random find the probability that it will be of a yellow flower.
(b) Find P(S′).
(c) Write down $n(S \cap T \cap Y)$.
(d) Find the probability that a seed picked at random is of a tall yellow flower.
(e) Given that a seed picked at random was of a tall flower find the probability that it was also yellow.

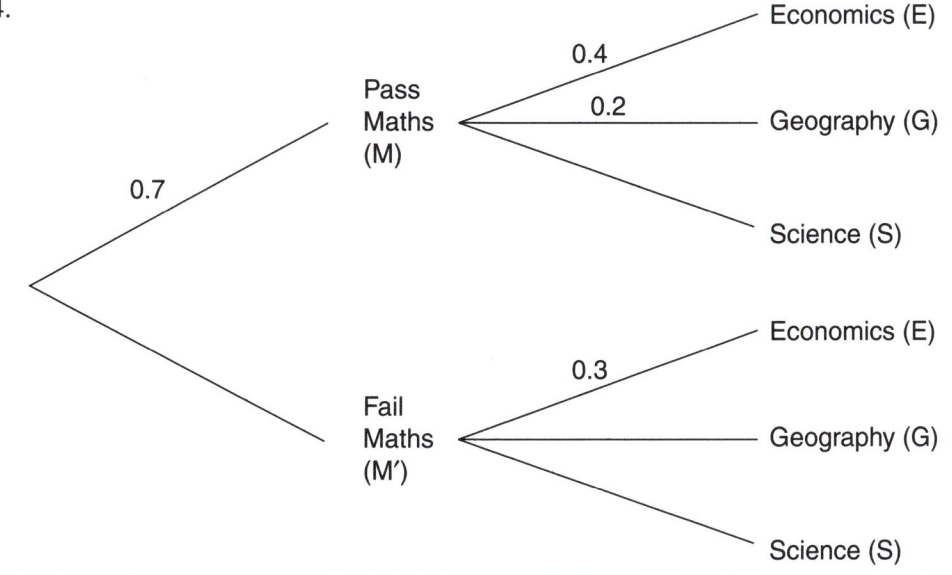

4.

Ashna is about to take a Maths examination. When she gets her results she will decide whether to study Economics (E), Geography (G) or Science (S). If she passes her Maths examination (M) the probability that she will study Economics is 0.4. If she fails her Maths examination (M′) the probability that she will study Geography (G) is 0.6.

(a) Copy and complete the tree diagram above, and use it to answer the following questions.
(b) What is the probability that she will pass her Maths examination and study Economics?
(c) Calculate P(M′ ∩ E).
(d) Calculate the probability that she will study Science.

Examination Questions

5. An ordinary unbiased die has faces numbered 1, 2, 3, 4, 5 and 6.
   Sarah and Terry each threw this die once.
   Expressing each answer as a fraction **in its lowest terms**, find the probability that
   (i) Sarah threw a 7,
   (ii) they both threw a 6,
   (iii) neither threw an even number,
   (iv) Sarah threw exactly four more than Terry.

   (4024/02 May/June 2003 q 5 (b))

6.

| A | D | A | M | | D | A | N | I | E | L |

   Adam writes his name on four red cards and Daniel writes his name on six white cards.
   (a) One of the ten cards is chosen at random.
      Find the probability that
      (i) the letter on the card is **D**,
      (ii) the card is red,
      (iii) the card is red **or** the letter on the card is **D**,
      (iv) the card is red **and** the letter on the card is **D**,
      (v) the card is red **and** the letter on the card is **N**.
   (b) Adam chooses a card at random and then Daniel chooses one of the remaining 9 cards at random.
      Giving your answers as fractions, find the probability that the letters on the two cards are
      (i) both **D**,
      (ii) both **A**,
      (iii) the same,
      (iv) different.

   (0580/04 Oct/Nov 2003 q 5)

7. Two unbiased spinners are used in a game.
   One spinner is numbered form 1 to 6 and the other is numbered from 1 to 3.
   The scores on each spinner are multiplied together. The table below shows the possible outcomes.

		First Spinner					
		1	2	3	4	5	6
Second Spinner	1	1	2	3	4	5	6
	2	2	4	6	8	10	12
	3	3	6	9	12	15	18

   (a) Find the probability that the outcome is even.
   (b) When the outcome is even, find the probability that it is also greater than 11.

   (0580/02 May/June 2007 q 15)

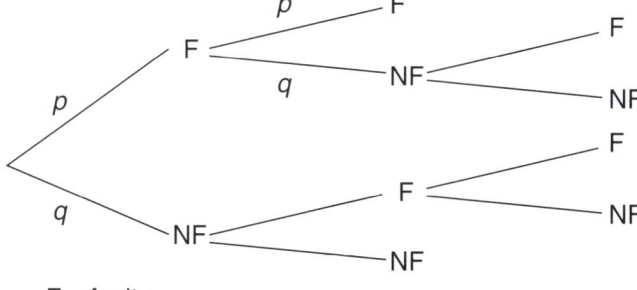

8.
   F = faulty
   NF = not faulty

   The tree diagram shows a testing procedure on calculators, taken from a large batch.
   Each time a calculator is chosen at random, the probability that it is faulty (F) is $\frac{1}{20}$.
   (a) Write down the values of p and q.
   (b) Two calcuators are chosen at random.
       Calculate the probability that
       (i) both are faulty,
       (ii) **exactly one** is faulty.
   (c) If **exactly one** out of two calculators tested is faulty, then a third calculator is chosen at random. Calculate the probability that exactly one of the first two calculators is faulty **and** the third one is faulty.

(d) The whole batch of calculators is rejected
   either   if the first two chosen are both faulty
   or   if a third one needs to be chosen and if is faulty.
   Calculate the probability that the whole batch is rejected.
(e) In one month, 1000 batches of calculators are tested in this way.
   How many batches are expected to be rejected?

(0580/04 May/June 2009 q 8)

9. In a servey, 100 students are asked if they like basketball (B), football (F) and swimming (S). The Venn diagram shows the results.

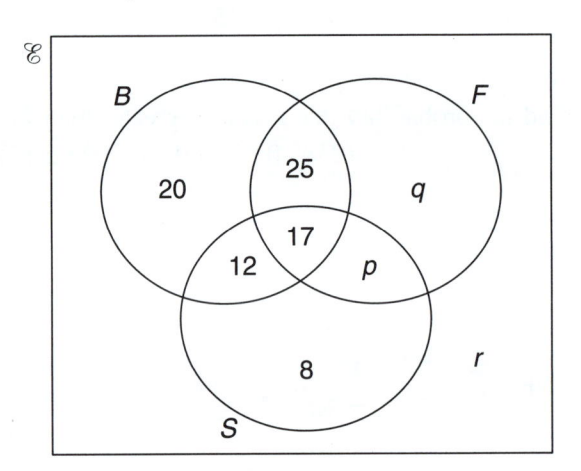

42 students like swimming.
40 students like exactly one sport.

(a) Find the values of p, q and r.
(b) How many students like
   (i) all three sports,
   (ii) basketball and swimming but not football?
(c) Find
   (i) $n(B')$,
   (ii) $n((B \cup F) \cap S')$.
(d) One student is chosen at random from the 100 students. Find the probability that the student
   (i) only likes swimming,
   (ii) likes basketball but not swimming.
(e) Two Students are chosen at random from those who like basketball. Find the probability that they each like exactly one other sport.

(0580/04 Oct/Nov 2008 q 9)

10. **Give your answers to this question as fractions.**

   (a) the probability that it rains today is $\frac{2}{3}$.

   If it rains today, the probability that it will rain tomorrow is $\frac{3}{4}$.

   If it does not rain today, the probability that it will rain tomorrow is $\frac{1}{6}$.

   The tree diagram below shows this information.

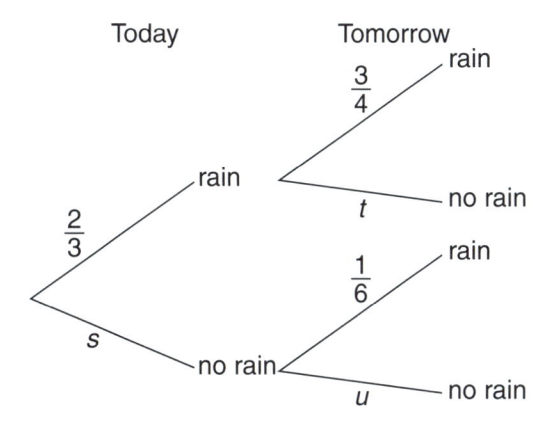

   (i) Write down, as fractions, the values of s, t and u.
   (ii) Calculate the probability that it rains on both days.
   (iii) Calculate the probability that it will not rain tomorrow.

   (b) Each time that Christina throws a ball at a target, the probability that she hits the target is $\frac{1}{3}$.

   She throws the ball three times.
   Find the probability that she hits the target
   (i) three times,   (ii) at least once.

   (c) Each time Eduardo throws a ball at the target, the probability that he hits the target is $\frac{1}{4}$.

   He throws the ball until he hits the target.
   Find the probability that he **first** hits the target with his
   (i) 4th throw,   (ii) nth throw.

   (0580/04 Oct/Nov 2006 q 5)

11. (a)

Grade	1	2	3	4	5	6	7
Number of students	1	2	4	7	4	8	2

The table shows the grades gained by 28 students in a history test.
- (i) Write down the mode.
- (ii) Find the median.
- (iii) Calculate the mean.
- (iv) Two students are chosen at random.
  Calculate the probability that they both gained grade 5.
- (v) From all the students who gained grades 4 or 5 or 6 or 7, two are chosen at random.
  Calculate the probability that they both gained grade 5.
- (vi) Students are chosen at random, one by one, from the original 28, until a student chosen has a grade 5.
  Calulate the probability that this is the third student chosen.

(b) Claude goes to school by bus.
The probability that the bus is late is 0.1.
If the bus is late, the probability that Claude is late to school is 0.8.
If the bus is not late, the probability that Claude is late to school is 0.05.
- (i) Calculate the probability that the bus is late and Claude is late to school.
- (ii) Calculate the probability that Claude is late to school.
- (iii) The school term lasts 56 days.
  How many days would Claude expect to be late?

(0580/04 Oct/Nov 2007 q 2)

12.

| 1 | 1 | 6 | 7 | 11 | 12 |

Six cards are numbered 1, 1, 6, 7, 11 and 12.
In this question, give all probabilities as fractions.
(a) One of the six cards is chosen at random.
- (i) Which number has a probability of being chosen of $\frac{1}{3}$?
- (ii) What is the probability of choosing a card with a number which is smaller than **at least three of the other numbers**?

(b) Two of the six cards are chosen at random, without replacement.
Find the probability that
- (i) they are both numbered 1,
- (ii) the total of the two numbers is 18,
- (iii) the first number **is not** a 1 and the second number **is** a 1.

(c) Cards are chosen, without replacement, until a card numbered 1 is chosen.
Find the probability that this happens before the third card is chosen.

(d) A seventh card is added to the six cards shown in the diagram.
The mean value of the seven numbers on the cards is 6.
Find the number on the seventh card.

(0580/04 Oct/Nov 2009 q 3)

13. (a)

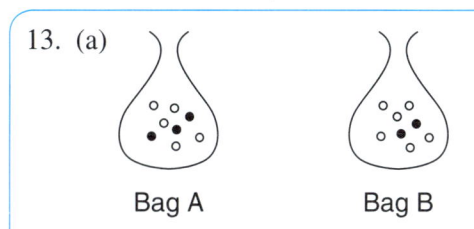

Bag A   Bag B

Nadia must choose a ball from Bag A or from Bag B.

The probability that she chooses Bag A is $\frac{2}{3}$.

Bag A contains 5 white and 3 black balls.
Bag B contains 6 white and 2 black balls.
The tree diagram below shows some of this information.

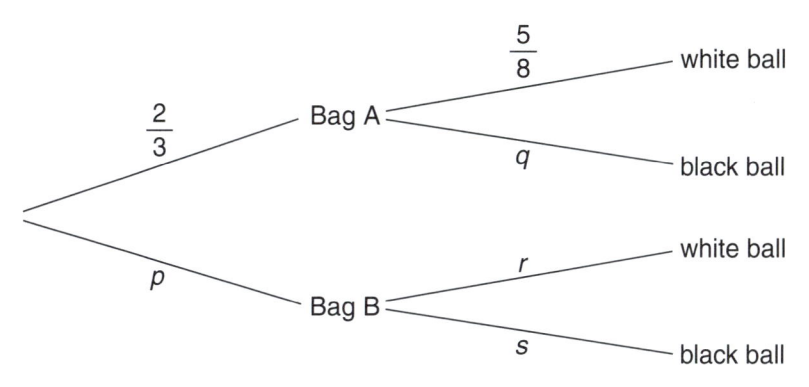

```
              5/8  ── white ball
        Bag A ─
  2/3  ╱       q  ── black ball
      ╱
      ╲        r  ── white ball
   p   ╲ Bag B ─
              s   ── black ball
```

 (i) Find the values of *p*, *q*, *r* and *s*.
 (ii) Find the probability that Nadia chooses Bag A and then a white ball.
 (iii) Find the probability that Nadia chooses a white ball.

(b) Another bag contains 7 green balls and 3 yellow balls.
Sani takes three balls out of the bag, without replacement.
 (i) Find the probability that all three balls he chooses are yellow.
 (ii) Find the probability that at least one of the three balls he chooses is green.

(0580/04 May/June 2008 q 3)

14. (a) There are 30 students in a class.
20 study Physics 15 study Chemistry and 3 study neither physics nor chemistry.

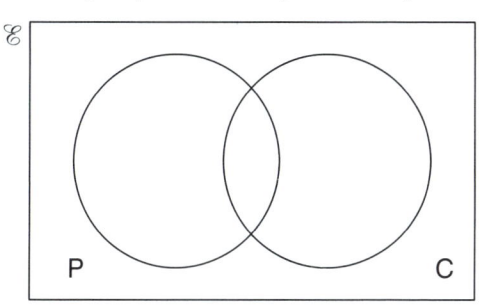

(i) **Copy and complete** the Venn diagram to show this information.
(ii) Find the number of students who study both Physics and Chemistry.
(iii) A student is chosen at random. Find the probability that the student studies Physics but not Chemistry.
(iv) A student who studies Physics is chosen at random. Find the probability that this student does not study Chemistry.

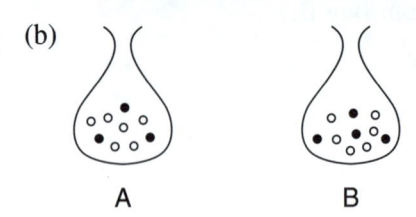

(b)

A   B

Bag A contains 6 white beads and 3 black beads.
Bag B contains 6 white beads and 4 black beads.
One bead is chosen at random from each bag.
Find the probability that
 (i) both beads are black,
 (ii) at least one of the two beads is white.
The beads are not replaced.
A second bead is chosen at random from each bag.
Find the probability that
(iii) all four beads are white,
(iv) the beads are not all the same colour.

(0580/04 May/June 2004 q 7)

Revision and Examination Hints

These hints are in addition to those given in the Core Book, which you should read carefully.

For your examination you need to understand and be familiar with the whole of the Core course and the Extended course. The items outlined below are those which can be overlooked, or which sometimes have been given insufficient attention, and can therefore lead to unnecessary loss of marks.

In particular you should remember the following.

- The number of marks available for each part of the question gives an indication of the amount of work which needs to be shown for that part.
- Avoid using rough or other extra paper. If you really have run out of space make sure that you indicate in your answer booklet where the rest of the working will be found, and number that work carefully so that the examiner can find it. You cannot get marks for working that cannot be found!
- Copy down numbers from the questions carefully, and transfer your answer carefully to the answer space.
- Make sure that your calculator is set in degrees, not 'rads' or 'grads'.
- Do not lose marks for 'premature approximation' by not working to sufficient numbers of significant figures. To be on the safe side use your calculator value for subsequent working. Learn to use your calculator memory.
- Read the question carefully to make sure that you are giving your answer in the required form. For example, which units are required, or what degree of accuracy is expected.
- In questions which require you to 'show that' some statement or result is true make sure that you write down sufficient working to show that you know the method that is required, and give your answer in a clear and logical order.
- Be careful how you write fractions. A quarter of x should be written $\frac{1}{4}x$, or with the x level with the 1 in the numerator, <u>not</u> level with the 4 in the denominator. Careless use of fractions can all too easily lead to wrong working, thus losing marks unnecessarily. Remember that, while examiners try to award 'follow through' marks for subsequent working, it can happen that the subsequent working is not actually possible if a fundamental mistake has been made.
- Remember to indicate vectors by underlining the lower case letter you are using to represent the vector (for example, <u>u</u>), and directed line segments by using an arrow over the capital letters (for example, $\overrightarrow{AB}$).
- Do not waste time using the Sine and Cosine Rules in a right-angled triangle; the sine, cosine and tangent ratios are quicker and simpler.

It is recommended that you make sure that you understand the following.

- Locus
- Reverse percentages.

- The equations of straight lines on graphs including those parallel to the axes. (Be aware of the possibility of different scales on each axis.)
- The general shapes of different curves, including whether a parabola has a maximum point (negative x^2 term) or a minimum point (positive x^2 term).
- How to find the total surface area and volume of a prism.
- Position vectors.
- How to draw histograms, including the use of frequency density.
- The mean from a grouped frequency table.
- Cumulative frequency curves and how to read them. (For example, using a cumulative frequency curve to find the number of students getting *more* than a certain mark in an examination.)
- The difference between distance/time graphs (gradient = speed) and speed/time graphs (gradient = acceleration, and area under graph = distance gone).
- The notation for successive transformations and composite functions. (For example, fg(*x*) means do g before f.)
- Fractional and negative indices.

LEARN the following.

- The quadratic formula.
- All circle facts.
- How to find gradients, midpoints and lengths of straight line segments on graphs.
- The formulae for the area and circumference of a circle.
- How to work out the length of an arc and the area of a sector of a circle.
- How to find a determinant and an inverse matrix.
- The Sine Rule.
- The Cosine Rule, in both forms.
- The formula for the area of a non-right angled triangle.
- The basic matrices associated with transformations. (It is useful to learn these, but not essential if you know how to find them.)

PRACTISE the following.

- Drawing Venn diagrams and using them. (For example, to find the intersection and union of sets.)
- Factorising algebraic expressions (common factors, difference of squares and quadratics), and completing the square.
- The procedure for Variation. (For example, if *m* is inversely proportional to the square of *n* then $m \propto \dfrac{1}{n^2}$ and $m = \dfrac{k}{n^2}$. Then use the given information to find *k*.)
- Using the length, area and volume ratios of similar shapes. (For example, if the length ratio is 2 : 3, the area ratio is 4 : 9 and the volume ratio is 8 : 27.)
- Finding and using the sines and cosines of obtuse angles, and the correct use of the calculator with the Sine Rule and Cosine Rule.

- Finding centres and invariant lines in transformations and giving sufficient information when describing transformations.
- Working with algebraic fractions.

Finally, make sure that you are very familiar with your calculator before you go into your examination. You can now be confident that you have done all you can to prepare and should be able to achieve your best possible grade in the examination.

Answer Key

NOTE: In some of these answers a range of values has been given. Answers within these ranges should obtain full marks. They represent the different values that might be obtained in the following ways.

- Correct working, but using either rounded or calculator values from previous (numbered) parts of the question. (Remember never to round <u>within</u> the working for part of a question.)
- Using different values for π. (The calculator value is best.)
- Reading from graphs.
- Measurements from accurate drawings.

The graphs and diagrams in these answers are here to guide you, but they may have been reduced in scale for reasons of space. Yours should be full size. Where working has to be shown some of the key steps are sometimes included in the given answer.

Chapter 1

Core Skills

1. (a) 100, 1, 9, 18, 24, 6, $\sqrt{25}$, 2, 49
   (b) 100, 1, 9, 18, 24, 6, 0, $\sqrt{25}$, 2, 49
   (c) 100, 22.5, 1, 9, 18, 24, 6, 0, $\sqrt{25}$, 2, $\frac{2}{7}$, 49
   (d) π, $\sqrt{2}$ (e) $\sqrt{25}$, 2   (f) 1, 9, $\sqrt{25}$
   (g) 18, 24, 6   (h) 2   (i) 49   (j) 9
2. (a) {1, 2, 3, 4, 6, 8, 9, 12, 16, 18, 24, 36, 48, 72, 144}
   (b) {2, 3}   (c) $144 = 2^4 \times 3^2$
3. 22
4. 90
5. 0.5006, 0.5111, 0.513, 0.52
6. $\sqrt{64} = 2^3$ $\sqrt{64} < 3^2$ $19 > 18$
7. (a) 6.01371×10^4 (b) 5.401×10^{-3}
8. (a) 0   (b) 5   (c) 1
9. {23, 29, 31, 37}
10. $12\frac{1}{3}$ 11. $\frac{28}{5}$ 12. $\frac{25}{75} = \frac{1}{3} = \frac{4}{12}$
13. (a) $1\frac{5}{7}$ (b) $1\frac{15}{49}$ (c) 4
14. (a) 71%   (b) $\frac{3}{20}$ (c) 0.14
15. 8.64
16. 44.4% to 3 s.f.
17. $\frac{9}{14}$

Exercise 1.1

1. (a) {16, 25, 36}   (b) {March, May}
   (c) {1, 2, 3, 4, 5, 6, 7, 8, 9, 10}
2. (a) The set of odd numbers less than 10
   (b) The set of days of the week.
   (c) The set of letters of the English alphabet.
3. (a) and (d) are empty sets
4. (a) $1 \notin$ {prime numbers}
   (b) $1000 \in$ {even numbers}
5. (a) {a, c, e, h, i, m, s, t}   (b) 8

Exercise 1.2

1. (a) {3, 5, 7}   (b) {5}
   (c) {3, 4, 5, 6, 7}   (d) {3, 7, 8, 9}
   (e) {4, 6, 8, 9}   (f) {3, 7}
   (g) {8, 9}
2. ∅
3. {4, 5, 6, 9}
4. 2   5. 4   6. 4

Exercise 1.3

1. (a)

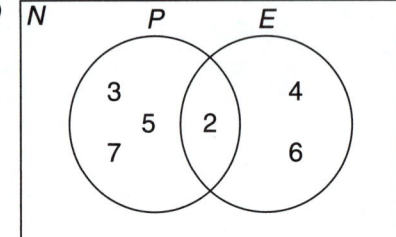

Answer Key 369

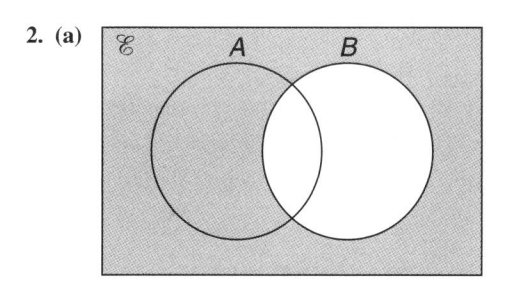

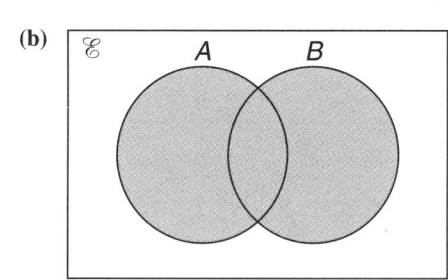

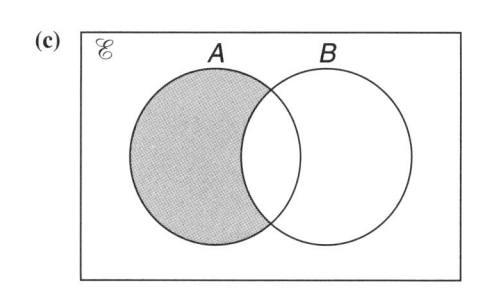

(b) Venn diagram with universal set Z, sets S and E overlapping.

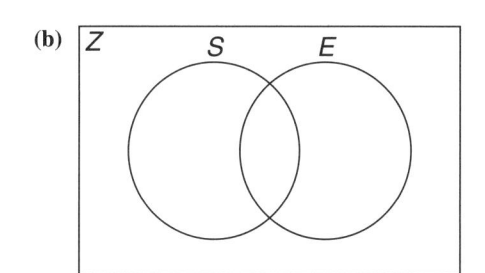

2. (a) Venn diagram with universal set $\mathscr{E}$, sets A and B overlapping; everything outside A shaded (i.e., A').

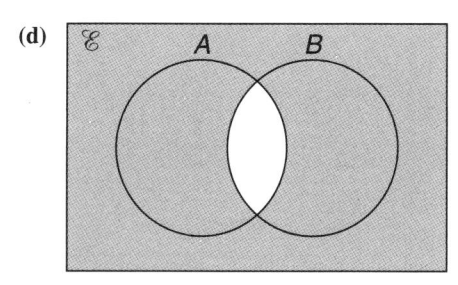

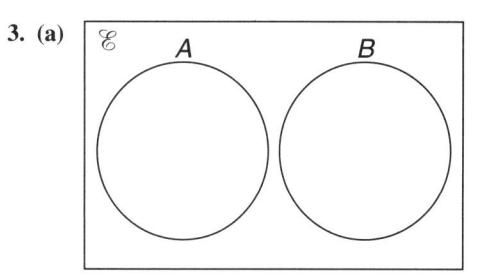

(b) Venn diagram with A and B overlapping; A shaded entirely including intersection ($A \cup (A\cap B)$ = A region plus... actually A and its parts shaded).

(c) Venn diagram with A only (excluding intersection) shaded, i.e., $A \setminus B$.

(d) Venn diagram with everything shaded except the intersection of A and B.

3. (a) Venn diagram with $\mathscr{E}$, disjoint sets A and B.

(b) Venn diagram with $\mathscr{E}$, A inside B.

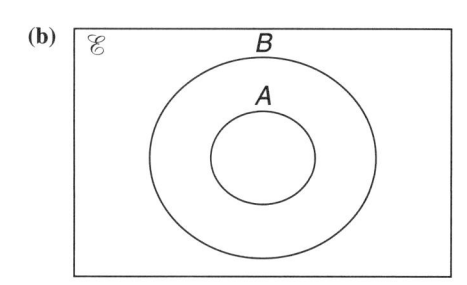

4. Venn diagram with universal set $\mathscr{E}$, set P containing R and S (R and S overlap inside P):
- Outside P: 9, 10
- Inside P only: 7, 5
- Inside R only: 6, 8
- R ∩ S: 2, 4
- Inside S only: 1, 3

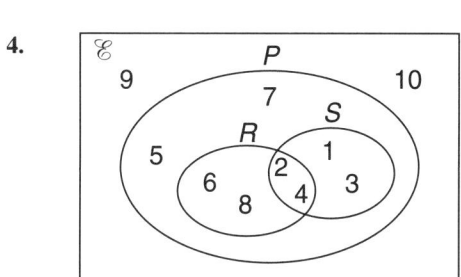

5. (a) Venn diagram with A containing B:
- Outside A: 23
- A only: 15
- B: 10

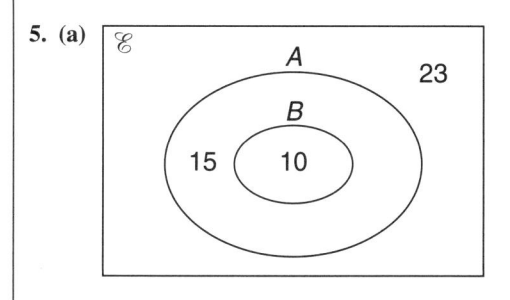

(b) (i) 15   (ii) 23

6. (a) Venn diagram with P and S overlapping:
- P only: 4
- P ∩ S: 3
- S only: 13

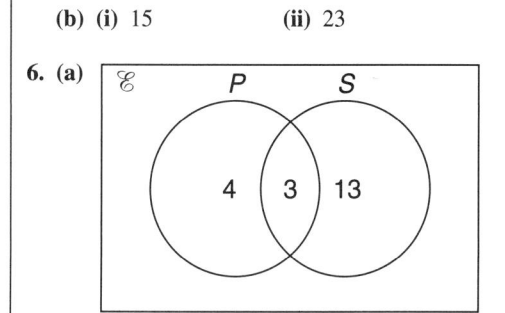

(b) 3

370 Extended Mathematics for Cambridge IGCSE

Exercise 1.4

1. (a) $\sqrt{784} = \sqrt{4 \times 4 \times 49} = 4 \times 7 = 28$

   (b) $\sqrt{1600} = \sqrt{16 \times 100} = 4 \times 10 = 40$

   (c) $\sqrt{625} = \sqrt{5 \times 5 \times 25} = 5 \times 5 = 25$

2. (a) $\sqrt{180} = \sqrt{9 \times 20} = \sqrt{9 \times 4 \times 5} = 6\sqrt{5}$

   (b) $\sqrt{98} = \sqrt{2 \times 49} = 7\sqrt{2}$

   (c) $\sqrt{192} = \sqrt{2 \times 96} = \sqrt{2 \times 2 \times 48}$
   $= \sqrt{4 \times 3 \times 16} = 2 \times 4 \times \sqrt{3} = 8\sqrt{3}$

3. (a) $\sqrt{50} = \sqrt{2 \times 25} = 5\sqrt{2}$ irrational

   (b) $\sqrt{144} = 12$ rational

   (c) $\sqrt{45} = \sqrt{5 \times 9} = 3\sqrt{5}$ irrational

Exercise 1.5

1. (a) $2\dfrac{23}{100}$ (b) $\dfrac{223}{100}$

2. 2×10^{-3} 0.21% $\dfrac{1}{250}$ $\dfrac{21}{500}$

3. (a) 130%   (b) 95.5%   (c) 0.5%

4. $\dfrac{47}{99}$ $\dfrac{22}{41}$ $\dfrac{53}{79}$

5. (a) $\dfrac{7}{8}$ (b) $\dfrac{49}{60}$

Exercise 1.6

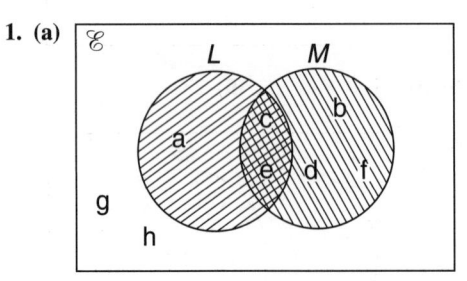

1. (a) [Venn diagram with sets L and M: a, c in L only; c, e in L∩M; b, d, f in M only; g, h outside]

   (b) (i) {c, e}   (ii) {a, b, c, d, e, f}
   (iii) {b, d, f, g, h}   (iv) {a, g, h}
   (v) {a, b, d, f, g, h}

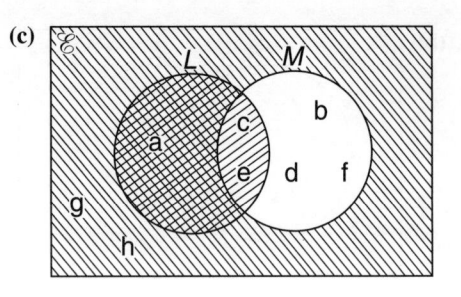

   (c) [Venn diagram with sets L and M showing shaded regions]

   (i) {a}   (ii) {a, c, e, g, h}

2. {−2, −1, 0, 1, 2}

3. (a) The set of the first four cube numbers.
   (b) The set of the first five square numbers.

4. (a) $\sqrt{4096} = \sqrt{4 \times 1024} = \sqrt{4 \times 4 \times 256}$
   $= \sqrt{16 \times 4 \times 64} = 4 \times 2 \times 8 = 64$

   (b) $\sqrt{2450} = \sqrt{2 \times 1225} = \sqrt{2 \times 5 \times 245}$
   $= \sqrt{2 \times 5 \times 5 \times 49} = 5 \times 7 \times \sqrt{2} = 35\sqrt{2}$

5. (a) is irrational

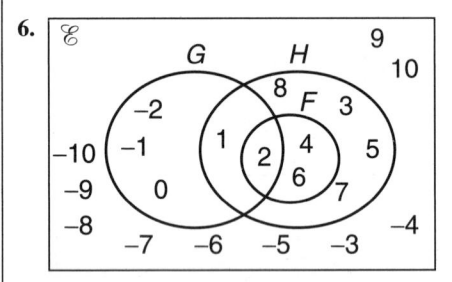

6. [Venn diagram with sets G, H, F and various numbers]

   (a) 1   (b) 7   (c) 1
   (d) 11   (e) 10

7. (a) $\sqrt{16}$ or $\dfrac{65}{13}$ (b) π or $\sqrt{14}$

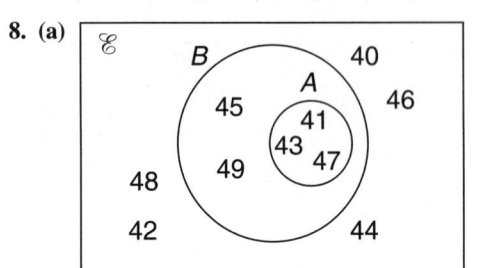

8. (a) [Venn diagram with sets B and A containing numbers 40–49]

   (b) 2

9. (a) $-1, \sqrt{36}$ (b) $\sqrt{2}, \sqrt{30}$

10. (a) $A \cap (B \cup C)'$ or $A \cap B' \cap C'$
(b) 3
11. (a) 8 (b) 18
12. $\dfrac{3}{2500}$, $\dfrac{1}{8}$%, 0.00126
13. 150%
14. $\dfrac{1}{125000}$, 8×10^{-5}, 0.0008, 0.8%

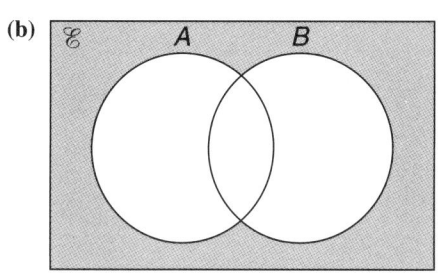

15. (a) [Venn diagram: A ∩ B shaded]
(b) [Venn diagram: everything shaded]
(c) [Venn diagram: A and outside shaded, B only unshaded]

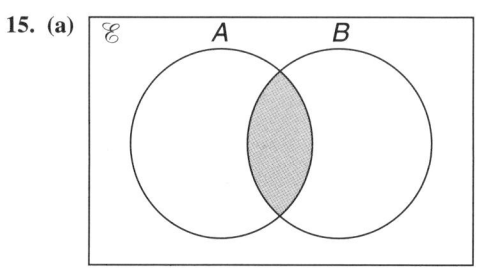

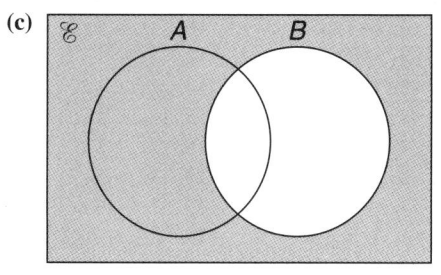

16. 97

17. (a) [Venn diagram with B inside A: 0 outside, 7 in A only, 11 in B]

(b) [Venn diagram with A and C disjoint: 18 in A, 11 in C]

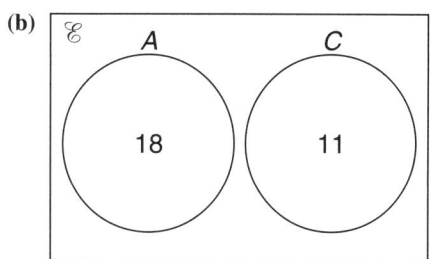

18. (a) [Venn diagram A, B, C with everything except A only and C only shaded]
(b) [Venn diagram A, B, C: A∩B and B∩C shaded]

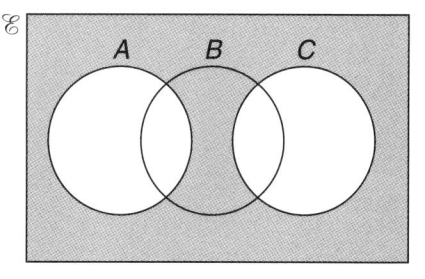

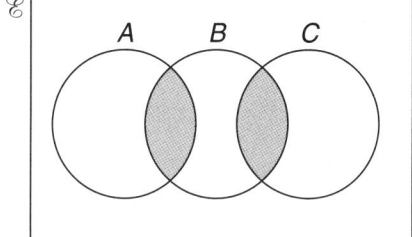

19. [Venn diagram A, B: 2 outside, 8 in A only, 4 in intersection, 7 in B only]

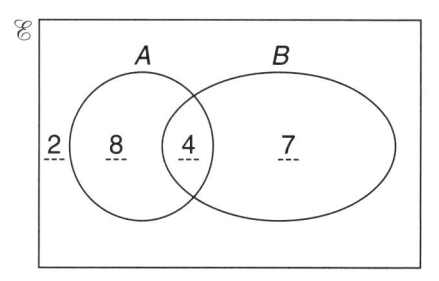

20. (a) $p = 5$, $q = 12$, $r = 1$
(b) (i) 17 (ii) 57
(c) (i) 26 (ii) 57

21. $\dfrac{598}{601}$, $\dfrac{399}{401}$, $\dfrac{698}{701}$

22. (a) $\varnothing$ (b) $\mathscr{E}$ (c) A

23. $1\dfrac{22}{37}$ or $\dfrac{59}{37}$

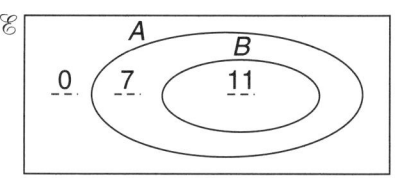

Chapter 2

Core Skills

1. (a) $a^2 = b^2 + c^2$ (b) $y = 2x + 3$
   (c) $2x + 5y - x - 1$
2. (a) $2x$ (b) -6
   (c) -1 (d) $3xy$ and $-4xy$
3. (a) -3 (b) -9
   (c) -2 (d) 7
4. (a) -8 (b) 30
   (c) $-\dfrac{1}{2}$ or -0.5 (d) $\dfrac{5}{6}$
   (e) -1 (f) 16
5. (a) a^5 (b) x^9
   (c) b^3 (d) c^{10} (e) a^3b^6
6. (a) $x^2 - x^3$ (b) $\dfrac{3x}{y}$
   (c) $xy + x^2y^2$ (d) b
7. (a) x^7y^7 (b) $6x^6$
   (c) x^{12} (d) $x^{12}y^{30}$
8. (a) $-3x - 2y$ (b) $2a + 2c$
   (c) 0 (d) $-ab + 4a$
9. (a) 25 (b) 29
   (c) -8 (d) 17
10. (a) $\dfrac{1}{3}$ (b) $\dfrac{1}{16}$
    (c) 4 (d) $\dfrac{4}{25}$
11. (a) $n = 0$ (b) $n = -1$
    (c) $n = -2$ (d) $n = 1$
12. (a) $-4y$ (b) $-pq$
    (c) $a^3 + 2a^2b^2 + b^3$ (d) $1 - x^2 - y^2$
13. (a) a^2bc (b) $3x^2y$
14. (a) $5xy(z + 2x)$ (b) $7xy(2x - 3y)$
    (c) $3a^2(1 - 2a)$ (d) $x(2x - 1)$

Exercise 2.1

1. $a^2 + 2a + 1$ 2. $x^2 + 9x + 20$
3. $x^2 - x - 20$ 4. $x^2 + x - 20$
5. $x^2 - 9x + 20$ 6. $2b^2 + 3b + 1$
7. $5c^2 - 12c + 4$ 8. $12x^2 + 28x + 15$
9. $x^2 + 2xy + y^2$ 10. $x^2 - 2xy + y^2$
11. $x^2 - y^2$ 12. $4d^2 - 9e^2$
13. $14z^2 - 5z - 1$ 14. $8 + 6x + x^2$
15. $4 - x^2$ 16. $a^3 + a^2b + ab + b^2$
17. $x^3 + x^2 - x - 1$ 18. $2b^2 + 4bc + b + 2c^2 + c$
19. $x^3 - 1$ 20. $x^4 - 1$
21. $4x^2 + 12x + 9$ 22. $4x^2 - 9$
23. $4x^2 - 12x + 9$ 24. $9x^2 - 16$
25. $4b^2 - 1$ 26. $2ac + 2ad + 3bc + 3bd$

Exercise 2.2

1. $(x + 1)(x + 3)$ 2. $(x + 1)(x + 5)$
3. $(x + 1)(x + 12)$ 4. $(x + 2)(x + 6)$
5. $(x + 3)(x + 4)$ 6. $(x + 4)(x + 4)$ or $(x + 4)^2$
7. $(x + 1)(x + 16)$ 8. $(x + 2)(x + 8)$
9. $(x + 1)(x + 17)$ 10. $(x + 1)(x + 1)$ or $(x + 1)^2$

Exercise 2.3

1. $(x + 1)(x + 5)$ 2. $(x - 1)(x + 5)$
3. $(x + 1)(x - 5)$ 4. $(x - 1)(x - 5)$
5. $(x + 1)(x - 8)$ 6. $(x - 2)(x - 4)$
7. $(x - 2)(x + 4)$ 8. $(x - 1)(x + 36)$
9. $(x + 2)(x - 18)$ 10. $(x - 4)(x - 9)$

Exercise 2.4

1. $(x + 1)(3x + 1)$ 2. $(2x + 1)(3x + 1)$
3. $(x + 1)(3x + 2)$ 4. $(x + 2)(3x + 1)$
5. $(x - 1)(3x + 1)$ 6. $(x + 1)(3x - 1)$
7. $(x - 1)(3x - 1)$ 8. $(x + 1)(6x + 1)$
9. $(4x + 1)(x - 2)$ 10. $(2x + 1)(2x + 3)$
11. $(4x + 1)(x + 3)$ 12. $(2x + 1)(2x - 3)$
13. $(2x - 1)(2x + 3)$ 14. $(2x - 1)(2x - 3)$
15. $(4x + 1)(2x + 3)$ 16. $(4x - 1)(2x + 3)$
17. $(4x + 1)(2x + 3)$ 18. $(2x - 1)(4x + 3)$
19. $(2x + 5)(4x + 3)$ 20. $(3x + 5)(3x + 5)$
21. $(3x - 5)(2x - 5)$ 22. $(2x + 5y)(4x + 3y)$
23. $(3x + 5y)(3x + 5y)$ 24. $(3xy - 5)(2xy - 5)$

Exercise 2.5

1. $(x - y)(x + y)$ 2. $(a - 1)(a + 1)$
3. $(x - 3)(x + 3)$ 4. $(2y - 3)(2y + 3)$
5. $(5 - a)(5 + a)$ 6. $(6a - 7b)(6a + 7b)$
7. $(ab - xy)(ab + xy)$ 8. $(1 - 2c)(1 + 2c)$

Exercise 2.6

1. $3(3a + 5b)$ 2. $(1 - x)(1 + x)$
3. $2(3 - x)(3 + x)$ 4. $(2x + 1)(x - 1)$

Answer Key

5. $6x(x-3)$
6. $2(x-3)(x-3)$
7. $5x(2x-1)(2x+1)$
8. $3(x+2)(x-1)$
9. $2y(y-1)(y-1)$
10. $4xy(4y-x)$
11. $4xy(2y-x)(2y+x)$
12. $3(2a+x)(2b+y)$

Exercise 2.7

1. 3
2. $\dfrac{1}{3}$
3. $\dfrac{1}{27}$
4. $7x^{\frac{2}{5}}$
5. $10x^{\frac{4}{5}}$
6. $\dfrac{2}{5}$
7. 10
8. $\dfrac{10}{x^{\frac{4}{5}}}$
9. $\dfrac{2}{5}x^{\frac{4}{5}}$
10. $2y^{\frac{5}{4}}$
11. $\dfrac{2}{y^{\frac{1}{4}}}$
12. $x-1$
13. $\dfrac{1}{x^{\frac{1}{2}}y}$
14. $2+2x^{\frac{1}{2}}$
15. $4x^{\frac{4}{5}}$
16. $x-2x^{\frac{1}{2}}+1$
17. $x-2+\dfrac{1}{x}$
18. (a) $x=\dfrac{1}{5}$ (b) $x=\dfrac{1}{4}$ (c) $x=\dfrac{1}{3}$

Exercise 2.8

1. $\dfrac{1}{7}$
2. $\dfrac{20}{11}$
3. $\dfrac{17}{32}$
4. $2x-4$
5. $\dfrac{1}{2x+3}$
6. $y+z$
7. $\dfrac{x}{2}$
8. $\dfrac{1}{3y+5z}$
9. $\dfrac{3x+2}{5x-4}$
10. $\dfrac{3yz}{y-4z}$
11. $\dfrac{1}{x-y}$
12. $2x+3$
13. $\dfrac{1}{y+1}$
14. $\dfrac{x+1}{x-1}$
15. $\dfrac{x-1}{x+1}$
16. $\dfrac{1}{9}$
17. $\dfrac{y+x^2}{x+y^2}$
18. a^2+b^2

Exercise 2.9

1. $\dfrac{8}{27}$
2. $\dfrac{2}{3}$
3. $\dfrac{xy}{x^2+y^2}$
4. $\dfrac{x+1}{y(x-1)}$
5. $\dfrac{x^2y}{(x-1)(x+1)}$
6. $\dfrac{1}{x^2+y^2}$
7. $\dfrac{x+y}{xyz}$
8. $\dfrac{1}{x}$

Exercise 2.10

1. $\dfrac{19}{15}$
2. $\dfrac{x+1}{x}$
3. $\dfrac{7x}{12}$
4. $\dfrac{x^2-y^2}{xy}$
5. $\dfrac{2a}{(a-b)(a+b)}$
6. $\dfrac{2b}{(a-b)(a+b)}$
7. $\dfrac{6x-3}{x(x-1)}$
8. $\dfrac{3}{x(x-1)}$
9. $\dfrac{-x-6}{12}$
10. $\dfrac{17x-18}{12}$
11. $\dfrac{x+5y}{(x+y)(x+2y)}$
12. $\dfrac{x^2+y^2}{(x-y)(x+y)}$
13. $\dfrac{x^2+2xy-y^2}{(x-y)(x+y)}$
14. $\dfrac{2(a^2+b^2)}{(a-b)(a+b)}$
15. $\dfrac{4ab}{(a-b)(a+b)}$
16. $\dfrac{-3x^2+5xy-y^2}{(2x-y)(x+y)}$
17. $\dfrac{x^2-1}{xy}$
18. $\dfrac{2x-1}{x(x+1)(x-1)}$

Exercise 2.11

1. (a) $\dfrac{1}{3}$ (b) 2 (c) $\dfrac{1}{81}$
    (d) 6 (e) 65 (f) 16
    (g) $\dfrac{7}{6}$ (h) $\dfrac{1}{42}$ (i) 42

2. (a) $\dfrac{1}{y^2}$ (b) $\dfrac{1}{y^{\frac{1}{3}}}$ (c) $4x^{\frac{5}{2}}$
    (d) $24a^{\frac{7}{2}}b^{\frac{17}{6}}$ (e) $\dfrac{x^{\frac{1}{2}}}{y^2}$

3. (a) $6x^2-7x-49$ (b) x^2-2x+1
    (c) $x^4-2x^2y^2+y^4$ (d) x^2y^2-1
    (e) $xy+xb+ay+ab$
    (f) $6ac-10bc+9ad-15bd$

4. (a) $(3x-2)(x+1)$ (b) $(x+5)(x-10)$
    (c) $(2x-7)(x-7)$ (d) $2(5x-3)(5x+3)$
    (e) $xy(x-1)(x+1)$ (f) $3(2x+3)(x-4)$
    (g) $(a+b)(2c+d)$ (h) $(2c-d)(a-b)$
    (i) $2ab(2a-3b)(2a+3b)$

5. (a) $\dfrac{1}{ax-1}$ (b) $\dfrac{x+1}{x+6}$
    (c) $\dfrac{2}{x}$ (d) $\dfrac{2+x}{2-x}$
    (e) $\dfrac{3x^2}{2}$ (f) $\dfrac{y+x^2}{x(y+x)}$

6. (a) $\dfrac{2a^2}{(a+b)(a-b)}$ (b) $\dfrac{4x}{(x-1)(x+1)}$
    (c) $\dfrac{(x+1)^2}{(x-1)^2}$ (d) 1

7. (a) $\dfrac{c^2+cd-1}{c^2-d^2}$ (b) $\dfrac{c(c+d+1)}{c^2-d^2}$

8. $\dfrac{1}{2}p^{20}$

9. 7

10. (a) $3(2x-y)(2x+y)$
 (b) (i) x^2-6x+9 (ii) $p=3, q=1$

11. (a) -1 (b) $5k$

12. $\dfrac{x+3}{x(x+1)}$

13. $\dfrac{x+11}{(x-3)(x+4)}$

14. (a) $(a-2b)(1-3c)$ (b) $5t^2+6$

15. $\dfrac{x+7}{(x-3)(x+2)}$

16. (a) 8 (b) $\dfrac{1}{3}$ (c) -3

17. (a) 2 (b) $\dfrac{1}{3}$

18. (a) x^3-1 (b) $(a-b)(x-3y)$

19. $\dfrac{4(x+1)}{x(x+2)}$

20. (a) $\dfrac{x^{18}}{9}$ (b) 2^x

21. (a) $(2x-3)(2x+3)$ (b) $x(4x-9)$
 (c) $(4x-1)(x-2)$

22. (a) 26 (b) (i) x^2 (ii) x^3

23. (a) (i) $3x^2-4$ (ii) $\dfrac{x^2(a-1)}{x(a-1)}=x$
 (b) $7(x-3)(x+3)$

24. $\dfrac{x^2-6x+25}{4(x-3)}$

25. (a) $\dfrac{5}{2}$ or 2.5 (b) -1

26. $\dfrac{11x}{18}$

27. $9x^2$

28. $\dfrac{-18}{(2x+3)(x-3)}$

29. (a) 0 (b) 0.2 or $\dfrac{1}{5}$ (c) 0.6 or $\dfrac{3}{5}$

30. (a) $3x^2$ (b) -6

31. $\dfrac{5}{7}, 72\%, \sqrt{\dfrac{9}{17}}, \left(\dfrac{4}{3}\right)^{-1}$

32. $\dfrac{2}{c}$

Chapter 3

Core Skills

1. (a) 10500 cm² or 1.05 m² (b) 3014 cm or 30.14 m
2. (a) 4.67 (b) 501 (c) 0.01
 (d) 0.0106 (e) 516cm (f) 100 kilograms
 (g) 9200 (h) 1000
3. (a) $438.5 \leqslant 439 < 439.5$
 (b) $5665 \leqslant 5670 < 5675$
4. (a) 2×10^{10} mm³ (b) 2×10^7 cm³
 (c) 2×10^1 m³
5. (a) 0.5 cm² (b) 0.5 litres (c) 12000 g
6. (a) $700 \times 0.7 \simeq 500$ (b) $\dfrac{80}{80} \div \dfrac{200}{400} = 2$
 (c) $\dfrac{1000}{500} + \dfrac{400}{30} \simeq 15$
7. (a) 1.425×10^{-2} (b) 2.49×10^{12} to 3 s.f.
 (c) 1.107×10^3 (d) 6.002×10^{-5}
8. (a) 1.93 (b) 4.07 (c) 6.32
 (d) 2.82 (e) 2.02

9. (a) 2 : 25 (b) 1 : 400
 (c) 4 : 1 (d) 48 : 85
 (e) 2 : 1 : 10 (f) 37 : 500
10. (a) 1 : 0.0002 (b) 5000 : 1
11. Ana : €255 Bette : €102
12. 112 blocks
13. 15 days
14. 55km/h
15. 20.8 m/s to 3 s.f.
16. 20 15
17. 11.25 am

Exercise 3.1

1. (a) (i) 1.0 (ii) 3.0
 (b) (i) 26 (ii) 78
 (c) (i) 155.25 (ii) 181.25
 (d) (i) 1.08 (ii) 1.26 to 3 s.f.

Answer Key 375

2. (a) (i) 702.4 (ii) 728.8
   (b) (i) 20.88 (ii) 21.67
   (c) (i) 69.2875 (ii) 71.0775

Exercise 3.2

1. 8.4
2. 0.73
3. 50 cm
4. 1200 kg
5. 90
6. 9

Exercise 3.3

1. $90
2. $12.50
3. $1388
4. (a) Rs 480 (b) Rs 408

Exercise 3.4

1. (a) 19.8 cm (b) 22.2525 cm²
2. (a) upper bound of short pieces = 10.5 cm
   lower bound of large piece = 99.5 cm
   $\frac{99.5}{10.5} = 9.476$ so it may only be possible to cut 9 whole lengths.
   (b) upper bound of large piece = 100.5 cm
   lower bound of short pieces = 9.5 cm
   lower bound of 10 short pieces = 95 cm
   maximum left over = 100.5 − 95 = 5.5 cm
3. (a) 264 (b) 147 (c) 112
4. (a) 43.2 (b) 4.6 (c) 0.0719
5. (a) $20493 (b) $5508
6. (a) €34 (b) €40
7. (a) 1 : 13 or 13 : 15 (b) 12600
   (c) 41.0% (d) 20000 (e) 14000
8. (a) (i) $132 (ii) 110%
   (b) $185 (c) 48 cm
   (d) (i) 19.8 km (ii) 414 km

9. 3.23%
10. (a) 115 125 (b) 2400 m²
11. 3 h 20 min
12. (a) 5.8 × 10⁸ (b) 98 (c) 10200
13. 50.1225 cm²
14. 201.25 m²
15. 38
16. $231.13
17. (a) 1045.28 (b) 1000
18. (a) $450 (b) (i) $120 (ii) $80
    (c) (i) $441 (ii) 5.125 or 5.12 or 5.13
19. 20
20. £3000
21. 62225000 or 6.2225 × 10⁷ or 62.225 million
22. (a) 2.67 × 10⁻² (b) 0.0267
23. 50.1225 cm²
24. (a) (i) $6000 (ii) 15%
    (b) $11200 (c) (i) $7500 (ii) $\frac{9}{80}$
    (d) $8640
25. (a) 06 41 (b) $204
26. (a) (i) $250 (ii) $2600 (iii) 6.12%
    (b) (i) 12 m (ii) 7 h 12 min
    (iii) 2.78 km/h (iii) 16 07
    (c) 25000 or 2.5 × 10⁴
27. 237.5 242.5
28. (a) $\frac{0.003 \times 3000}{(10+20)^2}$ (b) 0.01 or $\frac{1}{100}$
29. (a) 0.701 (b) £190
30. 135 165

Chapter 4

Core Skills

1. (a) $x = 2.5$ or $\frac{5}{2}$ (b) $x = 6$ (c) $x = 2$
   (d) $x = 3.5$ or $\frac{7}{2}$ (e) $x = 4$
   (f) $x = 3.5$ or $\frac{7}{2}$ (g) $a = 0$
2. (a) $x = sh$ (b) $x = \frac{A}{l}$
   (c) $x = \frac{V-c}{y}$ (d) $x = \frac{u^2 - v^2}{2a}$
   (e) $x = \frac{2A}{a+b}$ (f) $x = \frac{b}{c}$
3. (a) $5n - 3$ (b) $5n - 15$
   (c) $n^2 + 1$ (d) $-2n + 12$ (e) n^2
4. (a) (i) $\frac{9}{2}$ or 4.5 (ii) 5202
   (b) (i) −9 (ii) 9000
   (c) (i) −9 (iii) −801
5. (a) $x = 11, y = -4$
   (b) $x = 2, y = 3$

(c) $x = 0, y = 2$
(d) $x = 5, y = 4$

Exercise 4.1

1. $x = 3$
2. $x = 6$
3. $x = 14$
4. $x = 2$
5. $x = -21$
6. $x = \dfrac{5}{8}$
7. $x = -\dfrac{1}{2}$
8. $x = -1$

Exercise 4.2

1. $x = -2$ or -3
2. $x = -2$ or 1
3. $x = 5$ or 1
4. $x = -\dfrac{1}{2}$ or 1
5. $x = -1$ or 1
6. $x = 1$ or $\dfrac{5}{2}$
7. $x = -1$ or $\dfrac{4}{3}$
8. $x = \dfrac{1}{2}$ or $x = \dfrac{3}{2}$
9. $x = -\dfrac{3}{2}$ or $\dfrac{3}{2}$
10. $x = 3$ or 5
11. $x = -2$ or 2
12. $x = -1$
13. $x = 1$
14. $x = 1$ or $\dfrac{8}{3}$

Exercise 4.3

1. $x = -6.14$ or 1.14
2. $x = -1.14$ or 6.14
3. $x = -3.12$ or -0.214
4. $x = -0.618$ or 1.62
5. $x = -1.56$ or 2.56
6. $x = 0.314$ or 3.19

Exercise 4.4

1. $(x - 3)^2 - 8$
2. $\left(x + \dfrac{5}{2}\right)^2 - \dfrac{17}{4}$
3. $\left(x - \dfrac{3}{2}\right)^2 - \dfrac{21}{4}$
4. $2(x + 1)^2 - 7$
5. $3(x - 1)^2 - 7$
6. $2\left(x - \dfrac{3}{4}\right)^2 + \dfrac{7}{8}$
7. $2\left(x + \dfrac{5}{4}\right)^2 - \dfrac{65}{8}$

Exercise 4.5

1. $x = -1 \pm \sqrt{2}$
2. $x = \dfrac{1 \pm \sqrt{33}}{4}$
3. $x = 5$ or -1
4. $x = \dfrac{-2 \pm \sqrt{5}}{2}$ or $-1 \pm \dfrac{\sqrt{5}}{2}$
5. $x = \dfrac{7 \pm \sqrt{41}}{2}$
6. $x = \dfrac{1}{2}$ or 1

Exercise 4.6

1. $x = -0.618$ or 1.62
2. $x = -7.87$ or -0.127
3. $x = 1$
4. $x = -2$ or $\dfrac{3}{2}$
5. $x = 1.31$ or -0.306
6. 0.382 or 2.62
7. $x = -0.303$ or 3.30
8. $x = -1$ or $-\dfrac{1}{3}$
9. $x = -0.215$ or 1.55
10. $x = 1$
11. $x = -42.6$ or -0.103
12. $x = -\dfrac{5}{6}$ or 3

Exercise 4.7

1. $x = -1$ or -2
2. $x = \dfrac{1}{2}$ or 2
3. $x = \dfrac{1}{2}$ or 2
4. $x = -3$ or 2
5. $x = -2$ or 1
6. $x = \dfrac{1}{2}$ or 1
7. $x = 1$
8. $x = -\dfrac{1}{3}$ or 1
9. $x = 2$ or $-\dfrac{1}{2}$
10. $x = 1$ or 4
11. (a) $(x - 4)(x - 5) = 12$
    (b) $x = 1$ or 8
    (c) 4 cm by 3 cm
12. 8, 9
13. 12, 14
14. (a) $x(x + 3) = 40$
    $x^2 + 3x - 40 = 0$
    (b) $x = -8$ or 5
    (c) length 8 cm, breadth 5 cm

Exercise 4.8

1. (a) 1, 4, 9, 16 (b) −1, 2, 7, 14
   (c) −1, 0, 3, 8 (d) −3, −2, 1, 6
   (e) $\dfrac{1}{2}, \dfrac{2}{3}, \dfrac{3}{4}, \dfrac{4}{5}$ (f) $\dfrac{1}{2}, \dfrac{1}{5}, \dfrac{1}{10}, \dfrac{1}{17}$

Answer Key 377

(g) 0, 1, 4, 9   (h) 1, 8, 27, 64
(i) 3, 11, 30, 67   (j) 0, 6, 24, 60
(k) 3, 6, 9, 12
2. (a) $n(n+1)$ (b) $n^2 - n$ (c) $(n+1)^2$
(d) $2n^2$ (e) $\dfrac{n^2}{n+1}$ (f) $\dfrac{3n}{n+3}$
3. (a) 19, 99, 201
   (b) (i) $n = 4$ (ii) $n = 11$
4. (a) 24, 63, 10200
   (b) (i) 1   (ii) 9
       (iii) 10   (iv) 16

Exercise 4.9

1. $y = 2, x = -1$ 2. $y = \dfrac{1}{2}, x = \dfrac{1}{2}$
3. $y = -\dfrac{1}{6}, x = \dfrac{5}{6}$ 4. $y = 2, x = 2$

Exercise 4.10

1. $x = 1, y = 3$ 2. $x = -1, y = -1$
3. $x = \dfrac{29}{3}, y = \dfrac{5}{3}$ 4. $x = -2, y = 10$
5. $x = \dfrac{3}{10}, y = \dfrac{9}{10}$ 6. $x = \dfrac{9}{14}, y = \dfrac{5}{7}$

Exercise 4.11

1. $x = 0, y = 4$ 2. $x = 10, y = -1$
3. $x = 3, y = 3$ 4. $x = -10, y = -5$
5. $x = 18, y = -1$ 6. $x = 5, y = -\dfrac{2}{13}$
7. $x = -\dfrac{3}{5}, y = \dfrac{19}{5}$ 8. $x = 7, y = -2$
9. $x = -4, y = 2$ 10. $x = -\dfrac{9}{7}, y = \dfrac{52}{7}$

Exercise 4.12

1. (a) $x > -\dfrac{1}{2}$

(b) $x > 1$

(c) $x \leqslant -\dfrac{5}{2}$

(d) $x \geqslant 3$

(e) $x > -6.4$

(f) $-\dfrac{7}{2} < x < 2$

(g) $2 \leqslant x < 7$

(h) $-5 < x < -3$

2. (a) $x < -2$ $\{\ldots -5, -4, -3\}$
   (b) $x \leqslant -2$ $\{\ldots -5, -4, -3, -2\}$
   (c) $x < \dfrac{3}{2}$ $\{\ldots -2, -1, 0, 1\}$
   (d) $-3 < x < \dfrac{1}{2}$ $\{-2, -1, 0\}$
   (e) $-1 \leqslant x < 4$ $\{-1, 0, 1, 2, 3\}$
   (f) $-4 \leqslant x \leqslant -1$ $\{-4, -3, -2, -1\}$
   (g) $-3 < x < 1$ $\{-2, -1, 0\}$

Exercise 4.13

1. (a) $T = 33N$ (b) the cost of one item
2. $y = \dfrac{4}{9}x^2$ 3. (a) $\dfrac{10}{7}$ (b) $D = \dfrac{21}{V}$
4. (a) $a = \dfrac{4}{3}\sqrt{b}$ (b) $\dfrac{16}{3}$ (c) 3.52
5. (a) $m = 3l^3$ (b) 681 g
6. (a) $F = \dfrac{10^{38}}{d^2}$ (b) $F = 10^8$ Newtons

Exercise 4.14

1. $b = \sqrt{12V} + a$
2. $r = \dfrac{S-a}{S}$ or $r = 1 - \dfrac{a}{S}$
3. $z = \sqrt{x^2 + y^2 - 2xyC}$
4. $a = \dfrac{y}{4x}$
5. $a = \dfrac{1}{2}\left(\dfrac{2s}{n} - (n-1)d\right)$
6. $V = \sqrt{\dfrac{2E}{m} + u^2}$
7. (a) $w = \dfrac{W(1-e)}{e}$ (b) $W = \dfrac{ew}{1-e}$
8. $u = \sqrt{v^2 - 2as}$

9. $l = \dfrac{t^2 g}{4\pi^2}$

10. $A = \dfrac{b^2 + c^2 - a^2}{2bc}$

Exercise 4.15

1. (a) $x = \dfrac{10}{7}$ (b) $x = -7$

2. (a) $x = -\dfrac{5}{3}$ or $\dfrac{5}{3}$ (b) $x = -4.5$ or 3
   (c) $x = -3.10$ or 0.431

3. $(x-1)^2 - \dfrac{21}{4}$

4. (a) nth term $= \dfrac{2n-1}{3n}$ (b) nth term $= \dfrac{(n-1)^2}{n^2}$

5. (a) $y = \dfrac{30}{\sqrt{x}}$ (b) (i) $y = 4.74$ (ii) $x = 2.25$

6. (a) $B = A - Dc$ (b) $b = \dfrac{aB}{A}$

7. $-2.5 < x \leq 1$

8. (a) $x = 4, y = -1$ (b) $x = -\dfrac{44}{3}, y = \dfrac{1}{3}$

9. $\dfrac{1}{9}, 0, \dfrac{1}{25}, \dfrac{1}{9}, \dfrac{9}{49}$

10. (a) (i) $k > 2$ (ii) $t = 30$
    (b) $x = 18.5, y = 10.5$

11. $x = 8, y = 6$

12. (a) $(2x - 3)(2x + 3)$ (b) $x(4x - 9)$
    (c) $(4x - 1)(x - 2)$

13. $x = -5.2$

14. (a) (i) $\dfrac{40}{x}$ (ii) $\dfrac{40}{x+2} = \dfrac{40}{x} - 1$
    (iii) $-10, 8$ (iv) \$8
    (b) (i) $m = n + 2.55, 2m = 5n$
    (ii) $m = 4.25, n = 1.7$

15. (a) $wf = 300000$ (b) 500

16. (b) (i) $p = 25, q = 40$
    (ii) $x = n^2, y = (n+1)^2, z = (n+1)^2 + n^2 - 1$,
    or $z = 2n(n+1)$
    (c) (i) $\dfrac{2}{3} + f + g = 4$
    (ii) $\dfrac{2}{3} \times 2^3 + f \times 2^2 + g \times 2 = 12$
    (iii) $f = 2, g = \dfrac{4}{3}$ (iv) 880

17. (a) $x = 3.6$ (b) $x = -11.7$ or -0.3

18. $x > -\dfrac{4}{7}$ or $x > -0.571$

19. (a) $x = -12$ (b) $x > -1$

20. 3200

21. 7.5

22. (a) $y = \dfrac{k}{x^2}, k = 4.8 \times 5^2 = 120$ (b) $y = 30$
    (c) $x = 3.46$ (d) $x = 4.93$
    (e) y is divided by 4 (f) x increased by 25%
    (g) $x = \sqrt{\dfrac{120}{y}}$

23. $c = \dfrac{b^2 + 5}{3}$

24. 210

25. $d = \sqrt[3]{2(c - 5)}$

26. (a) p varies inversely as the square root of q
    (b) $q = 9$

27. 5×10^4

28. (a) $p = -7, q = 512, r = \dfrac{8}{9}, s = 81, t = 2187, u = -2106$
    (b) (i) $9 - 2n$ (ii) n^3 (iii) $\dfrac{n}{n+1}$
    (iv) $(n+1)^2$ (v) 3^{n-1} (vi) $(n+1)^2 - 3^{n-1}$
    (c) 393 (d) 12

29. (a) $x = 13.5$ (b) $x = -1$ or 4

30. (a) 2870 (b) $(n + 3)^2 + 1$

31. $x < -23.5$

32. (a) 3×10^{11} (b) 5×10^6

33. (a) (i) $(x + 4)(x - 5)$ (ii) $x = -4$ or 5
    (b) $x = -0.55$ or 1.22
    (c) (i) $(m - 2n)(m + 2n)$ (ii) -12
    (iii) $y = 20x + 5$ (iv) $n = \sqrt{\dfrac{m^2 - y}{4}}$
    (d) (i) $k = 4$ or -4
    (ii) $n(m - 2n)(m + 2n)(m^2 + 4n^2)$

34. $x = 4, y = -3$ 35. $t = 0.128$

36. $x > -\dfrac{4}{25}$

37. $\sqrt{\left(\dfrac{6}{T}\right)^2 - 1}$ or $\sqrt{\dfrac{36}{T^2} - 1}$

38. (a) $x^2(a + b)$ (b) $x = \sqrt{\dfrac{p^2 + d^2}{a + b}}$

39. 1.25

40. $y = (9(1 - x))^2$ 41. $x = 2, y = -6$

42. $c = 2$

43. (a) $m = -13$
    (b) (i) $y = 0.5$
    (ii) $\dfrac{x + 11}{(x - 1)(x + 3)}$ (iii) $x = -\dfrac{1}{3}$
    (c) $q = \dfrac{p + t}{p}$

Answer Key 379

Chapter 5

The answers to questions in this chapter which require drawing and measuring have been given as a range of values. You should get your answers within the given range, or very close; if you do not, sharpen your pencil and try again with more attention to accuracy! The answers to many of the questions can be arrived at by different routes.

Where reasons are required one explanation has been given to help you. Yours could be different and still be correct.

Core Skills

1. (a) $x = 20$ (b) $x = 36$ (c) $x = 22.5$
2. 55–56°, 41–42°, 82–83°
3. 140°   4. 1980°
5. (a) rectangle or parallelogram (b) kite
   (c) trapezium   (d) rhombus
6. (a) b and d   (b) c and e
7. (a) rotational symmetry, order 2.
       2 lines of symmetry
   (b) rotational symmetry, order 2.
       No lines of symmetry
   (c) rotational symmetry, order 5.
       5 lines of symmetry
   (d) rotational symmetry, order 4.
       4 lines of symmetry
8. Circle, centre A, radius 5 cm.
9. (regular) tetrahedron
10. (a) for example, $\angle TAP$ and $\angle ABR$
    (b) for example, $\angle PAB$ and $\angle ABC$
    (c) for example, $\angle ACB$ and $\angle SCV$
    (d) $\angle PAB = \angle RBU = 50°$ (corresponding angles)
        $\angle BAC = 180 - 50 - 55 = 75°$ (angles on a straight line)

Exercise 5.1

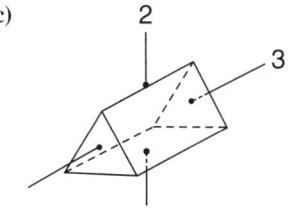

1. (a), (c)
   (b) 4
   (d) 4

2. (a), (b) and (c)

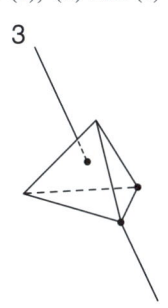

3. (a) One axis of rotational symmetry, order infinity.
       Infinite number of planes of symmetry.
   (b) One axis of rotational symmetry, order 4.
       4 planes of symmetry.
   (c) Rotational symmetry, order 6 about one axis.
       6 more axes, each with rotational symmetry order 2.
       7 planes of symmetry.

Exercise 5.2

1. $a = 20$ (isosceles triangle)
   $b = 90$ (angles between radius and tangent)
   $c = 140$ (isosceles triangle)
   $d = 40$ (angles on a straight line)
   $e = 50$ (angle sum of a triangle)

2. Let $BS = x$
   $AU = AS = 4$ cm (tangents from a point outside the circle are equal)
   $CT = CU = 6$ cm (tangents from a point outside the circle are equal)
   $BT = BS = x$ cm (tangents from a point outside the circle are equal)
   $2x + 2 \times 4 + 2 \times 6 = 29$
   $x = 4.5$ $a = 10.5$ cm

3. $a = 35$ (isosceles triangle)
   $b = 110$ (angle sum of a triangle)

$c = 55$ ($\angle ABC = 90°$, angle in a semicircle)
$d = 55$ (isosceles triangle)
$e = 25$ (isosceles triangle)
$f = 25$ (alternate angles)
$g = 45$ (angle sum of a triangle)

Exercise 5.3

1. (a) $\angle BAO = 40°$ (isosceles triangle)
   $\angle AOB = 100°$ (angle sum of a triangle)
   $x = 80$ (angles on a straight line)
   (b) $\angle OAB = \angle OCB = 90°$ (angle between radius and tangent)
   $\angle AOC = 130°$ (angle sum of a quadrilateral)
   $x = 65$ (the angle at the centre is twice the angle at the circumference)

2. $a = 70$ (cyclic quadrilateral)
   $b = 70$ (angles in the same segment)
   $c = 20$ (angle sum of a triangle)
   $d = 20$ (angle sum of a triangle)

3. $a = 20$ (isosceles triangle)
   $b = 20$ (alternate angles)
   $c = 180 - (180 - 2 \times 20)$ (angle sum of an isosceles triangle and angles on a straight line)
   $c = 40$
   $d = 40$ (angles in the same segment)
   $e = 180 - (20 + 40)$ (angle sum of a triangle and vertically opposite angles)
   $e = 120$
   $f = 20$ (angles in the same segment)

4. $a = 35$ (corresponding angles)
   $b = 90$ (angle in a semicircle)
   $c = 55$ (angle sum of a triangle)
   $d = 90$ (angle in a semicircle)
   $e = 45$ (isosceles triangle)
   $f = 45$ (alternate angles)

5. $\angle ADC = 50°$ (angles on a straight line)
   $\angle ABC = 130°$ (angles on a straight line)
   $\angle ADC + \angle ABC = 180°$
   therefore $ABCD$ is a cyclic quadrilateral (opposite angles add up to $180°$)

Exercise 5.4

1. 13   2. $x = 74$ 3. $a = 20, b = 70$

Exercise 5.5

1. $x = 60$ $y = 65$
2. $w = 30$
   $x = 75$
   $y = 60$
   $z = 50$
3. (a) $a = 70$
   $b = 140$
   $c = 70$
   $d = 40$
   (b) $b + d = 180$
   therefore $AOCT$ is a cyclic quadrilateral (opposite angles of a cyclic quadrilateral are supplementary)
   (c) On the midpoint of OT
4. One axis of symmetry, order 6.
   6 planes of symmetry
5. (a) $AE = BD$ (equidistant from centre of circle)
   (b) All equal and half AE and BD (perpendicular bisector of a chord passes through centre of circle)
6. $\angle OCB = 50°$ (vertically opposite angles)
   $\angle CBO = 90°$ (angle between tangent and radius)
   $\angle COB = 40°$ (angle sum of a triangle)
   $\angle AOB = 140°$ (angles on a straight line)
   $x = 20$ (isosceles triangle)
7. $a = 55$ (isosceles triangle)
   $b = 70$ (angle sum of a triangle)
   $c = 125$ (cyclic quadrilateral)
   $d = 110$ (cyclic quadrilateral)
   $e = 27.5$ (isosceles triangle)
8. (a) $x = 78$ (alternate angles)
   $y = 144$ (opposite angles of a cyclic quadrilateral)
   $z = 102$ (opposite angles of a cyclic quadrilateral)
   (b) Draw DE produced to F.
   $\angle AEF = 78°$ (angles on a straight line)
   $\angle AEF \neq \angle EAC$, therefore ED and AC are not parallel
   (c) Reflex angle $EOC = 2 \times 144 = 288$ (angle at the centre is twice the angle at the circumference)
   Therefore $\angle EOC = 72°$ (angles round a point)
   (d) $51°$

Answer Key 381

9. (a)

[Diagram: semicircle with diameter PL, point M on arc, point Q inside, arcs constructed]

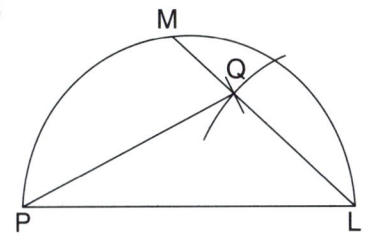

(b) 108–111° (c) 3.2–3.5 cm
(d) The angle in a semicircle is a right angle
10. (a) 44°   (b) 158°
11. (a), (b) (i)

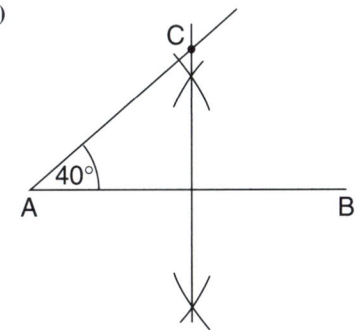

[Diagram: angle of 40° at A with ray to C, perpendicular bisector/construction with B on horizontal ray]

(ii) 45.0 to 46.5 m
(iii) 320°
12. $p = 54$ $q = 51$ $r = 78$
13. $w = 30$ $x = 22$
    $y = 30$ $z = 52$
14. (i) 33°   (ii) 24°
    (iii) 57° (iv) 123°
15. (a) $w = 26$ $x = 128$
    (b) $\angle OQT = \angle OPT = 90°$ (angle between tangent and radius)
    $y = 360 - 90 - 90 - 128$ (angle sum of a quadrilateral)
    $y = 52$
16. (a) 54°   (b) 42°   (c) 78°
17. (a) (i) 69   (ii) 57
        (iii) 72   (iv) 15
    (b) 135
18. (a) 65°   (b) 25°
    (c) 58°   (d) 206°
19. Circle, centre B, radius BD
20. (a) 58°   (b) 32°
    (c) 58°   (d) 24°

Chapter 6

Core Skills

1. (a) [Graph: line $y = x + 1$ through (-1, 0) and (0, 1)]

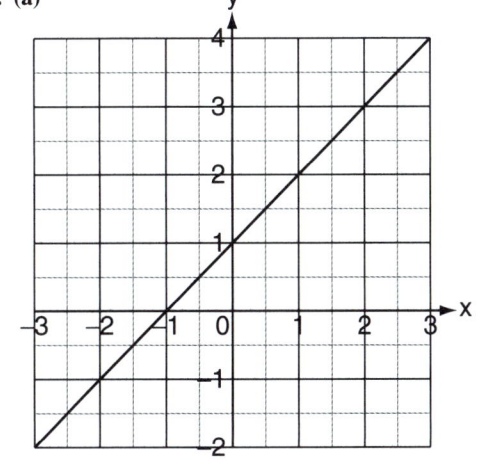

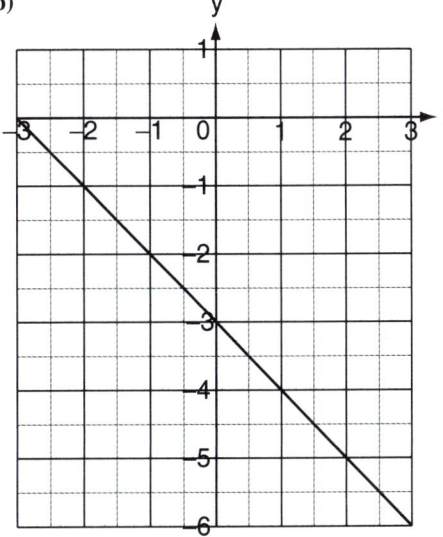

   (b) [Graph: line $y = -2x - 3$ with negative slope through (0, -3)]

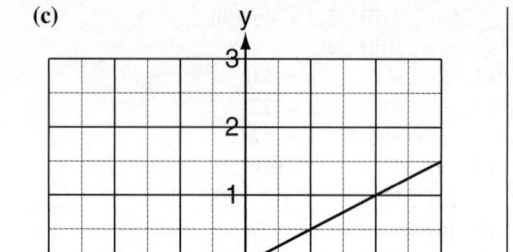

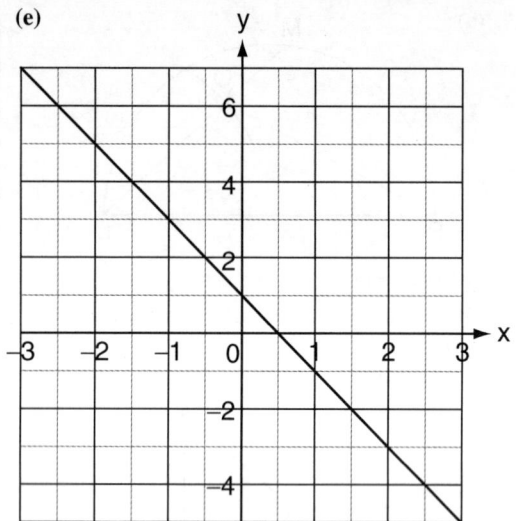

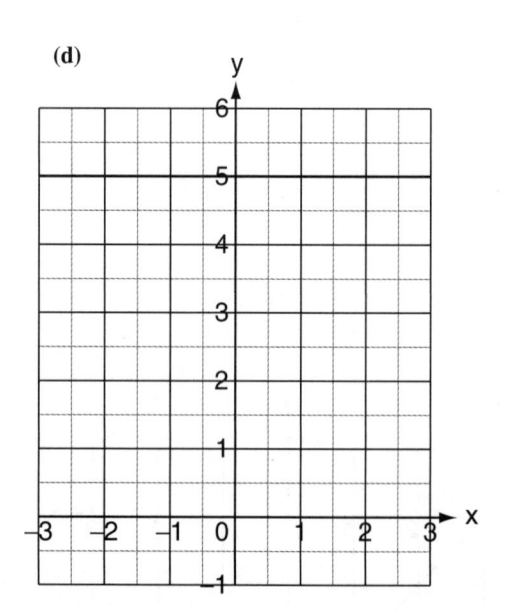

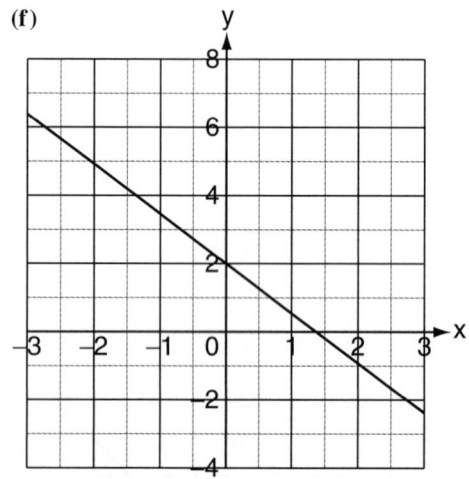

(c), (d), (e), (f) [graphs]

2. y-intercept $= \dfrac{3}{5}$, gradient $= -\dfrac{2}{5}$

3. (a) (i) $m = 2$ (ii) $c = -3$
   (b) (i) $m = -1$ (ii) $c = 1$

4. (a) $y = -2$ (b) $x = 3$
   (c) $y = -x$ (d) $y = -2x - 5$

5.

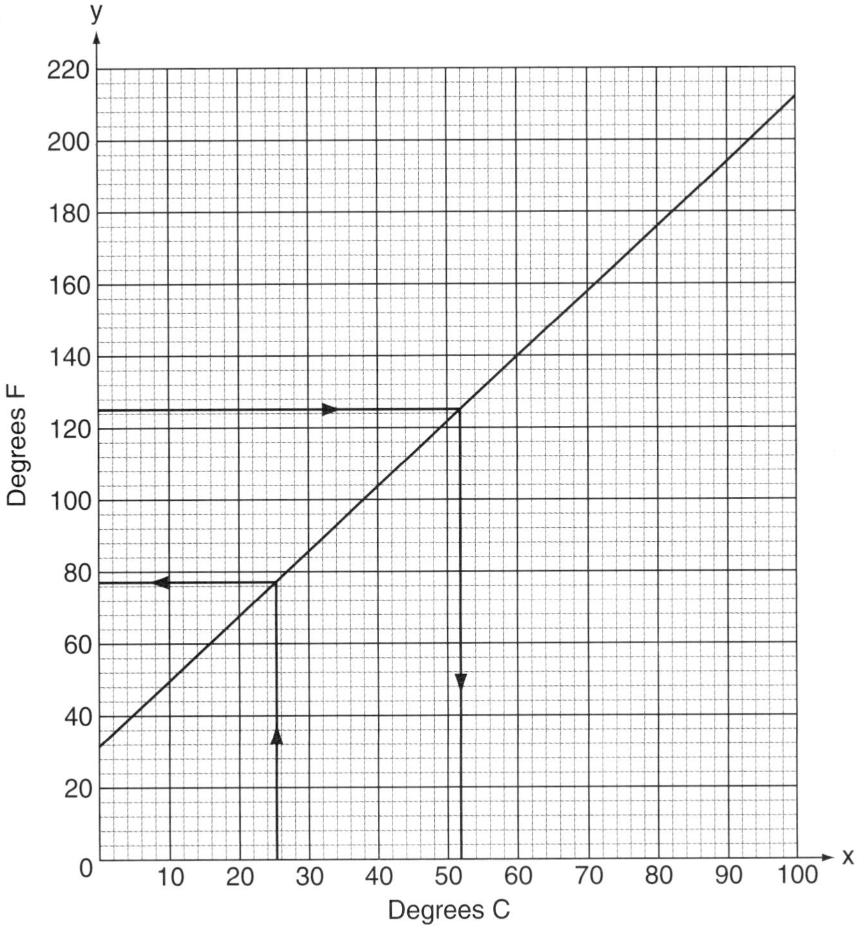

(b) **(i)** 76° to 78°F **(ii)** 51° to 53°C

6. **(a)** 0, −2, 0 **(b)**

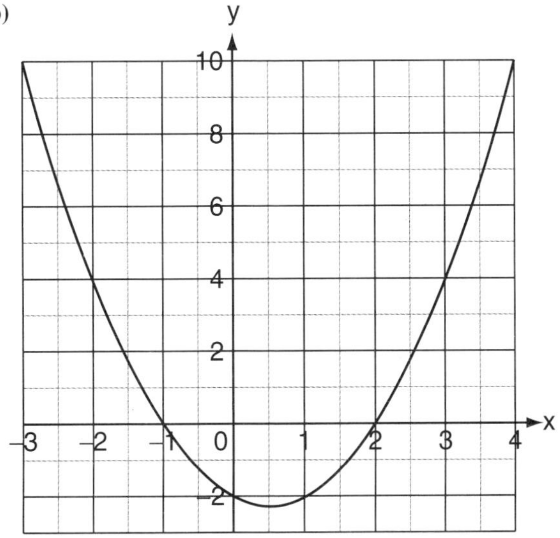

(c) $x = 0.5$

7.

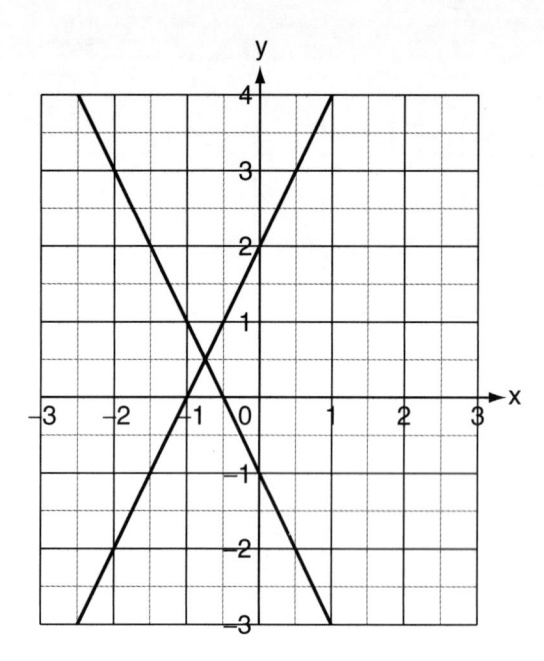

$x = -0.75, y = 0.5$

Exercise 6.1

1. (a) 7.75 cm²   (b) 9 cm²
2. (a) 45 minutes   (b) $2\frac{2}{3}$ km/h
   (c) before visit: 9 km/h
   after visit: 7.5 km/h
   (d) approximately 7 27 am
   (e) 1.7 km/h
   (f) approximately 7 12 am and 7 51 am
3. (a) 1050 km/h²   (b) 890 km

Exercise 6.2

1.

Question	Function	x	−3	−2	−1	0	1	2	3
(a)	$y = x^2$	y	9	4	1	0	1	4	9
(b)	$y = -x^2$	y	−9	−4	−1	0	−1	−4	−9
(c)	$y = (x + 1)^2$	y	4	1	0	1	4	9	16
(d)	$y = x^2 - 4$	y	5	0	−3	−4	−3	0	5
(e)	$y = x^2 - 2x - 8$	y	7	0	−5	−8	−9	−8	−5

(a) (ii)

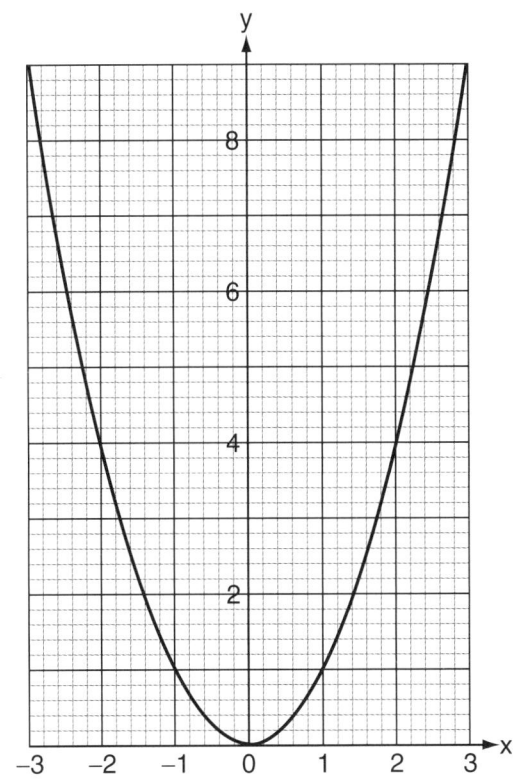

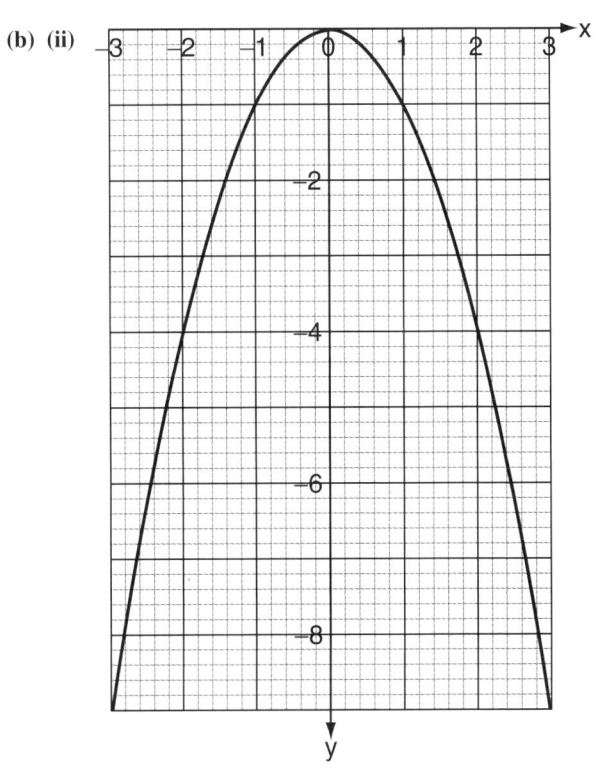

(b) (ii)

(c) (ii)

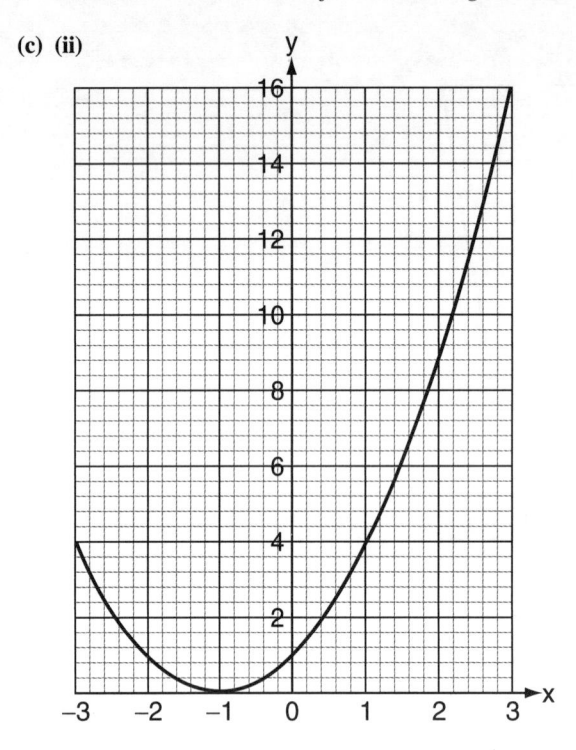

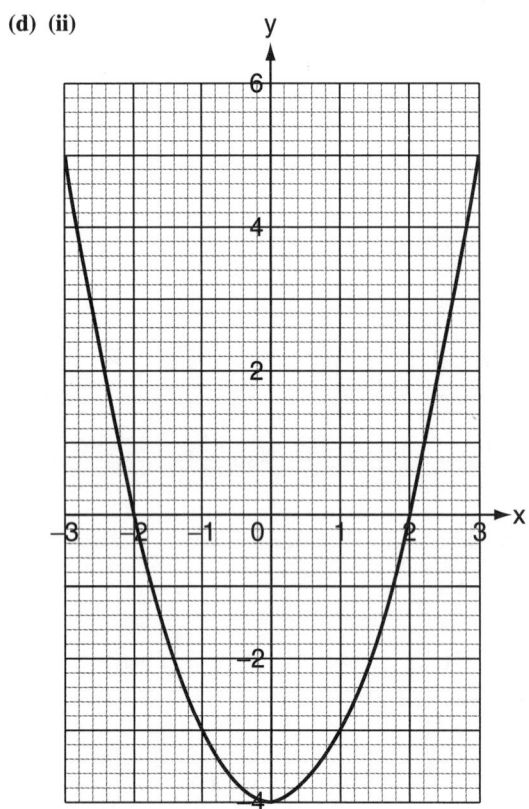

(d) (ii)

(e) (ii)

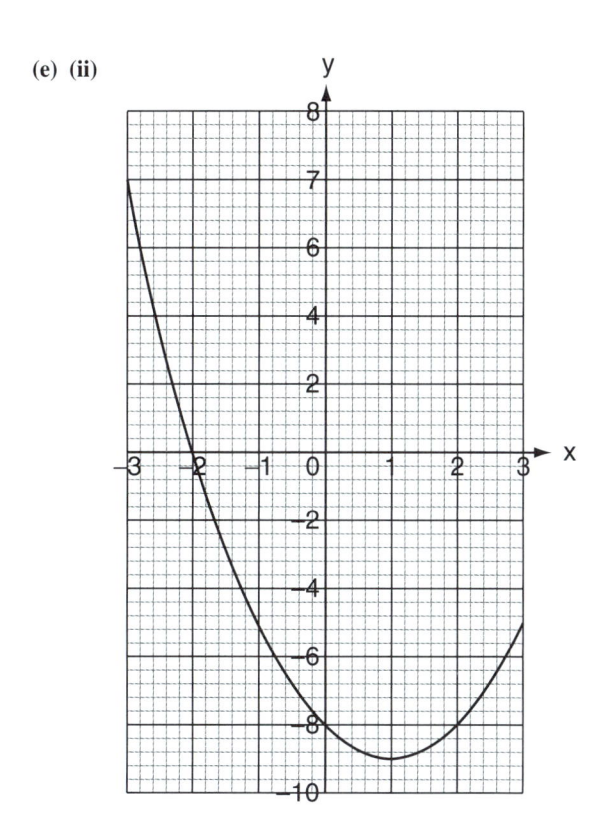

2.

Question	Function	x	−3	−2	−1	0	1	2	3
(a)	$y = x^3$	y	−27	−8	−1	0	1	8	27
(b)	$y = -x^3$	y	27	8	1	0	−1	−8	−27
(c)	$y = (x+1)^3$	y	−8	−1	0	1	8	27	
(d)	$y = x^3 - 4$	y	−31	−12	−5	−4	−3	4	23

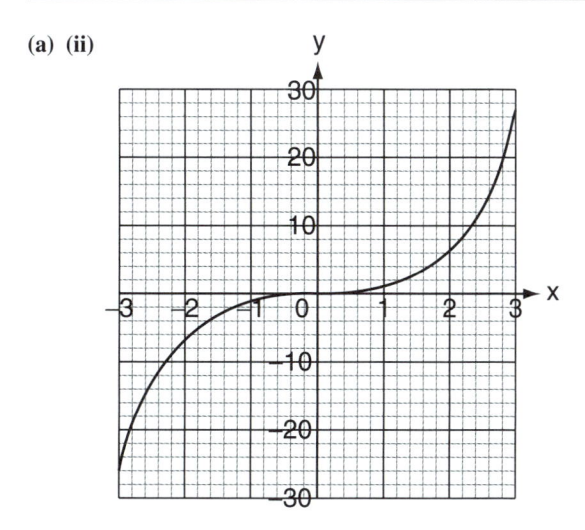

(a) (ii)

(b) (ii)

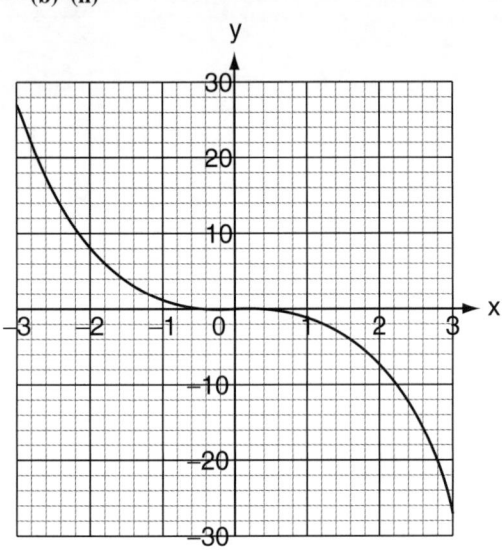

(d) (ii)

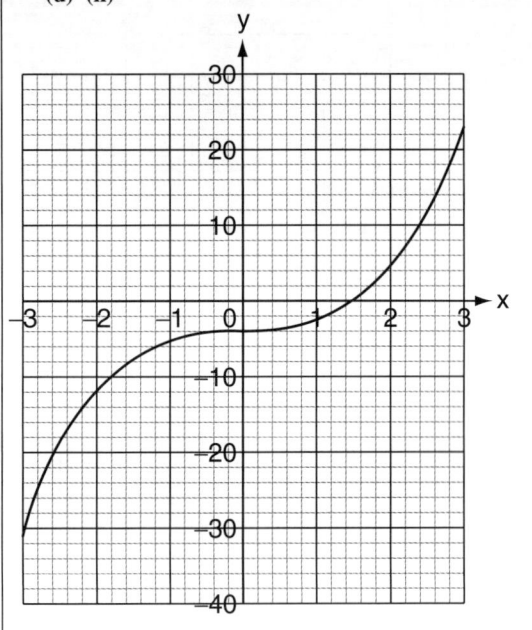

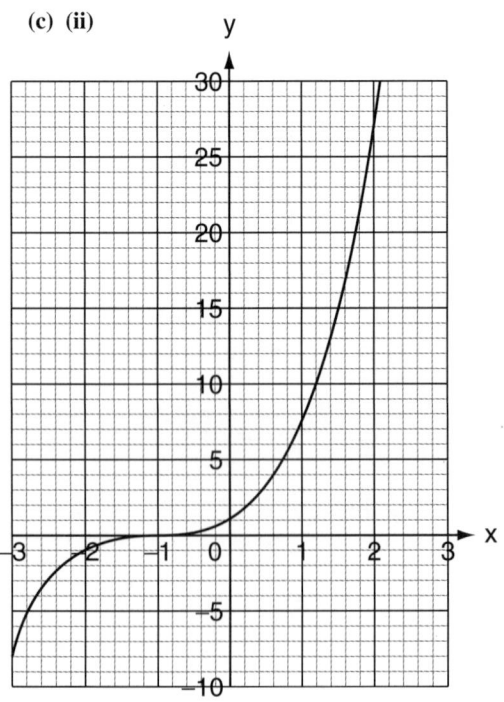

(c) (ii)

Answer Key 389

3. (a) (i)

x	−4.5	−4	−3	−2.5	−2	−1.5	−1	−0.5	0	0.5	1
$y = x^3 + 4x^2 + 1$	−9.1	1	10	10.4	9	6.6	4	1.9	1	2.1	6

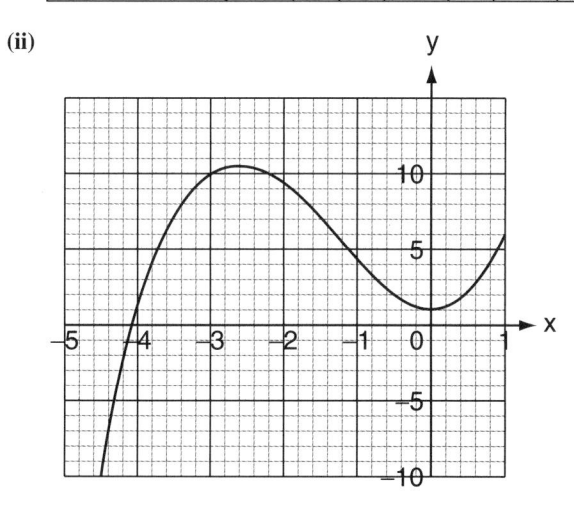

(ii)

(b) (i)

x	−2.5	−2	−1.5	−1	−0.5	0	0.5	1	2
$y = x^3 + x^2 - 2x$	−4.4	0	1.9	2	1.1	0	−0.6	0	8

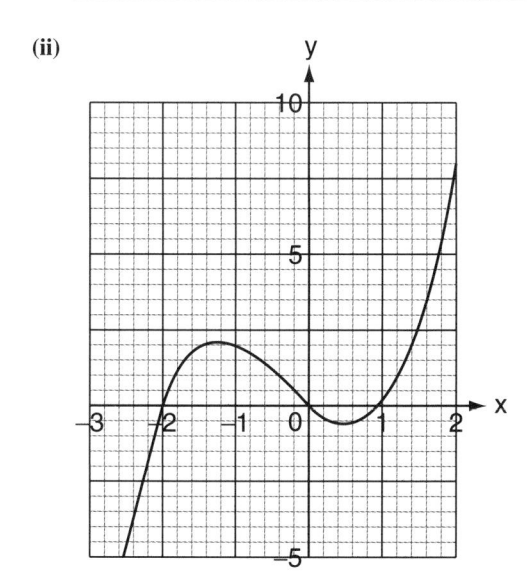

(ii)

(c) (i)

x	−2.5	−2	−1.5	−1	−0.5	0	0.5	1
$y = x^2(x + 2)$	−3.1	0	1.1	1	0.4	0	0.6	3

(ii)

(d) (i)

x	−2.5	−2	−1.5	−1	−0.5	0	0.5	1
$y = -x^2(x+2)$	3.1	0	−1.1	−1	−0.4	0	−0.6	−3

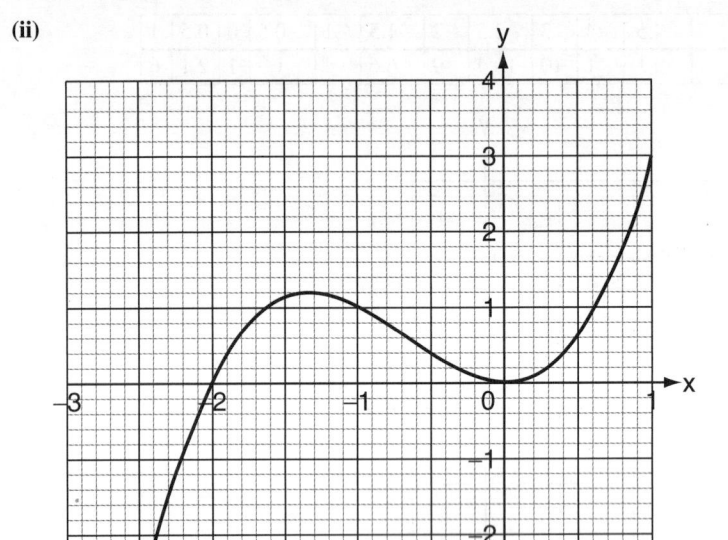

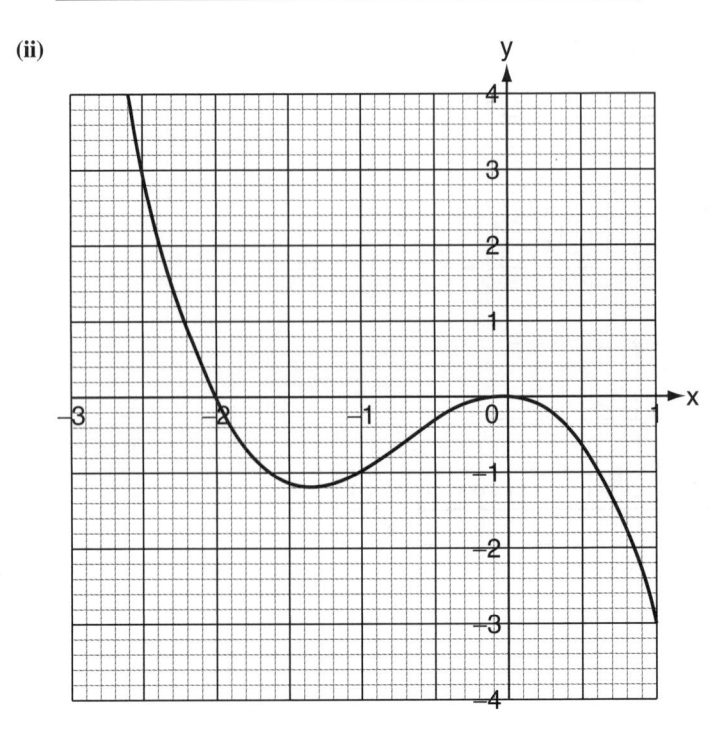

(ii)

4.

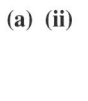

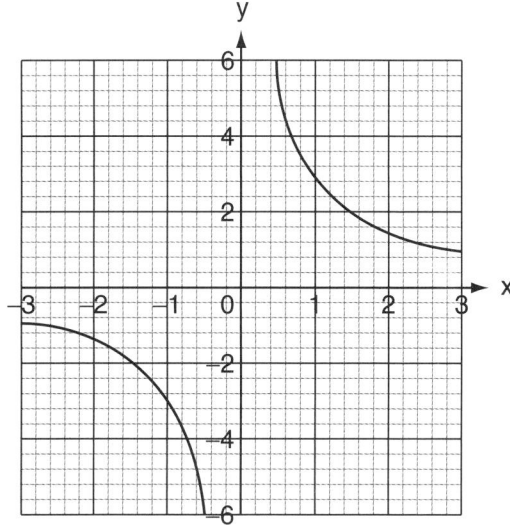

Question	Function	x	−3	−2.5	−2	−1.5	−1	−0.5	0.5	1	1.5	2	2.5	3
(a)	$y = \dfrac{3}{x}$	y	−1	−1.2	−1.5	−2	−3	−6	6	3	2	1.5	1.2	1
(b)	$y = \dfrac{9}{x^2}$	y	1	1.4	2.3	4	9	36	36	9	4	2.3	1.4	1
(c)	$y = x^2 + \dfrac{3}{x}$	y	8	5.1	2.5	0.3	−2	−5.8	6.3	4	4.3	5.5	7.5	10

(a) (ii)

(b) (ii)

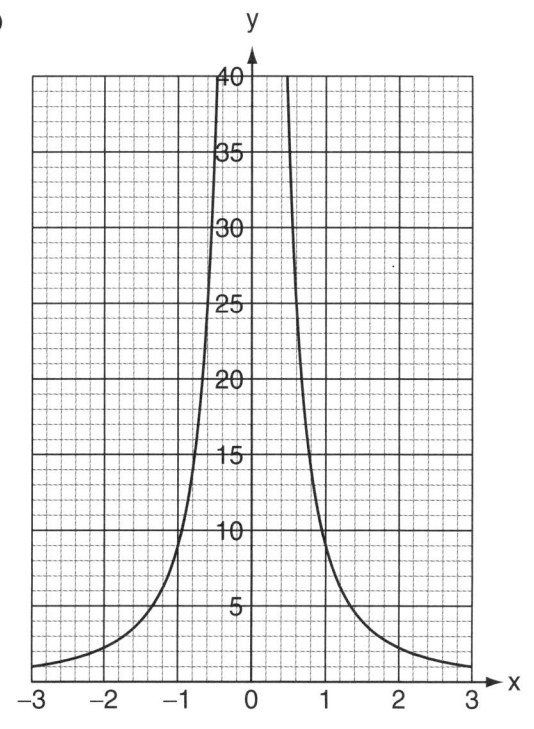

(c) (ii)

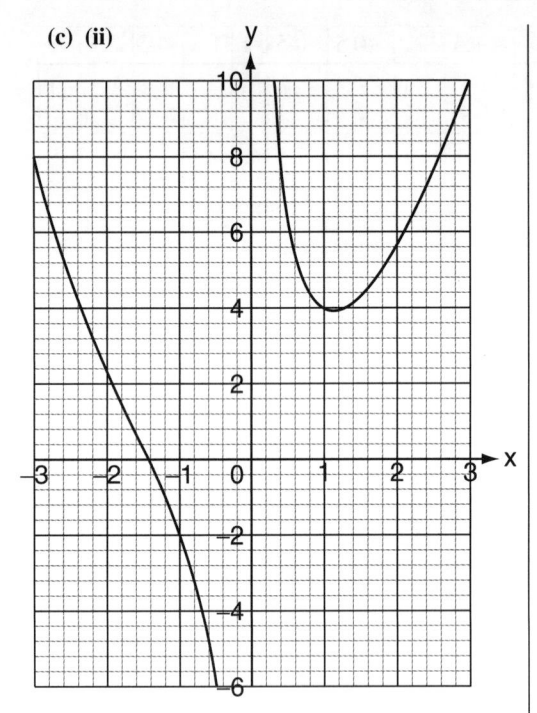

Exercise 6.3

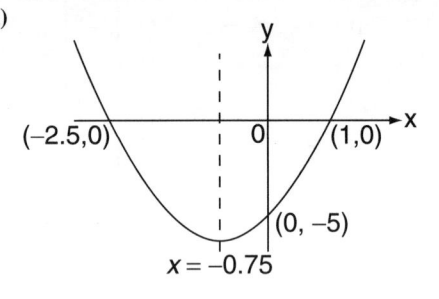

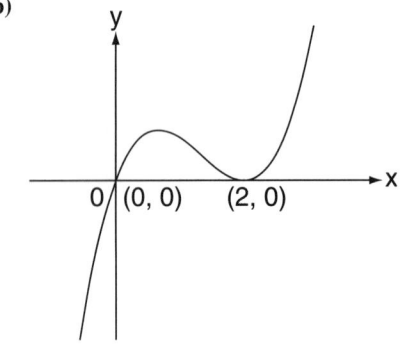

1. (a) parabola with roots $(-2.5, 0)$ and $(1, 0)$, y-intercept $(0, -5)$, axis $x = -0.75$

(b) cubic through $(0, 0)$ and $(2, 0)$

(c) downward parabola with roots $(1, 0)$ and $(2, 0)$, y-intercept $(0, -2)$, axis $x = 1.5$

(d) cubic with roots $(-2, 0)$, $(1, 0)$, $(3, 0)$, passing through $(6, 0)$

5.

Function	x	−0.5	−0.25	0	0.25	0.5	0.75	1
$y = 10^x$	y	0.3	0.6	1	1.8	3.2	5.6	10

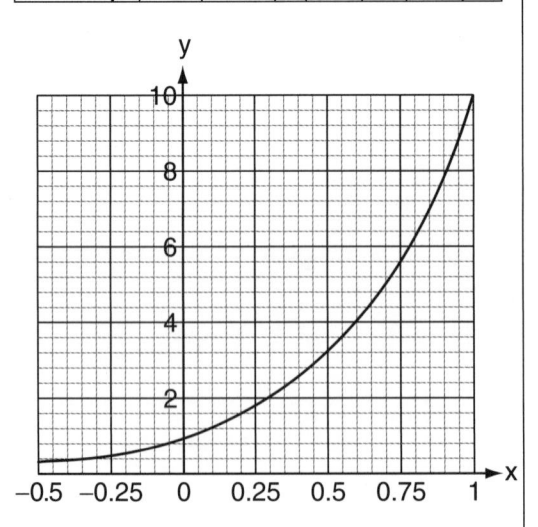

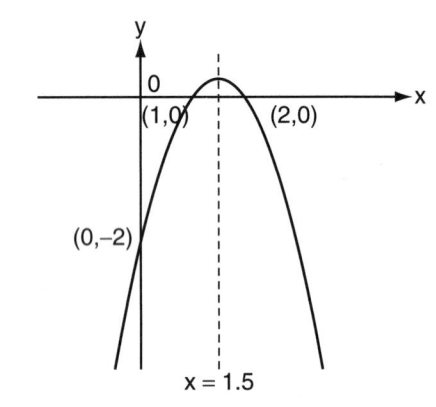

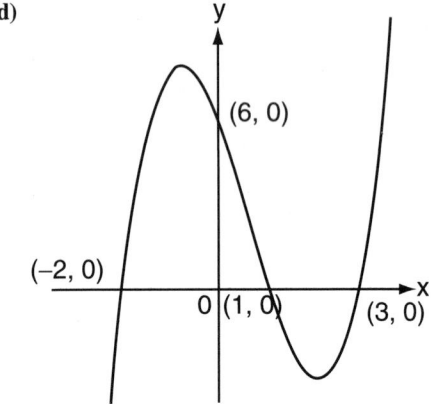

2. (a) (iii) **(b)** (v) **(c)** (vi)

3. (a)

x	0	1	2	3	3.5	4
y	0	2.5	2.2	1.9	1.7	1.6

(b)

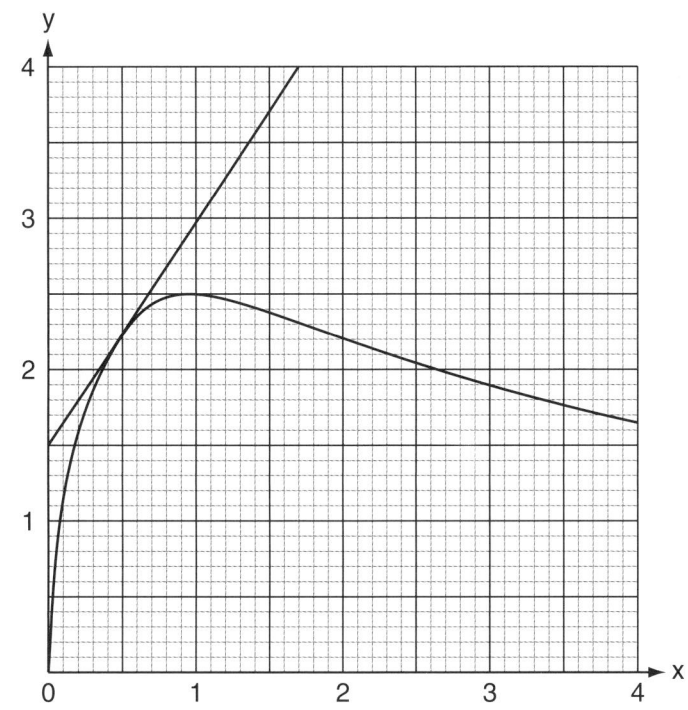

(c) 1.4 to 1.5

Exercise 6.4
1. **(a)** 4.12 **(b)** 4 **(c)** (0.5, 4)
2. **(a)** 1.41 **(b)** 1 **(c)** (−1.5, −1.5)
3. **(a)** 22.4 **(b)** 2 **(c)** (0, 0)
4. **(a)** 24.1 **(b)** $\frac{9}{8}$ **(c)** (5, 5)
5. **(a)** 10.6 **(b)** $-\frac{8}{7}$ **(c)** (0.5, −3)
6. **(a)** 5.66 **(b)** −1 **(c)** (1, 1)
7. **(a)** 9.22 **(b)** $-\frac{9}{2}$ **(c)** (−2, 0.5)
8. **(a)** 6.40 **(b)** $\frac{5}{4}$ **(c)** (3, 0.5)

Exercise 6.5
1. (a)

t (seconds)	0	0.25	0.5	0.75	1	1.25	1.5	1.75	2	2.25	2.5	2.75	3
h (metres)	0	3.4	6.3	8.4	10	10.9	11.3	10.9	10	8.4	6.3	3.4	0

(b)

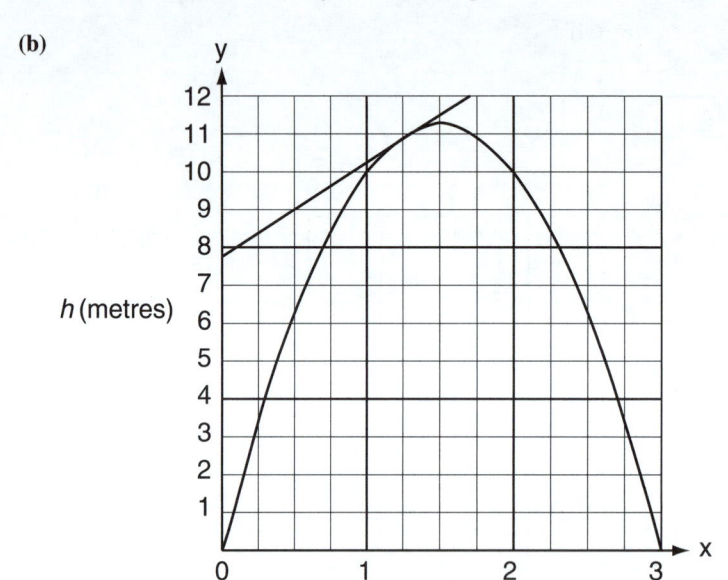

h (metres)

t (seconds)

NOTE: This graph is reduced in size to show you the general shape. Yours should be to the scale stated in the question.

(c) 11.1 to 11.5 metres
(e) It stops momentarily before changing direction
(d) 2 to 3 m/s
(f) −16 to −14 m/s

2. (a)

t (seconds)	0	0.25	0.5	0.75	1	1.25	1.5
speed (m/s)	15	12.5	10	7.5	5	2.5	0

(b)

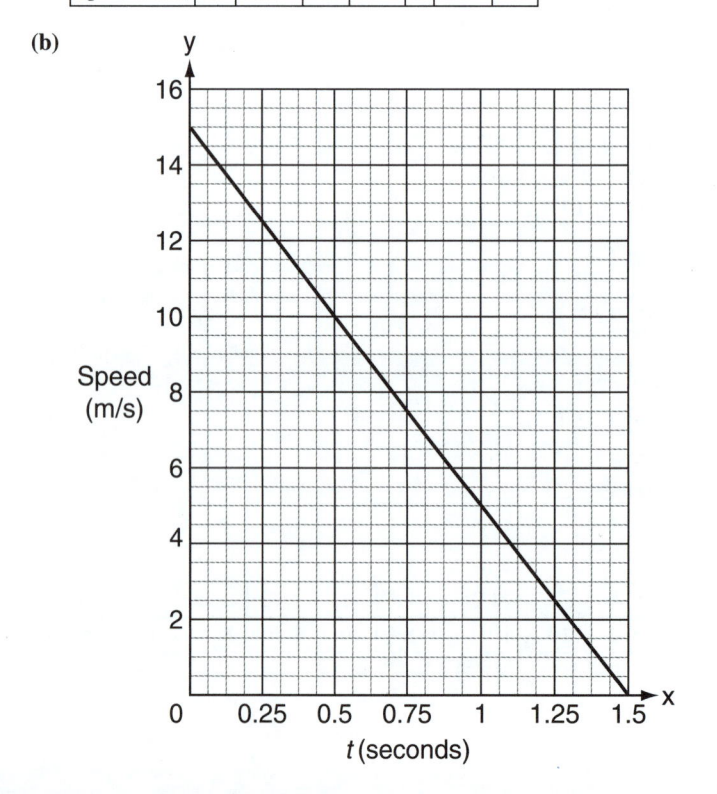

Speed (m/s)

t (seconds)

(c) (i) The graph would continue in a straight line downwards so that the speed would correspond to negative numbers on the vertical axis

(ii) The ball has reached its maximum height, where the speed is momentarily zero, and then is coming back down, so travelling in the opposite direction

(d) 10 metres

(e) $-\dfrac{15}{1.5} = -10$ m/s² **NOTE: The acceleration is negative because the ball is slowing down as it travels upwards.**

3. $\sqrt{52}$ or $2\sqrt{13}$ or 7.21 units

4. (a)

x	0.1	0.2	0.4	0.6	0.8	1	1.5	2	2.5	3
y	10	5.0	2.7	2.0	1.9	2	2.9	4.5	6.7	9.3

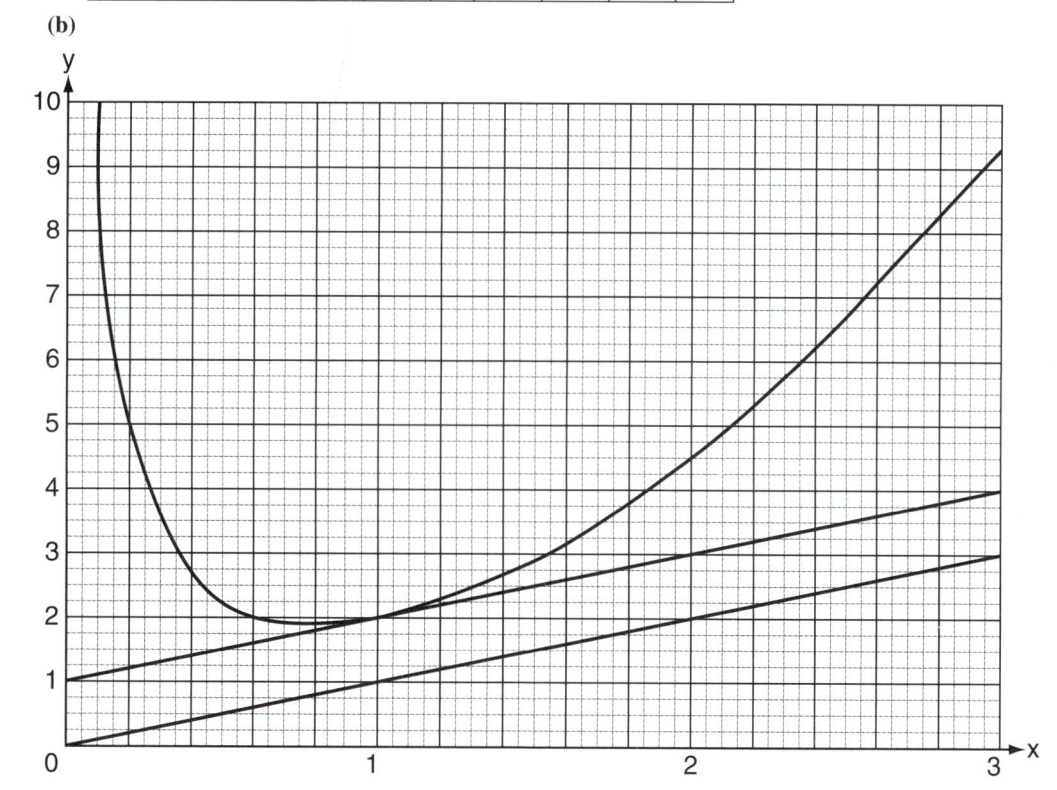

(b)

(d) (1, 2)

(e) $y = x + 1$

5.

x	−3	−2	−1	0	1	2	3	4
$y = x^2 - x - 6$	6	0	−4	−6	−6	−4	0	6
$y = -x^2 - x + 6$	0	4	6	6	4	0	−6	−14

$x = -2.4$ to -2.5, $y = 2.2$ to 2.4
$x = 2.4$ to 2.5, $y = -2.2$ to -2.4

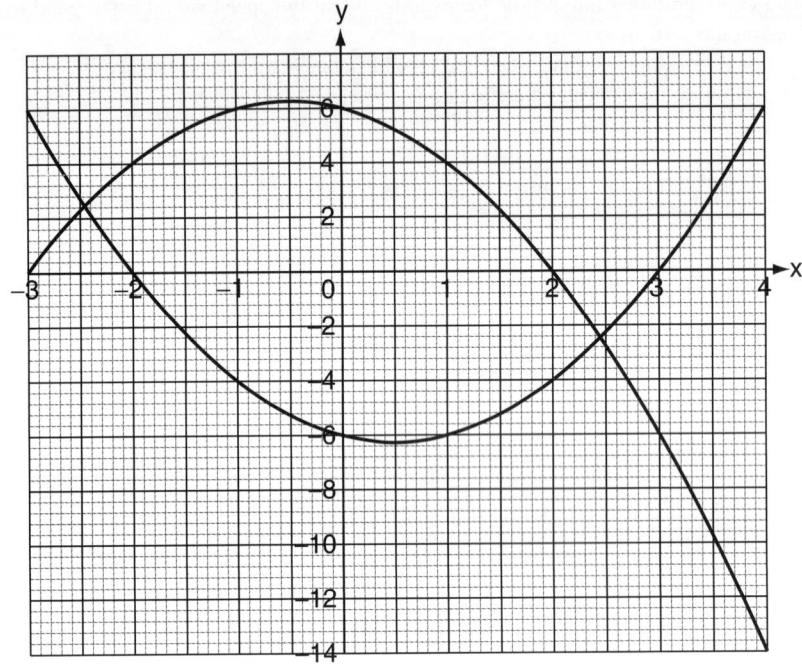

6. (a) 1.8 m/s²   (b) 450 m   (c) 13 m/s
7. (a) A: (2, 0) B: (0, -6)   (b) 6.32   (c) (1, -3)
8. (a) 0.1   0.3   0.6   1   1.7   3
   (b)   (c) 1.6 to 1.65

Answer Key 397

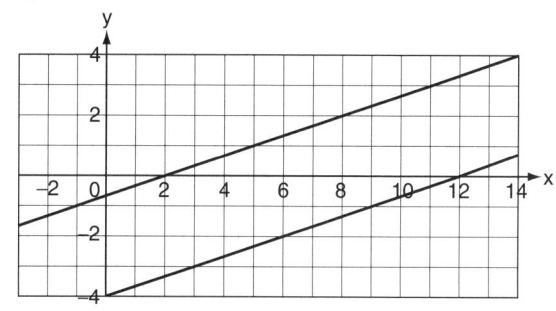

9. (a) (5, 0.5)
   (b) (i)
   (ii) $k = 12$
10. (a), (b)

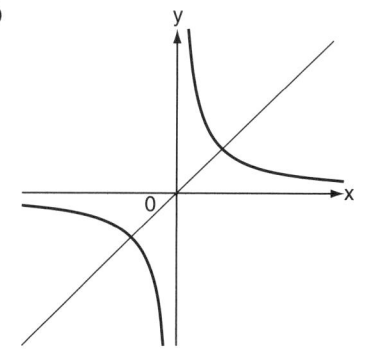

   (c) $k = -\sqrt{3}$ or $\sqrt{3}$
11. (a) $p = 2.5$
    (b), (d), (e)

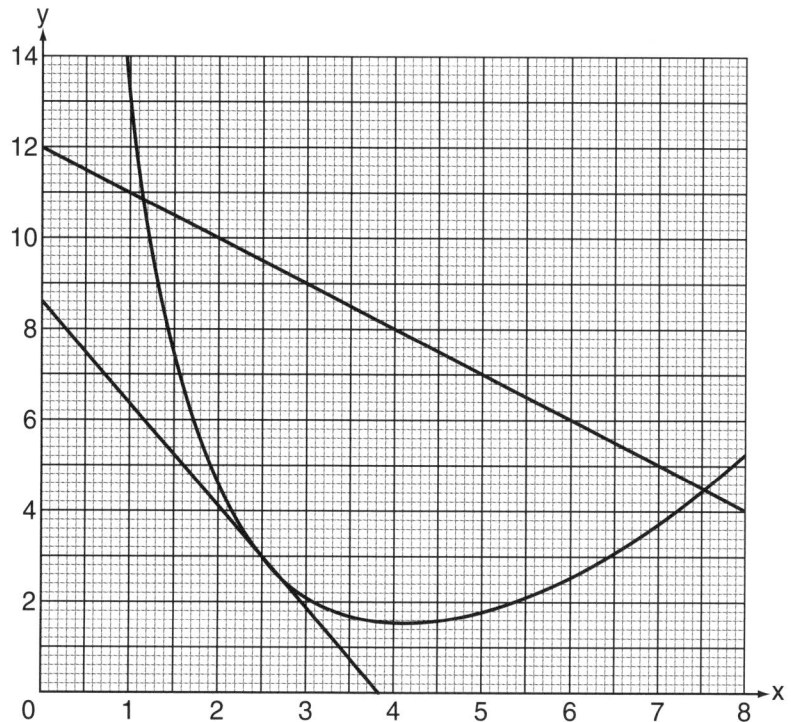

(c) (i) 1.4 to 1.5   (ii) 6.4 to 6.5
(d) 2.0 to 2.5
(f) $A = 8$ $B = -136$

12. (a) (i) $p = 12$, $q = 1.5$, $r = 1.2$
    (ii), (iii), (b)

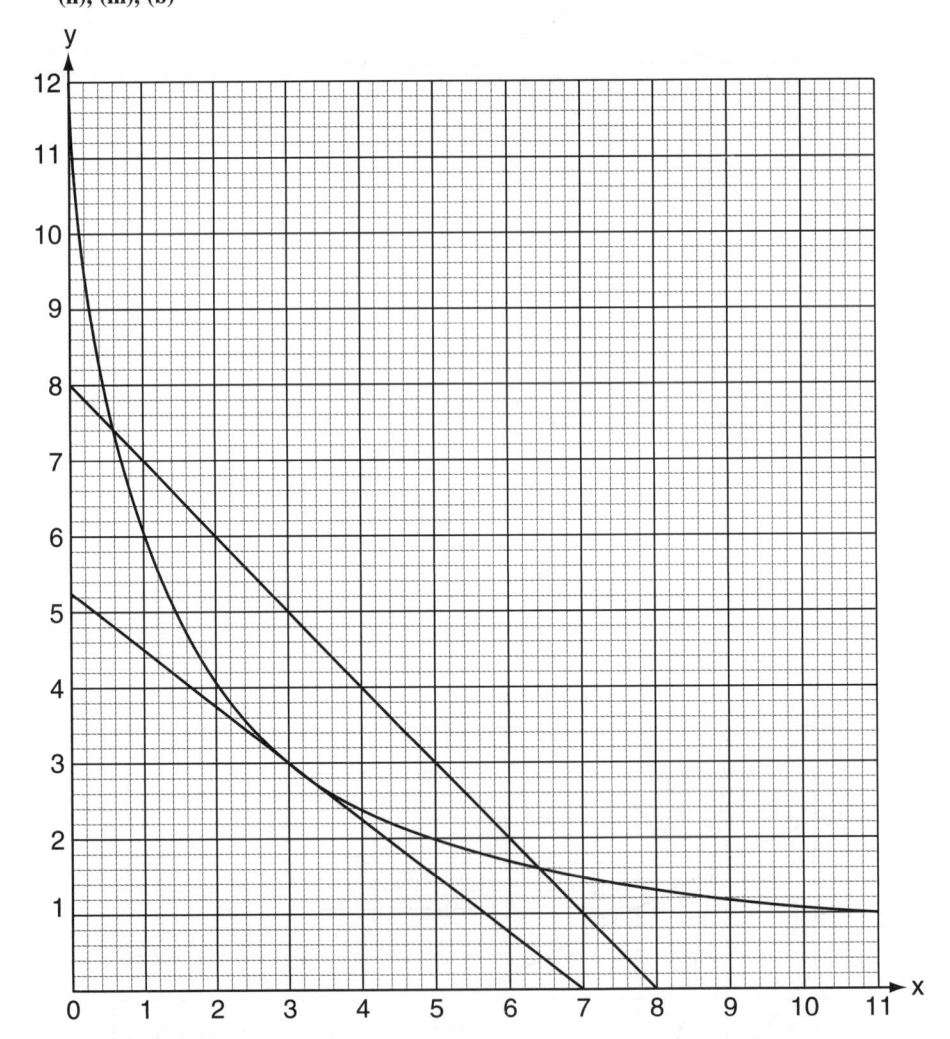

gradient: -0.6 to -1.0

(c) (i) $\dfrac{12}{x+1} = 8 - x$

$12 = (8-x)(x+1)$

$x^2 - 7x + 4 = 0$

(ii) 0.5 to 0.8, 6.2 to 6.5

13. (a) (i) $y = 5$ (ii) $K: (-1, 0)$, $L: (3, 0)$ (iii) $(1, -4)$
    (b) (i) The graph would be the other way up, with a maximum point
        (ii) The minimum point would be $(0, 0)$
    (c) (i) $c = 0$ (ii) $a = 2$, $b = -6$

14. (i) 2 (ii) 1.1 to 1.3
15. (a) 0.6 m/s² (b) 1170 m
16. (a), (b), (c) (i), (d) (ii)

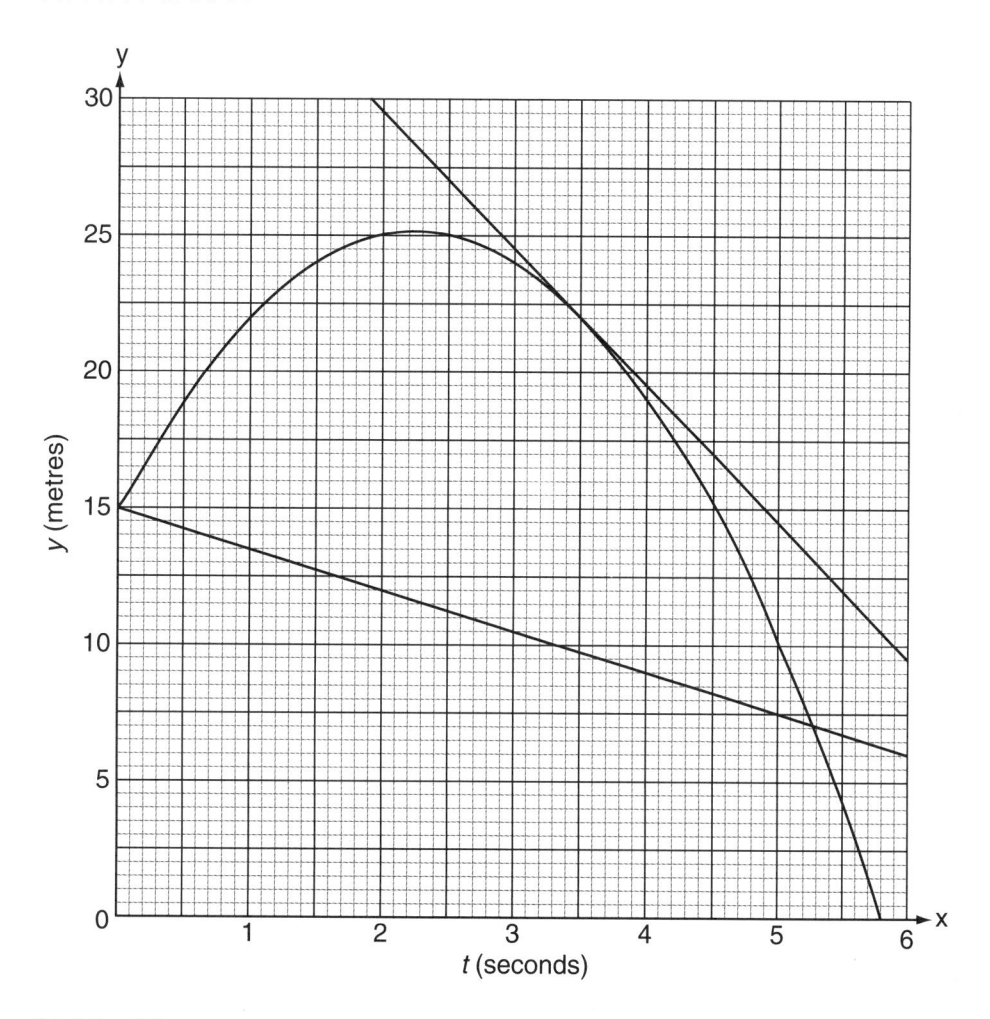

(b) 5.7 to 5.9
(c) (i) −6 to −4
   (ii) speed (or velocity)
(d) (i) (a) 15 m   (b) 9 m
   (iii) 7 to 7.4 m
17. (a) (i) $k = 1$ (ii) $m = -1$ (iii) $n = \dfrac{3}{2}$
    (b) (i) $r = 0.25$ $s = 1$ $t = 8$

(c) (i), (ii), (iii)

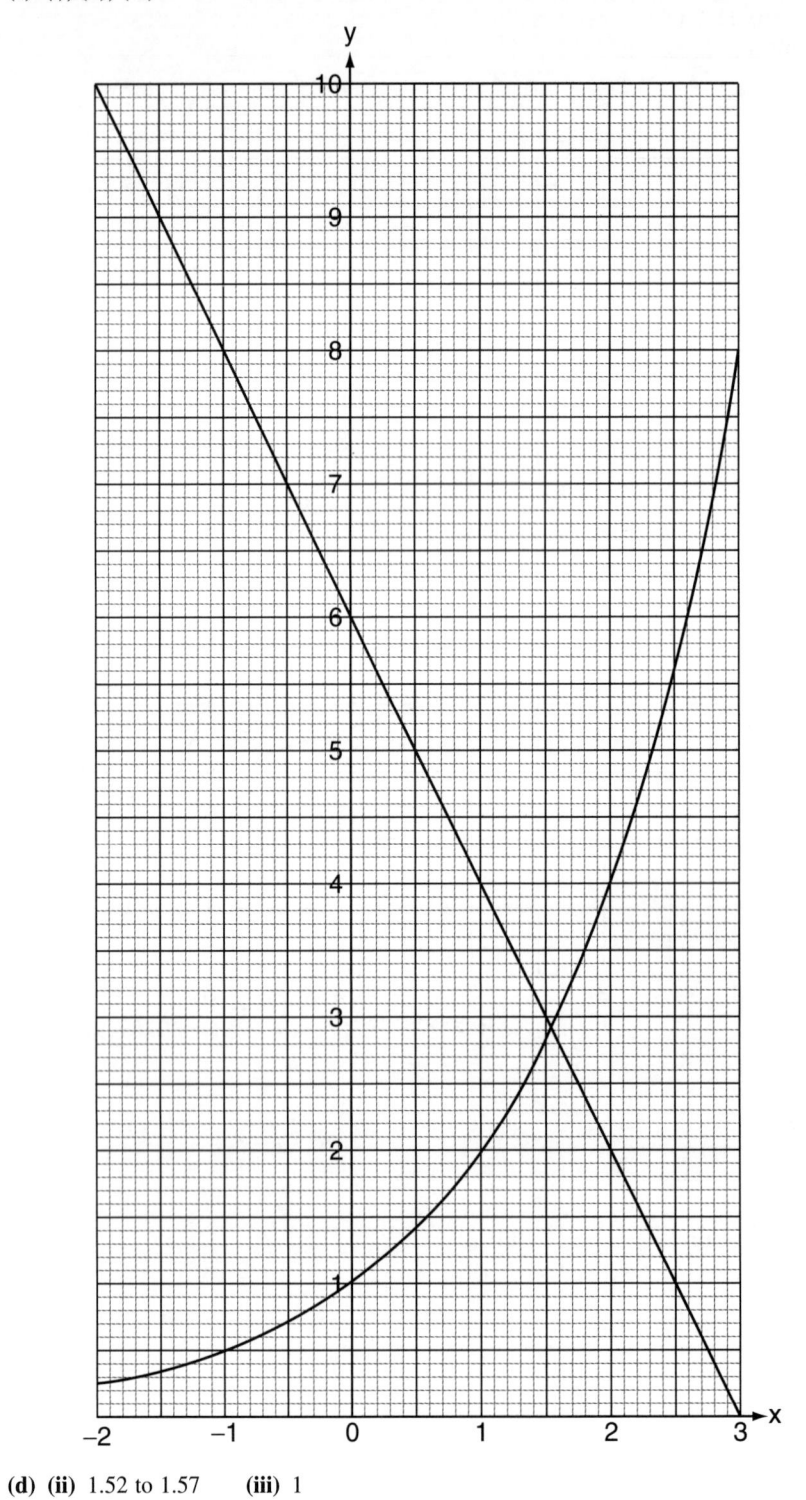

(d) (ii) 1.52 to 1.57   **(iii)** 1

18. (a) $p = 0.89$ $q = -10.1$
(b), (d), (f) (i)

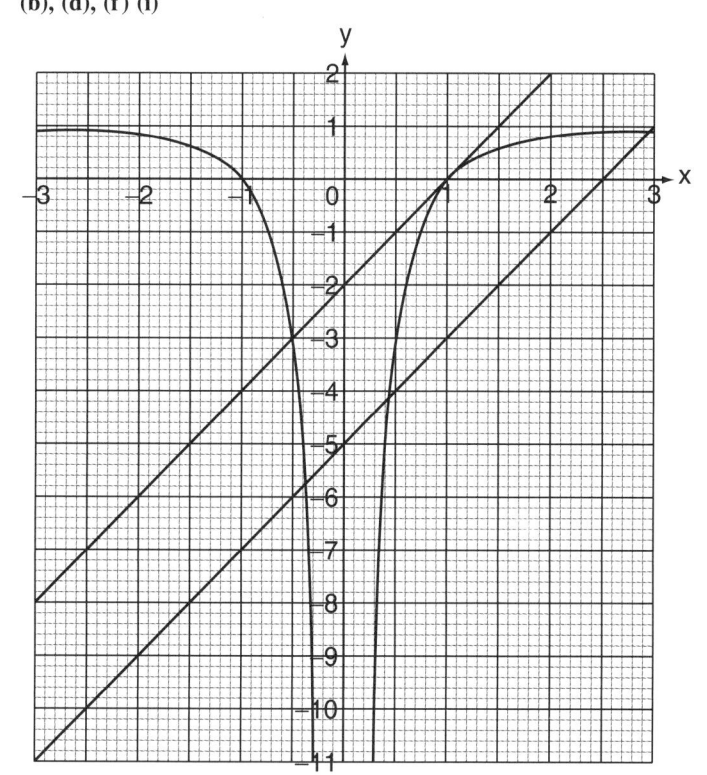

(c) for example, $k = 2$
(e) (i) -0.45 to -0.3, 0.4 to 0.49, 2.9 to 2.99
  (ii) $2x^3 - 6x^2 + 1 = 0$
(f) (ii) $y = 2x - 2$

19. (a), (b) (i), (c) (ii)

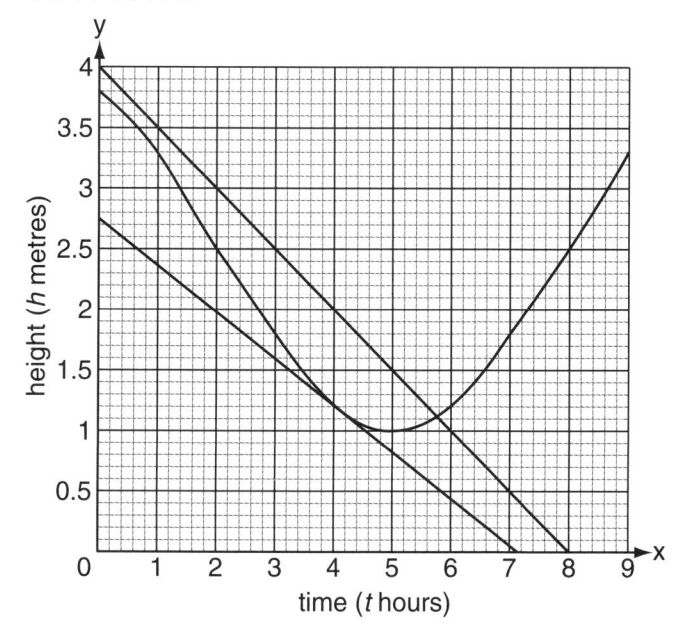

(b) (i) −0.45 to −0.32
   (ii) This is the rate at which the water level is changing
   (c) (i) 4 m   (iii) 75 to 85 cm   (iv) 5.7 to 5.9 hours
20. (a) (2.5, 4)   (b) 5   (c) $\dfrac{4}{3}$
    (d) $3y = 4x + 2$ (e) (7, 2)
21. (a) figure 6   (b) figure 4   (c) figure 2
22. $y = \dfrac{1}{2}x + 5$
23. (a) $108.16 or $108   (b) $p = 148, q = 324$
    (c), (e) (iii)

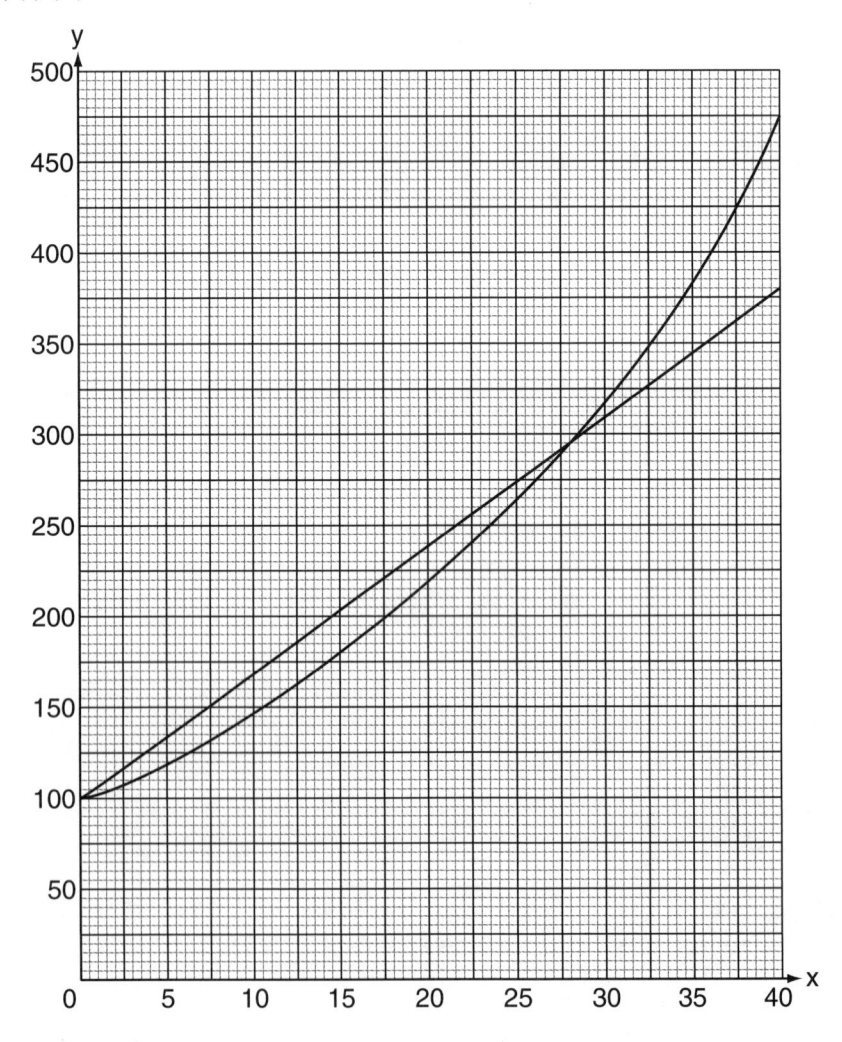

(d) (i) $265 to $270   (ii) 17 or 18
(e) (i) $100 + 7 \times 20 = 140$ (ii) $380
(f) 27 to 29
24. (4, 2)

25. (a) (i) 1.5, 3.75, −1.5
  (ii)

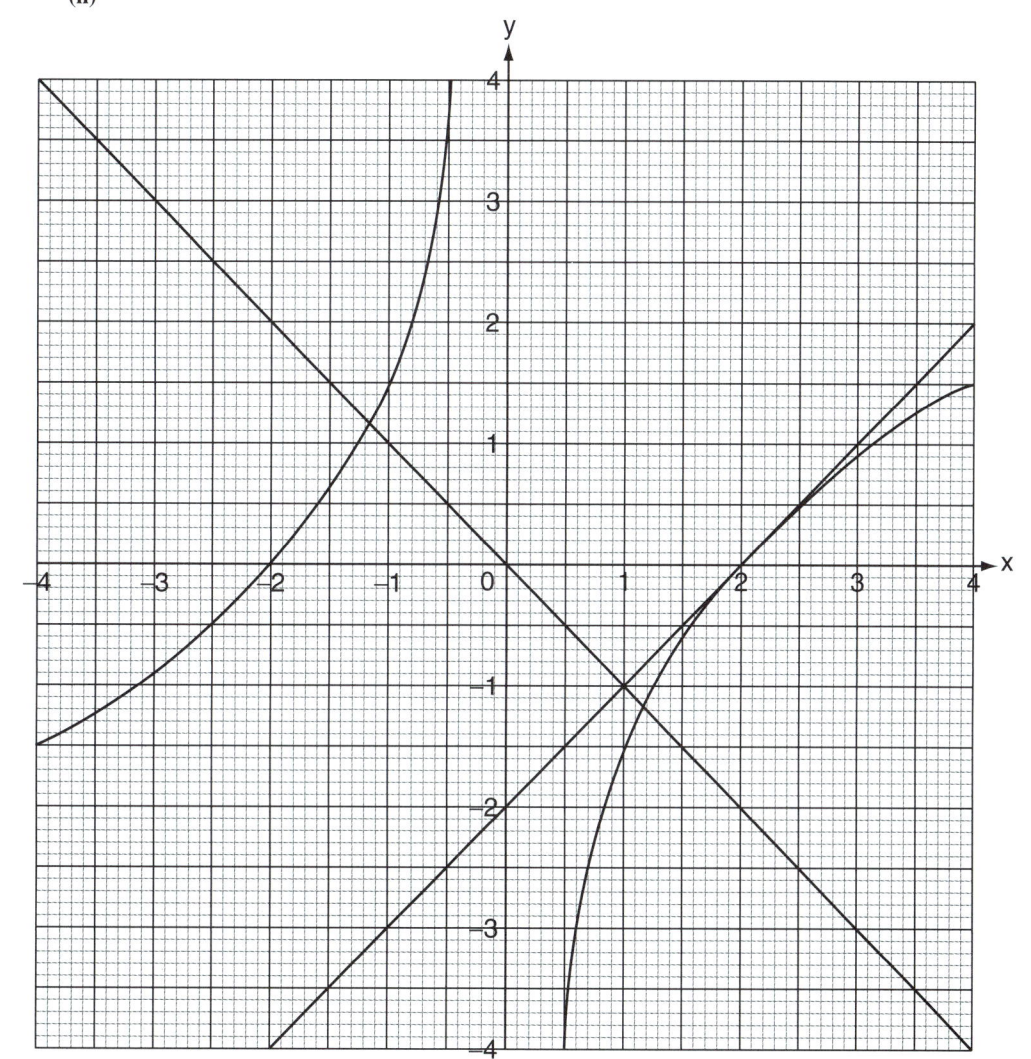

(b) −1.4 to −1.1 and 3.1 to 3.4
(c) (i) 0.8 to 1.2 (ii) 0.8 to 1.2
(d) (ii) −1.3 to −1.05 and 1.05 to 1.3
(e) for example, $y = x$ or $y = x - 1$

Chapter 7

Core Skills

1. (a) 0.25 m   (b) 0.00002 m³ or 2×10^{-5} m³
   (c) 150000 cm²   (d) 1700 ml
2. (a) 56.5 cm   (b) 254 cm²
3. (a) (i) 20 cm   (ii) 21 cm²
   (b) (i) 19.1 cm   (ii) 21.9 cm²
   (c) (i) 18 cm   (ii) 19 cm²
   (d) (i) 19 cm   (ii) 20 cm²
   (e) (i) 15.6 cm   (ii) 13.5 cm²

4. (a) 460 cm² **(b)** 600 cm³
5. (a) 471 cm² **(b)** 785 cm³
6. 707 ml

Exercise 7.1

(a) 5.24 cm **(b)** 26.2 cm² **(c)** 19.5 cm
(d) 78.2 cm² **(e)** 17.5 cm **(f)** 43.6 cm²
(g) 191.0° **(h)** 15 cm² **(i)** 11.5 cm
(j) 68.8 cm² **(k)** 15.3° **(l)** 4 cm
(m) 15.1 cm **(n)** 5.28 cm **(p)** 35.4°
(q) 5.56 cm **(r)** 8.29 cm **(s)** 14.5 cm

Exercise 7.2

1. (a) 113 cm² **(b)** 113 cm³ **(c)** 84.8 cm²
(d) 56.5 cm² **(e)** 616 cm² **(f)** 1440 cm³
(g) 462 cm² **(h)** 718 cm³ **(i)** 1.26 cm
(j) 8.41 cm³ **(k)** 15 cm² **(l)** 4.21 cm³
(m) 1.93 cm **(n)** 46.7 cm² **(p)** 35.0 cm²
(q) 15 cm³ **(r)** 2.88 cm **(s)** 104 cm²
(t) 100 cm³ **(u)** 78.1 cm²

2. (a) 47.1 cm² **(b)** 264 cm³ **(c)** 1.91 cm
(d) 30.6 cm³ **(e)** 1.20 cm **(f)** 48.9 cm²

3. (a) 339 cm² **(b)** 467 cm² **(c)** 75.6 cm³
4. (a) 3.00 cm **(b)** 245 cm²

Exercise 7.3

1. (a) (i) 1:25 **(ii)** 1:5
(b) (i) 0.4 cm **(ii)** 424 cm²
(iii) 100 cm **(iv)** 10000 cm³
2. 1.26 cm
3. (a) 2:3 **(b)** 4:9 **(c)** 45 cm²
4. (a) 125:343 **(b)** 2000 ml, or 2.00 litres
5. 3.125×10^5 m²

Exercise 7.4

1. (a) In triangle ABC
  $\angle ABC = 53.1°$ (angles on a straight line)
  $\angle BAD = 36.9°$ (angle sum of a triangle)
  $\angle DAC = 45°$ (isosceles triangle)
  $\angle BAC = 36.9 + 45 = 81.9°$
  In triangle PQR
  $\angle PQR = 53.1°$ (angle sum of a triangle)
  $\angle SPR = 45°$ (isosceles triangle)
  $\angle QPR = 36.9 + 45 = 81.9°$
  In triangles ABC and PQR
  $\angle ABC = \angle PQR = 53.1°$
  $\angle BAC = \angle QPR = 81.9°$
  $\angle ACB = \angle PRQ = 45°$
  the triangles are similar (equiangular)
(b) 5.6 cm **(c)** 7.65 cm² **(d)** 27.44 cm²

2. (a) In triangles ABC and CDE
  $\angle ABC = \angle CDE$ (alternate angles)
  $\angle BAC = \angle DCE$ (alternate angles)
  $\angle ACB = \angle DCE$ (vertically opposite angles)
  the triangles are similar (equiangular)
(b) 5.71 cm **(c)** 18.375 cm²

3. (a) In triangles PRT and QRS
  $\angle PTR = \angle SQR$ (cyclic quadrilateral and angles on a straight line)
  $\angle RPT = \angle RSQ$ (cyclic quadrilateral and angles on a straight line)
  $\angle R$ is common to both
  The triangles are similar (equiangular)
(b) 30 cm

Exercise 7.5

1. 144°
2. 1250 m²
3. (a) $AB:DF = 3:4.5 = 1:1.5$
  $BC:EF = 8:12 = 1:1.5$
  $AC:DE = 6:9 = 1:1.5$
  The triangles are similar (sides in the same ratio)
(b) $w = 121.9°$, $x = 39.5°$, $y = 18.6°$, $z = 39.5°$
(c) 1:2.25
4. square, side 6 cm
  rectangle, length 9 cm, breadth 4 cm
5. (a) $\frac{1}{3}\pi x^3$ **(b)** $\frac{2}{3}\pi x^3$
(c) The conical bowl has twice as much wood as the hemispherical bowl
6. (a) $6\pi x$ cm **(b)** $3\pi x^2$ cm²
7. (a) 42 cm² **(b)** 16 cm
8. (a) (i) 462 cm³ **(ii)** 216 cm²
  (iii) (a) 118 cm² **(b)** 81.2 cm²
(b) (i) 71 to 72 cm³ **(ii)** 15.7 to 16.4 cm

9. (i) $\dfrac{y}{y+2} = \dfrac{y+1}{2y-1}$
  $y(2y-1) = (y+1)(y+2)$
  $2y^2 - y = y^2 + 3y + 2$
  $y^2 - 4y - 2 = 0$

Answer Key   405

(ii) $y = \dfrac{4 \pm \sqrt{16+8}}{2}$
   $y = -0.45$ or $y = 4.45$
(iii) 7.90 cm
10. (a) 64.8 m³   (b) 1230%
    (c) 22.1 m   (d) 150 m
11. (a) 14.1 cm²   (b) 24.8 cm
12. (a) (i) 20°   (ii) 70°
    (b) (i) 3.49 cm   (ii) 8.73 cm²
13. (a) 74.8 to 74.9 cm
    (b) 365 cm²
    (c) 14600 cm³
    (d) 3720 to 3730 cm²
14. (i) $(x^2 - 40) + (x + 2) + (2x + 4) + x = 62$
        $x^2 + 4x - 96 = 0$
    (ii) $x = -12$ or 8 (iii) 8
    (iv) 176 square units

15. (a) (i) 63 m²   (ii) 1512 m³
       (iii) 1512000 litres
    (b) (i) $1891.62   (ii) $1900
    (c) (i) 6870 cm³   (ii) 2 days 13 hours
16. radius = 18 cm   height = 42 cm
17. (a) 45498 km   (b) 7240 km
18. (a) 320 cm³   (b) 567 cm²
19. (a) 40 cm³   (b) 0.00004 m³
20. (a) 440 cm²
    (b) $h = \dfrac{A - 2\pi r^2}{2\pi r}$
    (c) 3.99 to 4.01 cm²   (d) 9.77 to 9.78 cm
21. (a) (i) 2400 km   (ii) 520000 km²
    (b) (i) 1 : 5000000 (ii) 738 to 742 km/h
22. 28.2, 28.6
23. 314 cm²

Chapter 8

1. (a)

	Black	White
Laptops	10	20
Desktops	4	15

$\begin{pmatrix} 10 & 20 \\ 4 & 15 \end{pmatrix}$

(b)

	Win	Draw	Lose
Reds	3	2	1
Blues	4	0	2
Greens	0	5	1
Yellows	0	3	3

$\begin{pmatrix} 3 & 2 & 1 \\ 4 & 0 & 2 \\ 0 & 5 & 1 \\ 0 & 3 & 3 \end{pmatrix}$

2. (a) $\begin{pmatrix} 3 & 10 \\ -8 & 3 \end{pmatrix}$ (b) $\begin{pmatrix} -1 & 1 \\ 1 & -7 \end{pmatrix}$
   (c) $\begin{pmatrix} 2 & 0 \\ 6 & -16 \end{pmatrix}$ (d) $\begin{pmatrix} 2 & -5 \\ 0 & -1 \end{pmatrix}$

3. (a) **A** 2 × 3   **B** 2 × 2   **C** 2 × 3
       D 2 × 2   **E** 3 × 2   **F** 1 × 3

   (b) $\mathbf{A}' = \begin{pmatrix} 5 & 4 \\ 1 & 5 \\ -3 & 2 \end{pmatrix}$ $\mathbf{B}' = \begin{pmatrix} 3 & 7 \\ 9 & 1 \end{pmatrix}$

   $\mathbf{C}' = \begin{pmatrix} 1 & 4 \\ 2 & 9 \\ 5 & 3 \end{pmatrix}$ $\mathbf{D}' = \begin{pmatrix} 7 & 3 \\ 0 & 1 \end{pmatrix}$

   $\mathbf{E}' = \begin{pmatrix} 5 & 4 & 9 \\ 1 & 6 & 3 \end{pmatrix}$ $\mathbf{F}' = \begin{pmatrix} -1 \\ 2 \\ 6 \end{pmatrix}$

   (c) (i) **D′** 2 × 2   (ii) **E′** 2 × 3   (iii) **F′** 3 × 1
   (d) (i) $\begin{pmatrix} 6 & 3 & 2 \\ 8 & 14 & 5 \end{pmatrix}$ (ii) $\begin{pmatrix} 4 & -9 \\ -4 & 0 \end{pmatrix}$
       (iii) $(-4 \ \ 8 \ \ 24)$

Exercise 8.2

1. **FG** not   2. **GH** not   3. **HK** 2 × 2
4. **KH** 4 × 4   5. **LG** not   6. **GL** 3 × 2
7. **LF** 3 × 2   8. **FL** not

Exercise 8.3

1. $a = 34$ 2. $b = -8 \ \ c = 8$
3. (a) not possible   (b) $\begin{pmatrix} 5 & 3 & 5 \\ 15 & 7 & 9 \end{pmatrix}$
   (c) $(12 \ \ 25)$ (d) $\begin{pmatrix} 30 & 13 & 15 \\ 20 & 6 & 2 \\ -5 & 2 & 10 \end{pmatrix}$
4. $x = 2 \ \ y = 0 \ \ z = 2$
5. (a) $\begin{pmatrix} -3 & 9 \\ 0 & -6 \end{pmatrix}$ (b) $\begin{pmatrix} 1 & -3 \\ 0 & 2 \end{pmatrix}$ (c) $\begin{pmatrix} 1 & -9 \\ 0 & 4 \end{pmatrix}$
   (d) $\begin{pmatrix} -2 & 0 \\ 0 & -2 \end{pmatrix}$ (e) $\begin{pmatrix} -2 & 3 \\ 3 & -4 \end{pmatrix}$

Exercise 8.4

1. det **P** = 5
2. det **Q** = 2
3. det **R** = 18
4. det **S** = −19

Exercise 8.5

1. $\mathbf{A}^{-1} = -\dfrac{1}{2}\begin{pmatrix} 10 & -8 \\ -9 & 7 \end{pmatrix}$ 2. $\mathbf{B}^{-1} = -\dfrac{1}{3}\begin{pmatrix} -2 & -1 \\ 1 & 2 \end{pmatrix}$

3. $\mathbf{C}^{-1} = \dfrac{1}{8}\begin{pmatrix} 10 & 6 \\ -3 & -1 \end{pmatrix}$ 4. $\mathbf{D}^{-1} = -\begin{pmatrix} 0 & -1 \\ -1 & 0 \end{pmatrix} = \mathbf{D}$

5. $\mathbf{I}^{-1} = \begin{pmatrix} 1 & 0 \\ 0 & 1 \end{pmatrix} = \mathbf{I}$

Exercise 8.6

1. **E**: 2 × 2   **F**: 2 × 3   **G**: 2 × 2
   H: 2 × 1   **J**: 1 × 2   **K**: 3 × 2
2. (a) 2 × 2   (b) 3 × 3   (c) 2 × 2
   (d) 1 × 1   (e) 3 × 1

3. $\mathbf{F}' = \begin{pmatrix} 7 & 3 \\ 1 & 5 \\ 6 & 2 \end{pmatrix}$ **F**′ : 3 × 2

4. (a) $\begin{pmatrix} 7 & 1 & 6 \\ 10 & 6 & 4 \end{pmatrix}$ (b) $\begin{pmatrix} 6 & 10 & 4 \\ 14 & 2 & 12 \end{pmatrix}$

   (c) $\begin{pmatrix} 4 & 6 \\ 8 & 12 \end{pmatrix}$ (d) (16)

5. $\begin{pmatrix} 1 & 0 \\ 0 & 1 \end{pmatrix}$

6. (a) $\dfrac{1}{2}\begin{pmatrix} 2 & 0 \\ 0 & 1 \end{pmatrix}$ (b) $-\dfrac{1}{4}\begin{pmatrix} 0 & -2 \\ -2 & 0 \end{pmatrix}$

   (c) $\begin{pmatrix} 1 & 0 \\ 0 & 1 \end{pmatrix}$ (d) $\begin{pmatrix} 1 & 0 \\ 0 & 2 \end{pmatrix}$ (e) $\begin{pmatrix} 0 & 2 \\ 2 & 0 \end{pmatrix}$

7. (a) $\begin{pmatrix} a \\ b \end{pmatrix} = \begin{pmatrix} 5 \\ -7 \end{pmatrix}$ (b) $a = 5$, $b = -7$

8. (a) $\begin{pmatrix} p & q \\ r & s \end{pmatrix} = \begin{pmatrix} -3 & 4 \\ -2 & 4 \end{pmatrix}$

   (b) $p = -3$, $q = 4$, $r = -2$, $s = 4$

9. (a) $\begin{pmatrix} 17 \\ 10 \\ 18 \end{pmatrix}$ (b) $\begin{pmatrix} x \\ 10x \\ 5x \end{pmatrix}$

10. (a) $a = 1.5$, $b = 1$, $c = -2.5$, $d = 2$
    (b) $x = 4$, $b = 6$ (c) $y = \dfrac{1}{3}$
    (d) $p = \dfrac{1}{4}$, $q = 3$, $r = 7$

11. (a)

	Reliables	Gofasters
Retail Value	$5000	$6000

	Blue	Red	Black
Reliables	10	5	2
Gofasters	5	16	3

(b) **R** = (5000 6000) $\mathbf{N} = \begin{pmatrix} 10 & 5 & 2 \\ 5 & 16 & 3 \end{pmatrix}$

(c) **RN** = (8000 121000 28000) (d) 229000
(e) The total value of the cars sold by the salesperson

12. **A** is 2 × 3   **B** is 1 × 4
    They are not conformable for multiplication either as **AB** or **BA**.

13. The determinant is zero. The inverse does not exist

Exercise 8.7

1. (a) $\dfrac{1}{3}$ (b) $-\dfrac{1}{3}$ (c) $\dfrac{2}{7}$ (d) 0
2. $x = -5$ or $x = -1$
3. (a) $\dfrac{1}{5}$ (b) 0.25   (c) $\dfrac{11}{13}$
4. $x = -3$
5. (a) 0   (b) 0   (c) 0.5
   (d) −1 or 3   (e) −2 or 2   (f) −1

Exercise 8.8

1. $f^{-1}(x) = \dfrac{x-1}{3}$ 2. $f^{-1}(x) = \dfrac{2-x}{5}$

3. $f^{-1}(x) = 2(x-1)$ 4. $f^{-1}(x) = \dfrac{4}{3}x - 2$

5. $f^{-1}(x) = 5 - 2x$ 6. $g^{-1}(x) = x^2 + 3$

7. $g^{-1}(x) = (x+1)^{\frac{1}{3}}$ **8.** $g^{-1}(x) = \dfrac{1-x}{x}$

9. $f^{-1}(x) = \dfrac{x}{1-x}$ **10.** $g^{-1}(x) = \dfrac{x}{3} + \dfrac{1}{2}$

Exercise 8.9

1. (a) $fg(x) = 15x + 5$ (b) $gf(x) = 15x - 3$
   (c) 35   (d) −18
2. (a) 203   (b) 442
3. (i) (a) $fg(x) = x^2 + 4$ (b) $gf(x) = (x+4)^2$
   (c) 104
   (ii) $x = -1.5$
4. (a) $2\dfrac{1}{9}$ (b) $\dfrac{1}{11}$
5. (a) $fg(x) = 3(x + 3)$ (b) $gf(x) = 3x + 3$
   (c) $ff(x) = 9x$ (d) $gg(x) = x + 6$
   (e) $g^{-1}(x) = x - 3$ (f) $gg^{-1}(x) = x$
   (g) $f^{-1}(x) = \dfrac{x}{3}$ (h) $f^{-1}f(x) = x$
6. (a) $ff(x) = (x^2 + 1)^2 + 1$
   (b) $fff(x) = \left[\left(x^2+1\right)^2 + 1\right]^2 + 1$

Exercise 8.10

1. (a) −1   (b) −2   (c) $-\dfrac{1}{2}$ or −1
2. (a) $gf(x) = \dfrac{1}{x^2 + 1}$ (b) $gg(x) = \dfrac{x+1}{x+2}$
   (c) $\dfrac{5}{4}$ (d) $\dfrac{13}{9}$ (e) $g^{-1}(x) = \dfrac{1-x}{x}$
3. (a) $\dfrac{5}{7}$ (b) $\dfrac{13}{18}$ (c) $h^{-1}(x) = \dfrac{3x-2}{1-x}$
   (d) $x = -2$ (e) $-\dfrac{8}{3}$ (f) −2   (g) $-\dfrac{8}{3}$

Exercise 8.11

1. (a) $x + y \leqslant 5, y > x + 3, x \geqslant 0$
   (b) $x + y \leqslant 5, x \geqslant 3, y \geqslant 0$
   (c) $y \geqslant 0, x \geqslant 0, x + y \leqslant 5, y < x + 3, x \leqslant 3$
   (d) $y \leqslant 0, x \geqslant 3, x + y \leqslant 5$
   (e) $x \geqslant 3, x + y \geqslant 5, y < x + 3$
   (f) $x + y \geqslant 5, y > x + 3$
   (g) $x + y \leqslant 5, y > x + 3, x \leqslant 0$

2. (a)

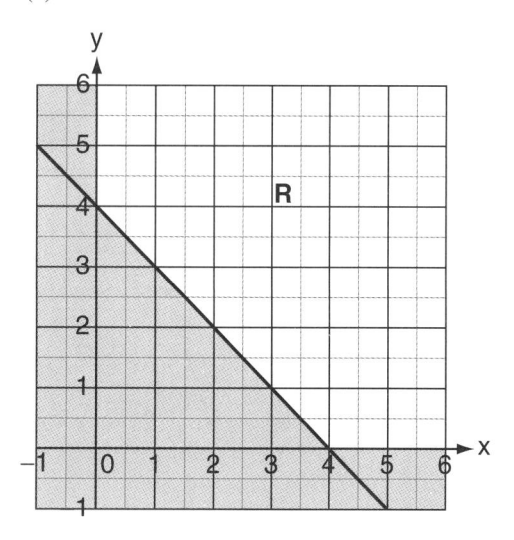

(b)

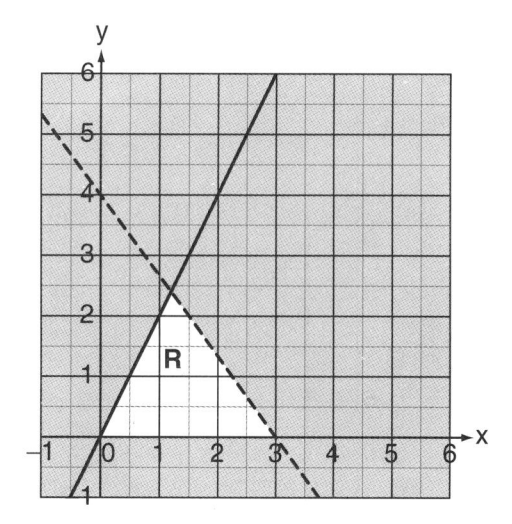

408 *Extended Mathematics for Cambridge IGCSE*

3. (a)

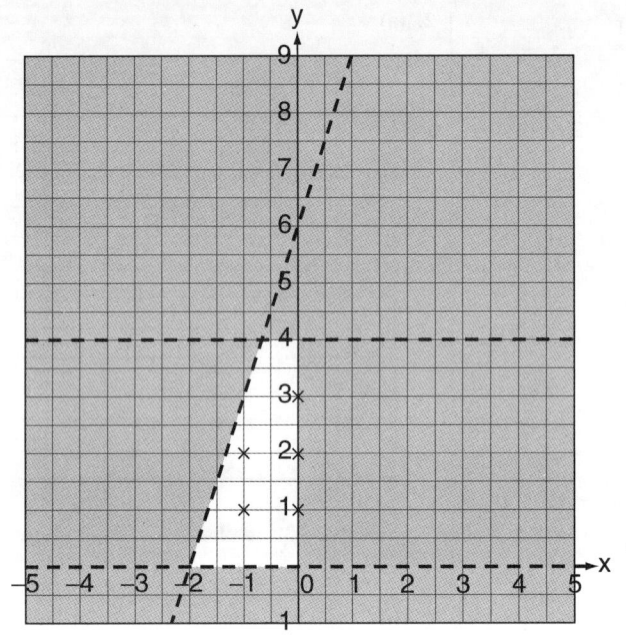

(c) (−1, 2), (−1, 1), (0, 1), (0, 2), (0, 3)

4. (a) $x + y \leqslant 10, x \geqslant 1, y \geqslant 4, y > x$

(b)

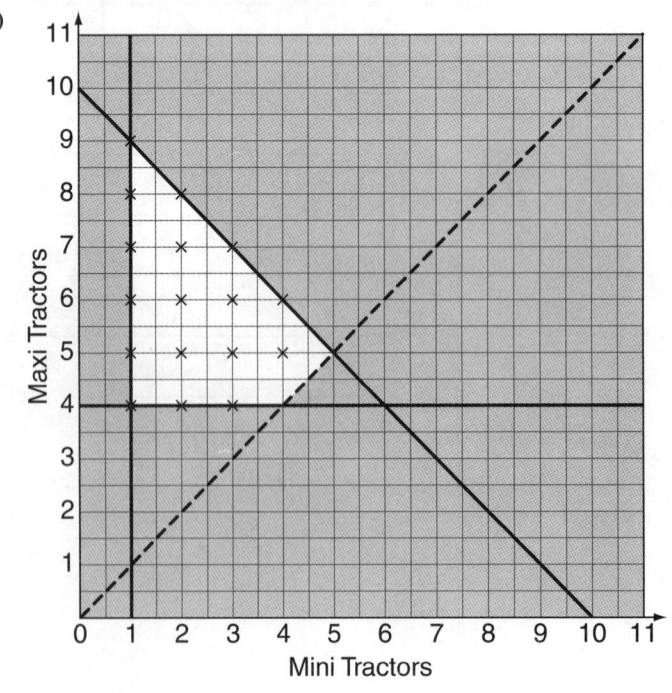

(d) 1 Mini and 9 Maxi Tractors. Total Profit = $19500

5. (a) $x + y \leqslant 10$, $y \geqslant 3$, $y < x$

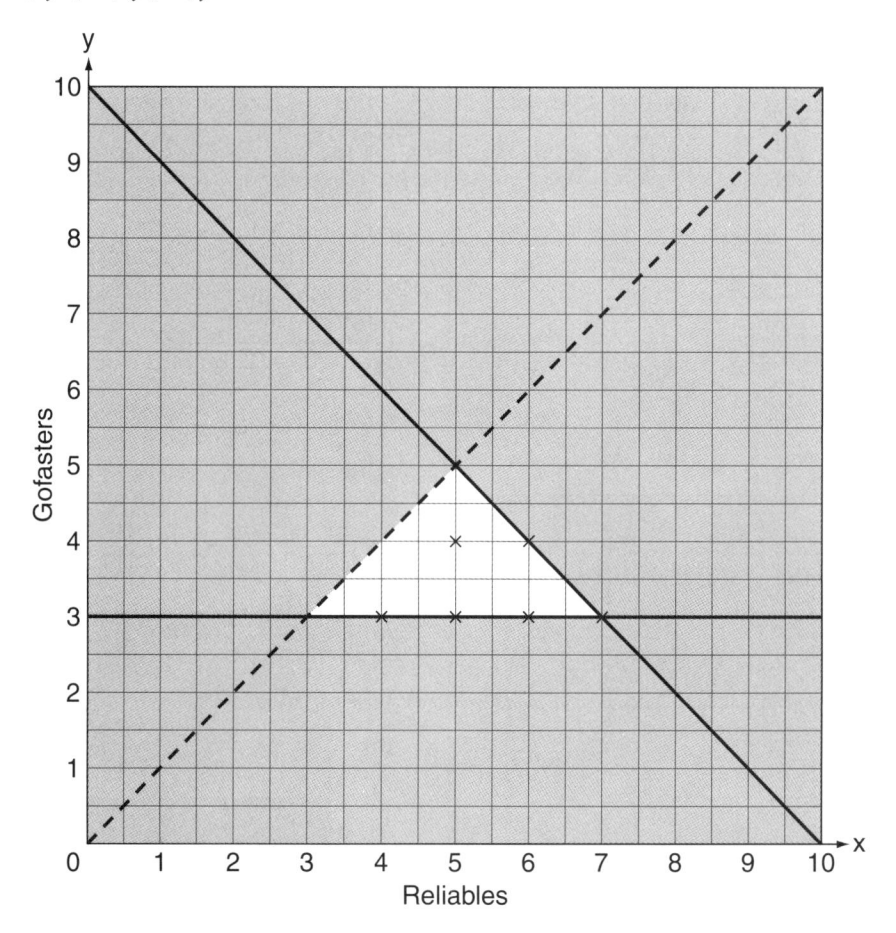

(d) (i)
$$\begin{matrix} x & y \end{matrix}$$
$$\begin{pmatrix} 5 & 4 \\ 4 & 3 \\ 5 & 3 \\ 6 & 4 \\ 6 & 3 \\ 7 & 3 \end{pmatrix} \quad \begin{pmatrix} 1500 \\ 1800 \end{pmatrix}$$
$$6 \times 2 \qquad 2 \times 1$$

(ii)
$$\begin{pmatrix} 14700 \\ 11400 \\ 12900 \\ 16200 \\ 14400 \\ 15900 \end{pmatrix}$$

6 Reliables and 4 Gofasters would make the biggest profit ($16200)

Exercise 8.12

1. (a) **A**: 2×2 **B**: 2×2 **C**: 2×3 **D**: 1×2
   E: 3×2 **F**: 2×1

(b) (i) 2×2 **(ii)** 3×3
   (iii) 1×1 **(iv)** 2×2

(c) $\mathbf{C}' = \begin{pmatrix} 1 & 2 \\ 3 & 0 \\ -5 & 4 \end{pmatrix}$

(d) (i) det $\mathbf{A} = 1$ **(ii)** $\begin{pmatrix} 2 & 1 \\ 3 & -1 \end{pmatrix}$

(iii) $\begin{pmatrix} 2 & 1 \\ 3 & -1 \end{pmatrix}$ **(iv)** $\begin{pmatrix} 1 & 0 \\ 0 & 1 \end{pmatrix}$

(v) $\mathbf{B}^{-1} = -\dfrac{1}{5}\begin{pmatrix} -1 & -1 \\ -3 & 2 \end{pmatrix}$

2. (a) (i) 1   (ii) 1   (iii) $x = 4$

(iv) $g^{-1}(x) = \dfrac{1}{x}$ (v) 4

(b) (i) $fg(x) = x^2$ (ii) $gf(x) = x^2$

(iii) $\dfrac{1}{100}$ (iv) $fgh(x) = (x^2 + 1)^2$

3. (a) $f^{-1}(x) = \dfrac{1 - 3x}{2x}$

(b) $g^{-1}(x) = \dfrac{3x}{1 - 2x}$

4. (a)

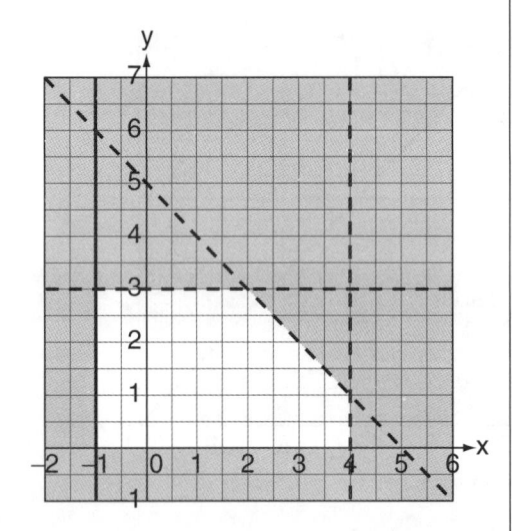

(b) (−1, 0), (−1, 1), (−1, 2), (0, 0), (0, 1), (0, 2), (1, 0), (1, 1), (1, 2), (2, 0), (2, 1), (2, 2), (3, 0), (3, 1)

5. (a) $15x + 25y \leqslant 2000$

(b) $y \leqslant x$

(c) $y \geqslant 35$

(d)

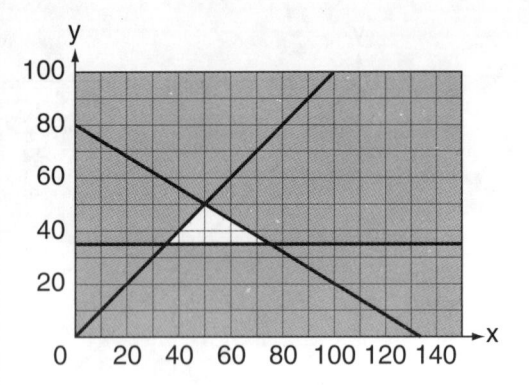

(e) 38

(f) $6.20

6. (a) $4\dfrac{1}{2}$ (b) $\dfrac{1}{2}\begin{pmatrix} 2 & 1 \\ 4 & 3 \end{pmatrix}$

7. (a) $\begin{pmatrix} 10 & 17 & 4 \\ -6 & -9 & 0 \end{pmatrix}$ (b) $\dfrac{1}{2}\begin{pmatrix} -2 & -4 \\ 3 & 5 \end{pmatrix}$

8. (a) $m = -1$ $c = 8$

(b)

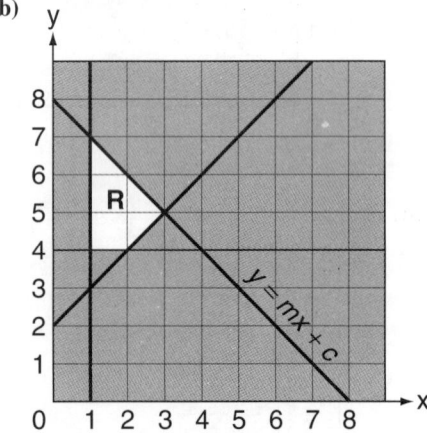

9. (a) $k = \dfrac{h}{9}$ (b) $\begin{pmatrix} 8 & 4 \\ -6 & 0 \end{pmatrix}$

Answer Key   411

10. (a) 13   (b) −4
11. (a) $x + y \leqslant 12$ (b) $y \geqslant 4$

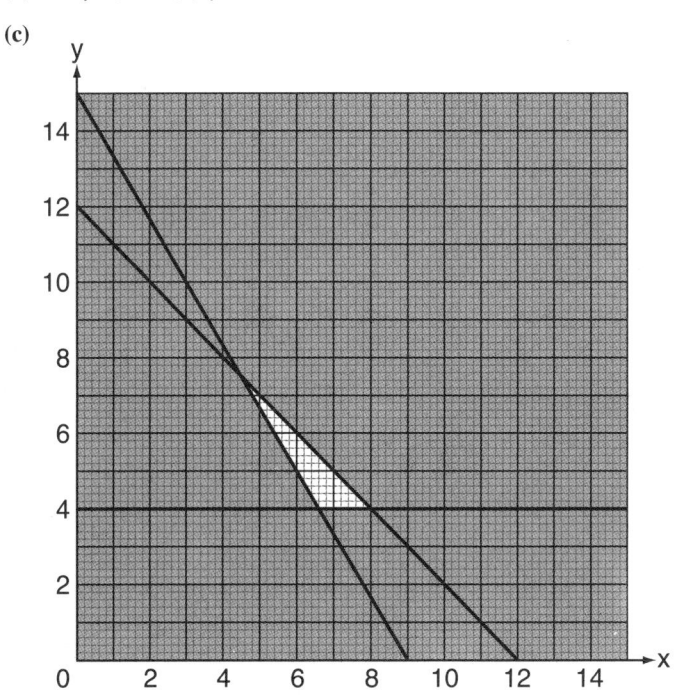

(c)

(e) 6 super and 5 mini, or 5 super and 7 mini

(f) (i) $274, $260   (ii) $94

12. (a) $x = 1$ or 3 (b) $g^{-1}(x) = \dfrac{x+1}{2}$
(c) $x = 0.76$ or 5.24 (d) 29
(e) $fg(x) = 4x^2 - 12x + 8$

13. (a) $y \geqslant \dfrac{1}{2}x$ (b) $x = -4$ or -3

14. (a) $(-1, 3)$ (b) $y < 3$, $y > \dfrac{1}{2}x$

15. (a) $\dfrac{2}{3}$ (b) $f^{-1}(x) = \dfrac{3x+4}{5}$

16. (a) (i) 3   (ii) −4.25 to −4
(b) (i) −1.6 or 2 or 8.6   (ii) 9.2
(c) $k = -9$ or 3 (d) $0 < x < 6$
(e) (i) $y = -x + 1$ (ii) 3

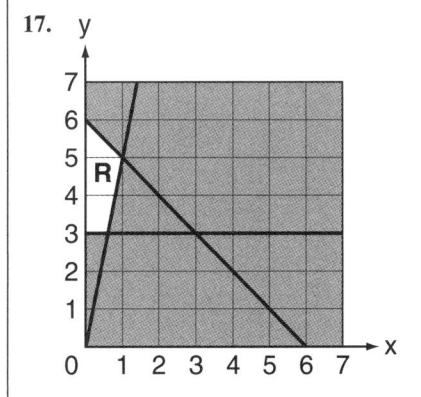

17.

18. (a) $f^{-1}f(x) = x$ (b) $ff(x) = 9x - 4$
19. (a) 3.16   (b) 0
20. (a) **BA**   (b) $\begin{pmatrix} 38 & 0 \\ 0 & 38 \end{pmatrix}$ (c) $\dfrac{1}{38}\begin{pmatrix} 4 & 6 \\ 5 & -2 \end{pmatrix}$

21. (a) $\begin{pmatrix} 0 & 1 \\ -1 & 2 \\ 0 & -3 \end{pmatrix}$ (b) $(1\ -1)$

22. (a) $\begin{pmatrix} 2 & 0 \\ 0 & 2 \end{pmatrix}$ (b) $\frac{1}{2}\begin{pmatrix} 4 & 2 \\ 1 & 1 \end{pmatrix}$

23. $\begin{pmatrix} -11 \\ -11 \\ -14 \end{pmatrix}$

24. (a) $\begin{pmatrix} 0 & 0 \\ 0 & 0 \end{pmatrix}$ (b) I

25. (a) $|A| = x^2 - 16$ (b) $x = -5$ or 5

26. (a) -14 (b) $gf(x) = 2x^3 - 6x^2 + 12x - 9$
(c) $g^{-1}(x) = \dfrac{x+1}{2}$

27. $a = 3, b = 4$

28. $\dfrac{1}{2}\begin{pmatrix} 5 & -3 \\ 4 & -2 \end{pmatrix}$

29. (a) $\begin{pmatrix} 2x+12 & 3x+6 \\ 14 & 15 \end{pmatrix}$ (b) $x = 5$

30. (a) (i) -2 (ii) 26 (iii) $\dfrac{1}{8}$
(b) $f^{-1}(x) = \dfrac{x+1}{2}$ (c) $x = \sqrt{z-1}$
(d) $gf(x) = 4x^2 - 4x + 2$ (e) $x = 9$
(f) $x = -4.24$ or 0.24
(g) (i)

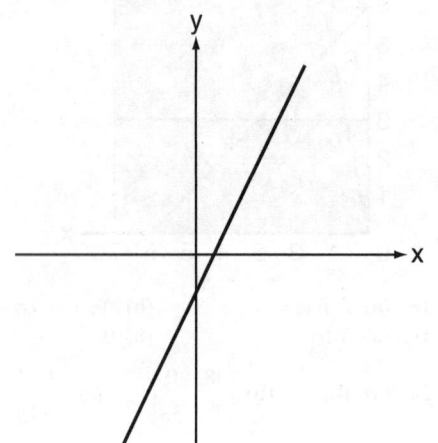

(ii)

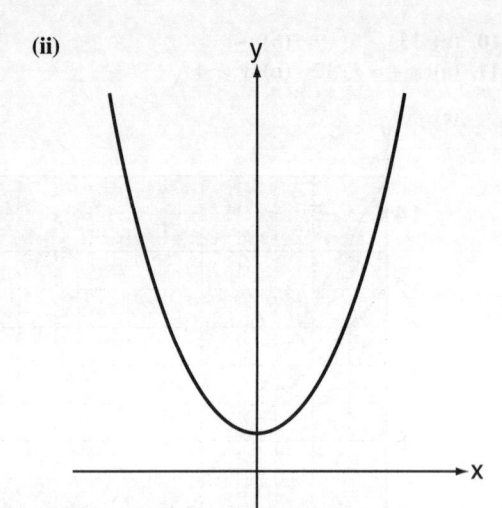

31. (a) $x + y \leqslant 12, x \geqslant 4$
(b)

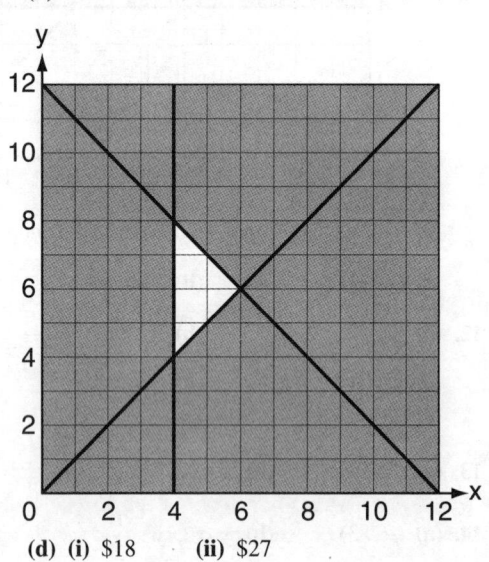

(d) (i) $18 (ii) $27

32. $\begin{pmatrix} 13 & 21 \\ 21 & 34 \end{pmatrix}$

33. (a) 8 (b) $f^{-1}(x) = \dfrac{5-x}{3}$ (c) 8

34. (a) $600x + 1200y \geqslant 720000$ (b) $x + y \leqslant 900$
(d) 300

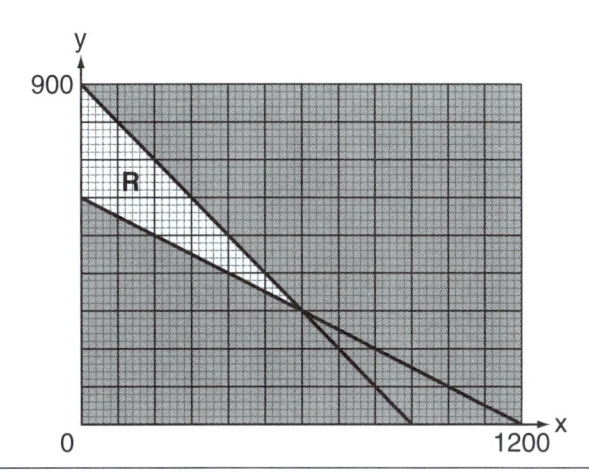

Chapter 9

Core Skills

1. (a) $a = 53.1°$
(b) $b = 16.6$ cm, $c = 25.0°$
(c) $d = 9.96$ cm, $e = 8.36$ cm
(d) $f = 15.7$ cm
(e) $g = 13.3$ cm, $h = 33.6°$
(f) $j = 4.04$ cm
(g) $k = 11.8$ cm, $l = 10.7$ cm
(h) $m = 6.93$ cm, $n = 9.17$ cm

2. 245°

3. 020°

4. (a) 135° (b) 337.5°

Exercise 9.1

1. 31.5 m
2. (a) 32.6 m (b) 73.1 m

Exercise 9.2

1. (a) 0.643 (b) 0.643 (c) 0.5
(d) 0.966 (e) 0.766 (f) −0.766
(g) −0.866 (h) 0.259

2. (a) 74.3° (b) 53.8°
(c) 105.7° (d) 126.2°

3. (a) 163.7° (b) 123.9°
(c) 145.2° (d) 129.6°

4. (a) 34.3°, 145.6° (b) 7.2°, 172.8°
(c) 57.5°, 122.5° (d) 32.2°, 147.8°

Exercise 9.3

1. $a = 36.3$ cm   **2.** $b = 23.8°$
3. $c = 14.4$ cm   **4.** $d = 20.4°$
5. $e = 60.4°$ **6.** $f = 12.5$ cm
7. $g = 18.8$ cm   **8.** $j = 65.8°$ or 114.2°

Exercise 9.4

1. $a = 95.7°$ **2.** $b = 14.9$ cm
3. $c = 8.55$ cm   **4.** $d = 28.1°$
5. $e = 52.4°$ **6.** $f = 142.6°$
7. $g = 7.57$ cm   **8.** $h = 13.7$ cm
9. $i = 111.8°$ **10.** $j = 46.2$ cm

Exercise 9.5

1. (a) $c = 3.42$ cm (b) $b = 12.7$ cm
(c) $d = 5.12$ cm (d) $e = 9.86$ cm
(e) $f = 137.0°$ (f) $a = 56.9°$
(g) $g = 8.09$ cm, $h = 31.4°$

2. (a) $a = 128°$ (b) $b = 30.8$ km
(c) $c = 32.5°$ (d) 273° or 272.5°

Exercise 9.6
1. (a) 26.6 cm² (b) 45.1 cm²
   (c) 1000 km² (d) 32.4 m²
   (e) 43.3 cm²
2. (a) (i) 50.6° (ii) 99.4° (iii) 23.1 cm²
   (b) (i) 129.4° (ii) 20.6° (iii) 8.22 cm²

Exercise 9.7
1. (a) 7.07 cm (b) 6.96 cm
   (c) 59.5° (d) 67.4°
2. 13.9 cm
3. (a) 4.79 (b) 15.7 cm
   (c) 16.6 cm (d) 24.9°

Exercise 9.8
1. (a) 22.1 cm (b) 43.6 cm
   (c) 553 cm² (d) 128.4°
2. (a) 100° (b) 15.3 cm (c) 66.3 cm²
3. (a) 7.37 cm (b) 14.0 cm
   (c) 136.8° (d) 17.7 cm²
4. (a) 6.46 cm (b) 3.77 cm
   (c) 79.2° (d) 27.8 cm²
5. (a) $(b+c)^2$ (b) a^2
   (c) $\frac{1}{2}bc$ (d) $\frac{(b+c)^2 - a^2}{4}$
   (e) $(b+c)^2 - a^2 = 2bc$
   $b^2 + 2bc + c^2 - a^2 = 2bc$
   $b^2 + c^2 = a^2$
   (f) Pythagoras' Theorem
6. (a) 37.9° (b) 0.6 to 0.61 m
   (c) 41.59 to 42°
7. (a) 43.1 cm
   (b) $\frac{1}{2} \times 61 \times 30 \times \sin 41 = 600$
   (c) The two triangles have the same height, and bases in the ratio
   $45 : 30 = 3 : 2$.

Note: Drop a perpendicular from *B* to the line *CA*. This is the height of both triangles.

   (d) 900 m²
   (e) 41.7 to 41.9 m (f) 21.0 to 21.1°

8. (a) $\frac{1}{2} \times 40.3 \times 26.8 \times \sin 92 = 540 \, \text{cm}^2$

   (b) $\frac{AB}{\sin 92} = \frac{40.3}{\sin 55}$

   $AB = \frac{40.3 \times \sin 92}{\sin 55} = 49.2 \, \text{cm}$

   (c) 55° (angles in the same segment)
   (d) 33°
   (e) similar
   (f) $\frac{XD}{40.3} = \frac{20.1}{26.8}$
   $XD = 30.2$ cm
9. (a) 0.176 (b) 1.76×10^{-1}
10. tan 100° cos 100° sin 100°
11. 7.31
12. 1
13. (a) 0 (b) −1.5
    (c) It is below the height at midday
14. (a) 232° (b) 175.4°
15. (a) 12 cm (b) 192 cm² (c) 67.4°
    (d) 36.8 to 36.9° (e) (i) 76.7° (ii) 6.40 cm
16. (i) $(2y-1)^2 = y^2 + (y+2)^2$
    $4y^2 - 4y + 1 = y^2 + y^2 - 4y + 4$
    $2y^2 - 8y - 3 = 0$
    (ii) −0.35 or 4.35 (iii) 13.8
17. (a) (i) 60° (ii) 13 km
    (b) (i) 145° (ii) 61.4° (iii) 15.3 km
    (c) 139 or 140 km²
18. (a) 2 (b) 30 cm³ (c) 45°
    (d) 37.5° (e) 4.92 to 4.93 cm
19. 7.94
20. 60, 120
21. (a) 0.5 (b) −1
    (c) $g^{-1}(f(x)) = \frac{\cos x - 4}{2}$
22. (a) (i) $\frac{1}{2}x(x+4) = 48$
    (ii) −12 or 8 (iii) 12 cm
    (b) $\frac{4}{5}$ or 0.8
    (c) (i) $(x+4)^2 + x^2 = 9^2$
    $x^2 + 8x + 16 + x^2 = 81$
    (ii) $x = \frac{-8 \pm \sqrt{64 + 4 \times 2 \times 65}}{4}$
    $x = -8.04$ or 4.04
    (iii) 21.1 cm
23. (a) $\frac{1}{2} \times 10 \times 14 \times \sin P = 48$
    $\sin P = \frac{48}{70}$, $P = 43.29 = 43.3°$ to 1 dp
    (b) 9.60 cm

Answer Key 415

24. (a) $(7x)^2 + (24x)^2 = 150^2$
$625x^2 = 22500, x^2 = 36$
(b) 336 cm
25. (a) 24.7 m (b) 11.5 m
26. (a) 393 to 393.5 km (b) 1210 km
(c) 820900 to 822000 km²
(d) (i) 073° (ii) 289°
(e) 1 : 20 000 000

27. (a) 14 46 or 2 46 pm
(b) (i) 260° (ii) 145°
(c) 85.0 km (d) 39.8° (e) 73.8 km
28. (a) 11.3° (b) 233°
29. (a) (i) 5.74 cm (ii) 6.32 cm
(b) 132 cm²
(c) (i) $PX^2 = 5.74^2 - 3^2$
(ii) 50.7 to 50.8° (iii) 78.3 to 79°
(iv) 44.4° (v) *PHN* or *PGM*

Chapter 10

NOTE: Diagrams have been drawn for some of the answers in this chapter. For others, enough information has been given for you to check your own diagrams.

Core Skills

1. (a) Reflection in the line $x = 3$
(b) Rotation 180° about the point (2, 3)
(c) Translation $\begin{pmatrix} -2 \\ -3 \end{pmatrix}$
(d) Enlargement, centre the origin, scale factor 2
2. (a) $\begin{pmatrix} -1 \\ 5 \end{pmatrix}$ (b) $\begin{pmatrix} 1 \\ 0 \end{pmatrix}$
3. (a) $x = -6, y = -3$
(b) $x = 3, y = 2$
4. *b* and *c* and *e*
5. They are parallel, but pointing in opposite directions, and *v* is 3 times longer than *w*.

Exercise 10.1

1. (a) $\sqrt{90} = 3\sqrt{10}$ (b) $\sqrt{50} = 5\sqrt{2}$
(c) $\sqrt{52} = 2\sqrt{13}$ (d) $\sqrt{29}$
2. (a) (i) $\begin{pmatrix} 8 \\ 2 \end{pmatrix}$ (ii) $\begin{pmatrix} 20 \\ -17 \end{pmatrix}$ (iii) $\begin{pmatrix} 6 \\ 0 \end{pmatrix}$
(b) (i) $\sqrt{149}$ (ii) $\sqrt{41}$
(iii) $\sqrt{104} = 2\sqrt{26}$
3. (a) $\overrightarrow{AC} = v + u$ (b) $\overrightarrow{BD} = -u + v$ or $v - u$
4. (a) Because *PQRS* is not a parallelogram (or $\overrightarrow{QR} \neq b$)
(b) $\overrightarrow{QS} = -a + b$ or $b - a$ (c) $\overrightarrow{PM} = \frac{1}{2}(a + b)$
5. (b) (i) $\overrightarrow{BA} = -p$
(ii) $\overrightarrow{BD} = q + r$

(iii) $\overrightarrow{AD} = p + q + r$ or $2q$
(iv) $\overrightarrow{FM} = p$
(v) $\overrightarrow{ME} + \overrightarrow{CB} = \overrightarrow{CD} + \overrightarrow{CB} = r - q$
(vi) $\overrightarrow{CF} - \overrightarrow{AB} = \overrightarrow{CF} + \overrightarrow{BA} = 2p - p = p$
(vii) $\overrightarrow{ED} + \overrightarrow{MA} + \overrightarrow{FM} = p - q + p = 2p - q$

Exercise 10.2

1. (a) $-a + b$ or $b - a$ (b) $-2a + b + c$
(c) $a + b - 2c$ (d) $-a + 2b - c$
2. (a) $-a + b$ or $b - a$ (b) $\frac{1}{2}(b - a)$
(c) $\frac{1}{2}(a + b)$

3.

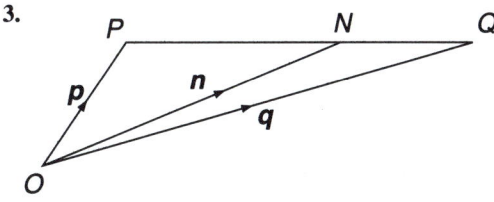

NOTE: Your diagram may not look like this because it depends on where you put *O*.

(a) $q - p$ (b) $\frac{2}{3}(q - p)$
(c) $\frac{1}{3}(p + 2q)$ (d) $\frac{1}{3}(q - p)$

4.

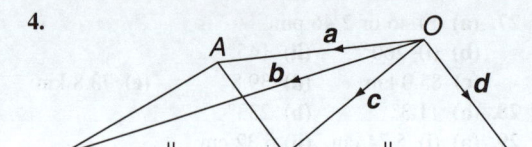

NOTE: As in question 3, the diagram depends on where you put O.

(a) $-b + a$ (b) $2(-b + c)$
(c) $-b + c$ (d) $a + b - 2c$

5. (a) $\begin{pmatrix} 7 \\ -1 \end{pmatrix}$

(b) $|\overrightarrow{AB}| = 3\sqrt{2}$, $|\overrightarrow{BC}| = 4\sqrt{2}$, $|\overrightarrow{AC}| = 5\sqrt{2}$

(c) $|\overrightarrow{AB}|^2 = 18$, $|\overrightarrow{BC}|^2 = 32$, $|\overrightarrow{AC}|^2 = 50$

so $|\overrightarrow{AC}|^2 = |\overrightarrow{BC}|^2 + |\overrightarrow{AB}|^2$, $\angle ABC = 90°$,

(by Pythagoras)

6. (a) (i) $\begin{pmatrix} 4 \\ 0 \end{pmatrix}$ (ii) $\begin{pmatrix} -1 \\ -9 \end{pmatrix}$ (iii) $\begin{pmatrix} 3 \\ -9 \end{pmatrix}$

(iv) $\begin{pmatrix} -7 \\ -9 \end{pmatrix}$ (v) $\begin{pmatrix} 0 \\ 0 \end{pmatrix}$

(b) (i) $3\sqrt{10}$ (ii) $\sqrt{82}$ (iii) 4

Exercise 10.3

1. (a) $(2, -1)$ (b) $(-6, -1)$
(c) $(11, 6)$ (d) $(8, 0)$

2. (a) $\begin{pmatrix} -1 \\ -20 \end{pmatrix}$ (b) $\begin{pmatrix} -9 \\ -7 \end{pmatrix}$ (c) $\begin{pmatrix} -2 \\ 8 \end{pmatrix}$

Exercise 10.4

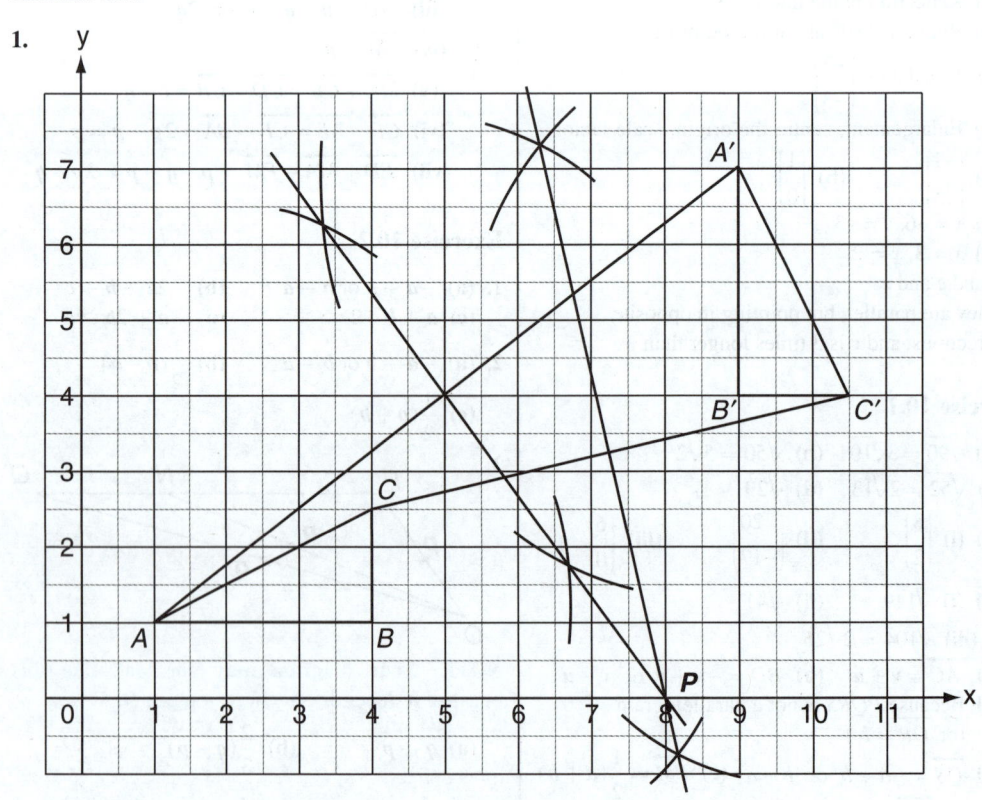

1.

(f) P is the centre of the rotation which maps triangle ABC on to triangle $A'B'C'$. AA' and CC' are chords of two circles which have the same centre. The perpendicular bisectors of the chords pass through the centre of the circles, so where the bisectors meet is the centre of rotation.

(g) rotation, 90° clockwise, centre $(8, 0)$.

2. (a)

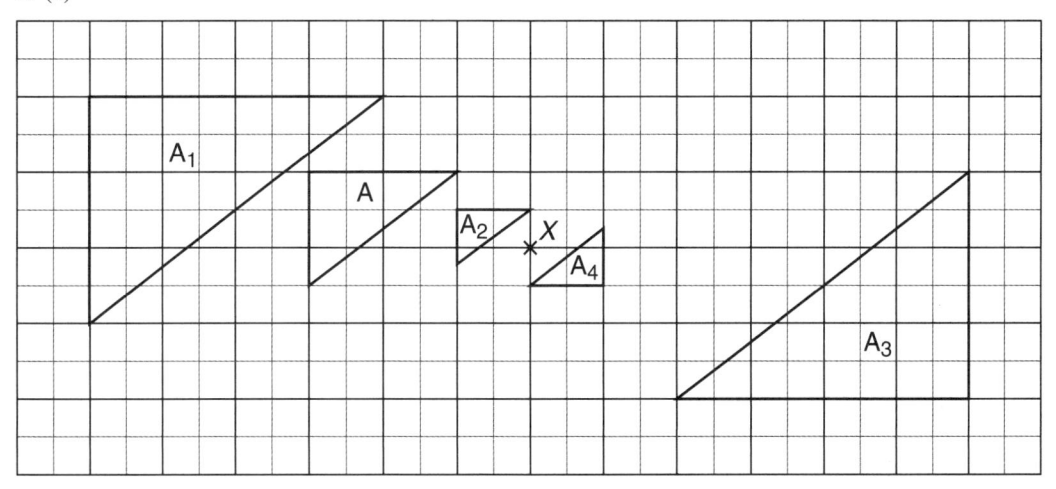

(b) enlargement centre X, scale factor $-\dfrac{1}{4}$

3.

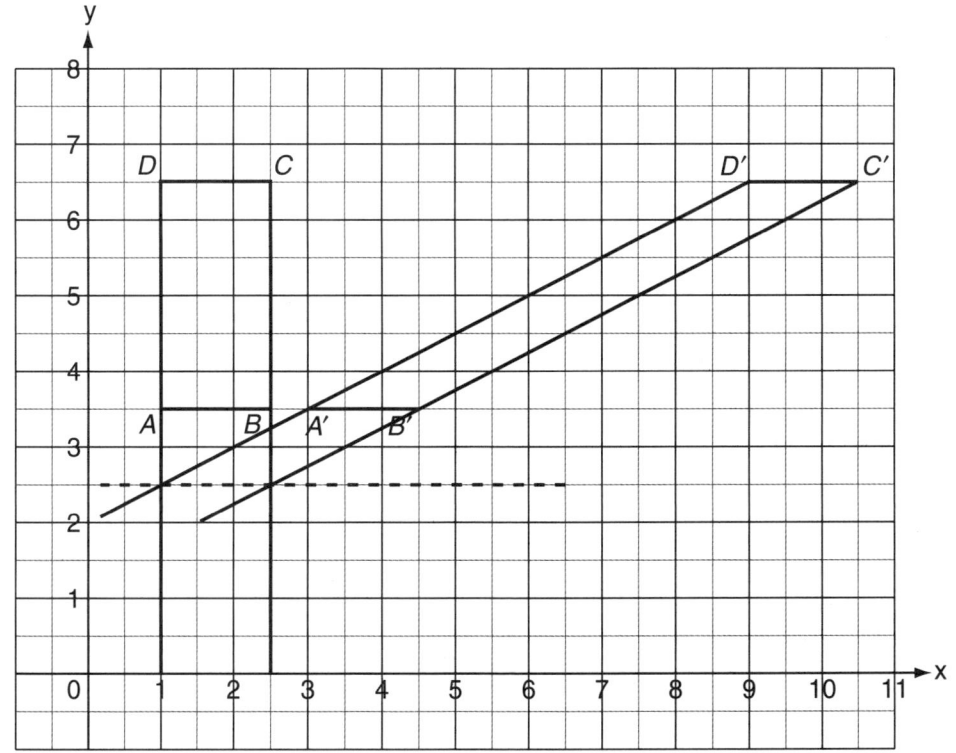

(c) shear, factor 2, invariant line $y = 2.5$

4. (a) (i) $f^{-1}(x) = \dfrac{2x}{6-x}$
   (ii), (iii), (iv)

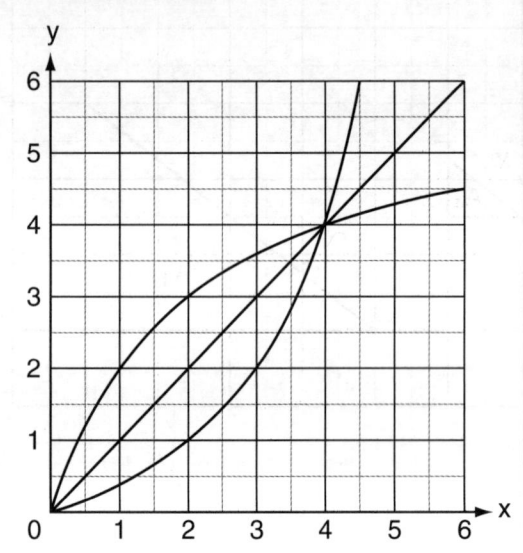

   (v) reflection in the line $y = x$

(b) (i) $f^{-1}(x) = \dfrac{3}{2}(x-1)$
   (ii), (iii), (iv)

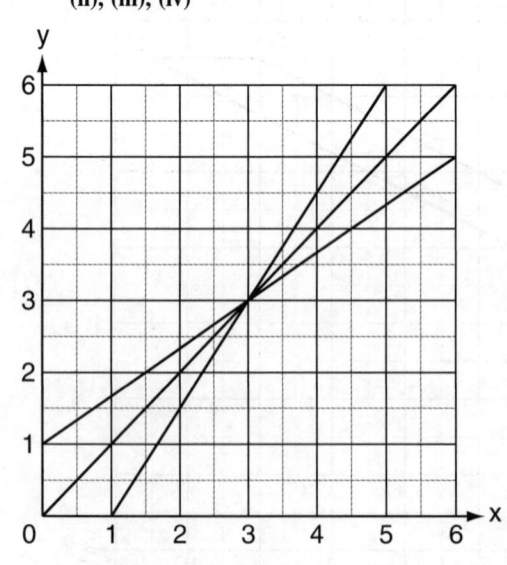

   (v) reflection in the line $y = x$

5. (a) The sides of triangles A and A' produced should meet on the line $x = 1.5$.
   Stretch, factor 3, invariant line $x = 1.5$
(b) The diagonals of rectangles B and B' produced meet on the y-axis.
   Stretch, factor 2, invariant line y-axis

Exercise 10.5

1. enlargement, centre (0, 0), scale factor $-\dfrac{1}{2}$
2. (a) vertices at (2, −1), (5, −4), (7, −4)
   (b) vertices at (1, 3), (1, −3), (3, −3)
   (c) (i) vertices at (−1.5, 1.5), (−4.5, 1.5), (−1.5, 6)
       (ii) enlargement, centre (0, 0), scale factor $-\dfrac{2}{3}$
3. shear, factor 2, invariant line $y = 2$

Exercise 10.6

1. (a) $\begin{pmatrix} 0 & -1 \\ -1 & 0 \end{pmatrix}$ (b) $\begin{pmatrix} 1 & 0 \\ 2 & 1 \end{pmatrix}$

   (c) $\begin{pmatrix} 3 & 0 \\ 0 & 1 \end{pmatrix}$ (d) $\begin{pmatrix} -2 & 0 \\ 0 & -2 \end{pmatrix}$

2. (a) $\begin{pmatrix} 2 & 0 \\ 0 & 2 \end{pmatrix}$ (b) enlargement, centre (0, 0), scale factor 2

3. (a) $\begin{pmatrix} 0 & -1 \\ -1 & 0 \end{pmatrix}$ (b) reflection in the line $y = -x$

   (c) $\begin{pmatrix} 0 & -1 \\ -1 & 0 \end{pmatrix} \begin{pmatrix} -1 & 0 \\ 6 & -4 \end{pmatrix} = \begin{pmatrix} -6 & 4 \\ 1 & 0 \end{pmatrix}$

4. (a) $\begin{pmatrix} 2 \\ -2 \end{pmatrix}$ (b) $\begin{pmatrix} -2 \\ 0 \end{pmatrix}$ (c) $\begin{pmatrix} 3 \\ -3 \end{pmatrix}$

5. $A' = (1, -3)$, $B' = (-5, 7)$, $C' = (0, 1)$

Exercise 10.7

1. (i) reflection in the line $y = x$
   Enlargement, centre (0, 0), scale factor 2.5

   (ii) $\begin{pmatrix} 0 & 0 \\ 1 & 0 \end{pmatrix}$, $\begin{pmatrix} 2.5 & 0 \\ 0 & 2.5 \end{pmatrix}$, $\begin{pmatrix} 0 & 2.5 \\ 2.5 & 0 \end{pmatrix}$

   (iii) No. $\begin{pmatrix} 0 & 1 \\ 1 & 0 \end{pmatrix} \begin{pmatrix} 2.5 & 0 \\ 0 & 2.5 \end{pmatrix} = \begin{pmatrix} 2.5 & 0 \\ 0 & 2.5 \end{pmatrix} \begin{pmatrix} 0 & 1 \\ 1 & 0 \end{pmatrix}$
   $= \begin{pmatrix} 0 & 2.5 \\ 2.5 & 0 \end{pmatrix}$

2. (a) $\begin{pmatrix} 2 & 1 \\ 2 & 3 \end{pmatrix}$
   (b) vertices at (0, 0), (2, 2), (1, 3), (3, 5)
   (c) $-\dfrac{1}{4}\begin{pmatrix} 3 & 1 \\ -2 & -2 \end{pmatrix}$

3. (b) $A_2(-2, -1)$, $B_2(-4, -1)$, $C_2(-2, -2)$
   (c) reflection in $y = x$

4. (a) $I = \begin{pmatrix} 1 & 0 \\ 0 & 1 \end{pmatrix}$ (b) $R_1 = \begin{pmatrix} 0 & 1 \\ -1 & 0 \end{pmatrix}$

   (c) $R_2 = \begin{pmatrix} -1 & 0 \\ 0 & -1 \end{pmatrix}$ (d) $M_1 = \begin{pmatrix} 1 & 0 \\ 0 & -1 \end{pmatrix}$

Answer Key 419

(e) $M_2 = \begin{pmatrix} -1 & 0 \\ 0 & 1 \end{pmatrix}$ (f) $M_3 = \begin{pmatrix} 0 & 1 \\ 1 & 0 \end{pmatrix}$

(g) $M_4 = \begin{pmatrix} 0 & -1 \\ -1 & 0 \end{pmatrix}$

5. $R_1^{-1} = \begin{pmatrix} 0 & -1 \\ 1 & 0 \end{pmatrix}$ Rotation 90 anticlockwise about (0, 0)

6.   Second transformation

	R_1	R_2	M_1	M_2	M_3	M_4
R_1	R_2	R_1^{-1}	M_3	M_4	M_2	M_1
R_2	R_1^{-1}	I	M_2	M_1	M_4	M_3
M_1	M_4	M_2	I	R_2	R_1^{-1}	R_1
M_2	M_3	M_1	R_2	I	R_1	R_1^{-1}
M_3	M_1	M_4	R_1	R_1^{-1}	I	R_2
M_4	M_2	M_3	R_1^{-1}	R_1	R_2	I

First transformation (row labels)

Exercise 10.8

1. $x = 0, y = 8$
2. (a) $A'(10.5, 3), B'(3, 7.5)$
   (b) similar
   (c) 40 square units
3. (a), (b)

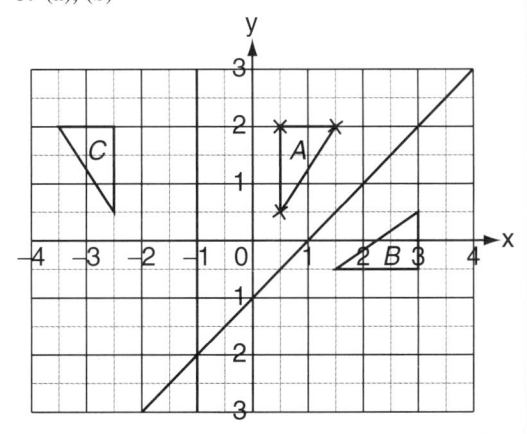

(c) Rotation, 90° anticlockwise, centre $(-1, -2)$
4. (b) shear factor 1.5
5. shear, factor 2, invariant line $y = 0.5$
6. (a) $D = (2, 1), B' = (7, 3), D' = (4, 1)$
   (b) scale factor = 3
7. (a) $\begin{pmatrix} 0 & 5 & 7 & 2 \\ 0 & 1 & 5 & 4 \end{pmatrix}$
   (b) (ii) rhombus
   (c) 18 square units
   (d) (i) det $M = 18$ (ii) They are the same

8. (a) (i) translation $\begin{pmatrix} -3 \\ 0 \end{pmatrix}$
   (ii) rotation, 90° anticlockwise, centre (0, 1)
   (iii) $\begin{pmatrix} -1 & 0 \\ 0 & 1 \end{pmatrix}$
   (b) (i) -2 (ii) (3, 1)
   (c) (i) 2 (ii) $\frac{1}{2}\begin{pmatrix} 4 & -3 \\ 2 & -1 \end{pmatrix}$
   (iii) $\begin{pmatrix} -1 & 3 \\ -2 & 4 \end{pmatrix}\begin{pmatrix} x \\ y \end{pmatrix} = \begin{pmatrix} 4 \\ -2 \end{pmatrix}$, $x = 11, y = -2$
   $X = \begin{pmatrix} 11 \\ 5 \end{pmatrix}$

9. (c) (i) $A_1 = (5, -7), B_1 = (8, -7), C_1 = (8, -5)$
   (ii) $A_2 = (-4, 2), B_2 = (-7, 2), C_2 = (-7, 4)$
   (iii) $A_3 = (-2, -2), B_3 = (-5, -2), C_3 = (-5, -4)$
   (d) (i) $A_4 = (3, 2), B_4 = (7.5, 2), C_4 = (7.5, 4)$
   (ii) $\frac{1}{15}\begin{pmatrix} 1 & 0 \\ 0 & 1.5 \end{pmatrix}$
   (iii) stretch, y-axis invariant, factor $\frac{2}{3}$

10. $\frac{1}{2}a - \frac{1}{2}c$

11. (c) $A_1 = (1, 2), B_1 = (3, 3), C_1 = (5, 1)$
    (d) $A_2 = (-2, 1), B_2 = (-3, 3), C_2 = (-5, 1)$
    (e) reflection in the y-axis
    (f) (i) $A_3 = (2, -1), B_3 = (3, 0), C_3 = (5, -4)$
    (ii) shear, y-axis invariant, factor 1
    (iii) $\begin{pmatrix} 1 & 0 \\ 1 & 1 \end{pmatrix}$

12. (a) (i) (5, 3) (ii) (3, 5)
    (b) $\begin{pmatrix} 0 & 1 \\ 1 & 0 \end{pmatrix}$
    (c) $M(Q) = (k - 3, k - 2)$
    $TM(Q) = (k - 3 + 3, k - 2 + 2) = (k, k)$
    this point lies on the line $y = x$
    (d) $\begin{pmatrix} 0 & 1 \\ 1 & 0 \end{pmatrix}$
    (e) (i) $\begin{pmatrix} 0 & 1 \\ -1 & 0 \end{pmatrix}$ (ii) rotation, centre (0, 0), 90° clockwise

13. (a) (i) $-p + q$ (ii) $\frac{2}{3}(-p + q)$
    (iii) $-\frac{2}{3}p - \frac{1}{3}q$ (iv) $\frac{1}{3}p + \frac{2}{3}q$
    (b) (i) (4, -2) (ii) $\begin{pmatrix} -3 \\ 4 \end{pmatrix}$
    (c) (i) rotation, 90° clockwise about (0, 0)

(ii) (3, −5)

(d) $\begin{pmatrix} 0 & 1 \\ 1 & 0 \end{pmatrix}$

14. (a) (i) a (ii) $-a + b$ (iii) $a + b$
(b) triangle OAB is equilateral, so the lengths of OA, OB and AB are equal
(c) (i) (a) b (b) $3b$
(ii) Y, A and X lie on a straight line
(d) $3a$
(e) $\overrightarrow{XZ} = -3a, \overrightarrow{YX} = 3b, \overrightarrow{YZ} = 3(b - a)$,
since $|a| = |b| = |b - a|$, then $|\overrightarrow{XZ}| = |\overrightarrow{YX}| = |\overrightarrow{YZ}|$
The triangle is equilateral
(f) $\dfrac{1}{9}$

15. (a) R is 1 cm (2 small squares) below P
(b) $h = -\dfrac{3}{4}$

16. (a) (0, −1)   (b) reflection in $y = -x$
(c) (2, 1), (2, −1), (3, −3)
(d) (5, 3), (5, 4), (7, 4)

17. (a) stretch, factor 2, y-axis invariant
(b) (i) $\begin{pmatrix} 8 \\ 0 \end{pmatrix}$
(ii) (a) (4, 0), (−7, −2), (9, 2)   (b) 4

18. (a) (i) translation $\begin{pmatrix} -6 \\ 1 \end{pmatrix}$
(ii) reflection in $y = -x$
(iii) enlargement, centre (0, 6), scale factor 3
(iv) shear, x-axis invariant, factor 0.5
(b) (i) $\begin{pmatrix} 0 & -1 \\ -1 & 0 \end{pmatrix}$ (ii) $\begin{pmatrix} 1 & 0.5 \\ 0 & 1 \end{pmatrix}$

19. (a) (i) rotation 90° anticlockwise about (0, 0)
(ii) translation $\begin{pmatrix} -2 \\ -5 \end{pmatrix}$
(iii) reflection in $y = -x$
(iv) 180° rotation about (1, −1)
(v) enlargement, centre (0, 0), scale factor 2
(vi) shear y-axis invariant, scale factor 1
(b) B
(c) (i) $\begin{pmatrix} 1 & 0 \\ 0 & -1 \end{pmatrix}$ (ii) $\begin{pmatrix} 1 & 0 \\ 1 & 1 \end{pmatrix}$

20. (c) (i) (6, 8), (10, 6), (12, 10)
(ii) $\begin{pmatrix} 0 & 1 \\ 1 & 0 \end{pmatrix}$
(d) (i) (4, 3), (3, 5), (5, 6)
(ii) enlargement, centre (0, 0), scale factor 0.5
(e) (4, 6), (3, 10), (5, 12)

21. (a) (i) $p + r$ (ii) $-p + r$
(iii) $-p + \dfrac{2}{3}r$

(iv) $p + \dfrac{1}{2}r$

(b) (i) $-\dfrac{3}{2}p + r$ (ii) $-\dfrac{3}{2}p$
(c) Q, R and S lie on a straight line

22. (a) $2\sqrt{25^2 + 7^2} = 48$
(b) (i) 147°   (ii) 33°
(c) (i) $p + q$ (ii) $-p + q$
(d) $p + 3q$ (e) $0.5p + 2.5q$
(f) (i) $\begin{pmatrix} 0 \\ 24 \end{pmatrix}$ (ii) $\begin{pmatrix} 7 \\ -24 \end{pmatrix}$
(g) 50

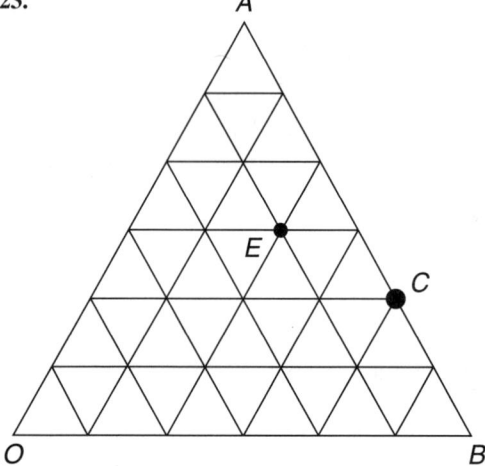

23.
(b) $\dfrac{1}{2}a + \dfrac{1}{3}b$ (c) $-\dfrac{1}{2}a + \dfrac{2}{3}b$

24. (a) (i) translation $\begin{pmatrix} 0 \\ -11 \end{pmatrix}$
(ii) reflection in $x = 1$
(iii) reflection in $y = -x$
(iv) enlargement, centre (2, 0), scale factor 0.5
(v) stretch, factor 2, x-axis invariant
(b) (i) $\begin{pmatrix} 0 & -1 \\ -1 & 0 \end{pmatrix}$ (ii) $\begin{pmatrix} 1 & 0 \\ 0 & 2 \end{pmatrix}$

25. (a) (i) $\begin{pmatrix} 2 \\ 1 \end{pmatrix}$ (ii) $\begin{pmatrix} 1 \\ 2 \end{pmatrix}$
(b) translation $\begin{pmatrix} 0 \\ -4 \end{pmatrix}$
(c) $y \geqslant 0$, $x \leqslant 2$, $y \geqslant \dfrac{1}{2}x$, $y \leqslant 2x + 4$

26. (a) (i) (−1, −2), (−1, −3), (−3, −2)
(ii) reflection in $y = -x$
(b) $\begin{pmatrix} 0 & -1 \\ 1 & 0 \end{pmatrix}$

Chapter 11

Core Skills

1. (a)

Data item	1	2	3	4	5	6	7	8	9
Frequency	5	4	5	3	6	4	4	4	4

(b) mean = 4.87
median = 5
mode = 5
range = 8

2.

3. $a = 120°$ $b = 6$ $c = 60°$

4. (a) no correlation
   (b) positive correlation

5.

Heights of students (h cm)	Lower class boundary	Upper class boundary
141 to 150	140.5	150.5
151 to 160	150.5	160.5
161 to 170	160.5	170.5

Exercise 11.1

1. (a)

Class (l cm)	Frequency (f)	Class boundaries		Class width (w)	Frequency density ($f \div w$)
$20 < l \leqslant 30$	1	20	30	10	0.1
$30 < l \leqslant 50$	10	30	50	20	0.5
$50 < l \leqslant 70$	15	50	70	20	0.75
$70 < l \leqslant 100$	20	70	100	30	0.67
$100 < l \leqslant 150$	3	100	150	50	0.06

Modal class = $50 < l \leqslant 70$

(b)

Class	Frequency (f)	Class boundaries		Class width (w)	Frequency density ($f \div w$)
0–30	5	0	30.5	31	0.16
31–50	25	30.5	50.5	20	1.25
51–70	41	50.5	70.5	20	2.05
71–100	50	70.5	100.5	30	1.67
101–150	7	100.5	150.5	50	0.14
151–250	2	150.5	250.5	100	0.02

Modal class = 51–70

2. (a)

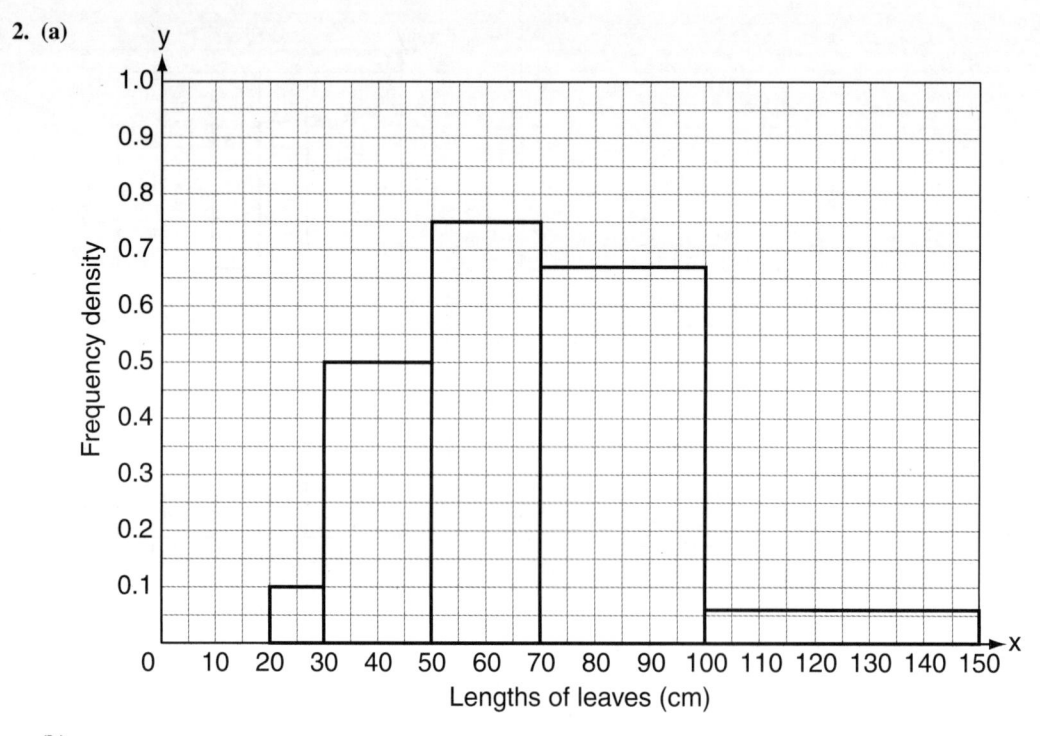

Frequency density vs Lengths of leaves (cm)

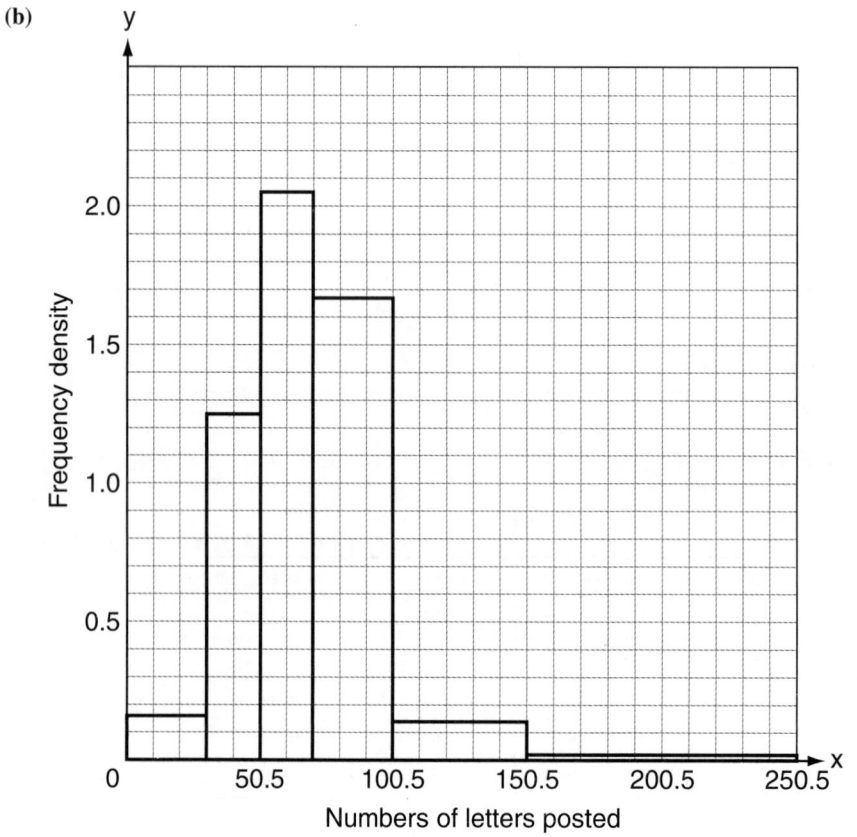

(b)

Frequency density vs Numbers of letters posted

3. (a)

Age (a)	Class boundaries		Class width	Frequency density	Frequency
$0 \leqslant a < 20$	0	20	20	0.2	4
$20 \leqslant a < 40$	20	40	20	0.5	10
$40 \leqslant a < 50$	40	50	10	1.0	10
$50 \leqslant a < 65$	50	65	15	0.8	12
$65 \leqslant a < 85$	65	85	20	0.7	14
$85 \leqslant a < 110$	85	110	25	0.4	10

(b)

Number of peas	Class boundaries		Class width	Frequency density	Frequency
1–3	0.5	3.5	3	2	6
4–5	3.5	5.5	2	5	10
6–8	5.5	8.5	3	3	9
9–10	8.5	10.5	2	0.5	1

4. (a), (c)

Class	$0 < s \leqslant 1000$	$1000 < s \leqslant 5000$	$5000 < s \leqslant 10000$	$10000 < s \leqslant 20000$
Frequency	20	100	105	100
Class width	1000	4000	5000	10000
Frequency density	0.02	0.025	0.021	0.01

(b)

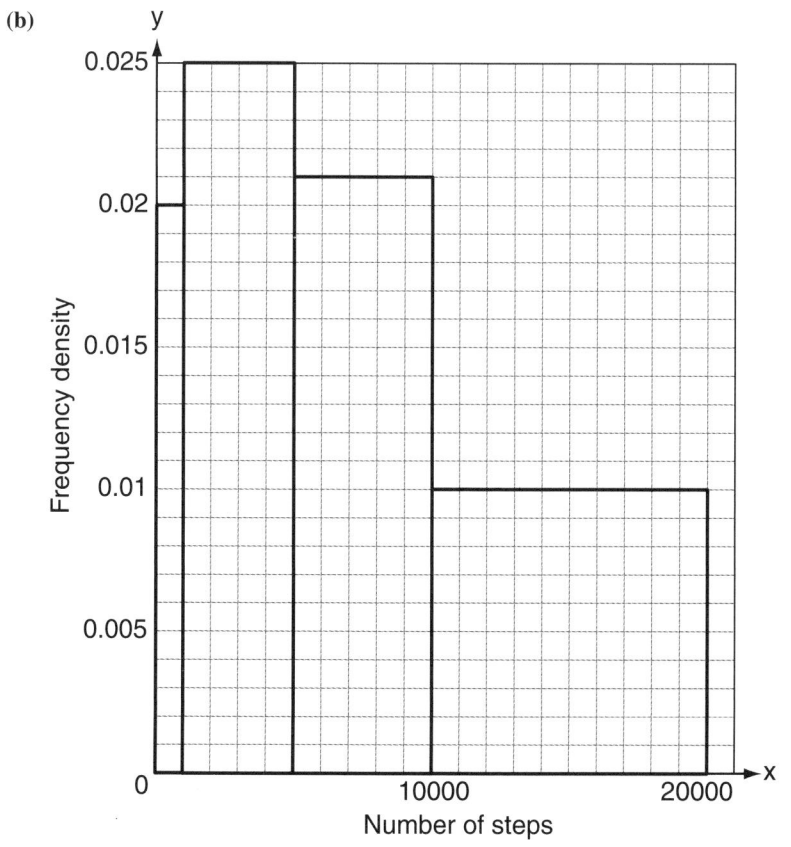

(d) $1000 < s \leqslant 5000$

424 *Extended Mathematics for Cambridge IGCSE*

Exercise 11.2
1. 30.1 2. 5.17 3. 1.70

Exercise 11.3
1. (a)

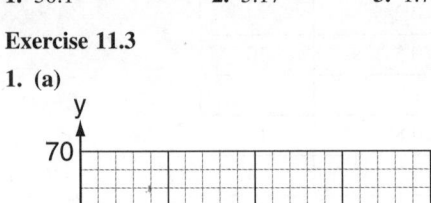

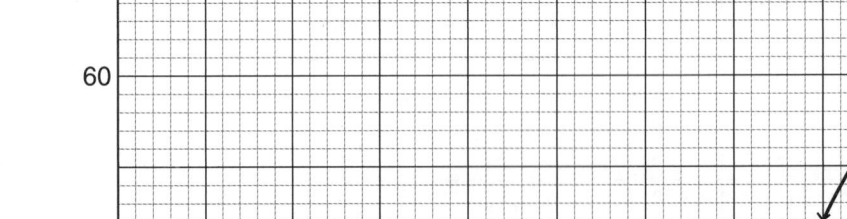

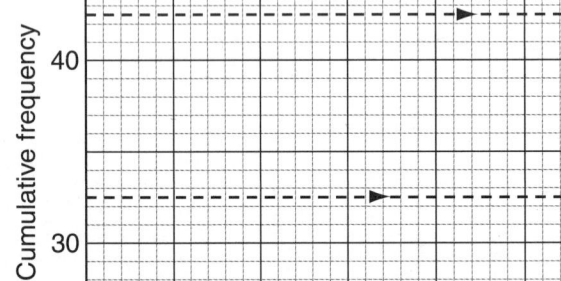

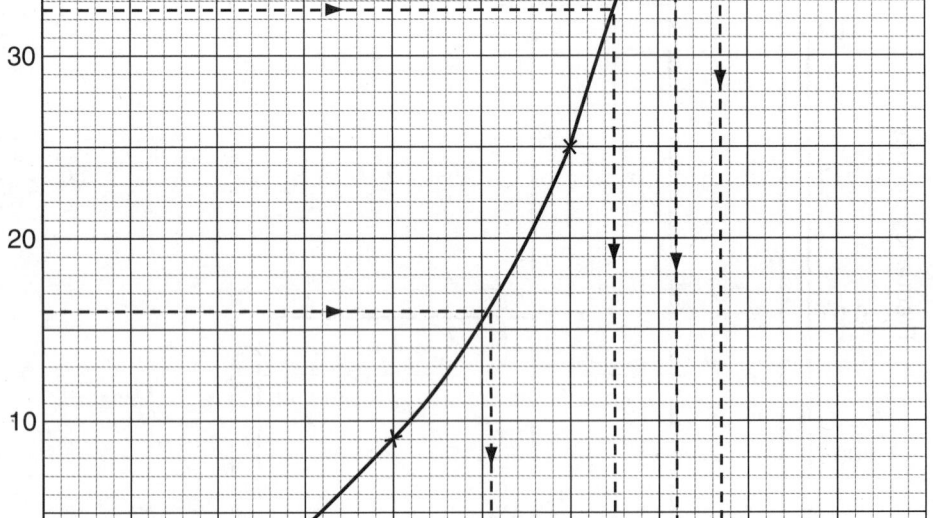

(b) median = 31.5 to 33.5
Q_1 = 24.5 to 26.5 Q_2 = 37.5 to 39.5
Inter-quartile range = 11 to 15
(c) 65th percentile = 35 to 37

2. (a)

Class	$65 < m \leqslant 70$	$70 < m \leqslant 75$	$75 < m \leqslant 80$	$80 < m \leqslant 85$	$85 < m \leqslant 90$	$90 < m \leqslant 100$
Frequency	3	5	6	7	4	3
Cumulative frequency	3	8	14	21	25	28

(b)

(c) (i) 79.5 to 80.5 kg
   (ii) 10 to 12 kg
   (iii) 77 to 78 kg

Exercise 11.4

1.

Time	Class boundaries		Midpoint	Class width	Frequency	Cumulative frequency	Frequency density
$70 < t \leqslant 80$	70	80	75	10	4	4	0.4
$80 < t \leqslant 90$	80	90	85	10	5	9	0.5
$90 < t \leqslant 100$	90	100	95	10	9	18	0.9
$100 < t \leqslant 120$	100	120	110	20	38	56	1.9
$120 < t \leqslant 160$	120	160	140	40	8	64	0.2

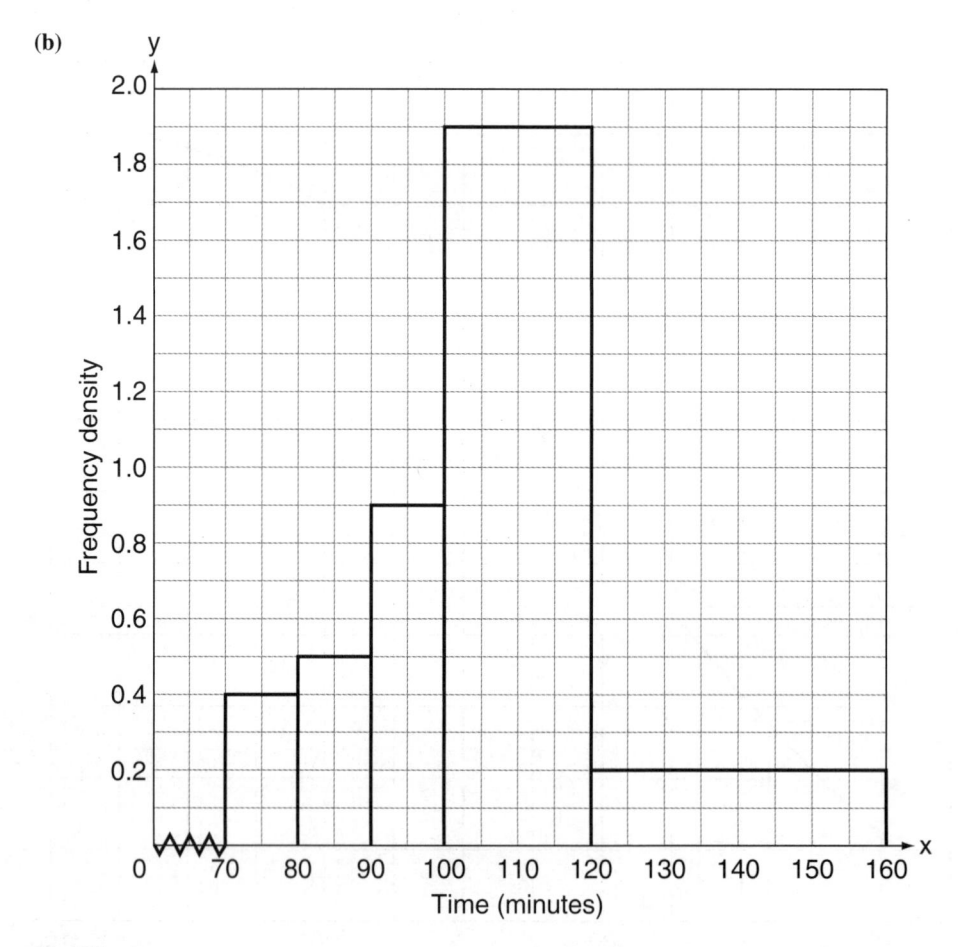

(b)

(c) 108 min

(d) $100 < t \leqslant 120$

(e)

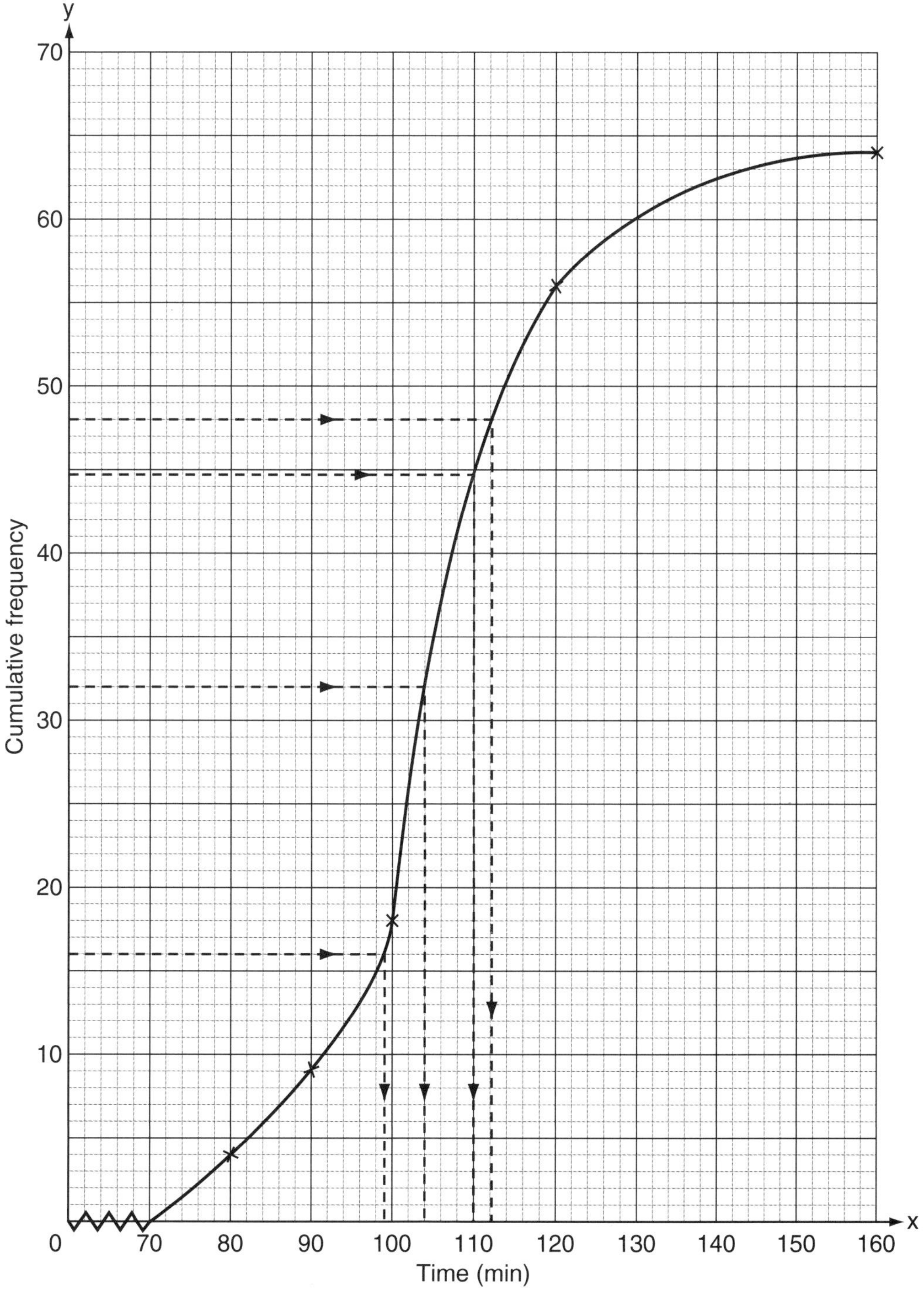

(f) median = 102 to 106 lower quartile = 97 to 101
upper quartile = 110 to 114 70th percentile = 108 to 112

2. (a)

Class	Class width	Midpoint	Frequency	Cumulative frequency	Frequency density
$70 < t \leqslant 80$	10	75	6	6	0.6
$80 < t \leqslant 90$	10	85	12	18	1.2
$90 < t \leqslant 100$	10	95	20	38	2.0
$100 < t \leqslant 110$	10	105	24	62	2.4
$110 < t \leqslant 120$	10	115	6	68	0.6

(b) 96.8 minutes

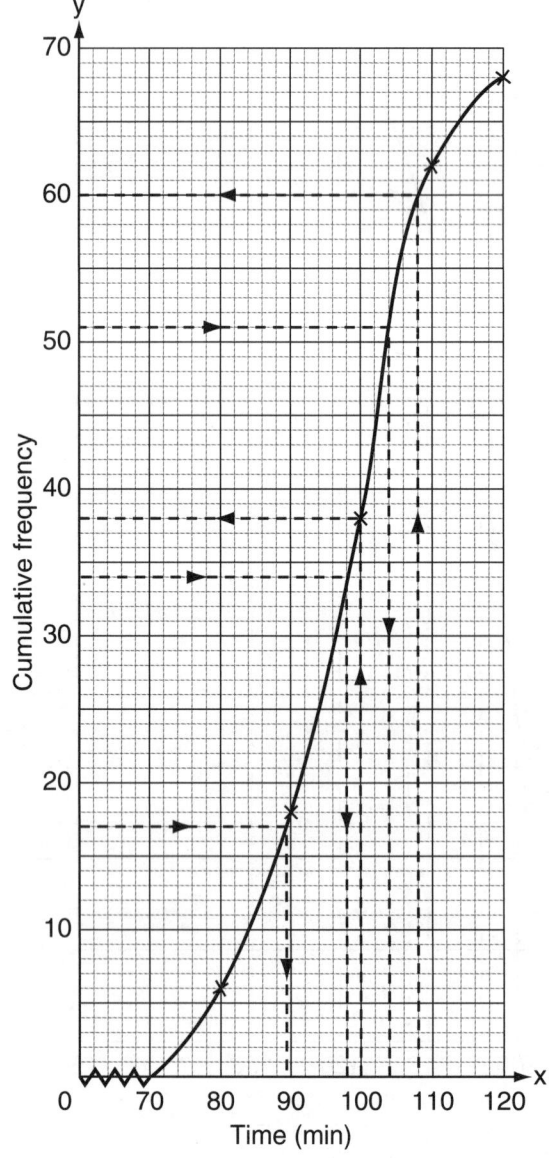

(c)

(d) median = 96 to 100 Q_1 = 88 to 92 Q_2 = 102 to 106
(e) one of 37, 38 or 39   (f) one of 7, 8 or 9

Answer Key 429

3. Total girls = 25.4 × 12 = 305 to 3 s.f.
   Total boys = 23.8 × 15 = 357
   Total students = 662
   Total classes = 22
   Mean number of students per class = $\frac{662}{22} = 30.1$
4. 71.8%
5. (a) 24 (b) 77.2%
6. (a) (i) $125 < h \leqslant 135$ (ii) 126.25 cm
   (b) (i) 11 cm (ii) 16
7. (a) $1.5 < x \leqslant 2$ (b) 1.73 litres

(c)

Amount of water (x litres)	Number of people	Cumulative frequency
$0 < x \leqslant 0.5$	8	8
$0.5 < x \leqslant 1$	27	35
$1 < x \leqslant 1.5$	45	80
$1.5 < x \leqslant 2$	50	130
$2 < x \leqslant 2.5$	39	169
$2.5 < x \leqslant 3$	21	190
$3 < x \leqslant 3.5$	7	197
$3.5 < x \leqslant 4$	3	200

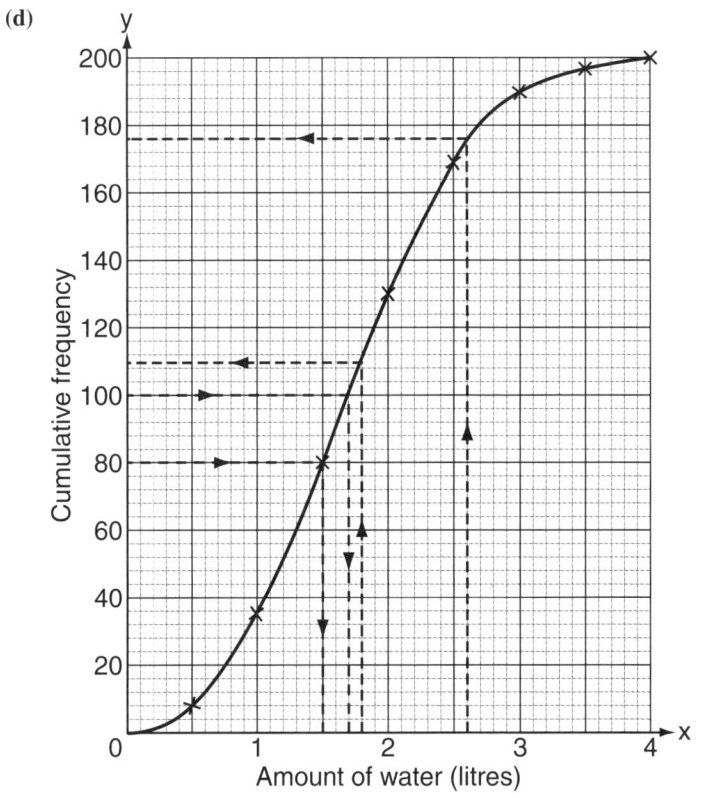

(d)

NOTE: Shown half-scale. Yours should be twice this scale.

(e) (i) 1.65 to 1.75 litres
 (ii) 1.5 litres
 (iii) one of 23, 24, 25, 26, 27, 28 or 29
(f) 54 to 56.5%

8. (a) (i) 1
 (ii) 3
 (iii) 4
(b) 38.2 seconds
(c) (i) $p = 20$, $q = 72$

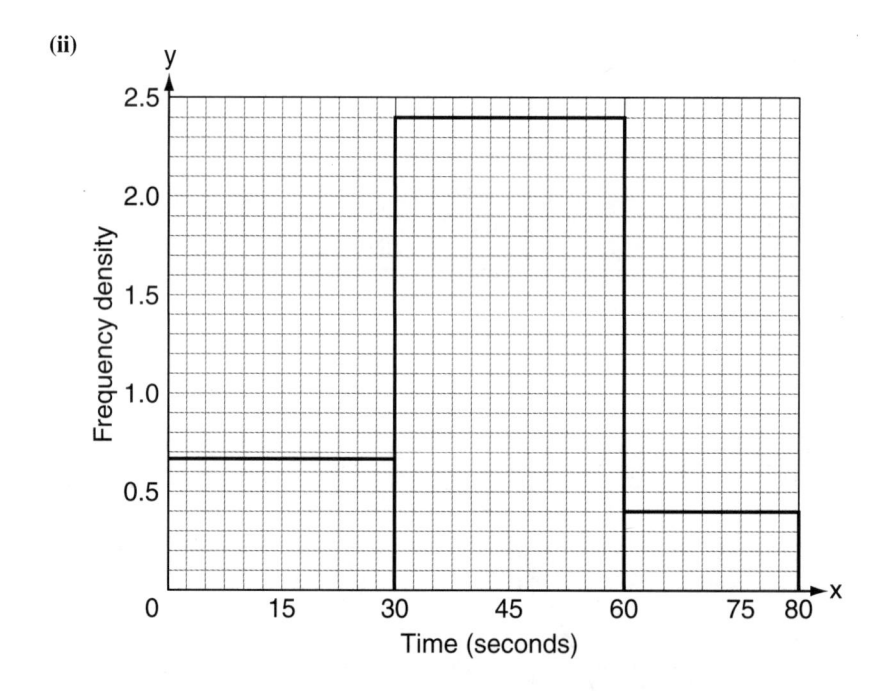

(ii)

9. (a) $160 < h \leqslant 170$
(b) (i) 162 cm
 (ii) The distribution of heights in each class interval is estimated by using the mean of each class
(c) $p = 15$ $q = 39$ $r = 75$

(d)

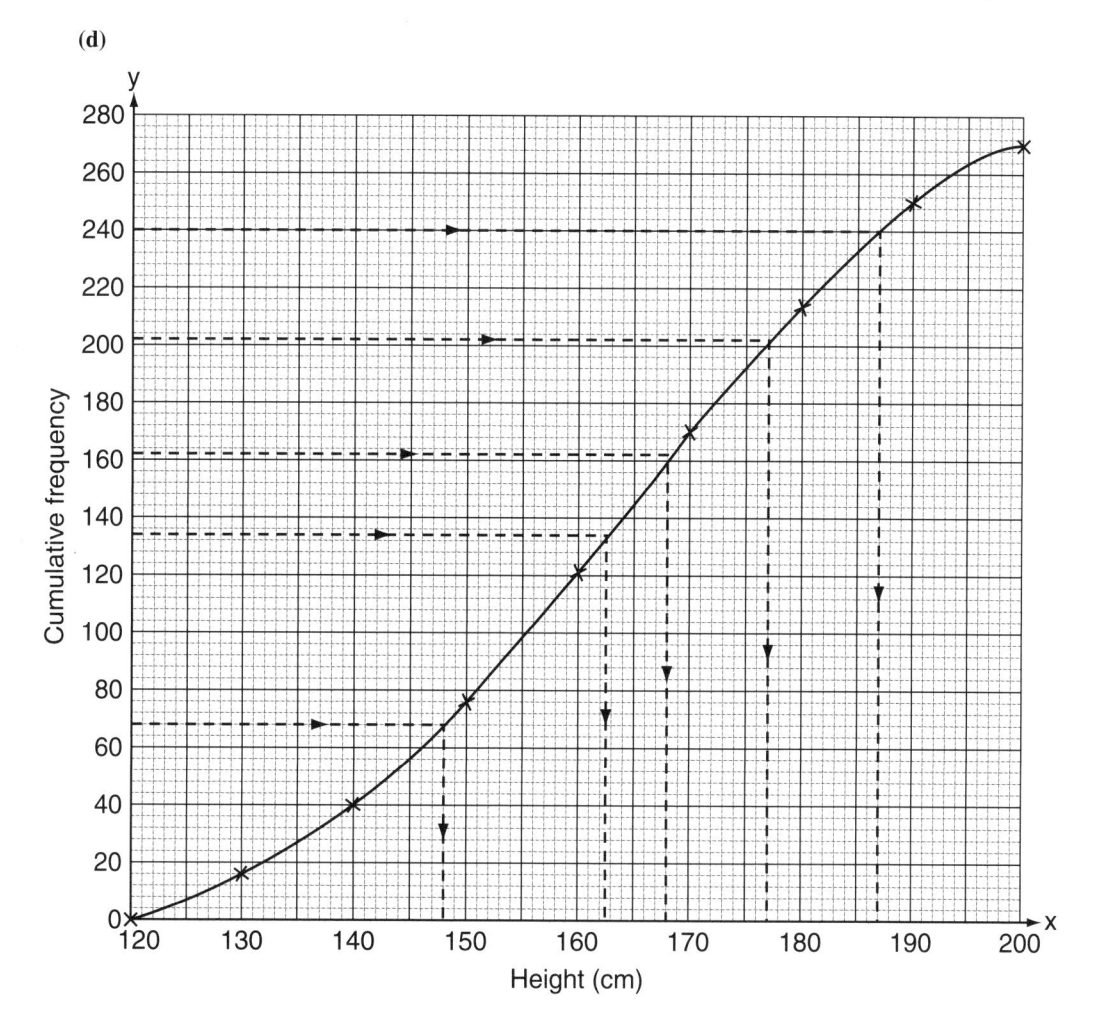

(e) (i) 162 to 164 cm
(ii) 176 to 178 cm
(iii) 28 to 30 cm
(iv) 167.5 to 168.5 cm
(f) 186.5 to 188 cm

10. (a) 74.4 to 74.7 kg
(b) 79.1 to 79.4 kg
(c) one of 23, 24 or 25

11. **(a)** one of 219, 220 or 221 plants

(b) 13

(c)

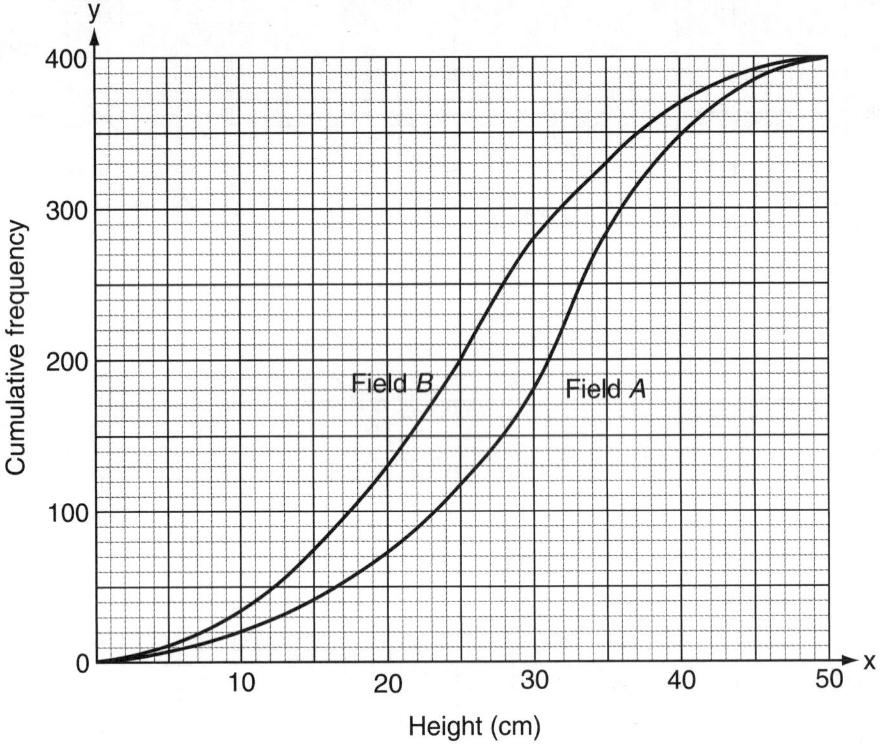

(d) For example, we can see that Field *A* produced the taller plants because the whole of curve for Field *A* is to the right of the curve for Field *B*. The median for Field *A* is 31 cm. The median for Field *B* is 25 cm

12. **(a)** **(i)** $60 < x \leqslant 80$ **(ii)** \$64.40

(b) **(i)**

upper class boundaries	20	40	60	80	100	140
cumulative frequency	10	42	90	144	180	200

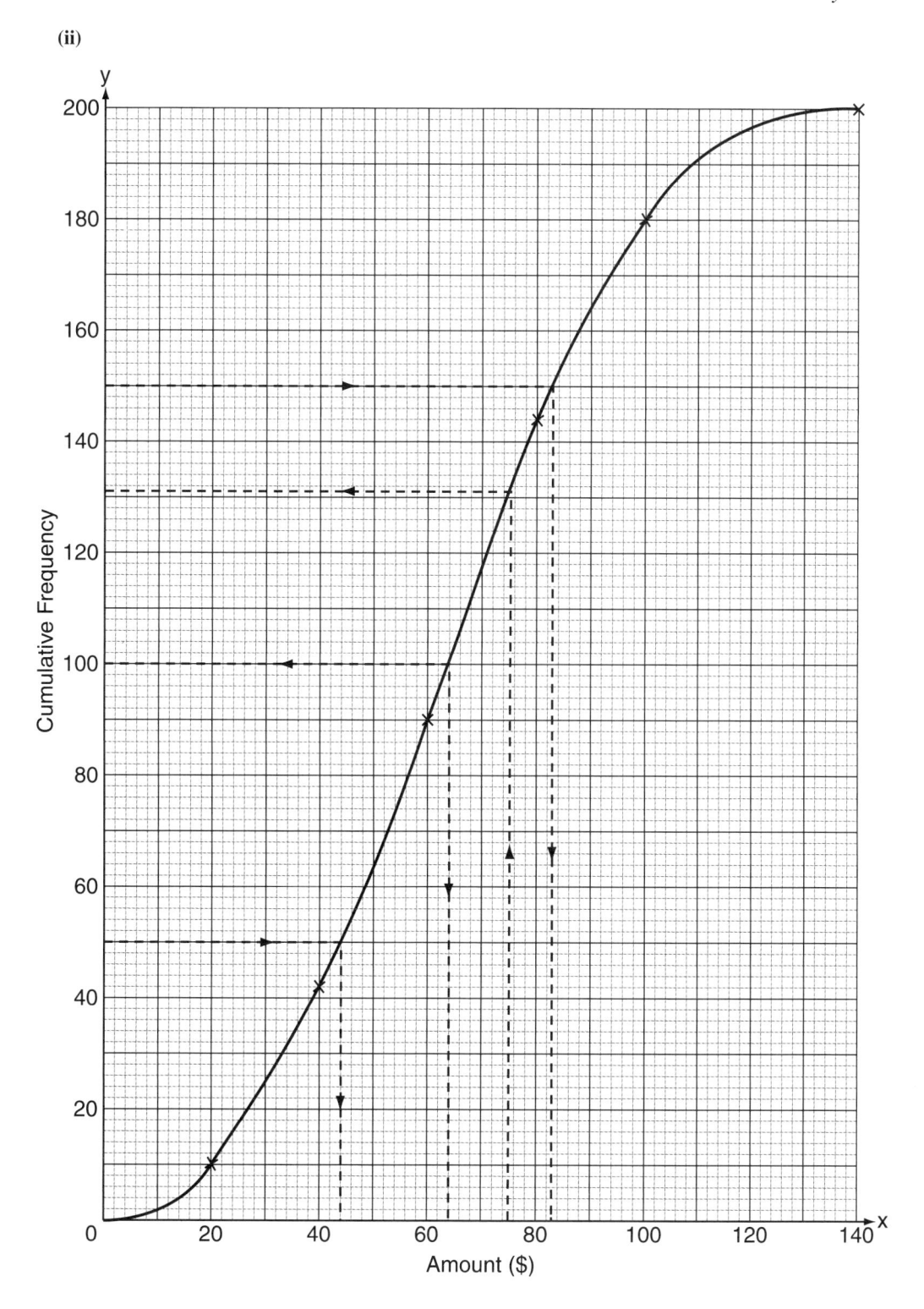

(ii)

(c) (i) \$63 to \$64 **(ii)** \$82 to \$84
(iii) \$38 to \$41 **(iv)** one of 67, 68, 69, 70, 71, 72

13. (a)

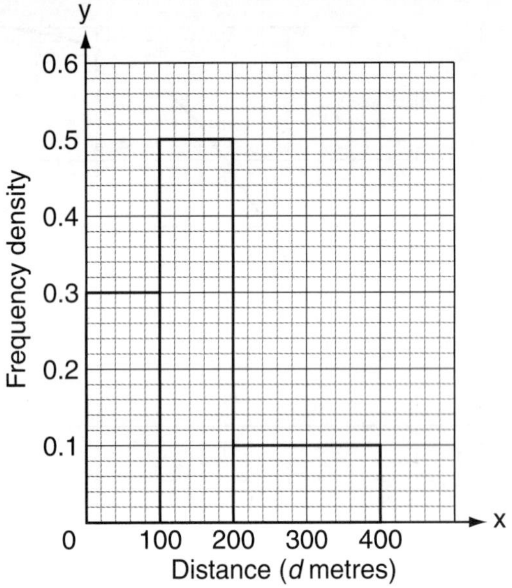
y
0.6
0.5
0.4
0.3
0.2
0.1
0
100
200
300
400
x
Frequency density
Distance (*d* metres)

(b) 72°

14. (a) (i) 40 **(ii)** 18

(b)

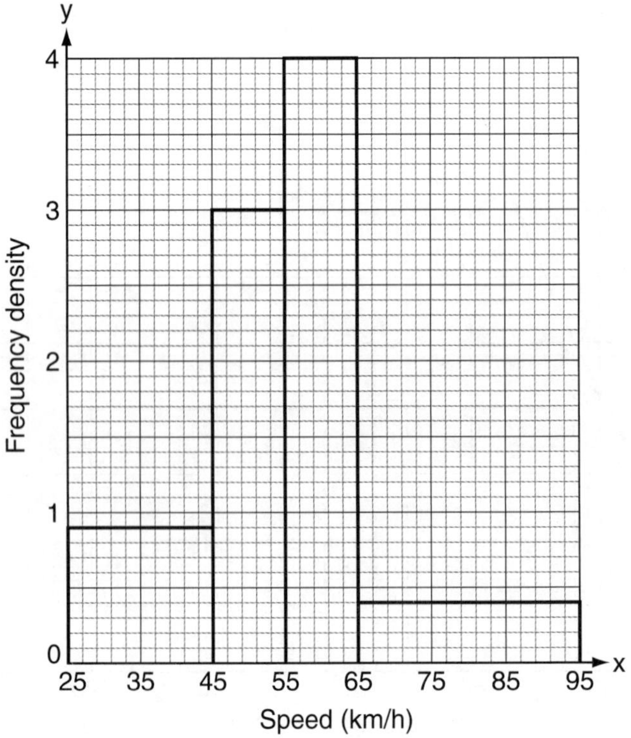
y
4
3
2
1
0
25
35
45
55
65
75
85
95
x
Frequency density
Speed (km/h)

15. (a) (i) 30 (ii) 30, 30.5, 31 (iii) 3
  (b) (i) 20.9 grams (ii) 2.6, 0.7, 0.8
16. (a) 3.365 to 3.375 grams
  (b) 0.26 to 0.27 grams
  (c) one of 55, 56 or 57
17. (a) (i) 3 (ii) 20 (iii) 7
  (b) (i) 14 to 14.2 (ii) 6 (iii) 28
  (iv) 22 (v) 31.5 to 32 (vi) 60
  (c) (i) 150 (ii) 125 (iii) 55.8
18. (a) 1 (b) 2.5
  (c) 2.96 (d) 2.9

Chapter 12

Core Skills

1. (a) 4 (b) $\dfrac{3}{10}$ (c) 0
2. (a) 6% (b) 90

Exercise 12.1

1. (a) 25 (b) $\dfrac{3}{25}$ (c) $\dfrac{10}{25}$
  (d) $\dfrac{5}{25}$ (e) 10 (f) 6

2. (a)

	R	B	G	P	W
R	RR	RB	RG	RP	RW
B	BR	BB	BG	BP	BW
G	GR	GB	GG	GP	GW
Y	YR	YB	YG	YP	YW

  (b) $\dfrac{3}{20}$

3. (a) $\dfrac{3}{12}$ (b) $\dfrac{2}{12}$
  (c) $\dfrac{1}{12}$ (d) 0

Exercise 12.2

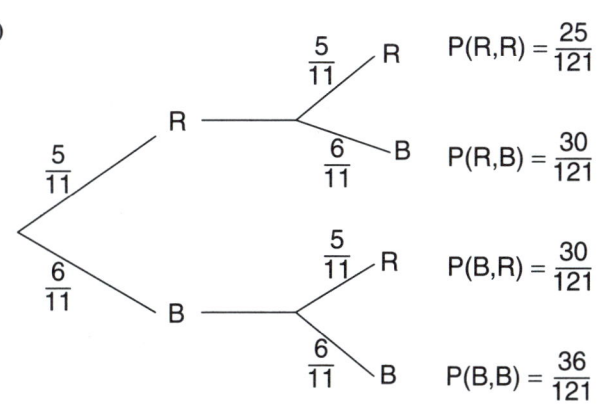

1. (a)

- $P(R,R) = \dfrac{25}{121}$
- $P(R,B) = \dfrac{30}{121}$
- $P(B,R) = \dfrac{30}{121}$
- $P(B,B) = \dfrac{36}{121}$

Tree diagram branches: R with $\dfrac{5}{11}$, B with $\dfrac{6}{11}$; from R: R $\dfrac{5}{11}$, B $\dfrac{6}{11}$; from B: R $\dfrac{5}{11}$, B $\dfrac{6}{11}$.

  (b) $\dfrac{60}{121}$

Answer Key 435

2. (a) (i) $\frac{13}{30}$ (ii) $\frac{17}{30}$

(b)

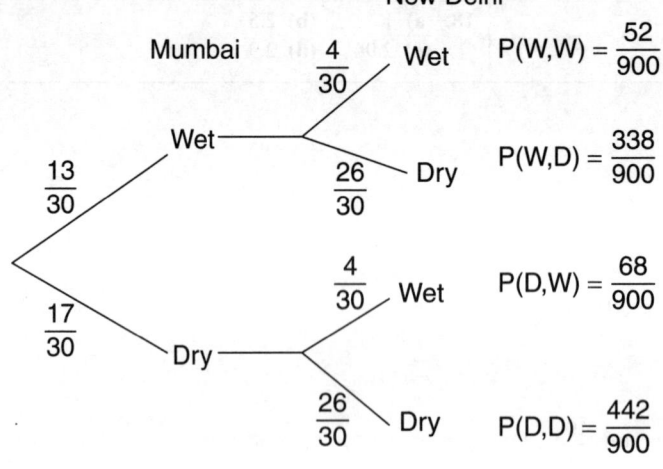

New Delhi: from Mumbai Wet ($\frac{13}{30}$) branches to Wet ($\frac{4}{30}$) giving $P(W,W) = \frac{52}{900}$ and Dry ($\frac{26}{30}$) giving $P(W,D) = \frac{338}{900}$; from Mumbai Dry ($\frac{17}{30}$) branches to Wet ($\frac{4}{30}$) giving $P(D,W) = \frac{68}{900}$ and Dry ($\frac{26}{30}$) giving $P(D,D) = \frac{442}{900}$.

(c) (i) $\frac{52}{900}$ (ii) $\frac{338}{900}$ (iii) $\frac{442}{900}$

(d) $\frac{26}{30}$

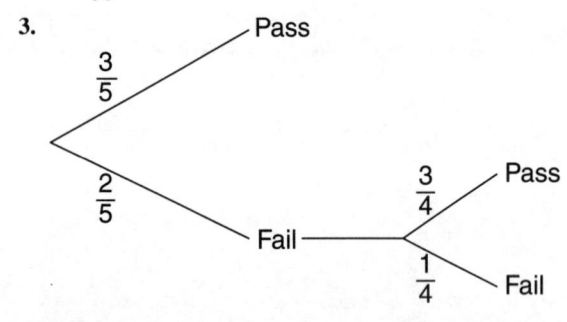

3. Tree diagram: Pass $\frac{3}{5}$; Fail $\frac{2}{5}$ then Pass $\frac{3}{4}$, Fail $\frac{1}{4}$.

Exercise 12.3

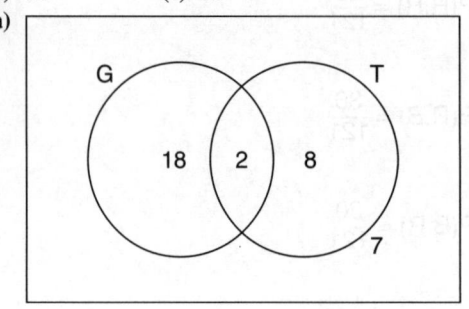

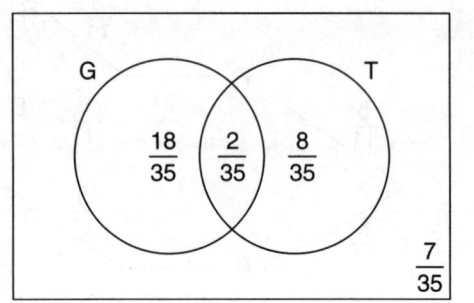

1. (a) Yes (b) 0.2 (c) 0
   (d) 0.3 (e) 0.9

2. (a) Venn diagrams: G and T with 18, 2, 8 and 7 outside; second diagram with $\frac{18}{35}$, $\frac{2}{35}$, $\frac{8}{35}$ and $\frac{7}{35}$ outside.

(b) $\frac{7}{35}$ (c) $\frac{2}{20}$

Exercise 12.4

1.

	Girls (G)	Boys (B)	Totals
Year Five (F)	10	13	23
Year Six (S)	14	12	26
Totals	24	25	49

(a) $\dfrac{24}{49}$ (b) $\dfrac{14}{49}$ (c) $\dfrac{23}{49}$ (d) $\dfrac{12}{26}$

2. (a)

1st Selection → 2nd Selection

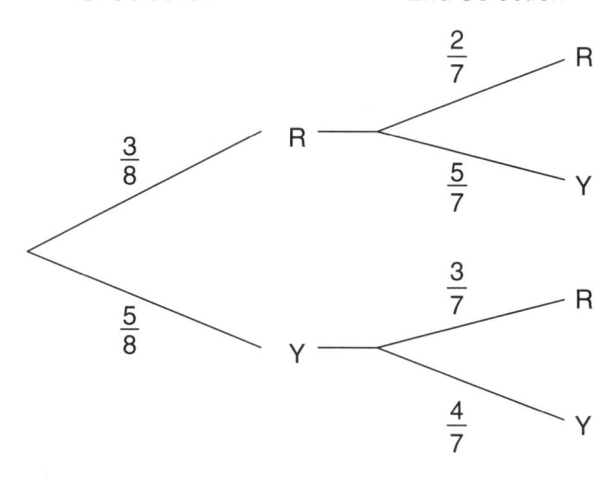

- $\dfrac{3}{8}$ R → $\dfrac{2}{7}$ R, $\dfrac{5}{7}$ Y
- $\dfrac{5}{8}$ Y → $\dfrac{3}{7}$ R, $\dfrac{4}{7}$ Y

(b) different (c) No (d) $\dfrac{5}{7}$ (e) $\dfrac{3}{8} \times \dfrac{5}{7} = \dfrac{15}{56}$ (f) $\dfrac{15}{56} + \dfrac{15}{56} = \dfrac{30}{56}$

3. (a) $\dfrac{20}{50}$ (b) $\dfrac{19}{50}$ (c) 8 (d) $\dfrac{9}{50}$ (e) $\dfrac{9}{25}$

4. (a)

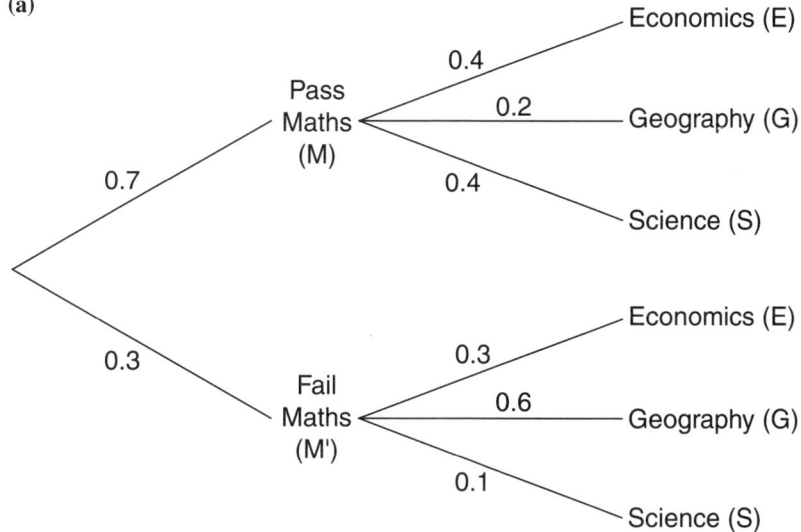

- 0.7 Pass Maths (M) → 0.4 Economics (E), 0.2 Geography (G), 0.4 Science (S)
- 0.3 Fail Maths (M') → 0.3 Economics (E), 0.6 Geography (G), 0.1 Science (S)

(b) 0.28 (c) 0.09 (d) 0.31

438 *Extended Mathematics for Cambridge IGCSE*

5. (i) 0   (ii) $\frac{1}{36}$ (iii) $\frac{1}{4}$ (iv) $\frac{1}{18}$
6. (a) (i) 0.2   (ii) 0.4   (iii) 0.5
   (iv) 0.1   (v) 0
   (b) (i) $\frac{1}{45}$ (ii) $\frac{1}{15}$ (iii) $\frac{4}{45}$ (iv) $\frac{41}{45}$
7. (a) $\frac{12}{18}$ (b) $\frac{3}{12}$
8. (a) $p = \frac{1}{20}$ $q = \frac{19}{20}$
   (b) (i) $\frac{1}{400}$ (ii) $\frac{38}{400}$
   (c) $\frac{38}{8000}$ (d) $\frac{58}{8000}$
   (e) 7 or 7.25 or 8
9. (a) $p = 5, q = 12, r = 1$
   (b) (i) 17   (ii) 12
   (c) (i) 26   (ii) 57
   (d) (i) $\frac{8}{100}$ (ii) $\frac{45}{100}$
   (e) $\frac{37}{74} \times \frac{36}{73} = \frac{18}{73}$
10. (a) (i) $s = \frac{1}{3}, t = \frac{1}{4}, u = \frac{5}{6}$
    (ii) $\frac{1}{2}$ (iii) $\frac{4}{9}$
    (b) (i) $\frac{1}{27}$ (ii) $\frac{19}{27}$
    (c) (i) $\frac{27}{256}$ (ii) $\left(\frac{3}{4}\right)^{n-1} \times \frac{1}{4}$
11. (a) (i) 6   (ii) 4.5   (iii) 4.54
    (iv) $\frac{1}{63}$ (v) $\frac{1}{35}$ (vi) $\frac{92}{819}$

(b) (i) 0.08   (ii) 0.125   (iii) 7
12. (a) (i) 1   (ii) $\frac{3}{6}$
    (b) (i) $\frac{2}{30}$ (ii) $\frac{4}{30}$ (iii) $\frac{8}{30}$
    (c) $\frac{18}{30}$ (d) 4
13. (a) (i) $p = \frac{1}{3}, q = \frac{3}{8}, r = \frac{6}{8}, s = \frac{2}{8}$
    (ii) $\frac{5}{12}$ (iii) $\frac{2}{3}$
    (b) (i) $\frac{1}{120}$ (ii) $\frac{119}{120}$
14. (a) (i)

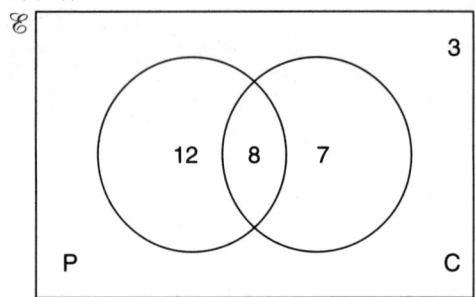

(ii) 8   (iii) $\frac{12}{30}$ (iv) $\frac{12}{20}$
(b) (i) $\frac{12}{90}$ (ii) $\frac{78}{90}$
(iii) $\frac{900}{6480}$ (iv) $\frac{5508}{6480}$

Index

NOTE: The page numbers given are for the principal pages on which the items occur.

acceleration, 78–79, 114–115
accuracy, limits of, 44
acute angle, 221
addition
    of algebraic fractions, 37–39
    of matrices, 169
    of vectors, 257–259
algebraic factors
    expanding products of, 21–23
algebraic fractions
    addition and subtraction, 37–39
    multiplication and division, 35–36
    simplifying, 32–34
ambiguous case, of Sine rule, 224–226
angle(s)
    acute, 221
    between 0 and 180, 217–222
    at centre of circle, 96
    of cyclic quadrilateral, 97
    of elevation and depression, 215–216
    obtuse, 221
    in opposite segments, 97
    in same segment, 97
    supplementary, 97–98
arc lengths, 147
area
    curved surface, 149–150
    of a sector, 148
    of a trapezium, 115
    of a triangle, 233–235
    ratio, 152–155
    total surface, 149
    axis of rotational symmetry, 91

bearing, three figure, 215
bisector
    perpendicular bisector of chord, 93

boundaries, class, 309, 312
bounds, upper and lower, 46–47
brackets, 21

cancelling, 35, 37
capacity, 154
chords
    equal, 93
    perpendicular bisector of , 93
circle facts
    angle at the centre, 96
    arc lengths, 147
    area of sector of, 148
    equal chords, 93
    finding the centre of, 98
    perpendicular bisector of chord, 93
    tangents from a point outside
        the circle, 94
class
    boundaries, 309, 312
    midpoint, 319
    width, 309
combined events, 342–345
combining vectors,
        257–260
complement, 7
completing the square, 62–65
composite functions, 190–191
conformable for multiplication, 171
constraints, 200
corresponding sides, 152
cosine
    acute and obtuse angles, 221
    curve, 220–221
    function, 221
    ratio, 218
    rule, 228–230

cumulative frequency, 320–329
   curve, 320–329
   polygon, 320
curved surface area, 149
   of cone, 149–150
   of cylinder, 149
curves, graphs of
   cosine, 219–221
   sine, 219–221
   sketching, 125
cyclic quadrilateral, 97

deceleration, 115
decimals, 13
denominator, 29, 32–33
dependent events, 349
det A, 178
determinant, 178
diagram
   mapping, 183–192
   possibility space, 343–345
   tree, 347–350
direct proportion, 76–78
directed line segment, 257
distance/time graph, 112–114
dividing
   algebraic fractions, 35–37
   a line in a given ratio, 261–262
domain, 183

element
   of a matrix, 169–171, 178, 181, 280
   of a set, 4
elimination, using to solve simultaneous equations, 70–71
empty set, 4
enlargement, 269–271
equal chords, 93
equations
   quadratic, 58–68
   simultaneous, 70–73
estimate the mean, 318
events
   combined, 342–351
   dependent, 349
   independent, 349
exclusive events, mutually, 352
expanding
   brackets, 22
   products of algebraic factors, 21–22
expression, quadratic, 23–29

factorising
   a difference of squares, 27
   a quadratic expression, 23–29
   by pairing, 27
   systematic, 28–29
feasible region, 200
finding
   the acceleration, 114–115
   the centre of a circle, 98
   the distance travelled, 115
   the gradient of a line segment, 129–130
   the length of a line segment, 130–131
   the midpoint of a line segment, 131–133
formula
   quadratic, 60–62
   rearranging, 78–81
fractional index, 29
fractions, 13
frequency
   cumulative, 320–330
   density, 309–315
function notation, using, 126
functions
   composite, 190–192
   inverse, 188–190

gradient, 112
   of a curve, 113–114
   of a line, 129–130
graphs
   distance/time, 112–114, 116
   of curves, 120–128
   of inequalities, 193–200
   of sine and cosine, 219–221
   sketching, 125–126
   speed/time, 114–115, 116
   using, 123–125
grouped frequency table, 318–319

head to tail, 258
histograms, 309–315

identity matrix, 177
increase and decrease by a given ratio, 47–48
independent events, 349
indices, fractional, 29
inequalities, 73–76
   graphs of, 193–200
integer values, 197, 200–201
inter-quartile range, 322
intersection of sets, 5

invariant line
    for a shear, 271
    for a stretch, 272–273
inverse
    functions, 188–190
    proportion, 76–78
irrational numbers, 12–13
irregular polygons, 100–101

length
    of arc, 147
    of line segment, 130
    of vector, 256–257
    ratio, 152–155
limits, 46
linear programming, 200–204
line segment, 129–133
    gradient of, 129–130
    length of, 130–131
    midpoint of, 131
lower quartile, 322–324
lowest common denominator, 37

magnitude (length) of a vector, 256–257
major segment, 97
many-to-one mapping, 185–186
mapping, 183–192
    diagram, 183–192
    one-to-many, 186
    one-to-one, 185–186
    many-to-one, 185–186
mark-up price, 49
matrix(ces), 168–183, 279–291
    addition and subtraction of, 169
    conformable for multiplication, 172
    identity, 177
    inverse, 177–180
    multiplication by a number, 170
    multiplication of, 171–176
    order of, 168
    and simultaneous equations, 181
    and transformations, 279–291
    transpose of, 169
    zero, 176
mean
    estimation of, 318
    from grouped frequency table, 318–319
median, 320–330
midpoint
    class, 319
    of line segment, 131–133

minor segment, 97
mod a, 257
modal class, 310
modulus, 257
multiplication
    of a matrix by a number, 170
    of a vector by a number, 260–261
    of matrices, 171–176
multiplying and dividing algebraic
        fractions, 35–37
multiplying or dividing an inequality
        by a negative number, 74
mutually exclusive events, 352

net, 149
notation
    function, 126, 184, 190
    set, 4–5, 352, 355
    transformation, 285–287
    vector, 269
nth term, 68
number line, 73–76
number machine, 80
numerator, 32–35, 38, 60, 66, 68,
        127, 129

obtuse angle, 218, 221
one-to-many mapping, 185–186
one-to-one mapping, 185–186

parabola, 125
parallel vectors, 262–263
percentages, 13
    reverse, 48–50
percentiles, 324–330
perpendicular
    bisector of a chord, 93
    height of a cone, 151
plane, 235
    of symmetry, 90–93
polygon, irregular, 100–101
position vectors, 265
possibility diagram, 343–345
power, 29–31
probability space diagram, 343–345
programming linear, 200–204
projection, 235
proper subset, 5–6
proportion
    direct, 76–78
    inverse, 76–78
Pythagoras' theorem, 130, 214, 228, 235

quadrants, 217–221
quadratic
    equations, 58–68
    expression, 23–29
    formula, 60–62
quadrilateral, cyclic, 97
quartile
    lower, 322–330
    upper, 322–330

radians, 229
range
    of a function, 183
    of data, 322
    inter-quartile, 322
ratio
    dividing line in given, 261–262
    increasing and decreasing by given, 47–48
    of areas, 152–155
    of lengths, 152–155
    of volumes, 152–155
rearranging (transformation of)
        formulae, 78–81
region, 193
    feasible, 200
resultant vector, 258
retail price, 49
retarding, 113
reverse percentages, 48–50
rotation, 275
rotational symmetry, 91–93

satisfy the equation, 58, 193
sector areas in circles, 148
segment
    of circle, 97
    major, 97
    minor, 97
    line, 129–133
    directed line, 257
selection
    without replacement, 349
    with replacement, 350
sequences, 68–70
set(s)
    complement of, 7
    defining a, 3
    empty, 5
    intersection of, 5
    notation, 4–5, 352, 355
    union, 6
    universal, 5

shear, 271–275
shear factor, 271
similar
    shapes, 152–155
    triangles, 155–159
simplifying algebraic fractions, 32–35
simultaneous equations
    and matrices, 181
    solving by elimination, 70–71
    solving by substitution, 71–73
sine
    acute and obtuse angles, 221
    curve, 219–220
    function, 221
    ratio, 218
    rule, 222–226
sketching graphs of curves, 125–126
slant height of cone, 150
solid shapes, 90–93, 235–239
solving quadratic equations
    by completing the square, 62–65
    by factorising, 58–60
    by the quadratic formula, 60–62
solving simultaneous equations, 70–73, 181
space diagram, probability, 343–345
speed, 78
speed/time graph, 114–115, 116
square
    completing the, 62–65
    unit, 281
straight line segment, 129–133
stretch, 272–275
subset, 6
substitution, in simultaneous equations, 71–73
subtended, 147
subtraction of vectors, 260
successive transformations, 286–291
supplementary angles, 97
surds, 13, 64
surface area, 149
symmetry
    in three dimensional shapes, 90–93
    line, 90
    plane of, 90
    rotational, 91

tangent
    to a curve, 114
    ratio, 221
tangents from a point outside the
        circle, 94
theodolite, 216